T0211302

Lecture Notes in Computer Science　　10159

Commenced Publication in 1973
Founding and Former Series Editors:
Gerhard Goos, Juris Hartmanis, and Jan van Leeuwen

More information about this series at http://www.springer.com/series/7410

Helena Handschuh (Ed.)

Topics in Cryptology – CT-RSA 2017

The Cryptographers' Track at the RSA Conference 2017
San Francisco, CA, USA, February 14–17, 2017
Proceedings

 Springer

Editor
Helena Handschuh
Cryptography Research Inc.
San Francisco, CA
USA

ISSN 0302-9743   ISSN 1611-3349   (electronic)
Lecture Notes in Computer Science
ISBN 978-3-319-52152-7   ISBN 978-3-319-52153-4   (eBook)
DOI 10.1007/978-3-319-52153-4

Library of Congress Control Number: 2016962026

LNCS Sublibrary: SL4 – Security and Cryptology

Printed on acid-free paper

This Springer imprint is published by Springer Nature
The registered company is Springer International Publishing AG
The registered company address is: Gewerbestrasse 11, 6330 Cham, Switzerland

Preface

The RSA conference has been a major international event for information security experts since its inception in 1991. It is an annual event that attracts several hundreds of vendors and close to ten thousand participants from industry, government, and academia.

Since 2001, the RSA conference has included the Cryptographer's Track (CT-RSA), which provides a forum for current research in cryptography.

CT-RSA has become a major publication venue in cryptography. It covers a wide variety of topics from public-key to symmetric-key cryptography and from cryptographic protocols to primitives and their implementation security. This year selected topics such as cryptocurrencies and white-box cryptography were added to the call for papers.

This volume represents the proceedings of the 2017 RSA Conference Cryptographer's Track, which was held in San Francisco, during February 14–17, 2017.

A total of 77 full papers were submitted for review, out of which 25 papers were selected for presentation. As chair of the Program Committee, I deeply thank all the authors who contributed the results of their innovative research. My appreciation also goes to the 33 members of the Program Committee and the numerous external reviewers who carefully reviewed these submissions. Each submission had at least three independent reviewers, and those co-authored by a member of the Program Committee had a least four reviewers. Together, Program Committee members and external reviewers generated close to 250 reviews. The selection process proved to be a very difficult task, as each contribution had its own merits. It was carried out with great professionalism and total transparency and generated a number of enthusiastic discussions among the members of the Program Committee. The submission process as well as the review process and the editing of the final proceedings were greatly simplified by the software written by Shai Halevi and we thank him for his kind and immediate support throughout the whole process.

In addition to the contributed talks, the program also included a panel discussion moderated by Bart Preneel on "Post-Quantum Cryptography: Is Time Running Out ?" including panelists Dan Boneh, Scott Fluhrer, Michele Mosca, and Adi Shamir.

November 2017   Helena Handschuh

Organization

CT-RSA 2017

RSA Conference Cryptographer's Track 2017

Moscone Center, San Francisco, California, USA
February 14–17, 2017

The RSA Cryptographer's Track is an independently managed component of the annual RSA conference.

Steering Committee

Helena Handschuh	Cryptography Research, USA and KU Leuven, Belgium
Kaisa Nyberg	Aalto University (retired), Finland
Ron Rivest	Massachusetts Institute of Technology, USA
Kazue Sako	NEC, Japan
Moti Yung	Snapchat, USA

Program Chair

Helena Handschuh	Cryptography Research, USA and KU Leuven, Belgium

Program Committee

Josh Benaloh	Microsoft Research, USA
Alex Biryukov	University of Luxembourg, Luxembourg
Chen-Mou Cheng	Osaka University, Japan
Jeremy Clark	Concordia University, Canada
Jean Paul Degabriele	Royal Holloway University of London, UK
Orr Dunkelman	University of Haifa, Israel
Junfeng Fan	Open Security Research, China
Henri Gilbert	ANSSI, France
Tim Güneysu	University of Bremen and DFKI, Germany
Stanislaw Jarecki	University of California at Irvine, USA
Thomas Johansson	Lund University, Sweden
Marc Joye	NXP Semiconductors, USA
Kwangjo Kim	KAIST, Republic of Korea
Susan Langford	Hewlett Packard Enterprise, USA
Tancrède Lepoint	SRI International, USA
David M'Raïhi	Symphony, USA
Stefan Mangard	Graz University of Technology, Austria

Mitsuru Matsui	Mitsubishi Electric, Japan
María Naya-Plasencia	Inria, France
Kaisa Nyberg	Aalto University (retired), Finland
Elisabeth Oswald	University of Bristol, UK
Raphael C.-W. Phan	Multimedia University, Malaysia
David Pointcheval	École Normale Supérieure, France
Bart Preneel	KU Leuven and iMinds, Belgium
Matt Robshaw	Impinj, USA
Reihaneh Safavi-Naini	University of Calgary, Canada
Kazue Sako	NEC, Japan
Palash Sarkar	Indian Statistical Institute, India
Nigel Smart	University of Bristol, UK
Marc Stevens	CWI, The Netherlands
Willy Susilo	University of Wollongong, Australia
Huaxiong Wang	Nanyang Technological University, Singapore
Brecht Wyseur	Nagra, Switzerland

External Reviewers

Hamza Abusalah	Paolo Gasti	Marcel Medwed
Jacob Alperin-Sheriff	Romain Gay	Bart Mennink
Ralph Ankele	Hannes Gross	Tarik Moataz
Florian Bache	Fuchun Guo	Amir Moradi
Timo Bartkewitz	Patrick Haddad	Fabrice Mouhartem
Sébastien Bellon	Qiong Huang	Michael Naehrig
Fabrice Benhamouda	Toshiyuki Isshiki	Toru Nakanishi
Elizabeth Berners-Lee	Chenglu Jin	Khoa Nguyen
David Bernhard	Antoine Joux	Léo Paul Perrin
Marc Blanc-Patin	Seny Kamara	Jiaxin Pan
Olivier Blazy	Sabyasachi Karati	Manuel San Pedro
Céline Blondeau	Pierre Karpman	Hervé Pelletier
Christina Brzuska	Takashima Katsuyuki	Peter Pessl
Zhenfu Cao	Marcel Keller	Duong Hieu Phan
Debrup Chakraborty	Dmitry Khovratovich	Jérôme Plût
Jie Chen	Handan Kilinç	Yogachandran
Rongmao Chen	Fuyuki Kitagawa	Rahulamathavan
Rakyong Choi	Thomas Korak	Somindu C. Ramanna
Yann Le Corre	Po-Chun Kuo	Oscar Reparaz
Jeroen Delvaux	Jianchang Lai	Vladimir Rozic
Daniel Dinu	Wai-Kong Lee	Pascal Sasdrich
Leo Ducas	Fuchun Lin	Tobias Schneider
Maria Eichlseder	Aaron Lye	Victor Servant
Ben Fisch	Dan Martin	Terence Spies
Jean-Bernard Fischer	Takahiro Matsuda	Douglas Stebila
Jean-Pierre Flori	Alexander May	Ron Steinfeld

Contents

Public Key Implementations

Choosing Parameters for NTRUEncrypt

Jeff Hoffstein[1], Jill Pipher[1], John M. Schanck[2,3], Joseph H. Silverman[1],
William Whyte[3], and Zhenfei Zhang[3(✉)]

[1] Brown University, Providence, USA
`{jhoff,jpipher,jhs}@math.brown.edu`
[2] University of Waterloo, Waterloo, Canada
[3] Security Innovation, Wilmington, USA
`{wwhyte,zzhang,jschanck}@securityinnovation.com`

Abstract. We describe a method for generating parameter sets, and calculating security estimates, for NTRUEncrypt. Our security analyses consider lattice attacks, the hybrid attack, subfield attacks, and quantum search. Analyses are provided for the IEEE 1363.1-2008 product-form parameter sets, for the NTRU Challenge parameter sets, and for two new parameter sets. These new parameter sets are designed to provide $\geq$ 128-bit post-quantum security.

Keywords: Public-key cryptography/NTRUEncrypt · Cryptanalysis · Parameter derivation

1 Introduction and Notation

In this note we will assume some familiarity with the details and notation of NTRUEncrypt. The reader desiring further background should consult standard references such as [11,12,16]. The key parameters are summarized in Table 1. Each is, implicitly, a function of the security parameter λ.

NTRUEncrypt uses a *ring of convolution polynomials*; a polynomial ring parameterized by a prime N, and an integer q, of the form $R_{N,q} = (\mathbb{Z}/q\mathbb{Z})[X]/(X^N - 1)$. The subscript will be dropped when discussing generic properties of such rings. We denote multiplication in R by $*$. An NTRUEncrypt public key is a generator for a cyclic R-module of rank 2, and is denoted $(1, h)$. The private key is an element of this module which is "small" with respect to a given norm and is denoted (f, g). Ring elements are written in the monomial basis. When an element of $\mathbb{Z}/q\mathbb{Z}$ is lifted to $\mathbb{Z}$, or reduced modulo p, it is identified with its unique representative in $[-q/2, q/2) \cap \mathbb{Z}$. The aforementioned norm is the 2-norm on coefficient vectors:

$$\left\| \sum_{i=0}^{N-1} a_i x^i \right\|^2 = \sum_{i=0}^{N-1} a_i^2.$$

An extended version of the paper is available at [10].

© Springer International Publishing AG 2017
H. Handschuh (Ed.): CT-RSA 2017, LNCS 10159, pp. 3–18, 2017.
DOI: 10.1007/978-3-319-52153-4_1

Table 1.

Primary NTRUEncrypt parameters	
N, q	Ring parameters $R_{N,q} = \mathbb{Z}_q[X]/(X^N - 1)$.
p	Message space modulus.
d_1, d_2, d_3	Non-zero coefficient counts for product form polynomial terms.
d_g	Non-zero coefficient count for private key component g.
d_m	Message representative Hamming weight constraint

This norm is extended to elements $(a, b) \in R \oplus R$ as

$$\|(a, b)\|^2 = \|a\|^2 + \|b\|^2.$$

There is a large degree of freedom in choosing the structure of the private key. In previous parameter recommendations [9,16] the secret polynomials f and g have been chosen uniformly from a set of binary or trinary polynomials with a prescribed number of non-zero coefficients. These are far from the only choices. The provably secure variant of NTRUEncrypt by Stehlé and Steinfeld [17], samples f and g from a discrete Gaussian distribution, and the NTRU-like signature scheme BLISS [6] samples its private keys from a set of polynomials with a prescribed number of ± 1s and ± 2s. The reasons for such choices are varied: binary polynomials were believed to allow for a small q parameter, but the desire to increase resistance against the hybrid combinatorial attack of [15] motivated the use of larger sample spaces in both NTRUEncrypt and BLISS. In the provably secure variant the public key must be computationally indistinguishable from an invertible ring element chosen uniformly at random. The discrete Gaussian distribution has several nice analytic properties that simplify the proof of such a claim, and sampling from such a distribution is reasonably efficient.

Our parameter choices use product-form polynomials for f and for the blinding polynomial, r, used during encryption. First introduced to NTRUEncrypt in [13], product form polynomials allow for exceptionally fast multiplication in R without the use of the Fourier transform.

An extended version of the paper is available at [10] which includes the following:

- a more detailed description of NTRUEncrypt algorithms;
- a survey of other known attacks and the security level against those attacks;
- tables that list suggested q parameter; and parameters for the NTRU challenge [2];
- some additional analysis for the hybrid attack.

2 General Considerations

2.1 Ring Parameters

The only restrictions on p and q are that they generate coprime ideals of $\mathbb{Z}[X]/(X^N - 1)$. In this document we will fix $p = 3$ and only consider q that are a power of 2. This choice is motivated by the need for fast arithmetic modulo q, and by the impact of p on decryption failure probability (see Sect. 6).

For NTRUEncrypt we take N to be prime. Many ideal lattice cryptosystems use the ring $\mathbb{Z}_q[X]/(X^{2^n} + 1)$ primarily because $X^{2^n} + 1$ is irreducible over the rationals. Some complications arise from using a reducible ring modulus, but these are easily remedied.

For prime N the ring modulus factors into irreducibles over $\mathbb{Q}$ as

$$X^N - 1 = (X - 1)\Phi_N(X)$$

where $\Phi_N(X)$ is the Nth cyclotomic polynomial. To maximize the probability that a random f is invertible in $R_{N,q}$ we should ensure that $\Phi_N(X)$ is irreducible modulo 2, i.e. we should choose N such that (2) is inert in the Nth cyclotomic field. Such a choice of N ensures that f is invertible so long as $f(1) \neq 0 \pmod 2$. It is not strictly necessary that $\Phi_N(X)$ be irreducible modulo 2, and one may allow a small number of high degree factors while maintaining a negligible probability of failure. Reasonable primes is provided in the full version of the paper [10]. Similar considerations apply for other choices of q.

2.2 Private Key, Blinding Polynomial, and Message Parameters

The analysis below will be considerably simpler if we fix how the values d_1, d_2, d_3, and d_g will be derived given N and q.

We set the notation:

$$\mathcal{T}_N = \{\text{trinary polynomials}\}$$

$$\mathcal{T}_N(d, e) = \left\{ \begin{array}{l} \text{trinary polynomials with exactly} \\ d \text{ ones and } e \text{ minus ones} \end{array} \right\}$$

$$\mathcal{P}_N(d_1, d_2, d_3) = \left\{ \begin{array}{l} \text{product form polynomials} \\ A_1 * A_2 + A_3 : A_i \in \mathcal{T}_N(d_i, d_i) \end{array} \right\}.$$

If N is fixed we will write $\mathcal{T}$, $\mathcal{T}(d, e)$, and $\mathcal{P}(d_1, d_2, d_3)$ instead.

A *product form private key* is of the form $(f, g) = (1 + pF, g)$ with $F \in \mathcal{P}_N(d_1, d_2, d_3)$ and $g \in \mathcal{T}_N(d_g + 1, d_g)$. Note that f must be invertible in $R_{N,q}$ for the corresponding public key $(1, h) = (1, f^{-1}g)$ to exist. The parameters recommended in this document ensure that, when F is sampled uniformly from $\mathcal{P}_N(d_1, d_2, d_3)$, the polynomial $1 + pF$ will always be invertible. One may optionally check that g is invertible, although this is similarly unnecessary for appropriately chosen parameters.

In order to maximize the size of the key space, while keeping a prescribed number of ± 1s in g, we take $d_g = \lfloor N/3 \rfloor$. The expected number of non-zero coefficients in f is $4d_1d_2 + 2d_3$. In order to roughly balance the difficulty of the search problems for f and g (Sect. 4), we take $d_1 \approx d_2 \approx d_3$ with $d_1 = \lfloor \alpha \rceil$ where α is the positive root of $2x^2 + x - N/3$. This gives us $2d_1d_2 + d_3 \approx N/3$.

A Hamming weight restriction is placed on message representatives to avoid significant variation in the difficulty of message recovery. Message representatives are trinary polynomials; we require that the number of +1s, −1s, and 0s each be greater than d_m. The procedure for choosing d_m is given in Sect. 5.

3   Review of the Hybrid Attack

We consider the hybrid attack [15] to be the strongest attack against NTRUEncrypt, and believe that cost estimates for the hybrid attack give a good indication of the security of typical NTRUEncrypt parameter sets. Information on other attacks can be found from the full version of the paper [10].

Suppose one is given an NTRU public key $(1, h)$ along with the relevant parameter set. This information determines a basis for a lattice $\mathcal{L}$ of rank $2N$ generated by the rows of

$$L = \left(\begin{array}{c|c} qI_N & 0 \\ \hline H & I_N \end{array} \right) \tag{1}$$

wherein the block H is the circulant matrix corresponding to h, i.e. its rows are the coefficient vectors of $x^i * h$ for $i \in [0, N-1]$. The map $(1, h)R_{N,q} \to \mathcal{L}/q\mathcal{L}$ that sends $(a, b) \mapsto (b_0, \ldots, b_{N-1}, a_0, \ldots, a_{N-1})$ is an additive group isomorphism that preserves the norm defined in Eq. 1. As such, if one can find short vectors of $\mathcal{L}$ one can find short elements of the corresponding NTRU module.

The determinant of L is $\Delta = q^N$, giving us a Gaussian expected shortest vector of length $\lambda_1 \approx \sqrt{qN/\pi e}$, though the actual shortest vector will be somewhat smaller than this. A pure lattice reduction attack would attempt to solve Hermite-SVP[1] with factor $\lambda/\Delta^{1/2N} = \sqrt{N/\pi e}$, which is already impractical for N around 100. The experiments of [8] support this claim, they were able to find short vectors in three NTRU lattices with $N = 107$ and $q = 64$ that were generated using binary private keys. Only one of these was broken with BKZ alone, the other two required a heuristic combination of BKZ on the full lattice and BKZ on a projected lattice of smaller dimension with block sizes between 35 and 41.

Consequently the best attacks against NTRUEncrypt tend to utilize a combination of lattice reduction and combinatorial search. In this section we will review one such method from [15], known as the hybrid attack.

The rough idea is as follows. One first chooses $N_1 < N$ and extracts a block, L_1, of $2N_1 \times 2N_1$ coefficients from the center of the matrix L defined in Eq. 1.

---

[1] In practice q has a strong impact on the effectiveness of pure lattice reduction attacks as well. For large q the relevant problem becomes Unique-SVP which appears to be somewhat easier than Hermite-SVP. Conservative parameter generation should ensure that it is difficult to solve Hermite-SVP to within a factor of $q/\Delta^{1/2N} = \sqrt{q}$.

The rows of L_1 are taken to generate a lattice $\mathcal{L}_1$.

$$\left(\begin{array}{c|c} qI_N & 0 \\ \hline H & I_N \end{array}\right) = \left(\begin{array}{c|c|c} qI_{r_1} & 0 & 0 \\ \hline * & L_1 & 0 \\ \hline * & * & I_{r_2} \end{array}\right)$$

A lattice reduction algorithm is applied to find a unimodular transformation, U', such that $U'L_1$ is reduced, and an orthogonal transformation, Y', is computed such that $U'L_1Y' = T'$ is in lower triangular form. These transformations are applied to the original basis to produce a basis for an isomorphic lattice:

$$T = ULY = \left(\begin{array}{c|c|c} I_{r_1} & 0 & 0 \\ \hline 0 & U' & 0 \\ \hline 0 & 0 & I_{r_2} \end{array}\right) \left(\begin{array}{c|c|c} qI_{r_1} & 0 & 0 \\ \hline * & L_1 & 0 \\ \hline * & * & I_{r_2} \end{array}\right) \left(\begin{array}{c|c|c} I_{r_1} & 0 & 0 \\ \hline 0 & Y' & 0 \\ \hline 0 & 0 & I_{r_2} \end{array}\right) = \left(\begin{array}{c|c|c} qI_{r_1} & 0 & 0 \\ \hline * & T' & 0 \\ \hline * & * & I_{r_2} \end{array}\right).$$

Notice that $(g,f)Y$ is a short vector in the resulting lattice.

In general it is not necessary for the extracted block to be the central $2N_1 \times 2N_1$ matrix, and it is sometimes useful to consider blocks shifted s indices to the top left along the main diagonal. Let $r_1 = N - N_1 - s$ be the index of the first column of the extracted block and $r_2 = N + N_1 - s$ be the index of the final column. The entries on the diagonal of T will have values $\{q^{\alpha_1}, q^{\alpha_2}, \ldots, q^{\alpha_{2N}}\}$, where $\alpha_1 + \cdots + \alpha_{2N} = N$, and the α_i, for i in the range $[r_1, r_2]$, will come very close to decreasing linearly. That is to say, L_1 will roughly obey the geometric series assumption (GSA). The rate at which the α_i decrease can be predicted very well based on the root Hermite factor achieved by the lattice reduction algorithm used.[2] Clearly $\alpha_i = 1$ for $i < r_1$ and $\alpha_i = 0$ for $i > r_2$. By the analysis in [10] we expect

$$\alpha_{r_1} = \frac{1}{2} + \frac{s}{2N_1} + 2N_1 \log_q(\delta) \tag{2}$$

$$\alpha_{r_2} = \frac{1}{2} + \frac{s}{2N_1} - 2N_1 \log_q(\delta), \tag{3}$$

and a linear decrease in-between. The profile of the basis will look like one of the examples in Fig. 1.

By a lemma of Furst and Kannan (Lemma 1 in [15]), if $y = uT + x$ for vectors u and x in $\mathbb{Z}^{2N}$, and $-T_{i,i}/2 < x_i \leq T_{i,i}/2$, then reducing y against T with Babai's nearest plane algorithm will yield x exactly. Thus if v is a shortest vector in $\mathcal{L}$ and $\alpha_{r_2} > \log_q(2\|v\|_\infty)$, it is guaranteed that v can be found by enumerating candidates for its final $K = 2N - r_2$ coefficients. Further knowledge about v can also diminish the search space. For example, if it is known that there is a trinary vector in $\mathcal{L}$, and $\alpha_{r_2} > \log_q(2)$, then applying Babai's nearest plane algorithm to some vector in the set $\{(0|v')T - (0|v') : v' \in T_K\}$ will reveal it.

The optimal approach for the attacker is determined by the balancing the cost of combinatorial search on K coordinates against the cost of lattice reduction that results in a sufficiently large α_{2N-K}. Unsurprisingly, naïve enumeration of the possible v' is not optimal.

---

[2] A lattice reduction algorithm that achieves root Hermite factor δ returns a basis with $\|b_1\|_2 \approx \delta^n \det(\Lambda)^{1/n}$.

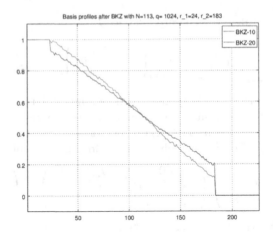

Fig. 1. Log length of ith Gram-Schmidt vector, $\log_q(\|b_i^*\|)$.

4   Meet in the Middle Search

The adaptation of meet-in-the-middle search algorithms to the structure of binary NTRU keys is due to Odlyzko and described in [14]. Generalizations to other private key types are described by Howgrave-Graham in [15]; this is the presentation we follow here. The key idea is to decompose the search space S as $S \subseteq S' \oplus S'$ for some set S' such that $|S'| \approx \sqrt{|S|}$. If s_1 and s_2 are elements of S' such that $s_1 + s_2 = f$, and (f, g) is an element with small coefficients in the NTRU module generated by $(1, h)$, then $(s_1, s_1 * h) = (f, g) - (s_2, s_2 * h)$. In particular, when the coefficients of g are trinary, this implies that $s_1 * h \approx -s_2 * h$ coordinate-wise.

Under the assumption that all approximate collisions can be detected, a meet in the middle search on the full product form NTRUEncrypt key space would require both time and memory of order $O(\sqrt{|\mathcal{P}_N(d_1, d_2, d_3)|})$.

A meet in the middle search is also possible on a basis that has been pre-processed for the hybrid attack as in Eq. 2. The assumption that all approximate collisions can be detected will turn out to be untenable in this case, however, in the interest of deriving conservative parameters we will assume that this complication does not arise. Let $\Pi : \mathbb{Z}^N \to \mathbb{Z}^K$ be a projection[3] onto K coordinates of $\mathbb{Z}^N$. Let $P_\Pi = \{v\Pi : v \in \mathcal{P}_N(d_1, d_2, d_3)\}$. The f component of the private key is guaranteed to appear in P_Π, so the expected time and memory required for the attack is $O(\sqrt{|P_\Pi|})$. That said, estimating the size of P_Π is non-trivial.

We may also consider an adversary who attempts this attack on the lattice corresponding to $(1, h^{-1})$ and searches for the g component of the private key instead. This may in fact be the best strategy for the adversary, because while $|\mathcal{P}_N(d_1, d_2, d_3)| < |\mathcal{T}_N(d_g + 1, d_g)|$ for parameters of interest to us, the presence

---

[3] We will abuse notation slightly and allow Π to act on elements of R by acting on their coefficient vectors lifted to $\mathbb{Z}^N$.

of coefficients not in $\{-1, 0, 1\}$ in product form polynomials leads to a large increase in the relative size of the projected set.

In either case we assume that it is sufficient for the adversary to search for trinary vectors, and that they may limit their search to a projection of $T_N(d, e)$ for some (d, e). When targeting g we have $d = d_g + 1$, $e = d_g$, and when targeting f we have that both d and e are approximately $2d_1 d_2 + d_3$. Clearly when $d = e = N/3$, and $N \gg K$, we should expect that the projection of a uniform random element of $T_N(d, e)$ onto K coordinates will look like a uniform random element of T_K. For such parameters, the size of the set that must be enumerated in the meet-in-the-middle stage is $\approx 3^{K/2}$.

For $d \neq N/3$, or for large K, not all trinary sequences are equally likely, and the adversary may choose to target a small set of high probability sequences. Consequently we must estimate the size of the set of elements that are typical under the projection. Fix N, K, Π, d, and e and let $\mathcal{S} = T_N(d, e)$. Let $p : T_K \rightarrow \mathbb{R}$ be the probability mass function on T_K induced by sampling an element uniformly at random from $\mathcal{S}$ and projecting its coefficient vector onto the set of K coordinates fixed by Π. We will estimate the size of the search space in the hybrid attack as, roughly, $2^{H(p)}$, where $H(p)$ is the Shannon entropy of p.

Let $\mathcal{S}_\Pi(a, b)$ be the subset of $\mathcal{S}$ consisting of vectors, v, such that $v\Pi$ has exactly a coefficients equal to $+1$ and b coefficients equal to -1. By the symmetry of $\mathcal{S}$ under coordinate permutations we have that $p(v\Pi) = p(v'\Pi)$ for all pairs $v, v' \in \mathcal{S}_\Pi(a, b)$. We choose a fixed representative of each type: $v_{a,b} = v\Pi$ for some $v \in \mathcal{S}_\Pi(a, b)$, and write

$$p(v_{a,b}) = \frac{1}{\binom{K}{a}\binom{K-a}{b}} \frac{|\mathcal{S}_\Pi(a,b)|}{|\mathcal{S}|} = \frac{\binom{N-K}{d-a}\binom{N-K-d+a}{d-b}}{\binom{N}{d}\binom{N-d}{d}}.$$

As there are exactly $\binom{K}{a}\binom{K-a}{b}$ distinct choices for $v_{a,b}$ this gives us:

$$H(p) = -\sum_{v \in T_K} p(v) \log_2 p(v) = -\sum_{0 \leq a, b \leq d} \binom{K}{a}\binom{K-a}{b} p(v_{a,b}) \log_2 p(v_{a,b}).$$

$$(4)$$

The size of the search space is further decreased by a factor of N since $x^i * g$ is likely to be a distinct target for each $i \in [0, N-1]$. Hence in order to resist the hybrid meet-in-the-middle attack we should ensure

$$\frac{1}{2}(H(p) - \log_2(N)) \geq \lambda.$$

The only variable not fixed by the parameter set itself in Eq. 4 is K. In order to fix K we must consider the cost of lattice reduction.

The block to be reduced is of size $(r_2 - r_1) \times (r_2 - r_1)$ where $r_2 = 2N - K$ and $r_1 = \lambda$. Recall that $s = N - (r_1 + r_2)/2$, and $N_1 = (r_2 - r_1)/2$. Having fixed these parameters we can use Eq. 3 to determine the strength of the lattice reduction needed to ensure that α_{r_2} is sufficiently large to permit recovery of a trinary vector. In particular, we need $\alpha_{r_2} = \frac{N_1 + s}{2N_1} - 2N_1 \log_q(\delta) \geq \log_q(2)$, which implies that

$$\log_2(\delta) \leq \frac{N_1 + s}{4N_1^2} \log_2(q) - \frac{1}{2N_1}. \tag{5}$$

Translating the required root Hermite factor, δ, into a concrete bit-security estimate is notoriously difficult. However there seems to be widespread consensus on the values that are currently out of reach for common security parameters. As such one might use the following step function as a first approximation:

$$\delta^*(\lambda) = \begin{cases} 1.009 & \text{if } \lambda \leq 60 \\ 1.008 & \text{if } 60 < \lambda \leq 80 \\ 1.007 & \text{if } 80 < \lambda \leq 128 \\ 1.005 & \text{if } 128 < \lambda \leq 256 \\ 1 & \text{otherwise.} \end{cases}$$

A more refined approach involving a BKZ simulator, from [4], is used in Sect. 8.1.

Rewriting Eq. 5 in terms of N, q, r_1 and K we define:

$$\log_2(\eta(N, q, r_1, K)) = \frac{(N - r_1) \log_2(q)}{4N^2 - 4N(K + r_1) + (K^2 + 2r_1K + r_1^2)} - \frac{1}{2N - (K + r_1)}.$$

Conclusion: A parameter set resists hybrid meet-in-the-middle attacks on private keys if Eq. 4 is satisfied and

$$1 < \eta(N, q, r_1, K) \leq \delta^*(r_1). \tag{6}$$

5   Rejecting Sparse (and Dense) Message Representatives

The parameter sets in this paper specify the exact number of 1's and -1's in each of r_1, r_2, r_3, which reveals the quantity $r(1)$, that is, the polynomial r evaluated at 1. As an encrypted message has the form $e = pr * h + m$, the value $m(1)$ modulo q is revealed by the known quantities $r(1), e(1), h(1)$. The value $m(1)$ in turn reveals the difference between the number of 1's and the number of -1's in the message representative.

We assume the message representative is uniformly distributed over $\mathcal{T}_N$. The expected value of $m(1)$ is zero, but for large $|m(1)|$, the size of the search space for m decreases, making a meet in the middle search for (r, m) easier. We assume that the adversary observes a very large number of messages and can freely condition their attack on the value of $m(1)$ regardless of the probability that a uniform random message representative takes that value.

In addition, we forbid the number of $+1$s, -1s, or 0s to be less than a given parameter, d_m. The choice of d_m depends primarily affects resistance against hybrid combinatorial attacks, but d_m also has an impact on decryption failure probability, as will be discussed in Sect. 6.

The calculation for determining resistance against hybrid combinatorial attacks is very similar to that leading up to Eq. 4, but there are two key differences. First, in Sect. 4 we were primarily concerned with validating the security of the obvious choice $d_g = \lfloor N/3 \rfloor$. Here we will need to search for d_m. Second, having fixed d_m we need to condition the distribution of projected elements on the value of $m(1)$.

The search space for d_m can be constrained by imposing an arbitrary upper bound on the probability of a failure. Such a failure is roughly as expensive as a full encryption, so d_m should be chosen to ensure that failures are rare.

Let $I(d_m) = \{(i,j) : d_m \le i < (N - 2d_m),\ d_m \le j < (N - d_m - i)\}$. We will only consider d_m satisfying:

$$2^{-10} \ge 1 - 3^{-N} \left(\sum_{(i,j) \in I(d_m)} \binom{N}{i}\binom{N-i}{j} \right)$$

Let K be the value derived in Sect. 4. Fix Π and let $\mathcal{S}(e_1, e_2; a, b)$ be the set of projections of elements of $\mathcal{T}(e_1, e_2)$ with a ones and b minus ones. Let $\mathcal{M}$ be the subset of $\mathcal{T}_N$ satisfying the d_m constraint. Let $p : \mathcal{T}_K \times \mathbb{Z} \times \mathbb{Z} \to \mathbb{R}$ be the probability mass function given by

$$p(v, e_1, e_2) = \mathrm{Prob}_{m \leftarrow_\$ \mathcal{M}} \left(m\Pi = v \text{ and } m \in \mathcal{T}_N(e_1, e_2) \right),$$

i.e. $p(v, e_1, e_2)$ is the probability that an m sampled uniformly from $\mathcal{M}$ has e_1 ones, e_2 minus ones, and is equal to v under projection. If the information leakage from $m(1)$ determined e_1 and e_2 then we could use essentially the same analysis as Sect. 4 and our security estimate would be

$$\frac{1}{2} \min_{(e_1, e_2) \in I(d_m)} H(p_{d_m} | e_1, e_2).$$

However the adversary only learns $m(1) = e_1 - e_2$, so we will account for their uncertainty about whether $m \in \mathcal{T}(e_1, e_2)$ given $m(1) = e_1 - e_2$.

The marginal distribution on e_1 and e_2 conditioned on the event $m(1) = y$ is

$$q(e_1, e_2 | m(1) = y) = \sum_{v \in \mathcal{T}_K} p(v, e_1, e_2 | m(1) = y)$$

$$= \binom{N}{e_1}\binom{N - e_1}{e_2} \left(\sum_{\substack{i - j = y \\ (i,j) \in I(d_m)}} \binom{N}{i}\binom{N-i}{j} \right)^{-1}.$$

Conclusion: As such we will consider a parameter set secure against hybrid meet-in-the-middle attacks on messages provided that:

$$\lambda \le \min_y \min_{\substack{e_1 - e_2 = y \\ (e_1, e_2) \in I(d_m)}} \frac{1}{2} H(p | e_1, e_2) - \log_2 q(e_1, e_2 | m(1) = y). \qquad (7)$$

Evaluating this expression is considerably simplified by noting that local minima will be found at the extremal points: $|e_1 - e_2| = N - 3d_m$ and $e_1 = e_2 \approx N/3$.

Note that unlike the estimate in Sect. 4 we do not include a $- \log_2(N)$ term to account for rotations of m.

6   Estimating the Probability of Decryption Failure

As remarked earlier, in order for decryption to succeed the coefficients of

$$a = p * (r * g + m * F) + m$$

must have absolute value less than $q/2$.

Assuming $p \in \mathbb{Z}$, and trinary g and m, the triangle inequality yields:

$$\|a\|_\infty \le p \left(\|r\|_1 \|g\|_\infty + \|F\|_1 \|m\|_\infty\right) + 1 = p \left(\|r\|_1 + \|F\|_1\right) + 1.$$

Thus with product form r and F decryption failures can be avoided entirely by ensuring $(q - 2)/2p > 8d_1 d_2 + 4d_3$. However, since ciphertext expansion scales roughly as $N \log_2(q)$, it can be advantageous to consider probabilistic bounds as well. The probability

$$\text{Prob}\,(\text{a given coefficient of } r * g + m * F \text{ has absolute value } \ge c)$$

can be analyzed rather well by an application of the central limit theorem. This was done for the case of trinary r, g, m, F in [9]. Here we provide a modified analysis for the case where the polynomials r and F take a product form. In particular, we assume that $r = r_1 * r_2 + r_3$, $F = F_1 * F_2 + F_3$, where each r_i and F_i has exactly d_i coefficients equal to 1, d_i coefficients equal to -1, and the remainder equal to 0.

Let X_k denote a coefficient of $r * g + m * F$. The spaces from which r and m are drawn are invariant under permutations of indices, so the probability that $|X_k| > c$ does not depend on the choice of k.[4] Note that X_k has the form

$$X_k = (r_1 * r_2 * g)_k + (r_3 * g)_k + (F_1 * F_2 * m)_k + (F_3 * m)_k,$$

and each term in the sum is itself a sum of either $4d_1 d_2$ or $2d_3$ (not necessarily distinct) coefficients of g or m. For instance, $(r_1 * r_2 * g)_k = \sum_{i,j} (r_1)_i (r_2)_j (g)_{(k-i-j)}$ and only the $4d_1 d_2$ pairs of indices corresponding to non-zero coefficients of r_1 and r_2 contribute to the sum. We can think of each index pair as selecting a sign $\epsilon(i)$ and an index $a(i)$ and rewrite the sum as $(r_1 * r_2 * g)_k = \sum_{i=1}^{4d_1 d_2} \epsilon(i)(g)_{a(i)}$. While the terms in this sum are not formally independent (since a may have repeated indices, and g has a prescribed number of non-zero coefficients) extensive experiments show that the variance of $(r_1 * r_2 * g)_k$ is still well approximated by treating $(g)_{a(i)}$ as a random coefficient of g, i.e. as taking a non-zero value with probability $(2d_g + 1)/N$:

$$\mathbb{E}\left[(r_1 * r_2 * g)_k^2\right] \approx \sum_{i=1}^{4d_1 d_2} \mathbb{E}\left[(\epsilon(i)(g)_{a(i)})^2\right] = \sum_{i=1}^{4d_1 d_2} \mathbb{E}\left[(g)_{a(i)}^2\right] = 4d_1 d_2 \cdot \frac{2d_g + 1}{N}$$

Nearly identical arguments can be applied to compute the variances of the other terms of Eq. 6, although some care must be taken with the terms involving m. While an honest party will choose m uniformly from the set of trinary

---
[4] The X_k for different k have the same distribution, but they are not completely independent. However, they are so weakly correlated as to not affect our analysis.

polynomials, m could be chosen adversarily to maximize its Hamming weight and hence the probability of a decryption failure. Due to the d_m constraint (Sect. 5), the number of non-zero coefficients of m cannot exceed $N - d_m$. As such we model the coefficients of m as taking ± 1 each with probability $(1 - d_m/N)$ and 0 with probability d_m/N.

With these considerations the variance of $(r_1 * r_2 * g)_k + (r_3 * g)_k$ is found to be $\sigma_1^2 = (4d_1 d_2 + 2d_3) \cdot \frac{2d_g + 1}{N}$, and the variance of $(F_1 * F_2 * m)_k + (F_3 * m)_k$ is found to be $\sigma_2^2 = (4d_1 d_2 + 2d_3) \cdot (1 - \frac{d_m}{N})$. Both terms are modeled as sums of i.i.d. random variables, and the d_i are chosen such that $4d_1 d_2 + 2d_3 \approx 2N/3$, so for sufficiently large N the central limit theorem suggests that each term will have a normal distribution. Finally X_k can be expected to be distributed according to the convolution of these two normal distributions, which itself is a normal distribution with variance

$$\sigma^2 = \sigma_1^2 + \sigma_2^2 = (4d_1 d_2 + 2d_3) \cdot \frac{N - d_m + 2d_g + 1}{N}.$$

The probability that a normally distributed random variable with mean 0 and standard deviation σ exceeds c in absolute value is given by the complementary error function, specifically $\texttt{erfc}(c/(\sqrt{2}\sigma))$. Applying a union bound, the probability that any of the N coefficients of $r * g + m * f$ is greater than c is bounded above by $N \cdot \texttt{erfc}(c/(\sqrt{2}\sigma))$.

Conclusion: To have negligible probability of decryption failure with respect to the security parameter, λ, we require

$$N \cdot \texttt{erfc}((q - 2)/(2\sqrt{2} \cdot p \cdot \sigma)) < 2^{-\lambda} \tag{8}$$

where $\sigma = \sigma(N, d_1, d_2, d_3, d_g, d_m)$ as in Eq. 6.

7   Product Form Combinatorial Strength

The search space for a triple of polynomials F_1, F_2, F_3 where each polynomial F_i has d_i 1's and d_i -1's is of size:

$$|\mathcal{P}_N(d_1, d_2, d_3)| = \binom{N}{d_1}\binom{N - d_1}{d_1}\binom{N}{d_2}\binom{N - d_2}{d_2}\binom{N}{d_3}\binom{N - d_3}{d_3}.$$

Thus a purely combinatorial meet-in-the-middle search on product form keys can be performed in time and space $O(\sqrt{|\mathcal{P}_N(d_1, d_2, d_3)|/N})$, where we have divided by N to account for the fact that rotations of a given triple are equivalent.

Finally, one could construct a $3N$ dimensional lattice attack by considering the lattice generated by linear combinations of the vectors $(1, 0, f_1 * h), (0, 1, h), (0, 0, q)$, where each entry corresponds to N entries in the lattice. The vector (f_2, f_3, g) will be a very short vector, but the increase of the dimension of the lattice by N, without any corresponding increase in the determinant of the lattice, leads to a considerably harder lattice reduction problem. As this attack also requires a correct guess of f_1 we will not consider it further.

8   Explicit Algorithm for Computing Parameters

Algorithm 1 determines the smallest recommended N that allows for k bit security. Additional details, such as recommendations on how to efficiently perform the search in Line 16, may be found in our implementation available at [1].

8.1   Sample Parameter Generation

We will ignore the implicit outer loop over security parameters and consider the case of $N = 401$ starting from Line 3.

Our recommendations for the key structure suggests taking $d_g = 134$, $d_1 = 8$, $d_2 = 8$, $d_3 = 6$. Taking $d_m = 102$ satisfies Eq. 5 with a probability of $2^{-10.4}$ of rejecting a message representative due to its coefficient sum. A direct meet-in-the-middle attack on the product form key space will involve testing approximately 2^{145} candidates. As this is an upper bound on the security of the parameter set we will ensure that our decryption failure probability is less than 2^{-145}. This implores us to take $q = 2048$, for which there is, by Eq. 8, a decryption failure probability of 2^{-217}.

In order to finish the parameter derivation we need a tighter estimate on its security. It may be significantly less than 145, in which case we may be able to reduce q.

We estimate the security of the parameter set by minimizing the adversary's expected cost over choices of the hybrid attack parameter K. Equation 4 specifies, for each K, the root Hermite factor, δ, that must be reached during the lattice reduction phase of the hybrid attack in order for the combinatorial stage to be successful. We use the BKZ-2.0 simulator of [4] to determine the blocksize and number of rounds of BKZ that will be required to reach root Hermite factor.

To turn the blocksize and iteration count into a concrete security estimate we need estimates on the number of nodes visited per call to the enumeration subroutine of BKZ. Table 2 summarizes upper bounds given by Chen and Nguyen in [4] and in the full version of the same paper [5]. The estimates of the full version are significantly lower than the original, and have perhaps not recieved the same scrutiny. In what follows we will consider the implications of both estimates.

Table 2. Upper bounds on $\log_2$ number of nodes enumerated in one call to enumeration subroutine of BKZ-2.0 as reported in the original and full versions of the paper.

β	100	110	120	130	140	150	160	170	180	190	200	210	220	230	240	250
LogNodes(β) [5]	39	44	49	54	60	66	72	78	84	96	99	105	111	120	127	134

To facilitate computer search for parameters we fit curves to the estimates in Table 2, and following [4] we estimate the per-node cost, as 2^7 operations. The resulting predictions for the cost of the lattice reduction stage, in terms of the

---

Algorithm 1. NTRUEncrypt parameter generation

Input: Desired security level k.

1: Let n_j be the jth value, ordered by magnitude, from the list of first 100 primes > 100 for which $\text{ord}_{(\mathbb{Z}/N\mathbb{Z})^*}(2) = (N-1)$, i.e. (2) is inert.

2: Set $j = 1$.

3: Set $N = n_j$.

4: Set $d_g = \lfloor \frac{N}{3} \rfloor$.

5: Set $d_1 = \left\lceil \frac{1}{4} \left(\sqrt{1 + \frac{8N}{3}} - 1 \right) \right\rceil$ {The next integer above the positive root of $2x^2 + x - N/3$.}

6: Set $d_2 = \left\lceil \left(\frac{N}{3} - d_1 \right) / (2d_1) \right\rceil$.

7: Set $d_3 = \max \left(\lceil \frac{d_1}{2} + 1 \rceil, \lceil \frac{N}{3} - 2d_1 d_2 \rceil \right)$.

8: Set d_m to be the largest value satisfying Eq. 5.

9: Set $k_1 = \lfloor \frac{1}{2} \log_2 (|\mathcal{P}_N(d_1, d_2, d_3)|/N) \rfloor$. {Cost of direct combinatorial search gives an upper bound on the security.}

10: **if** $k_1 < k$ **then**

11:    Increment j.

12:    Goto line 3.

13: **end if**

14: Set σ according to Eq. 6.

$$\sigma = \left((4d_1 d_2 + 2d_3) \cdot \frac{N - d_m + 2d_g + 1}{N} \right)^{1/2}.$$

15: Set q to be the smallest power of 2 satisfying

$$N \cdot \text{erfc} \left((q-2)/(6\sqrt{2}\sigma) \right) < 2^{-k_1}.$$

{Estimate security}

16: Search for a hybrid parameter K that minimizes the maximum of the cost estimates for hybrid attacks. Equation 4 gives the cost of the lattice reduction, and Eqs. 4 and 7 give the cost of combinatorial search for key- and message-recovery attacks respectively. Let k_2 be the corresponding security estimate.

17: **if** $k > \min(k_1, k_2)$ **then**

18:    Increment j.

19:    Go to Line 3.

20: **end if**

21: Let $q' = q/2$.

22: **if** $N \cdot \text{erfc} \left((q'-2)/(6\sqrt{2}\sigma) \right) < 2^{-k}$ **then**

23:    Set $q = q'$

24:    Repeat security estimate (Line 16) with modulus q' and set k_2 equal to the result.

25:    Go to Line 17.

26: **end if**

Output: $[N, q, d_1, d_2, d_3, d_g, d_m]$.

---

blocksize, the dimension of the sublattice to be reduced, and number of rounds are thus:

$$\text{LogNodes}(\beta) = 0.12081 \cdot \beta \log_2(\beta) - 0.42860 \cdot \beta$$

$$\text{BKZCost}(dim, \beta, rounds) = \text{LogNodes}(\beta) + \log_2(dimension \cdot rounds) + 7.$$

Finally our security estimate requires a search over K to balance the cost of lattice reduction against the cost of combinatorial search given by Eq. 4.

Fixing $K = 154$ the BKZ-2.0 simulator suggests that 10 rounds of BKZ-197 will achieve to the requisite $\delta = 1.0064$. The BKZCost estimate suggests that this reduction will require 2^{116} operations, matching the cost of 2^{116} given by Eq. 4 for the combinatorial search step.

We find that we cannot decrease q without violating the constraint on the decryption failure probability, and we are done.

The parameter set we have just (re-)derived originally appeared in the EESS #1 standard at the 112 bit security level. All four product-form parameter sets from EESS #1 are reviewed in Table 3 with security estimates following the above analysis. Note that while the algorithm in Sect. 8 rederives the $N = 401$ parameter set almost exactly (d_g is 133 in EESS #1), this is not true for the $N = 593$ and $N = 743$ parameter sets. In particular, all four of the published parameter sets take $q = 2048$, and this does not lead to a formally negligible probability of decryption failure for $N = 593$ or $N = 743$. Note also that the number of prime ideals lying above (2) is more than recommended for $N = 439$ and $N = 593$. Table 3 presents security estimates for the standardized parameters rather than those that would be output by the algorithm of Sect. 8.

Table 3.

EESS #1 Parameter sets and security estimates

Original security est	N	q	$(d_1, d_2, d_3, d_g, d_m)$	Hybrid attack parameters					Product form search cost	$\log_2$ dec. fail prob
				Dim	β	Rounds	K	Cost		
112	401	2048	(8, 8, 6, 133, 101)	532	197	10	154	**116**	145	-217
128	439	2048	(9, 8, 5, 146, 112)	571	221	10	174	**133**	147	-195
192	593	2048	(10, 10, 8, 197, 158)	732	316	8	261	201	**193**	-139
256	743	2048	(11, 11, 15, 247, 204)	880	407	8	350	272	**256**	-112

9 New Parameters

The parameter derivations above do not take quantum adversaries into consideration. The time/space tradeoff in the hybrid attack can be replaced (trivially) by a Grover search to achieve the same asymptotic time complexity as the hybrid attack with a space complexity that is polynomial in N. One may expect that a quantum time/space tradeoff could do even better, however this seems unlikely given the failure of quantum time/space tradeoffs against collision problems in

Table 4.

Classical security est	Quantum security est	N	q	$(d_1, d_2, d_3, d_g, d_m)$	Hybrid attack parameters					Product form search cost	$\log_2$ dec. fail prob
					Dim	β	Rounds	K	Cost		
128	128	443	2048	(9, 8, 5, 148, 115)	575	222	11	177	**133**	147	-196
192	128	587	2048	(10, 10, 8, 196, 157)	723	311	9	258	197	**193**	-139
256	128	743	2048	(11, 11, 15, 247, 204)	880	407	8	350	272	**256**	-112

Post-quantum parameter sets and security estimates

other domains [3]. Several proposals in this direction have been made, such as [7], however these assume unrealistic models of quantum computation. For now, it seems that the best quantum attack on NTRUEncrypt is the hybrid attack with meet-in-the-middle search replaced by Grover search in the Kth projected lattice.

Fluhrer has noted that there are weaknesses in the EESS #1 parameter sets assuming worst-case cost models for quantum computation [7]. In particular, if one Grover iteration is assigned cost equivalent to one classical operation, such as a multiplication in R, then attacks on the hash functions used in key generation and encryption can break the EESS #1 parameter sets.

Developing a realistic quantum cost model is outside the scope of this work. However we can easily provide parameter sets that are secure in Fluhrer's model. Since this model is in some sense a worst-case for quantum computation (it assigns the smallest justifiable cost to quantum operations) the quantum security estimates can be assumed to be quite conservative. In addition to using the parameters in Table 4 one must ensure that pseudorandom polynomial generation functions are instantiated with SHA-256, and that the message is concatenated with a random string b that is at least 256 bits. One should also ensure that any deterministic random bit generators used in key generation or encryption are instantiated with at least 256 bits of entropy from a secure random source.

The parameter sets for $N = 443$ and $N = 587$ in Table 4 are new, $N = 743$ is the same as ees743ep1 from EESS #1.

References

1. NTRU OpenSource Project.online. https://github.com/NTRUOpenSource Project/ntru-crypto
2. 2015. https://www.ntru.com/ntru-challenge/
3. Bernstein, D.J.: Cost analysis of hash collisions: will quantum computers make SHARCS obsolete? (2009). http://cr.yp.to/papers.html#collisioncost
4. Chen, Y., Nguyen, P.Q.: BKZ 2.0: better lattice security estimates. In: Lee, D.H., Wang, X. (eds.) ASIACRYPT 2011. LNCS, vol. 7073, pp. 1–20. Springer, Heidelberg (2011). doi:10.1007/978-3-642-25385-0_1
5. Chen, Y., Nguyen, P.Q.: BKZ 2.0: Better lattice security estimates (full version) (2011). http://www.di.ens.fr/~ychen/research/Full_BKZ.pdf

6. Ducas, L., Durmus, A., Lepoint, T., Lyubashevsky, V.: Lattice signatures and bimodal Gaussians. In: Canetti, R., Garay, J.A. (eds.) CRYPTO 2013. LNCS, vol. 8042, pp. 40–56. Springer, Heidelberg (2013). doi:10.1007/978-3-642-40041-4_3

7. Fluhrer, S.R.: Quantum cryptanalysis of NTRU. IACR Cryptology ePrint Archive, 2015:676 (2015)

8. Gama, N., Nguyen, P.Q.: Predicting lattice reduction. In: Smart, N. (ed.) EURO-CRYPT 2008. LNCS, vol. 4965, pp. 31–51. Springer, Heidelberg (2008). doi:10. 1007/978-3-540-78967-3_3

9. Hirschhorn, P.S., Hoffstein, J., Howgrave-Graham, N., Whyte, W.: Choosing NTRUEncrypt parameters in light of combined lattice reduction and MITM approaches. In: Abdalla, M., Pointcheval, D., Fouque, P.-A., Vergnaud, D. (eds.) ACNS 2009. LNCS, vol. 5536, pp. 437–455. Springer, Heidelberg (2009). doi:10. 1007/978-3-642-01957-9_27

10. Hoffstein, J., Pipher, J., Schanck, J.M., Silverman, J.H., Whyte, W., Zhang, Z.: Choosing Parameters for NTRUEncrypt (full version). IACR Cryptology ePrint Archive 2015:708 (2015)

11. Hoffstein, J., Pipher, J., Silverman, J.H.: NTRU: a ring-based public key cryptosystem. In: Buhler, J.P. (ed.) ANTS 1998. LNCS, vol. 1423, pp. 267–288. Springer, Heidelberg (1998). doi:10.1007/BFb0054868

12. Hoffstein, J., Silverman, J.H.: Optimizations for NTRU (2000)

13. Hoffstein, J., Silverman, J.H.: Random small hamming weight products with applications to cryptography. Discrete Appl. Math. **130**(1), 37–49 (2003)

14. Hoffstein, J., Silverman, J.H., Whyte, W.: Provable Probability Bounds for NTRU-Encrypt Convolution (2007). http://www.ntru.com

15. Howgrave-Graham, N.: A hybrid lattice-reduction and meet-in-the-middle attack against NTRU. In: Menezes, A. (ed.) CRYPTO 2007. LNCS, vol. 4622, pp. 150–169. Springer, Heidelberg (2007). doi:10.1007/978-3-540-74143-5_9

16. Howgrave-Graham, N., Silverman, J.H., Whyte, W.: Choosing parameter sets for NTRUEncrypt with NAEP and SVES-3. In: Menezes, A. (ed.) CT-RSA 2005. LNCS, vol. 3376, pp. 118–135. Springer, Heidelberg (2005). doi:10.1007/ 978-3-540-30574-3_10

17. Stehlé, D., Steinfeld, R.: Making NTRU as secure as worst-case problems over ideal lattices. In: Paterson, K.G. (ed.) EUROCRYPT 2011. LNCS, vol. 6632, pp. 27–47. Springer, Heidelberg (2011). doi:10.1007/978-3-642-20465-4_4

Encoding-Free ElGamal-Type Encryption Schemes on Elliptic Curves

Marc Joye[1]([⊠]) and Benoît Libert[2]

[1] NXP Semiconductors (USA), San Jose, USA
marc.joye@nxp.com
[2] CNRS, Laboratoire LIP (CNRS, ENSL, U. Lyon, Inria, UCBL),
ENS de Lyon, Lyon, France

Abstract. At PKC 2006, Chevallier-Mames, Paillier, and Pointcheval proposed a very elegant technique over cyclic subgroups of $\mathbb{F}_p^*$ eliminating the need to encode the message as a group element in the ElGamal encryption scheme. Unfortunately, it is unclear how to adapt their scheme over elliptic curves. In a previous attempt, Virat suggested an adaptation of ElGamal to elliptic curves over the ring of dual numbers as a way to address the message encoding issue. Advantageously the resulting cryptosystem does not require encoding messages as points on an elliptic curve prior to their encryption. Unfortunately, it only provides one-wayness and, in particular, it is not (and was not claimed to be) semantically secure.

This paper revisits Virat's cryptosystem and extends the Chevallier-Mames *et al.*'s technique to the elliptic curve setting. We consider elliptic curves over the ring $\mathbb{Z}/p^2\mathbb{Z}$ and define the underlying class function. This yields complexity assumptions whereupon we build new ElGamal-type encryption schemes. The so-obtained schemes are shown to be semantically secure and make use of a very simple message encoding: messages being encrypted are viewed as elements in the range $[0, p-1]$. Further, our schemes come equipped with a partial ring-homomorphism property: anyone can add a constant to an encrypted message –or– multiply an encrypted message by a constant. This can prove helpful as a blinding method in a number of applications. Finally, in addition to practicability, the proposed schemes also offer better performance in terms of speed, memory, and bandwidth.

Keywords: Public-key encryption · ElGamal encryption · Elliptic curves · Class function · Standard model

1 Introduction

Encryption is one of the most fundamental cryptographic primitives. It allows parties to exchange data privately. In the *asymmetric* setting, a (certified) public encryption key is made publicly available and the matching decryption key is kept private. Anyone can encrypt messages with the public key but only the intended recipient (possessing the private key) is able to decrypt ciphertexts. We refer the reader to Appendix A for background on public-key encryption.

© Springer International Publishing AG 2017
H. Handschuh (Ed.): CT-RSA 2017, LNCS 10159, pp. 19–35, 2017.
DOI: 10.1007/978-3-319-52153-4_2

ElGamal Encryption. The classical ElGamal public-key encryption scheme [12] readily extends to any group $\mathbb{G}$ wherein computing discrete logarithms is assumed to be intractable. In order to avoid sub-group attacks using the Pohlig-Hellman algorithm [25], the underlying group is usually restricted to a prime-order group $\mathbb{G} = \langle g \rangle$; see also [4]. We let q denote the order of $\mathbb{G}$.

The description of $\mathbb{G}$ and the generator g are made public. A random element $y = g^x \in \mathbb{G}$ is drawn for some randomly chosen $x \xleftarrow{R} \mathbb{Z}/q\mathbb{Z}$. The public-key/private-key pair is defined by (pk, sk) with $pk = \{\mathbb{G}, q, g\}$ and $sk = \{x\}$; the message space is $\mathcal{M} = \mathbb{G}$. The encryption of a message $m \in \mathbb{G}$ is given by the pair (c_1, c_2) where

$$c_1 = g^r \quad \text{and} \quad c_2 = m\, y^r$$

for a random integer $r \xleftarrow{R} \mathbb{Z}/q\mathbb{Z}$. Given the ciphertext $C = (c_1, c_2) \in \mathbb{G} \times \mathbb{G}$, message m is then recovered thanks to secret key x as $m = c_2/c_1{}^x$.

As described above, the ElGamal scheme is known to meet the IND-CPA security notion under the decisional Diffie-Hellman (DDH) assumption [29]. Loosely speaking, the *DDH assumption* states that no efficient algorithm can distinguish between the distributions (g, g^a, g^b, g^{ab}) and (g, g^a, g^b, g^c) where $a, b, c \xleftarrow{R} \mathbb{Z}/q\mathbb{Z}$.

Message Encoding. Elliptic curve cryptography [22, 24] benefits from the absence of sub-exponential algorithms to solve the underlying hard problem, the elliptic curve discrete logarithm problem. Elliptic curve cryptosystems therefore feature smaller key sizes, which results in significant gains in speed and memory. When applied to elliptic curves over a finite field, ElGamal encryption compels to express the plaintext message m as a point on an elliptic curve or, more precisely, as a point on a prime-order subgroup $\mathbb{G}$ thereof. This requires an injective encoding function mapping the message space to $\mathbb{G}$. Such encodings are provided in [2, 14, 15] for certain elliptic curves. Unfortunately they do not apply to prime-order elliptic curves as those recommended in most cryptographic standards.

Another option is to leverage the property that any element $w \in \mathbb{G} = \langle g \rangle$ is uniquely represented as $w = g^t$ for some $t \in \mathbb{Z}/q\mathbb{Z}$. This leads to the 'exponent' ElGamal scheme (see e.g. [10]). A message $m \subseteq \mathbb{Z}/q\mathbb{Z}$ is encoded as g^m. The corresponding ciphertext then becomes (c_1, c_2) with $c_1 = g^r$ and $c_2 = g^m\, y^r$ for some $r \xleftarrow{R} \mathbb{Z}/q\mathbb{Z}$. Unfortunately, decryption now involves the computation of a discrete logarithm in $\mathbb{G}$: m is the discrete logarithm of $c_2/c_1{}^x$ w.r.t. base g. Since discrete logarithms are supposed to be hard in $\mathbb{G}$, this limits the message space to a small subset of $\mathbb{Z}/q\mathbb{Z}$ so that discrete logarithms can be solved through, e.g., exhaustive search or Pollard's lambda method [26].

Yet another option is to modify the scheme by introducing a hash function. The resulting scheme is referred to as the hash-ElGamal scheme. In more details, let $h : \mathbb{G} \to \{0, 1\}^\ell, w \mapsto h(w)$ be a hash function that maps group elements to ℓ-bit strings. The message space is defined as $\mathcal{M} = \{0, 1\}^\ell$. The encryption of a message $m \in \mathcal{M}$ is given by (c_1, c_2) with $c_1 = g^r$ and $c_2 = m \oplus h(y^r)$. This variant elegantly solves the encoding problem. On the downside, unless one is willing to model h as a random oracle, the security analysis requires either additional assumptions on h – which should behave as a computationally secure (a.k.a. entropy-smoothing [27]) key derivation function – or larger key sizes [3,17,18]. Indeed, as observed in [18, Appendix A], using an information-theoretically secure key derivation function, the Leftover Hash Lemma [20,21] would require y^r to come from a distribution with about 300 bits of min-entropy in order to produce a 128-bit symmetric encryption key.

To overcome the message-encoding issue, Virat came with a different approach in [30]. Her idea consists in working with an elliptic curve over the ring $\mathbb{F}_p[\varepsilon]$, namely the ring of dual numbers over the prime field $\mathbb{F}_p$. Doing so, the message space becomes $\mathbb{F}_p$; i.e., messages are now viewed as integers in the set $\{0, \ldots, p-1\}$ rather than points on an elliptic curve.

Homomorphism Property. Malleability of ciphertexts is usually seen as an undesirable property. It proves nevertheless very useful in certain applications. Examples include electronic voting, electronic commerce or, more generally, privacy-preserving computations. The basic ElGamal scheme satisfies a homomorphism property with respect to the group law in $\mathbb{G}$. Namely, if $\cdot$ denotes the group law in $\mathbb{G}$ then given the ElGamal encryption of messages $m_1, m_2 \in \mathbb{G}$, anyone can derive the encryption of $m_1 \cdot m_2$. Indeed, letting $C_1 = (c_{1,1}, c_{1,2})$ and $C_2 = (c_{2,1}, c_{2,2})$ the respective encryption of m_1 and m_2, with $c_{i,1} = g^{r_i}$ and $c_{i,2} = m_i y^{r_i}$ ($i \in \{1, 2\}$), it is easily checked that

$$C_3 = (c_{1,1} \cdot c_{2,1}, c_{2,1} \cdot c_{2,2})$$

is the encryption of message $m_3 = m_1 \cdot m_2 \in \mathbb{G}$. For elliptic-curve ElGamal, including Virat's cryptosystem, this translates into the encryption of the (elliptic-curve) addition of two points. When the exponent variant is used, composing two ciphertexts yields the encryption of a message $m_3 = m_1 + m_2 \pmod{q}$, where messages m_1 and m_2 are viewed as elements in a small subset of $\mathbb{Z}/q\mathbb{Z}$.

Hash ElGamal is only *partially* homomorphic, w.r.t. the XOR operator. Given the encryption of a message m, anyone can compute the encryption of a message $m' = m \oplus K$ for any chosen value $K \in \{0, 1\}^\ell$. If $C = (c_1, c_2)$ with $c_1 = g^r$ and $c_2 = m \oplus h(y^r)$ then $C' = (c_1, c_2')$ with $c_2' = K \oplus c_2$ is the hash-ElGamal encryption of m'. This holds true, regardless of the underlying group. In particular, this is verified for elliptic curves.

Our Contribution. Compared to the classical elliptic-curve ElGamal encryption scheme, there are several drawbacks in Virat's cryptosystem. First it is computationally more demanding. Second it leads to an increased ciphertext expansion ratio. This is particularly damaging for elliptic curve cryptosystems as they are primarily designed to reduce the bandwidth. Third and more importantly, the security of the scheme is rather weak. It is only shown to be one-way; in particular, it does *not* provide semantic security.

We propose in this paper new ElGamal-type cryptosystems that enjoy the same advantage as Virat's cryptosystem (namely, no message encoding as points on elliptic curves) but without its drawbacks. In an earlier work, Chevallier-Mames *et al.* [9] astutely observe that certain mathematical properties of integers modulo p^2, where p is a prime number, allow getting rid of the message encoding from the classical ElGamal cryptosystem. Unfortunately, the solution of [9] is not known to be readily instantiable over elliptic curve subgroups. As a consequence, the Chevallier-Mames *et al.* [9] system loses the benefit of shorter keys enabled by elliptic curve cryptography. In this work, we solve a problem left open by Chevallier-Mames *et al.* [9] and provide an adaptation of their scheme [9] to the elliptic curve setting. The resulting encryption schemes features the same ciphertext expansion ratio as [9] and retains the partial homomorphism properties (additive or multiplicative). We prove that they are semantically secure in the standard model under a natural hardness assumption. We also describe a chosen-ciphertext secure extension of these schemes.

2   Encoding-Free ElGamal Schemes

2.1   Virat's Cryptosystem

Let $\mathbb{K}$ be a finite field of characteristic $p \neq 2, 3$. The *ring of dual numbers of* $\mathbb{K}$ is $\mathbb{K}[\varepsilon]$ with $\varepsilon^2 = 0$.

Consider the elliptic curve E over $\mathbb{K}[\varepsilon]$ given by the Weierstraß equation

$$E : y^2 = x^3 + ax + b \tag{1}$$

with $a, b \in \mathbb{K}[\varepsilon]$ and $4a^3 + 27b^2 \neq 0$. The set of points $(x, y) \in \mathbb{K}[\varepsilon] \times \mathbb{K}[\varepsilon]$ satisfying this equation together with the points at infinity, $\boldsymbol{O}_k = (k\varepsilon : 1 : 0)$ with $k \in \mathbb{K}$, form an Abelian group under the chord-and-tangent rule. Explicit addition formulæ are provided in [31, Table 2.1]. This group is denoted by $E(\mathbb{K}[\varepsilon])$ and its order by $\#E(\mathbb{K}[\varepsilon])$. Since $E(\mathbb{K}[\varepsilon])$ contains the p-torsion subgroup formed by the points at infinity, its order is a multiple of p.

Virat's cryptosystem relies on elliptic curves over $\mathbb{F}_p[\varepsilon]$ for some prime $p > 3$. Hence let E be an elliptic curve over $\mathbb{F}_p[\varepsilon]$ as per Eq. (1) of order pq for some prime $q \neq p$, and let $\hat{\boldsymbol{P}}$ be a generator of $E(\mathbb{F}_p[\varepsilon])$.

KeyGen(1^λ). On input security parameter λ, generate a cyclic group $E(\mathbb{F}_p[\varepsilon]) = \langle \hat{\boldsymbol{P}} \rangle$ of order pq as above. Next, choose a random integer $x \xleftarrow{R} \mathbb{Z}/q\mathbb{Z}$ and compute $\boldsymbol{Y} = [xp]\hat{\boldsymbol{P}}$.

The public key is $pk = \{E(\mathbb{F}_p[\varepsilon]), q, \hat{\boldsymbol{P}}, \boldsymbol{Y}\}$ and the private key is $sk = \{x\}$.

Encrypt(pk, m). The encryption of a message $m \in \mathbb{F}_p$ is given as follows:

1. Choose a random integer $r \xleftarrow{R} \mathbb{Z}/q\mathbb{Z}$;
2. Choose a random finite point $(x_0, y_0) \xleftarrow{R} E(\mathbb{F}_p)$;
3. Define $\hat{\boldsymbol{M}} = (x_0 + m\varepsilon, y_0 + y_1\varepsilon)$ where y_1 is the unique solution in $\mathbb{F}_p$ such that $\hat{\boldsymbol{M}} \in E(\mathbb{F}_p[\varepsilon])$;
4. Compute the points $\boldsymbol{C_1} = [rp]\hat{\boldsymbol{P}}$ and $\hat{\boldsymbol{C_2}} = \hat{\boldsymbol{M}} + [r]\boldsymbol{Y}$;
5. Output the ciphertext $C = (\boldsymbol{C_1}, \hat{\boldsymbol{C_2}})$.

Decrypt(sk, C). The decryption of $C = (\boldsymbol{C_1}, \hat{\boldsymbol{C_2}})$ is obtained as $\hat{\boldsymbol{M}} = \hat{\boldsymbol{C_2}} - [x]\boldsymbol{C_1}$ using secret key x, which in turn yields the value of m.

In a variant, Virat suggests to define the elliptic curve E over $\mathbb{F}_p[\varepsilon]$ but with curve parameters $a, b \in \mathbb{F}_p$. It is then shown that the scheme is one-way under the computational Diffie-Hellman assumption in $E(\mathbb{F}_p)$ [30, Theorem 6.4].

Given the x-coordinate of a finite point in $E(\mathbb{F}_p[\varepsilon])$, there are two possible values for its y-coordinate. So $2|p| + 1$ bits suffice to represent $\boldsymbol{C_1}$ or $\hat{\boldsymbol{C_2}}$, leading to a ciphertext expansion ratio of $4 \div 1$ [30, Sect. 5.2].

Remark 1. When the curve parameters $a, b \in \mathbb{F}_p$, Lemma 1 in [1] implies that for every finite point $\hat{\boldsymbol{P}} = (x_0 + x_1\varepsilon, y_0 + y_1\varepsilon) \in E(\mathbb{F}_p[\varepsilon])$ there exists a unique $k \in \mathbb{F}_p$ such that $\hat{\boldsymbol{P}} = \boldsymbol{P} + \boldsymbol{O}_k$ with $\boldsymbol{P} = (x_0, y_0) \in E(\mathbb{F}_p)$. It thus turns out that $[p]\hat{\boldsymbol{P}} = [p]\boldsymbol{P} + [p](k\varepsilon : 1 : 0) = [p]\boldsymbol{P} \in E(\mathbb{F}_p)$. In this case, it is interesting to define the public key as $pk = \{E(\mathbb{F}_p[\varepsilon]), q, \boldsymbol{Q}, \boldsymbol{Y}\}$ where $\boldsymbol{Q} = [p]\hat{\boldsymbol{P}} \in E(\mathbb{F}_p)$ and to evaluate $\boldsymbol{C_1}$ as $\boldsymbol{C_1} = [r]\boldsymbol{Q} \in E(\mathbb{F}_p)$. The ciphertext expansion ratio then drops to $3 \div 1$ using a compressed point representation (i.e., $\boldsymbol{C_1}$ is represented with $|p| + 1$ bits and $\hat{\boldsymbol{C_2}}$ with $2|p| + 1$ bits).

2.2 The Chevallier-Mames–Paillier–Pointcheval Scheme

The scheme of Chevallier-Mames *et al.* [9] is based on the class function over cyclic subgroups of $\mathbb{F}_p^*$. Specifically, for primes p and q such that $q \mid p - 1$, given a cyclic subgroup $\langle g \rangle \subseteq \mathbb{F}_p^*$ of order q, the class of $w = g^a \bmod p$ (w.r.t. $\hat{g}$) is denoted by $[\![w]\!]$ and is defined as the unique integer in $\mathbb{Z}/p\mathbb{Z}$ such that

$$\hat{g}^{\mathrm{CRT}([\![w]\!], a)} \bmod p^2 = w$$

for some $\hat{g} \in (\mathbb{Z}/p^2\mathbb{Z})^*$ of order pq and such that $\hat{g} \equiv g \pmod{p}$, and where $\mathrm{CRT}([\![w]\!], a)$ is an integer such that

$$\mathrm{CRT}([\![w]\!], a) \equiv [\![w]\!] \pmod{p} \quad \text{and} \quad \mathrm{CRT}([\![w]\!], a) \equiv a \pmod{q};$$

see [9, Sect. 4.1]. For example, if $\hat{g} = (1 - kp)\, g^p \bmod p^2$ with $k := \frac{(p-1)}{q}$ then

$$[\![w]\!] = \frac{(w^q \bmod p^2) - 1}{p} \bmod p.$$

Proof. Observe that $\hat{g} \equiv g^p \equiv g \pmod{p}$ as required. Remark also that, as elements in $(\mathbb{Z}/p^2\mathbb{Z})^*$, $1 - kp \pmod{p^2}$ is of order p and $g^p \pmod{p^2}$ is of order q. Hence, it follows that $w \equiv \hat{g}^{\mathrm{CRT}([\![w]\!], a)} \equiv (1 - kp)^{[\![w]\!]} (g^p)^a \pmod{p^2}$ and thus $w^q \equiv (1 - kp)^{[\![w]\!]q} \equiv 1 - (k\,[\![w]\!]\,q)p \equiv 1 + [\![w]\!]\,p \pmod{p^2}$. $\square$

Equipped with such an efficiently computable class function, the encryption scheme of Chevallier-Mames *et al.* goes as follows.

KeyGen(1^λ). On input security parameter λ, generate a prime p and an element $g \in \mathbb{F}_p^*$ of large prime order q. Next, compute $y = g^x \bmod p$ for some random integer $x \xleftarrow{R} \mathbb{Z}/q\mathbb{Z}$. The public key is $pk = \{\mathbb{F}_p^*, q, g, y\}$ and the private key is $sk = \{x\}$.

Encrypt(pk, m). The encryption of $m \in \mathbb{Z}/p\mathbb{Z}$ is given by the following algorithm:
1. Choose a random $r \stackrel{R}{\leftarrow} \mathbb{Z}/q\mathbb{Z}$. Compute $c_1 = g^r \bmod p$ and $d = y^r \bmod p$;
2. Define $c_2 = m + [\![d]\!] \pmod{p}$;
3. Output the ciphertext $C = (c_1, c_2)$.

Decrypt(sk, C). $C = (c_1, c_2)$ is decrypted as $m = c_2 - [\![c_1{}^x \bmod p]\!] \pmod{p}$ using the private key $sk = x$.

3 New Cryptosystems

Rather than considering elliptic curves over the ring $\mathbb{F}_p[\varepsilon]$, we work with elliptic curves defined over the ring $\mathbb{Z}/p^2\mathbb{Z}$. Borrowing the terminology of [9], this allows us to define a class function whereupon new ElGamal-type cryptosystems are derived. See also [16] for another family of cryptosystems making use of elliptic curves defined over a ring.

3.1 Class Function on Elliptic Curves

Since $\mathbb{F}_p = \mathbb{Z}/p\mathbb{Z} \subset \mathbb{Z}/p^2\mathbb{Z}$, we can view an elliptic curve given by a Weierstraß equation (with curve parameters $a, b \in \mathbb{F}_p$) over the ring $\mathbb{Z}/p^2\mathbb{Z}$. In order to deal with the points at infinity, we regard the projective form

$$Y^2 Z = X^3 + aX Z^2 + bZ^3.$$

The set of points on this elliptic curve over $\mathbb{Z}/p^2\mathbb{Z}$ is denoted by $E(\mathbb{Z}/p^2\mathbb{Z})$. The subset of points that reduce to $\boldsymbol{O} = (0 : 1 : 0)$ modulo p is denoted by $E_1(\mathbb{Z}/p^2\mathbb{Z})$; see [28, Sect. 2].

Proposition 1. *Using the previous notations, we have*

$$E_1(\mathbb{Z}/p^2\mathbb{Z}) = \{(\alpha p : 1 : 0) \mid 0 \le \alpha \le p - 1\}.$$

Proof. By definition, we have $E_1(\mathbb{Z}/p^2\mathbb{Z}) = \{(X : Y : Z) \in E(\mathbb{Z}/p^2\mathbb{Z}) \mid (X : Y : Z) \equiv (0 : 1 : 0) \pmod{p}\}$. Since $Y \equiv 1 \pmod{p}$ we obviously have $Y \not\equiv 0 \pmod{p^2}$ and so we can write $E_1(\mathbb{Z}/p^2\mathbb{Z}) = \{(\frac{X}{Y} : 1 : \frac{Z}{Y}) \in E(\mathbb{Z}/p^2\mathbb{Z}) \mid (X : Y : Z) \equiv (0 : 1 : 0) \pmod{p}\} = \{(\alpha p : 1 : \gamma p) \in E(\mathbb{Z}/p^2\mathbb{Z}) \mid 0 \le \alpha, \gamma \le p - 1\}$. Plugging $(\alpha p : 1 : \gamma p)$ into the Weierstraß equation yields $\gamma p = 0 \pmod{p^2} \iff \gamma = 0 \pmod{p}$. We therefore get $E_1(\mathbb{Z}/p^2\mathbb{Z}) = \{(\alpha p : 1 : 0) \mid 0 \le \alpha \le p - 1\}$. $\square$

The theory of formal groups [28, Proposition IV.3.2] implies that $E_1(\mathbb{Z}/p^2\mathbb{Z})$ is a group isomorphic to the additive group $(\mathbb{Z}/p\mathbb{Z})^+$. We have

$$\Gamma : E_1(\mathbb{Z}/p^2\mathbb{Z}) \xrightarrow{\sim} (\mathbb{Z}/p\mathbb{Z})^+, (\alpha p : 1 : 0) \longmapsto \alpha.$$

Hence, the sum of two elements $(\alpha_1 p : 1 : 0)$ and $(\alpha_2 p : 1 : 0)$ in $E_1(\mathbb{Z}/p^2\mathbb{Z})$ is given by $(\alpha_3 p : 1 : 0)$ with $\alpha_3 = (\alpha_1 + \alpha_2) \bmod p$. This also implies that $E_1(\mathbb{Z}/p^2\mathbb{Z})$ is a cyclic group of order p. Letting $\boldsymbol{U} = (p : 1 : 0)$, we can write $E_1(\mathbb{Z}/p^2\mathbb{Z}) = \langle \boldsymbol{U} \rangle$.

Given a finite point $P = (x, y) \in E(\mathbb{F}_p)$, with $y \neq 0$, we define

$$\Delta(P) = \frac{(x^3 + ax + b - y^2) \bmod p^2}{p} \quad \text{and} \quad \psi(P) = \frac{\Delta(P)}{2y} \bmod p.$$

[In the definition of $\Delta(P)$, point P is lifted; i.e., its coordinates x and y are viewed as integers.]

This gives rise to the map

$$\Psi : E(\mathbb{F}_p) \to E(\mathbb{Z}/p^2\mathbb{Z}), \begin{cases} O \mapsto O \\ (x, y) \mapsto (x, y + \psi(P)p). \end{cases}$$

To ease the notation, we will sometimes write $\tilde{P}$ for $\Psi(P)$.

We assume that E is not an anomalous curve (i.e., $\#E(\mathbb{F}_p) \neq p$) and we let $q = \text{ord}_E(P)$ denote the order of point $P \in E(\mathbb{F}_p)$. We define $V = [p]\tilde{P}$. Clearly, we have that V is of order q.

Consider now the subgroups $\mathbb{G} = \langle P \rangle \subseteq E(\mathbb{F}_p)$ of order q and $\hat{\mathbb{G}} = \langle U, V \rangle \subseteq E(\mathbb{Z}/p^2\mathbb{Z})$ of order pq. Any element $Q \in \hat{\mathbb{G}}$ can uniquely be written as

$$Q = [\beta]U + [\alpha]V \quad \text{for some } \alpha \in \mathbb{Z}/q\mathbb{Z} \text{ and } \beta \in \mathbb{Z}/p\mathbb{Z}. \tag{2}$$

We call integer β the *class of* Q and write $\beta = [\![Q]\!]$. The crucial observation is that $\Psi(\mathbb{G}) \subseteq \hat{\mathbb{G}}$. As a consequence, to any element $Q \in \mathbb{G}$, we similarly define its class as $[\![\tilde{Q}]\!]$. To ease the notation, we will sometimes omit the tilde and simply write $[\![Q]\!]$.

It is worth noticing that computing the class is easy. By definition, from the unique decomposition of a point $Q \in \hat{\mathbb{G}}$ as $Q = [\beta]U + [\alpha]V$ with $\beta = [\![Q]\!]$, it immediately follows that $[q]Q = [q\beta]U = (q\beta p : 1 : 0)$ and thus

$$[\![Q]\!] = \frac{\Gamma([q]Q)}{q} \bmod p. \tag{3}$$

3.2   An Additive Cryptosystem

With the above setting, we can now describe our first cryptosystem. The message space is $\mathbb{Z}/p\mathbb{Z}$ for some prime p.

KeyGen(1^λ). On input security parameter λ, generate an elliptic curve E over the prime field $\mathbb{F}_p$ and a point $P \in E(\mathbb{F}_p)$ of large prime order q. Next, compute the point $Y = [x]P \in E(\mathbb{F}_p)$ for some random integer $x \xleftarrow{R} \mathbb{Z}/q\mathbb{Z}$.

The public key is $pk = \{E(\mathbb{F}_p), q, P, Y\}$ and the private key is $sk = \{x\}$.

Encrypt(pk, m). The encryption of a message $m \in \mathbb{Z}/p\mathbb{Z}$ is given by the following algorithm:

1. Choose a random integer $r \xleftarrow{R} \mathbb{Z}/q\mathbb{Z}$;
2. Compute in $E(\mathbb{F}_p)$ the points $C_1 = [r]P$ and $C_2 = [r]Y$;
3. Compute $\beta = [\![\tilde{C_2}]\!]$;

4. Define $c_2 = m + \beta \pmod{p}$;
5. Output the ciphertext $C = (C_1, c_2)$.

Decrypt(sk, C). The decryption of $C = (C_1, c_2)$ is obtained as $m = c_2 - [\![\Psi([x]C_1)]\!] \pmod{p}$ using the secret key x.

The above cryptosystem presents a number of advantages. First, the ciphertexts are very compact. In their basic version, they feature a $3 \div 1$ ciphertext expansion ratio. This ratio can even be reduced to only $2 \div 1$ by using a compressed representation for C_1. Second, as will be shown in Sect. 4, it meets the standard IND-CPA security level in the standard model (while Virat's cryptosystem only satisfies one-wayness). Third, the proposed cryptosystem is to some extent malleable. More precisely, if (C_1, c_2) denotes the [additive] encryption of a message m then $(C_1, c_2 + K \pmod{p})$ is the encryption of message $m + K \pmod{p}$ for any $K \in \mathbb{Z}/p\mathbb{Z}$. Fourth, encryption is very fast. In an on-line/off-line mode [13], the encryption of a message m only requires a mere addition modulo p. Fifth, in contrast to classical ElGamal on elliptic curves over $\mathbb{F}_p$, no prior encoding of the message as a point on an elliptic curve is required.

3.3 A Multiplicative Cryptosystem

The previous cryptosystem is additive. As $\mathbb{Z}/p\mathbb{Z}$ is equipped with both addition and multiplication, we can define a multiplicative cryptosystem by replacing Step 3.2 in the encryption process accordingly.

KeyGen(1^λ) Idem.

Encrypt(pk, m). The encryption of a message $m \in \mathbb{Z}/p\mathbb{Z}$ is given by the following algorithm:
1. Choose a random integer $r \xleftarrow{R} \mathbb{Z}/q\mathbb{Z}$;
2. Compute in $E(\mathbb{F}_p)$ the points $C_1 = [r]P$ and $C_2 = [r]Y$;
3. Compute $\beta = [\![\tilde{C_2}]\!]$;
4. Define $c_2 = m \cdot \beta \pmod{p}$;
5. Output the ciphertext $C = (C_1, c_2)$.

Decrypt(sk, C). The decryption of $C = (C_1, c_2)$ is obtained as $m = c_2/[\![\Psi([x]C_1)]\!] \pmod{p}$ using the secret key x.

This multiplicative variant shares the advantages as its additive counterpart. The difference resides in that it is partially homomorphic w.r.t. multiplication; that is, if (C_1, c_2) is the [multiplicative] encryption of a message m then $(C_1, c_2 \cdot K \pmod{p})$ is the encryption of message $m \cdot K \pmod{p}$.

4 Security Analysis

4.1 Complexity Assumptions

Let $E(\mathbb{F}_p)$ be an elliptic curve over the prime field $\mathbb{F}_p$ and let $\mathbb{G} \subseteq E(\mathbb{F}_p)$ a cyclic subgroup thereof. Let also P be a generator of $\mathbb{G}$ and $\tilde{P} = \Psi(P) \in E(\mathbb{Z}/p^2\mathbb{Z})$.

We remind that the class of a point $Q \in \mathbb{G}$ (w.r.t. P), denoted $[\![Q]\!]$, is the unique integer $\beta \in \mathbb{Z}/p\mathbb{Z}$ such that $\Psi(Q) = [\beta]U + [\alpha]V$ where $U = (p : 1 : 0)$ and $V = [p]\tilde{P}$.

Given P and $[a]P, [b]P \overset{R}{\leftarrow} \mathbb{G} = \langle P \rangle \subseteq E(\mathbb{F}_p)$, the *elliptic curve class computational Diffie-Hellman (Class-CDH) problem* is to compute the class of $[ab]P$; i.e., $[\![ab]P]\!]$. Likewise, the *elliptic curve class decisional Diffie-Hellman (Class-DDH) problem* is to distinguish between the two distributions $(P, [a]P, [b]P, [\![ab]P]\!])$ and $(P, [a]P, [b]P, \vartheta)$ for $a, b \overset{R}{\leftarrow} [0, \#\mathbb{G})$ and $\vartheta \overset{R}{\leftarrow} \mathbb{Z}/p\mathbb{Z}$. We assume that these two problems are hard.

More formally, define an instance-generating algorithm $\mathcal{G}$ taking as input a security parameter λ and returning (the description of) a cyclic group $\mathbb{G} \subseteq E(\mathbb{F}_p)$, its order $q = \#\mathbb{G}$, and a generator P, as above. We consider the following experiment for an adversary $\mathcal{A}$.

$\mathsf{Class}_{\mathcal{A},\mathcal{G}}(\lambda)$:
1. Run $\mathcal{G}(1^\lambda)$ and obtain $(E(\mathbb{F}_p), q, P)$;
2. Choose $a, b \overset{R}{\leftarrow} \mathbb{Z}/q\mathbb{Z}$ and compute $[a]P$ and $[b]P$;
3. $\mathcal{A}$ is given $(E(\mathbb{F}_p), q, P, [a]P, [b]P)$ and outputs $\beta' \in \mathbb{Z}/p\mathbb{Z}$;
4. The output of the experiment is 1 if $\beta' = [\![C]\!]$ where $C = [ab]P \in E(\mathbb{F}_p)$, and 0 otherwise.

Definition 1. *The* Class-CDH *assumption says that for any probabilistic polynomial-time adversary $\mathcal{A}$ there exists a negligible function* negl *such that*

$$\Pr\big[\mathsf{Class}_{\mathcal{A},\mathcal{G}}(\lambda) = 1\big] \leq \mathrm{negl}(\lambda).$$

Definition 2. *The* Class-DDH *assumption says that for any probabilistic polynomial-time adversary $\mathcal{A}$ there exists a negligible function* negl *such that*

$$\bigg| \Pr\Big[\mathcal{A}\big(E(\mathbb{F}_p), q, P, [a]P, [b]P, [\![ab]P]\!]\big) = 1\Big] -$$

$$\Pr\Big[\mathcal{A}\big(E(\mathbb{F}_p), q, P, [a]P, [b]P, \vartheta\big) = 1\Big] \bigg| \leq \mathrm{negl}(\lambda),$$

where the probabilities are taken over the experiment of running $(E(\mathbb{F}_p), q, P) \leftarrow \mathcal{G}(1^\lambda)$ and choosing $a, b \overset{R}{\leftarrow} \mathbb{Z}/q\mathbb{Z}$ and $\vartheta \overset{R}{\leftarrow} \mathbb{Z}/p\mathbb{Z}$.

4.2   Semantic Security

Clearly the one-wayness of our cryptosystems is equivalent to the Class-CDH assumption.

We show below that the proposed cryptosystems are semantically secure under the Class-DDH assumption. We state:

Theorem 1. *The schemes of Sects. 3.2 and 3.3 are* IND-CPA *under the Class-DDH assumption.*

Proof. In order to deal with the two cryptosystems at the same time, we write the second part of the ciphertext, c_2, as $c_2 = m \star \beta \pmod{p}$ where $\star$ stands for addition modulo p or multiplication modulo p.

The goal is to construct a distinguisher $\mathcal{D}$ against the Class-DDH problem from an IND-CPA attacker $\mathcal{A}$ against the scheme. Consider the following algorithm $\mathcal{D}$ receiving as challenge the Class Diffie-Hellman triplet $([a]\boldsymbol{P}, [b]\boldsymbol{P}, \beta)$ for $(E(\mathbb{F}_p), q, \boldsymbol{P}) \leftarrow \mathcal{G}(1^\lambda)$, where either $\beta = [[ab]\boldsymbol{P}]$ or $\beta = \vartheta$, with $a, b \xleftarrow{R} \mathbb{Z}/q\mathbb{Z}$ and $\vartheta \xleftarrow{R} \mathbb{Z}/p\mathbb{Z}$:

1. Set $\boldsymbol{Y} = [a]\boldsymbol{P}$ and define $pk = \{E(\mathbb{F}_p), q, \boldsymbol{P}, \boldsymbol{Y}\}$;
2. Call $\mathcal{A}(pk)$ and receive two messages m_0 and m_1 in $\mathbb{Z}/p\mathbb{Z}$;
3. Choose a bit b at random and define $C = ([a]\boldsymbol{P}, m_b \star \beta)$;
4. Return ciphertext C to $\mathcal{A}$ and obtain its output bit b';
5. Output 1 if $b' = b$, and 0 otherwise.

When $\beta = [[ab]\boldsymbol{P}]$, C is a faithful ciphertext for message m_b. On the contrary, when $\beta = \vartheta$, C appears as a random value, independent of m_b. As a result, if $\epsilon(\lambda)$ denotes the probability that $\mathcal{A}$ wins the IND-CPA game, this means that

$$\Pr\left[\mathcal{D}\left(E(\mathbb{F}_p), q, \boldsymbol{P}, [a]\boldsymbol{P}, [b]\boldsymbol{P}, [[ab]\boldsymbol{P}]\right) = 1\right] = \epsilon(\lambda)$$

and

$$\Pr\left[\mathcal{D}\left(E(\mathbb{F}_p), q, \boldsymbol{P}, [a]\boldsymbol{P}, [b]\boldsymbol{P}, \vartheta\right) = 1\right] = \frac{1}{2}.$$

But the Class-DDH assumption says that their difference should be a negligible function in λ, that is, $\left|\epsilon(\lambda) - \frac{1}{2}\right| \leq \text{negl}(\lambda)$. □

5   Extension

5.1   Chameleon Hash Functions

Chameleon hash functions [23] are hash functions associated with a pair (hk, tk) of hashing/trapdoor keys. The name chameleon refers to the ability for the owner of the trapdoor key to modify the input without changing the output.

A chameleon hash function is defined by a tuple of three algorithms: (CMKg, CMhash, CMswitch). The key-generation algorithm CMKg, given a security parameter λ, outputs a key pair $(hk, tk) \leftarrow \text{CMKg}(1^\lambda)$. The hashing algorithm outputs $y = \text{CMhash}(hk, m, r)$ given the public key hk, a message m and random coins $r \in \mathcal{R}_{hash}$. On input of m, r, m' and the trapdoor key tk, the switching algorithm $r' \leftarrow \text{CMswitch}(tk, m, r, m')$ outputs $r' \in \mathcal{R}_{hash}$ such that

$$\text{CMhash}(hk, m, r) = \text{CMhash}(hk, m', r').$$

Collision-resistance mandates that it be infeasible to find pairs $(m', r') \neq (m, r)$ such that $\text{CMhash}(hk, m, r) = \text{CMhash}(hk, m', r')$ without knowing tk. Uniformity guarantees that the distribution of hashes is independent of the message m, in particular, for all hk and m, m', the distributions

$$\{r \leftarrow \mathcal{R}_{hash} : \text{CMhash}(hk, m, r)\} \quad \text{and} \quad \{r \leftarrow \mathcal{R}_{hash} : \text{CMhash}(hk, m', r)\}$$

are identical.

5.2   A Chosen-Ciphertext-Secure Construction

In this section, we describe an IND-CCA2-secure extension of our schemes which builds on the approach of Cash, Kiltz and Shoup [7] in its security analysis. We present below the additive variant. The multiplicative variant proceeds similarly.

KeyGen(1^λ). On input security parameter λ, generate an elliptic curve E over the prime field $\mathbb{F}_p$ and a point $\boldsymbol{P} \in E(\mathbb{F}_p)$ of large prime order q. Then, do the following.

1. Choose $y_0, y_1, z_0, z_1 \xleftarrow{R} \mathbb{Z}/q\mathbb{Z}$ and compute points $\boldsymbol{Y_0}, \boldsymbol{Y_1}, \boldsymbol{Z_0}, \boldsymbol{Z_1} \in E(\mathbb{F}_p)$ as

$$\boldsymbol{Y_0} = [y_0]\boldsymbol{P}, \qquad\qquad \boldsymbol{Y_1} = [y_1]\boldsymbol{P},$$
$$\boldsymbol{Z_0} = [z_0]\boldsymbol{P}, \qquad\qquad \boldsymbol{Z_1} = [z_1]\boldsymbol{P}.$$

2. Choose a chameleon hash function CMH = (CMKg, CMhash, CMswitch) that ranges over $\mathbb{Z}/q\mathbb{Z}$, with a key pair $(hk, tk) \leftarrow \mathsf{CMKg}(1^\lambda)$. We denote by $\mathcal{R}_{hash}$ the randomness space of the hashing algorithm.

The public key is $pk = \{E(\mathbb{F}_p), q, \boldsymbol{P}, \boldsymbol{Y_0}, \boldsymbol{Y_1}, \boldsymbol{Z_0}, \boldsymbol{Z_1}, hk\}$ and the matching private key is $sk = \{y_0, y_1, z_0, z_1\}$.

Encrypt(pk, m). To encrypt a message $m \in \mathbb{Z}/p\mathbb{Z}$, do the following.

1. Choose $r \xleftarrow{R} \mathbb{Z}/q\mathbb{Z}$ as well as $s \xleftarrow{R} \mathcal{R}_{hash}$;
2. Compute in $E(\mathbb{F}_p)$, $\boldsymbol{C_0} = [r]\boldsymbol{Y_0}$ and $\boldsymbol{C_1} = [r]\boldsymbol{P}$;
3. Compute $\beta = [\![\tilde{\boldsymbol{C_0}}]\!]$ and $c_0 = m + \beta \pmod{p}$;
4. Compute $t = \mathsf{CMhash}(hk, (c_0, \boldsymbol{C_1}), s_{hash}) \in \mathbb{Z}/q\mathbb{Z}$;
5. Compute

$$\boldsymbol{C_2} = [rt]\boldsymbol{Y_0} + [r]\boldsymbol{Z_0}, \qquad\qquad \boldsymbol{C_3} = [rt]\boldsymbol{Y_1} + [r]\boldsymbol{Z_1};$$

6. Output the ciphertext $C = (c_0, \boldsymbol{C_1}, \boldsymbol{C_2}, \boldsymbol{C_3}, s_{hash})$.

Decrypt(sk, C). Given the ciphertext $C = (c_0, \boldsymbol{C_1}, \boldsymbol{C_2}, \boldsymbol{C_3}, s_{hash})$ and the private key $sk = (y_0, y_1, z_0, z_1)$, conduct the following steps.

1. Compute $t = \mathsf{CMhash}(hk, (c_0, \boldsymbol{C_1}), s_{hash}) \in \mathbb{Z}/q\mathbb{Z}$;
2. Return $\perp$ if $\boldsymbol{C_2} \neq [ty_0 + z_0]\boldsymbol{C_1}$ or $\boldsymbol{C_3} \neq [ty_1 + z_1]\boldsymbol{C_1}$;
3. Compute $\boldsymbol{C_0} = [y_0]\boldsymbol{C_1}$ and return $m = c_0 - \beta \bmod p$, where $\beta = [\![\tilde{\boldsymbol{C_0}}]\!]$.

The above description follows a method suggested in [32] in that it makes use of a chameleon hash function to authenticate the message-carrying part c_0 of the ciphertext. We note that, instead of a chameleon hash function, the scheme could also use a strongly unforgeable one-time signature as in the Canetti-Halevi-Katz methodology [6]. However, this would incur longer ciphertexts. If we want to minimize the ciphertext overhead, the Boyen-Mei-Waters technique [5] can be used to eliminate the randomness s_{hash} of the chameleon hash function at the expense of introducing $O(\lambda)$ additional elliptic curve points in the public key.

Theorem 2. *The scheme is* IND-CCA2-*secure under the Class-DDH assumption, provided that the chameleon hash function is collision-resistant.*

Proof. The proof proceeds with a sequence of games. For each i, we denote by S_i the event that the adversary wins in Game i.

Game 0: This is the real game. In this game, the adversary $\mathcal{A}$ is given the public key pk and the challenger $\mathcal{B}$ answers all decryption queries by faithfully running the decryption algorithm. In the challenge phase, $\mathcal{A}$ chooses two distinct messages $m_0, m_1 \in \mathbb{Z}/p\mathbb{Z}$ and obtains a challenge ciphertext $C^\star = (c_0{}^\star, \boldsymbol{C_1}^\star, \boldsymbol{C_2}^\star, \boldsymbol{C_3}^\star, s_{hash}^\star)$ which encrypts m_d, for some random bit $d \xleftarrow{R} \{0,1\}$. In the second phase, the adversary $\mathcal{A}$ is granted further access to the decryption oracle. At the end of the game, $\mathcal{A}$ outputs a bit $d' \in \{0,1\}$ and we denote by S_0 the event that $d' = d$.

Game 1: This game is identical to Game 0 but the challenger $\mathcal{B}$ rejects all pre-challenge decryption queries $C = (c_0, \boldsymbol{C_1}, \boldsymbol{C_2}, \boldsymbol{C_3}, s_{hash})$ such that $\boldsymbol{C_1} = \boldsymbol{C_1}^\star$. Since $\boldsymbol{C_1}^\star$ is uniformly distributed in $\langle \boldsymbol{P} \rangle$ and independent of $\mathcal{A}$'s view before the challenge phase, the probability that $\mathcal{B}$ rejects a ciphertext that would not have been rejected in Game 0 is at most q_{dec}/q, where q_{dec} is the number of decryption queries. We have $|\Pr[S_1] - \Pr[S_0]| \leq q_{dec}/q$.

Game 2: In this game, the challenger $\mathcal{B}$ aborts if it realizes that, before or after the challenge phase, $\mathcal{A}$ has made a decryption query $C = (c_0, \boldsymbol{C_1}, \boldsymbol{C_2}, \boldsymbol{C_3}, s_{hash})$ such that

$$t = \mathsf{CMhash}(hk, (c_0, \boldsymbol{C_1}), s_{hash}) = \mathsf{CMhash}(hk, (c_0{}^\star, \boldsymbol{C_1}^\star), s_{hash}^\star) = t^\star.$$

Clearly, the latter event would contradict the collision-resistance property of the chameleon hash function. Moreover, Game 2 and Game 1 proceed identically until the latter event occurs, so that we obtain the inequality $|\Pr[S_2] - \Pr[S_1]| \leq \mathbf{Adv}^{\mathrm{CM\text{-}Hash}}(\lambda)$.

Game 3: This game is identical to Game 2 with the sole difference that the challenger $\mathcal{B}$ automatically rejects all post-challenge decryption queries of the form $C = (c_0{}^\star, \boldsymbol{C_1}^\star, \boldsymbol{C_2}, \boldsymbol{C_3}, s_{hash})$, where $(\boldsymbol{C_2}, \boldsymbol{C_3}) \neq (\boldsymbol{C_2}^\star, \boldsymbol{C_3}^\star)$. This change is only conceptual since these ciphertexts would be rejected in Game 2 as well. We thus have $\Pr[S_3] = \Pr[S_2]$.

Game 4: In this game, we modify the generation of the public key. At the outset of the game, $\mathcal{B}$ chooses a random value $t^\star \in \mathbb{Z}/q\mathbb{Z}$ in the range of the hashing algorithm CMhash, by hashing a random string R' using a random $s'_{hash} \xleftarrow{R} \mathcal{R}_{hash}$. It also picks $\gamma, \omega \xleftarrow{R} \mathbb{Z}/q\mathbb{Z}$ and sets $\boldsymbol{Y_1} = [\gamma]\boldsymbol{P} + [\omega]\boldsymbol{Y_0}$. It also picks $\gamma_0, \gamma_1 \xleftarrow{R} \mathbb{Z}/q\mathbb{Z}$ and sets

$$\boldsymbol{Z_0} = [-t^\star]\boldsymbol{Y_0} + [\gamma_0]\boldsymbol{P}, \qquad\qquad \boldsymbol{Z_1} = [-t^\star]\boldsymbol{Y_1} + [\gamma_1]\boldsymbol{P},$$

which implicitly defines the private key as $y_1 = \gamma + \omega y_0$, $z_0 = -t^\star y_0 + \gamma_0$ and $z_1 = -t^\star y_1 + \gamma_1$. In the challenge phase, $\mathcal{B}$ computes the challenge as

$$\boldsymbol{C_1}^\star = [r^\star]\boldsymbol{P}, \qquad \boldsymbol{C_2}^\star = [\gamma_0]\boldsymbol{C_1}^\star, \qquad \boldsymbol{C_3}^\star = [\gamma_1]\boldsymbol{C_1}^\star$$

while m_d is blinded as $c_0{}^\star = m_d + \beta^\star \pmod{p}$, where $\beta^\star = [\![\tilde{\boldsymbol{C}}_0{}^\star]\!]$, where $\boldsymbol{C_0}^\star = [y_0]\boldsymbol{C_1}^\star$. Finally, $\mathcal{B}$ uses the trapdoor key tk of the chameleon hash function to obtain $s_{hash}^{\star} = \mathsf{CMswitch}(tk, (R', s_{hash}'), (c_0{}^{\star}, \boldsymbol{C_1}{}^{\star}))$ such that $t^{\star} = \mathsf{CMhash}(hk, (c_0^{\star}, \boldsymbol{C_1}{}^{\star}), s_{hash}^{\star})$.

In Game 4, we remark that the public key pk and the challenge ciphertext $C^{\star} = (c_0{}^{\star}, \boldsymbol{C_1}{}^{\star}, \boldsymbol{C_2}{}^{\star}, \boldsymbol{C_3}{}^{\star}, s_{hash}^{\star})$ both have the same distribution as in Game 3, so that $\mathcal{A}$'s view has not changed. We have $\Pr[S_4] = \Pr[S_3]$.

Game 5: In this game, we modify the decryption oracle. Namely, at each decryption query $C = (c_0, \boldsymbol{C_1}, \boldsymbol{C_2}, \boldsymbol{C_3}, s_{hash})$, the challenger $\mathcal{B}$ computes the chameleon hash value $t = \mathsf{CMhash}(hk, (c_0, \boldsymbol{C_1}), s_{hash})$ as well as

$$\boldsymbol{W_1} = [(t - t^{\star})^{-1} \bmod q](\boldsymbol{C_2} - [\gamma_0]\boldsymbol{C_1})$$
$$\boldsymbol{W_2} = [(t - t^{\star})^{-1} \bmod q](\boldsymbol{C_3} - [\gamma_1]\boldsymbol{C_1})$$

At this point, $\mathcal{B}$ returns $\bot$ if $\boldsymbol{W_2} \neq [\gamma]\boldsymbol{C_1} + [\omega]\boldsymbol{W_1}$. Otherwise, $\mathcal{B}$ computes $\tilde{\boldsymbol{W}}_1 = \Psi(\boldsymbol{W_1})$, obtains $\beta = [\![\tilde{\boldsymbol{W}}_1]\!]$ and returns $m = c_0 - \beta \pmod{p}$.

It is easy to see that, in the adversary's view, Game 5 is identical to Game 4 until the event F_5 that $\mathcal{B}$ fails to reject a ciphertext that would have been rejected in Game 4. Using the same arguments as in [7,11], we can prove that $\Pr[F_5] \leq q_{dec}/q$. Specifically, event F_5 can only occur for a decryption query on an invalid ciphertext $C = (c_0, \boldsymbol{C_1}, \boldsymbol{C_2}, \boldsymbol{C_3}, s_{hash})$ where

$$\boldsymbol{C_1} = [r]\boldsymbol{P}, \qquad \boldsymbol{C_2} = [r + r']([t]\boldsymbol{Y_0} + \boldsymbol{Z_0}), \qquad \boldsymbol{C_3} = [r + r'']([t]\boldsymbol{Y_1} + \boldsymbol{Z_1})$$

and either $r' \neq 0$ or $r'' \neq 0$. This implies that $\boldsymbol{W_1} = [r + r_1]\boldsymbol{Y_0}$ and $\boldsymbol{W_2} = [r + r_2]\boldsymbol{Y_1}$, where $r_1 \neq 0$ (resp. $r_2 \neq 0$) if and only if $r' \neq 0$ (resp. $r'' \neq 0$). It is easy to see that, if $r_2 = 0$ and $r_1 \neq 0$ or $r_1 = 0$ and $r_2 \neq 0$, the equality $\boldsymbol{W_2} = [\gamma]\boldsymbol{C_1} + [\omega]\boldsymbol{W_1}$ never holds and we thus assume that $r_1 \neq 0$ and $r_2 \neq 0$. However, in this case $[\gamma]\boldsymbol{C_1} + [\omega]\boldsymbol{W_1}$ can be written $[r]\boldsymbol{Y_1} + [\omega r_1]\boldsymbol{Y_0}$, which is the sum of an information-theoretically fixed value $[r]\boldsymbol{Y_1}$ and another term $[\omega r_1]\boldsymbol{Y_0}$ that is completely undetermined in $\mathcal{A}$'s view: indeed, for a fixed $\boldsymbol{Y_1} = [\gamma]\boldsymbol{P} + [\omega]\boldsymbol{Y_0}$, we have q equally likely candidates for ω at the first decryption query such that $r' \neq 0$ or $r'' \neq 0$. For this query, we can only have the equality $\boldsymbol{W_2} = [\gamma]\boldsymbol{C_1} + [\omega]\boldsymbol{W_1}$ by pure chance, with probability $1/q$. Throughout the game, each invalid decryption query allows an unbounded adversary to eliminate one candidate for ω. Hence, after i queries, the adversary is left with a probability of $1/(q - i)$ of inferring the right ω. In the worst case, this probability is smaller than $1/(q - q_{dec})$ for a given decryption query. A union bound over all decryption queries gives the inequality $|\Pr[S_5] - \Pr[S_4]| \leq \Pr[F_5] \leq q_{dec}/(q - q_{dec})$. We remark that the private exponents (y_0, y_1, z_0, z_1) are not used any longer in Game 5 and we thus rely on the Class-DDH assumption to move to Game 6.

Game 6: This game is like Game 5 with the difference that, in the challenge ciphertext $C^{\star} = (c_0{}^{\star}, \boldsymbol{C_1}{}^{\star}, \boldsymbol{C_2}{}^{\star}, \boldsymbol{C_3}{}^{\star}, s_{hash}^{\star})$, $c_0{}^{\star}$ is chosen as a uniformly random element of $\mathbb{Z}/p\mathbb{Z}$. Under the Class-DDH assumption, this change should not be noticeable to $\mathcal{A}$ and we can write $|\Pr[S_6] - \Pr[S_5]| \leq \mathbf{Adv}^{\text{Class-DDH}}(\lambda)$.

In Game 6, we easily see that $\Pr[S_6] = 1/2$ since the challenge ciphertext can be seen as an encryption of a random message of $\mathbb{Z}/p\mathbb{Z}$, which is completely independent of m_0 and m_1. When counting probabilities throughout the sequence of games, we find that $|\Pr[S_0] - 1/2|$ is bounded by a sum of negligible functions under the aforementioned assumptions.   □

Acknowledgments. We thank Frederik Vercauteren for useful discussions and Antoine Joux for comments on an earlier version of this work. The second author's work has been supported in part by the "Programme Avenir Lyon Saint-Etienne de l'Université de Lyon" in the framework of the programme "Investissements d'Avenir" (ANR-11-IDEX-0007) and by the French ANR ALAMBIC project (ANR-16-CE39-0006).

A   Appendix

A.1   Public-Key Encryption

A *public-key encryption scheme* consists of three algorithms: (KeyGen, Encrypt, Decrypt).

Key generation. The key generation algorithm KeyGen is a randomized algorithm that takes on input some security parameter λ and returns a matching pair of public key and secret key for some user: $(pk, sk) \xleftarrow{R} \mathsf{KeyGen}(1^\lambda)$.

Encryption. Let $\mathcal{M}$ be the message space. The encryption algorithm Encrypt is a randomized algorithm that takes on input a public key pk and a plaintext $m \in \mathcal{M}$, and returns a ciphertext C. We write $C \leftarrow \mathsf{Encrypt}(pk, m)$.

Decryption. The decryption algorithm Decrypt takes on input secret key sk (matching pk) and a ciphertext C, and returns the corresponding plaintext m or a symbol $\perp$ indicating that the ciphertext is invalid. We write $m \leftarrow \mathsf{Decrypt}(sk, C)$ if C is a valid ciphertext and $\perp \leftarrow \mathsf{Decrypt}(sk, C)$ if it is not.

It is required that $\mathsf{Decrypt}\big(sk, \mathsf{Encrypt}(pk, m)\big) = m$ for any message $m \in \mathcal{M}$.

A.2   Security Notions

Beyond the basic property of one-wayness, data privacy in a public-key encryption scheme is captured by the notion of *semantic security*: An adversary should not learn any information whatsoever about a plaintext given its encryption beyond the length of the plaintext. This notion is known to be equivalent to the (easier to deal with) notion of *indistinguishability of encryptions* [19]. Furthermore, since the encryption key is public, an adversary can always encrypt messages of its choice; in other words, the adversary can mount chosen-plaintext attacks. It is therefore customary to let IND-CPA denote the security notion achieved by a semantically secure public-key encryption scheme.

The advantage of an adversary $\mathcal{A} = (\mathcal{A}_1, \mathcal{A}_2)$ in the IND-CPA experiment is defined as

$$\left| \Pr_{b \xleftarrow{R} \{0,1\}} \left[\begin{array}{l} (pk, sk) \leftarrow \mathsf{KeyGen}(1^\lambda), (m_0, m_1, s) \leftarrow \mathcal{A}_1(pk), \\ C^\star \leftarrow \mathsf{Encrypt}(pk, m_b) : \mathcal{A}_2(m_0, m_1, s, C^\star) = b \end{array} \right] - \frac{1}{2} \right| \qquad (*)$$

where the probability is taken over the random coins of the experiment according to the distribution induced by KeyGen(1^λ) as well as the ones of the adversary, and $m_0, m_1 \in \mathcal{M}$. An encryption is IND-CPA if the advantage of any polynomial-time adversary $\mathcal{A}$ is negligible a a function of λ.

The IND-CPA security notion offers an adequate security level in the presence of a *passive* adversary. The "right" security level against *active* attacks is that of IND-CCA2 security, or *security against chosen-ciphertext attacks*. The definition of the adversary's advantage as given by (*) extends to the IND-CCA2 model but the adversary $\mathcal{A} = (\mathcal{A}_1, \mathcal{A}_2)$ is now given an adaptive access to a decryption oracle to which it can submit any ciphertext of its choice with the exception that $\mathcal{A}_2$ may not query the decryption oracle on challenge ciphertext $C^\star$.

A.3   Consistent Lifting Problem

In this section, we extend the results of [8] to the elliptic curve setting.

Let $E(\mathbb{F}_p)$ be an elliptic curve over the prime field $\mathbb{F}_p$ and let $\mathbb{G} \subseteq E(\mathbb{F}_p)$ be a cyclic subgroup thereof. Let also $\boldsymbol{P}$ be a generator of $\mathbb{G}$ (i.e., $\mathbb{G} = \langle \boldsymbol{P} \rangle$) and $\tilde{\boldsymbol{P}} = \Psi(\boldsymbol{P}) \in E(\mathbb{Z}/p^2\mathbb{Z})$.

Given $\boldsymbol{P}$ and $\boldsymbol{Q} := [a]\boldsymbol{P} \xleftarrow{R} \mathbb{G}$, the *elliptic curve consistent lifting (ECCL) problem* is to compute $\boldsymbol{Q}' := [a]\tilde{\boldsymbol{P}}$. It is easily seen that this problem is equivalent to the discrete logarithm problem in $\mathbb{G}$. Indeed, given access to an ECCL solver, on input $\boldsymbol{Q}$, we receive $\boldsymbol{Q}'$ and then can obtain $\bar{a} := a \bmod p$ as $\bar{a} = \frac{[\boldsymbol{Q}']}{[\tilde{\boldsymbol{P}}]} \bmod p$. From Hasse's theorem, we know that $a = \bar{a}$ or $a = \bar{a}+p$; this can be easily decided by checking if $\boldsymbol{Q} = [\bar{a}]\boldsymbol{P}$ or $\boldsymbol{Q} = [\bar{a} + p]\boldsymbol{P}$. The other direction is straightforward. Given access to an ECDL solver, on input $\boldsymbol{Q}$, we obtain a and then can compute $\boldsymbol{Q}' = [a]\tilde{\boldsymbol{P}}$ where $\tilde{\boldsymbol{P}} = \Psi(\boldsymbol{P})$.

References

1. Belding, J.V.: A Weil pairing on the p-torsion of ordinary elliptic curves over $K[\epsilon]$. J. Number Theory **128**(6), 1874–1888 (2008)
2. Bernstein, D.J., Hamburg, M., Krasnova, A., Lange, T.: Elligator: elliptic-curve points indistinguishable from uniform random strings. In: ACM-CCS 2013, pp. 425–438. ACM Press (2013)
3. Boneh, D.: The decision Diffie-Hellman problem. In: Buhler, J.P. (ed.) ANTS 1998. LNCS, vol. 1423, pp. 48–63. Springer, Heidelberg (1998). doi:10.1007/BFb0054851
4. Boneh, D., Joux, A., Nguyen, P.Q.: Why textbook ElGamal and RSA encryption are insecure. In: Okamoto, T. (ed.) ASIACRYPT 2000. LNCS, vol. 1976, pp. 30–43. Springer, Heidelberg (2000). doi:10.1007/3-540-44448-3_3
5. Boyen, X., Mei, Q., Waters, B.: Direct chosen ciphertext security from identity based techniques. In: ACM-CCS 2005, pp. 320–329. ACM Press (2005)
6. Canetti, R., Halevi, S., Katz, J.: Chosen-ciphertext security from identity-based encryption. In: Cachin, C., Camenisch, J.L. (eds.) EUROCRYPT 2004. LNCS, vol. 3027, pp. 207–222. Springer, Heidelberg (2004). doi:10.1007/978-3-540-24676-3_13

7. Cash, D., Kiltz, E., Shoup, V.: The twin Diffie-Hellman problem and applications. In: Smart, N. (ed.) EUROCRYPT 2008. LNCS, vol. 4965, pp. 127–145. Springer, Heidelberg (2008). doi:10.1007/978-3-540-78967-3_8

8. Catalano, D., Nguyen, P.Q., Stern, J.: The hardness of hensel lifting: the case of RSA and discrete logarithm. In: Zheng, Y. (ed.) ASIACRYPT 2002. LNCS, vol. 2501, pp. 299–310. Springer, Heidelberg (2002). doi:10.1007/3-540-36178-2_19

9. Chevallier-Mames, B., Paillier, P., Pointcheval, D.: Encoding-free ElGamal encryption without random oracles. In: Yung, M., Dodis, Y., Kiayias, A., Malkin, T. (eds.) PKC 2006. LNCS, vol. 3958, pp. 91–104. Springer, Heidelberg (2006). doi:10.1007/11745853_7

10. Cramer, R., Gennaro, R., Schoenmakers, B.: A secure and optimally efficient multi-authority election scheme. In: Fumy, W. (ed.) EUROCRYPT 1997. LNCS, vol. 1233, pp. 103–118. Springer, Heidelberg (1997). doi:10.1007/3-540-69053-0_9

11. Cramer, R., Shoup, V.: A practical public key cryptosystem provably secure against adaptive chosen ciphertext attack. In: Krawczyk, H. (ed.) CRYPTO 1998. LNCS, vol. 1462, pp. 13–25. Springer, Heidelberg (1998). doi:10.1007/BFb0055717

12. ElGamal, T.: A public key cryptosystem and a signature scheme based on discrete logarithms. IEEE Trans. Inf. Theory 31(4), 469–472 (1985)

13. Even, S., Goldreich, O., Micali, S.: On-line/off-line digital schemes. J. Cryptol. 9(1), 35–67 (1996)

14. Farashahi, R.R.: Hashing into Hessian curves. In: Nitaj, A., Pointcheval, D. (eds.) AFRICACRYPT 2011. LNCS, vol. 6737, pp. 278–289. Springer, Heidelberg (2011). doi:10.1007/978-3-642-21969-6_17

15. Fouque, P.-A., Joux, A., Tibouchi, M.: Injective encodings to elliptic curves. In: Boyd, C., Simpson, L. (eds.) ACISP 2013. LNCS, vol. 7959, pp. 203–218. Springer, Heidelberg (2013). doi:10.1007/978-3-642-39059-3_14

16. Galbraith, S.D.: Elliptic curve Paillier schemes. J. Cryptol. 15(2), 129–138 (2002)

17. Gennaro, R., Krawczyk, H., Rabin, T.: Secure hashed Diffie-Hellman over non-DDH groups. In: Cachin, C., Camenisch, J.L. (eds.) EUROCRYPT 2004. LNCS, vol. 3027, pp. 361–381. Springer, Heidelberg (2004). doi:10.1007/978-3-540-24676-3_22

18. Gennaro, R., Shoup, V.: A note on an encryption scheme of Kurosawa and Desmedt. Cryptology ePrint Archive, Report 2004/194 (2004)

19. Goldwasser, S., Micali, S.: Probabilistic encryption. J. Comput. Syst. Sci. 28(2), 270–299 (1984)

20. Håstad, J., Impagliazzo, R., Levin, L.A., Luby, M.: A pseudorandom generator from any one-way function. SIAM J. Comput. 28(4), 1364–1396 (1999)

21. Impagliazzo, R., Levin, L.A., Luby, M.: A pseudorandom generator from any one-way function. In STOC 1989, pp. 12–24. ACM Press (1989)

22. Koblitz, N.: Elliptic curve cryptosystems. Math. Comput. 48, 203–209 (1987)

23. Krawczyk, H., Rabin, T.: Chameleon signatures. In: NDSS 2000. The Internet Society (2000)

24. Miller, V.S.: Use of elliptic curves in cryptography. In: Williams, H.C. (ed.) CRYPTO 1985. LNCS, vol. 218, pp. 417–426. Springer, Heidelberg (1986). doi:10.1007/3-540-39799-X_31

25. Pohlig, S.C., Hellman, M.E.: An improved algorithm for computing logarithms over GF(p) and its cryptographic significance. IEEE Trans. Inf. Theory 24(1), 106–110 (1978)

26. Pollard, J.M.: Monte Carlo methods for index computation mod p. Math. Comput. 32, 918–924 (1978)

27. Shoup, V., Sequences of games: a tool for taming complexity in security proofs. Cryptology ePrint Archive Report, 2004/332 (2004)
28. Silverman, J.H.: The Theory of Elliptic Curves, GTM 106. Springer-Verlag, Heidelberg (1986)
29. Tsiounis, Y., Yung, M.: On the security of ElGamal based encryption. In: Imai, H., Zheng, Y. (eds.) PKC 1998. LNCS, vol. 1431, pp. 117–134. Springer, Heidelberg (1998). doi:10.1007/BFb0054019
30. Virat, M.: A cryptosystem "à la" ElGamal on an elliptic curve over $\mathbb{F}_p[\varepsilon]$. In: WEWoRC 2005, LNI 74, pp. 32–44. Gesellschaft für Informatik e.V (2005)
31. Virat, M.: Courbes elliptiques sur un anneau et applications cryptographiques. Ph.D. thesis, Université de Nice-Sophia Antipolis (2009)
32. Zhang, R.: Tweaking TBE/IBE to PKE transforms with chameleon hash functions. In: Katz, J., Yung, M. (eds.) ACNS 2007. LNCS, vol. 4521, pp. 323–339. Springer, Heidelberg (2007). doi:10.1007/978-3-540-72738-5_21

Lattice-based Cryptanalysis

Gauss Sieve Algorithm on GPUs

Shang-Yi Yang[1], Po-Chun Kuo[1(✉)], Bo-Yin Yang[2], and Chen-Mou Cheng[1]

[1] Department of Electrical Engineering, National Taiwan University, Taipei, Taiwan
{ilway25,kbj,doug}@crypto.tw
[2] Institute of Information Science, Acamedia Sinica, Taipei, Taiwan
by@crypto.tw

Abstract. Lattice-based cryptanalysis is an important field in cryptography since lattice problems are among the most robust assumptions, and have been used to construct most cryptographic primitives. In this research, we focus on the Gauss Sieve algorithm, a heuristic lattice sieving algorithm proposed by Micciancio and Voulgaris. We propose the technique of *lifting* computations in prime-cyclotomic ideals into that in cyclic ideals. Lifting makes rotations easier to compute and reduces the complexity of inner products from $O(n^3)$ to $O(n^2)$. We implemented our Gauss Sieve on GPUs by adapting the framework of Ishiguro et al. in a single GPU, and the one of Bos et al. among multiple GPUs. We found a short vector at dimension 130 in the Darmstadt Ideal SVP Challenge (currently in first place in the Hall of Fame) using 8 GPUs in 824 h using our implementation.

Keywords: Lattice-based cryptography · Sieving algorithm · Gauss Sieve · GPU · Parallelization · Shortest vector problem · SVP · Ideal lattices

1 Introduction

Over the past two decades, lattice-based cryptosystems have attracted widespread interest. Not only are they among the group of PKCs that will potentially defend against the quantum threats, but they also provide the first constructions of many new cryptographic functionalities, e.g. fully-homomorphic encryption and multilinear maps [Gen09, GGH13]. Furthermore, in 1997 Ajtai and Dwork proved that some lattice problems possess worst-case to average-case reductions [AD97], which gives a strong guarantee on the security of lattice-based cryptosystems, and inspired the construction of many cryptographic primitives. Although many such constructions are in practice infeasible, ideal lattices have made both the keys shorter and algorithms faster, which brings many more new ideas closer to practicality.

Many lattice- and ideal-lattice-based schemes claim to base their security on the shortest vector problem (SVP): if the shortest vector in a lattice could be found, the lattice-based cryptosystems would be broken. However, it is unclear how to choose secure yet practical parameters for these schemes, and an accurate

© Springer International Publishing AG 2017
H. Handschuh (Ed.): CT-RSA 2017, LNCS 10159, pp. 39–57, 2017.
DOI: 10.1007/978-3-319-52153-4_3

assessment of their security levels would be indispensable if we are to select suitable parameters for them.

Several exact or approximate algorithms have been proposed for the SVP problem. Exact algorithms include enumeration, sieving, and ones based on Vonoroi cells [MV10]. The Vonoroi cell method, though single exponential both in time and space, proved to be impractical for dimensions higher than 10. On the other hand, lattice enumeration is an exhaustive search algorithm. The time complexity of lattice enumeration is $2^{O(n^2)}$ or $2^{O(n \log n)}$ and space complexity is polynomial [GNR10, KSD+11]. Lastly, sieving algorithms have $2^{O(0.52n)}$ time complexity and $2^{O(n)}$ space complexity [MV10]. In general, lattice enumeration has remained the fastest approach to solve SVP so far since almost all the data could be store in the CPU cache. However, the general approach is ill-suited for parallelizing on GPUs or similar wide vector architectures. For example, the speed-up in [KSD+11] is less than a factor of ten. It is also unclear how to make use of the special structure of ideal lattices when using enumeration.

In contrast, approximate algorithms run in polynomial time but output an approximate solution. Even though the output short vectors have length exponential in dimension, such vectors are in fact good enough for some applications or cryptanalysis. For example, the famous LLL algorithm can find, with high probability, the shortest vector for Goldstein-Mayer random lattices in the SVP challenge [Lat] for dimensions less than 30. For higher dimensions, however, the quality of vectors it outputs is insufficient. On the other hand, in the BKZ algorithm, which uses enumeration in sub-lattices as a subroutine, we can trade off execution time against the approximate factor. Whether sieving algorithms could be used as a sub-routine in the BKZ algorithm is still an open problem. Another open problem is whether there exists a poly-time algorithm which outputs a short vector with a polynomial approximation factor.

Although enumeration seems fastest so far in practice, sieving has a better complexity upper bound and may yet outperform enumeration in higher dimensions. Moreover, as far as we know sieving is currently the only way to use the ideal lattice structures. There are several papers on how to parallelize sieving and how to use the cyclic lattice structure, but how to do this on GPUs and use other ideal lattice structures is not clear yet. Previous works by Ishiguro et al. [IKMT14, MDB14] also seem to be limited to cyclic, anti-cyclic and trinomial ideal lattices. In this paper, we broaden the scope to include prime cyclotomic ideal lattices, and make the following contributions:

- We propose and implement the first lattice sieving algorithm for a single machine with multiple GPUs. Our variant includes two carefully designed layers of parallelism, both inter-GPU and intra-GPU (Sect. 5).
- We show that by *lifting* lattice vectors generated by the polynomial $x^n + \cdots + 1$ into ones generated by $x^{n+1} - 1$, not only do inner products (the critical path of Gauss Sieve) speed up, some register rotation problems on GPUs are mitigated (Sect. 4). Moreover, by heuristically applying *lazy rotation*, the complexity of reduction between two vectors with all their rotations goes down from $O(n^3)$ to $O(n^2)$ (only a constant times slower than the anticyclic lattice, cf. Sect. 4.3).

- We carefully crafted the reduction kernel to exploit both thread- and instruction-level parallelism (Sect. 6). Special care is taken with the layout of vectors in the register file, and some kernel-level heuristics are introduced that use the ideal lattice property.
- Incorporating these improvement into our implementation on GPUs, we were able to solve challenges of dimension 130 within 6583 GPU-hours. Our GPU implementation is 21.5 (resp. 55.8) times faster than a single-core CPU for general (resp. ideal) lattices (Sect. 7.2).
- We provide a lower bound complexity estimation for the SVP compared to the previous work (Sect. 7.4).

2 Preliminary

2.1 Definition and Notation

A *lattice* is a discrete additive group of all integer combinations of a basis $v_1, v_2, ..., v_m \in \mathbb{R}^n$, where $m \leq n$. In cryptography, integer lattices are often used, namely, the basis vectors are defined over $\mathbb{Z}^n$. The bases corresponding to a lattice are not unique, since multiplying a uni-modular matrix to a basis would not change the lattice spanned by the basis. We use $\mathcal{L}(B)$ to denote the lattice spanned by the basis B.

The first successive minimum $\lambda_1(\mathcal{L})$ is the length of the shortest nonzero vector of the lattice $\mathcal{L}$. The shortest vector problem (SVP) asks for the shortest nonzero vector in a given lattice. The SVP is NP-hard under randomized reduction [Ajt97]. The approximation shortest vector problem (SVP$_\alpha$) asks for a short vector of length shorter than $\alpha\lambda_1(L)$.

Extending the idea into rings, we have *ideal lattices*, a special class of lattices. Consider an ideal of a ring $\mathcal{I} = \langle g \rangle \subseteq \mathbb{Z}[x]/f(x)$, where f is a monic irreducible polynomial of degree n, an ideal lattice is $\mathcal{L}(B) \in \mathbb{Z}^n$ such that $B = \{g \mod f : g \in \mathcal{I}\}$. The polynomial of a ring affects its structure and computation cost. Thus, cryptographers are concerned with four type of ideal lattices defined by the polynomial $f(x)$:

- *Cyclic* ideal lattice, with $f_{cyclic}(x) = x^n - 1$, are the simplest ones and easy to compute. However, since the polynomials are always divided by $x - 1$, this kind of ideal lattice does not guarantee the worst-case collision resistance.
- *Anti-cyclic* ideal lattice, with $f_{anti-cyclic}(x) = x^n + 1$, are also eligible for easy multiplication and convolution. Such polynomials are irreducible over $\mathbb{Z}$ if n is a power of 2. This kind of ideal lattice is commonly used in cryptography.
- *Prime-cyclotomic* ideal lattices, with $f_{prime-cyclotomic}(x) = x^n + x^{n-1} + \cdots + 1$, are the main type we focus on. If $n + 1$ is prime, $f_{prime-cyclotomic}(x)$ is irreducible.
- *Trinomial* ideal lattices, with $f_{trinomial}(x) = x^n + x^{n/2} + 1$ where $n/2$ is a power of three, are the ones considered in [IKMT14].

By the definition of an ideal lattice, the vector $u = (u_0, u_1, \cdots, u_{n-1}) \in \mathbb{Z}^n$ also indicates a polynomial $u(x) = u_0 + u_1 x + \cdots + u_{n-1} x^{n-1} \in \mathbb{Z}[x]/f(x)$, the polynomial $x \cdot u(x)$ is still in the ideal. Thus, the vector corresponding to such polynomial is called the *(first) rotation* of u, denoted as $rot(u)$. For example, consider $f(x) = x^n - 1$, the rotation of $u = (u_0, u_1, \cdots, u_{n-1})$ is $\mathbf{rot}(u) = (u_{n-1}, u_0, u_1, \cdots, u_{n-2})$.

The central notation of the Gauss Sieve is Gauss reduction. Two vectors $u, v \in \mathcal{L}(B)$ satisfying $\|u \pm v\| \geq \max(\|u\|, \|v\|)$ are called *Gauss-reduced*. Given two arbitrary vectors u and v, we can reduce u with respect to v by $u \leftarrow u - \lfloor \frac{\langle u, v \rangle}{\langle v, v \rangle} \rceil v$. Thus, given two arbitrary vectors u and v, we can convert them into Gauss-reduced ones by repetitively applying the reduction procedure alternatingly, in a Euclidean algorithm-like manner, until the vectors no longer change. If any two vectors in a set are Gauss-reduced, it is *pairwise-reduced*.

Algorithm 1 shows the pseudo-code for reducing the list U with the list V. Algorithm 2 is the ideal lattice counterpart. In Algorithm 2, *times* represents the number of possible rotations in the input lattice. In other words, $x^{times} = \pm 1$. As concrete examples, for anti-cyclic lattices, $times = n$; for prime cyclotomic lattices, $times = n + 1$.

Algorithm 1. Gauss reduction between two lists for general lattices

Input : Lists U and V
Output: Reduced list U
1 **foreach** $u \in U$ **do**
2   **foreach** $v \in V$ **do**
3   **if** $2 \cdot |\langle u, v \rangle| > \langle v, v \rangle$ **then**
4             $u \leftarrow u - \lfloor \frac{\langle u, v \rangle}{\langle v, v \rangle} \rceil v$
5   Mark u as reduced.

Algorithm 2. Gauss reduction between two lists for ideal lattices

Input : Lists U and V
          Number of rotations: *times*
Output: Reduced list U, with all possible rotations
1 **foreach** $u \in U$ **do**
2   **foreach** $v \in V$ **do**
3   **for** $i \leftarrow 0$ **to** $times - 1$ **do**
4             $w \leftarrow x^i v$
5   **for** $j \leftarrow 0$ **to** $times - 1$ **do**
6                $(s, t) \leftarrow (x^j u, x^j w)$
7                $m \leftarrow \lfloor \langle s, t \rangle / \langle t, t \rangle \rceil$
8   **if** $m \neq 0$ **then**
9                   $u \leftarrow s - mt$
10   Mark u as reduced.

2.2 CUDA Programming

Here we provide a minimalist CUDA programming introduction, including only relevant information that our implementation takes into consideration. For more details, please refer to the CUDA C Programming Guide [CUD15].

Graphics processing units (GPUs) are high throughput, many-core architectures. Currently, the most widely used GPU development toolchain is CUDA by NVIDIA. CUDA supports writing fine-tuned programs for NVIDIA graphic cards. In this paper, we will especially focus on GPUs of the Maxwell architecture.

The CUDA programming model requires programmers to think in the *single instruction, multiple thread* (SIMT) programming model. The model exposes three key abstractions to programmers: a hierarchy of thread groups, shared memories, and barrier synchronization. Threads are first organized in blocks, and blocks are then organized in grids. A grid of GPU threads must run the same program (the kernel).

At the system level, blocks are independently dispatched to different processors. Since each block has a dedicated on-chip cache called the shared memory, threads within a block can only exchange data through the shared memory. However, this requires an explicit synchronization barrier that halts all the threads in a block, and thus can be a huge performance overhead for critical applications.

Fortunately, starting from the Kepler architecture, data exchange within a *warp* can be done using the *warp shuffle* instructions without any explicit synchronization barrier. A *warp*, consisting of 32 consecutive threads, is the smallest batch that can be scheduled and issued at once by a processor. For example, using the warp shuffle instructions, summing different values from threads with in a warp can be done relatively fast through the *parallel reduction* paradigm. If the threads in a warp are executing different instructions – most likely because of different branch conditions – severe *warp divergence* can occur, drastically lowering the warp utilization.

3 Background

3.1 Sieving Algorithms

The first sieving algorithm was proposed by Ajtai et al. in 2001 [AKS01]. They proved that the time/space complexity is $2^{O(n)}$, which is the first single-exponential time algorithm solving SVP. Following works either provided tighter theoretical bounds on the complexity [NV08, MV10, Sch11, Sch13], or improved the algorithm [MS11, MDB14, MBL15].

3.2 Gauss Sieve

The Gauss Sieve algorithm was proposed by Micciancio and Voulgrais in [MV10] and is the most practical version of sieving algorithms. The main idea of the algorithm is to mutually reduce samples with a list of vectors by Gauss reduction.

After Gauss reduction, the angle between any pair of two vectors is larger than $60°$. By the Kabatiansky-Levenshtein theorem, one can bound the number of such vectors, and thus obtain the time complexity of the algorithm.

The Gauss Sieve algorithm has been implemented on CPU in [IKMT14] and some improvements have been proposed by Bos et al. in [BNvdP14]. The work of Ishiquro et al. improved and implemented the parallel version of the Gauss Sieve algorithm proposed by Schneider [Sch13], and they also adapt to a specific ideal lattice called negacyclic ideal lattices. However, we promote this into more general polynomial ring and improve the performance. Later, Bos et al. proposed a different variant of the parallel Gauss Sieve algorithm which is more suited for high dimension lattice [BNvdP14]. We will expound on their ideas in Sect. 5. Moreover, Laarhoven incorporated locality-sensitive hashing into the algorithm [Laa15,BDGL16,BL16]. Instead of searching all the vector in the list, they group together near vectors using hash functions. Therefore, vectors are only reduced with more geometrically possible ones.

Prime Cyclotomic Rotation. First we state a nice property of anti-cyclic lattices.

Lemma 1 [BNvdP14]. *Let $a, b \in R = \mathbb{Z}[x]/(x^n + 1)$ with coefficient vector a, b. If $2\|\langle a, x^l \cdot b \rangle\| \leq \min\{\langle a, a \rangle, \langle b, b \rangle\}$ for all $0 \leq l < n$, then $x^i \cdot a$ and $x^j \cdot b$ are Gauss-reduced for all $i, j \in \mathbb{Z}$.*

In contrast to the anti-cyclic case described in Lemma 1, prime cyclotomic lattices do not possess this property. Therefore, all the rotations of two vectors can contribute to the global status. Thus, the list size might be even smaller.

However, prime cyclotomic lattices may have some disadvantages. To illustrate the computational overhead to find the norms of (all the) rotations of a vector, consider the vector $v = (5, 4, 3, 2, 1)$ in an ideal lattice generated by the polynomial $f(x) = x^5 + x^4 + x^3 + x^2 + x + 1$. The first rotation of v is

$$\mathbf{rot}(v) = (-1, 5 - 1, 4 - 1, 3 - 1, 2 - 1) = (-1, 4, 3, 2, 1).$$

Squaring and summing, we have the squared norm for $x \cdot v$:

$$\|\mathbf{rot}(v)\|^2 = (-1)^2 + 4^2 + 3^2 + 2^2 + 1^2 = 31.$$

Calculating norms like this can be slow, because only when the vector $\mathbf{rot}(v)$ is ready can we calculate the sum of squares. However, the value that is required in Gauss reduction is just $\|\mathbf{rot}(v)\|^2$, but not $\mathbf{rot}(v)$ per se. For processors with ADD, MUL and FMAD (fused multiply-add) instructions, it takes $2n$ operations to calculate the norm of an n-dimensional vector.

In this paper, we will see how to circumvent this by *lifting* a vector. By doing this, not only is the computation easier, but it also enables optimizations that are not possible without lifting.

4   Lifting Ideal Lattices

We now develop the properties for prime cyclic lattices and see how they can facilitate computation.

4.1   Lifting Prime Cyclotomic Polynomials

The idea behind lifting lattices is to supplement vectors with a bit of redundant information to ease computation. Specifically, we will express an n-dimensional vector with an $(n + 1)$-dimensional one. Let $\mathcal{L}$ be a lattice generated by $x^n + x^{n-1} + \cdots + 1$, and $\overline{\mathcal{L}}$ by $x^{n+1} - 1$. We wish to seek a way to connect the two lattices according to the following criteria:

- The conversion of vectors between the two lattices is simple.
- The rotation of vectors must be preserving, so that the complicated rotation in $\mathcal{L}$ can be done instead cyclically in $\overline{\mathcal{L}}$.

Technically speaking, we are looking for simple ring homomorphisms between $\mathbb{F}[x]/(x^n + x^{n-1} + \cdots + 1)$ and $\mathbb{F}[x]/(x^{n+1} - 1)$.

An intuitive clue to accomplish this comes from the observation that the polynomial $x^{n+1} - 1$ factorizes as

$$x^{n+1} - 1 = (x - 1)(x^n + x^{n-1} + \cdots + 1).$$

This suggests we connect u and its *lift* $\overline{u}$ by thinking of $\overline{u}$ as reduced modulo $x^n + x^{n-1} + \cdots + 1$:

$$u \equiv \overline{u} \pmod{x^n + x^{n-1} + \cdots + 1}.$$

Note that this choice also preserves rotation.

As an example, lifting directly $u = (1, 2, 3, 4, 5)$ in a lattice generated by $x^4 + x^3 + x^2 + x + 1$ gives $\overline{u} = (1, 2, 3, 4, 5, 0)$ in a lattice generated by $x^5 - 1$. This is not the only way to lift u. Another possibility is $\overline{u}' = (2, 3, 4, 5, 6, 1)$, since $(2 - 1, 3 - 1, 4 - 1, 5 - 1, 6 - 1) = (1, 2, 3, 4, 5)$. In general, to lift any u, we can choose p arbitrarily and lift u as $\overline{u} = (u_0 + p, u_1 + p, \cdots, u_n + p, p)$.

From now on, we will write a bar on top of a symbol to indicate that it is lifted from its underlying form. For example, $\overline{u}$ is a lift of the vector u and $\overline{L}$ is a lift of the lattice L. Whenever we see a lifted vector $\overline{u}$, we should keep in mind that it is merely a surface form representing its underlying original vector.

4.2   Norms and Inner Products

During the Gauss reduction, we are especially interested in the norms and inner products of rotations of vectors. Let us see how to derive these quantities for the underlying lattice directly, without converting from the lifted lattice.

Suppose $\bar{u} = (\bar{u}_0, \bar{u}_1, \cdots, \bar{u}_{n-1}, \bar{u}_n)$ in $\overline{\mathcal{L}}$. We first reduce $\bar{u}$ modulo the polynomial $x^n + x^{n-1} + \cdots + 1$ to get its underlying form:

$$u = (\bar{u}_0 - \bar{u}_n, \bar{u}_1 - \bar{u}_n, \cdots, \bar{u}_{n-1} - \bar{u}_n)$$
$$= (\bar{u}_0 - p, \bar{u}_1 - p, \cdots, \bar{u}_{n-1} - p).$$

Here, we rewrite $\bar{u}_n$ as p interchangeably, since $\bar{u}_n$ acts as a *pivot* for the vector. We can now calculate the norm of u:

$$\langle u, u \rangle^2 = \sum_{i=0}^{n-1} (\bar{u}_i - \bar{u}_n)^2$$

$$= \sum_{i=0}^{n} (\bar{u}_i - \bar{u}_n)^2 \qquad \text{since } \bar{u}_n - \bar{u}_n = 0.$$

$$= \sum_{i=0}^{n} \bar{u}_i^2 - 2\bar{u}_n \sum_{i=0}^{n} \bar{u}_i + (n+1)\bar{u}_n^2$$

$$= \boxed{\langle \bar{u}, \bar{u} \rangle^2} - 2p \boxed{\sum_{i=0}^{n} \bar{u}_i} + (n+1)p^2,$$

where the boxed terms remain constant throughout all cyclic rotations of $\bar{u}$, and thus can be saved beforehand. Note that we do not need to know what u is at all.

Similarly, the inner product of two vectors u and v is

$$\langle u, v \rangle^2 = \langle \bar{u}, \bar{v} \rangle^2 - p \sum_{i=0}^{n} \bar{v}_i - q \sum_{i=0}^{n} \bar{u}_i + (n+1)pq,$$

where q is the pivot of $\bar{v}$.

Simplifying Formulae. Although these formulae may look intimidating, we can always simplify them by choosing the "right" pivot. If we set $\sum_{i=0}^{n} \bar{u}_i = 0$, and solve for p:

$$0 = \bar{u}_0 + \bar{u}_1 + \cdots + \bar{u}_{n-1} + p \qquad \text{rewrite } \bar{u}_n \text{ as } p.$$
$$= (u_0 + p) + (u_1 + p) + \cdots + (u_{n-1} + p) + p$$
$$= (u_0 + u_1 + \cdots + u_{n-1}) + (n+1)p,$$

we can choose the pivot as

$$p = -\frac{\sum_{i=0}^{n-1} u_i}{n+1}.$$

This is our standard way to lift a vector.

Carrying out the same procedure for v, we can now write the inner product succinctly:

$$\langle u, v \rangle^2 = \langle \bar{u}, \bar{v} \rangle^2 + (n+1)pq.$$

Lifting in this manner, we amend Algorithm 2 into Algorithm 3. Note that the underlying vectors s and t are no longer needed on line 12. Since we do not have to track and update $\sum u_i$ anymore, simplifying in this manner eases some computational burden and memory overhead in the innermost loop for GPUs. However, integer vectors are now represented by floating points, which may lead to error accumulation after several rounds. We choose to rectify these vectors by unlifting and rounding the numbers when they are taken out from the stack for later rounds.

We could also eliminate the $n+1$ by "normalizing" and dividing vectors by $\sqrt{n+1}$, but this is less intuitive.

Algorithm 3. Gauss reduction between two lists for prime cyclotomic lattices (lifted)

> **Input**   : Lifted lists $\bar{U}$ and $\bar{V}$
> **Output** : Reduced, lifted list $\bar{U}$
>
> 1 **foreach** $\bar{u} \in \bar{U}$ **do**
> 2 $\quad$ **foreach** $\bar{v} \in \bar{V}$ **do**
> 3 $\quad\quad$ **for** $i \leftarrow 0$ **to** n **do**
> 4 $\quad\quad\quad$ $\bar{w} \leftarrow x^i \bar{v}$
> 5 $\quad\quad\quad$ $\langle \bar{w}, \bar{w} \rangle \leftarrow \langle \bar{v}, \bar{v} \rangle + (n+1)\bar{v}_{n-i}^2$
> 6 $\quad\quad\quad$ **for** $j \leftarrow 0$ **to** n **do**
> 7 $\quad\quad\quad\quad$ Calculate $\langle \bar{u}, \bar{w} \rangle$.
> 8 $\quad\quad\quad\quad$ $\langle s, t \rangle \leftarrow \langle \bar{u}, \bar{w} \rangle + (n+1)\bar{u}_{n-j}\bar{w}_{n-j}$
> 9 $\quad\quad\quad\quad$ $\langle t, t \rangle \leftarrow \langle \bar{w}, \bar{w} \rangle + (n+1)\bar{w}_{n-j}^2$
> 10 $\quad\quad\quad\quad$ $m \leftarrow \lfloor \langle s, t \rangle / \langle t, t \rangle \rceil$
> 11 $\quad\quad\quad\quad$ **if** $m \neq 0$ **then**
> 12 $\quad\quad\quad\quad\quad$ $(\bar{s}, \bar{t}) \leftarrow (x^j \bar{u}, x^j \bar{w})$
> 13 $\quad\quad\quad\quad\quad$ $\bar{u} \leftarrow \bar{s} - m\bar{t}$
> 14 $\quad\quad\quad\quad\quad$ Mark $\bar{u}$ as reduced.

4.3   Lazy Rotation

We now address two GPU performance bottlenecks in Algorithm 3, and provide two kernel-level heuristics to solve these problems.

First, on lines 12–14, whenever $\bar{u}$ is reduced, it is assigned as the difference of two rotated vectors $\bar{s}$ and $m\bar{t}$. However, such register indexing, unlike on CPUs, can cause spills on GPUs. Since $\bar{s} - m\bar{t} = x^j(\bar{u} - m\bar{w})$, we can instead write $\bar{u} \leftarrow \bar{u} - m\bar{w}$ and choose to rotate $\bar{u}$ back *lazily* after the kernel finishes. Now

---

Algorithm 4. Gauss reduction between two lists for prime cyclotomic lattices (lifted, with lazy rotation)

Input : Lifted lists $\overline{U}$ and $\overline{V}$
Output : Reduced, lifted list $\overline{U}$

1 **foreach** $\overline{u} \in \overline{U}$ **do**
2    $norm \leftarrow \langle \overline{u}, \overline{u} \rangle + (n+1)\overline{u}_n^2$
3   **foreach** $\overline{v} \in \overline{V}$ **do**
4   **for** $i \leftarrow 0$ **to** n **do**
5        $\overline{w} \leftarrow x^i \overline{v}$
6        $\langle \overline{w}, \overline{w} \rangle \leftarrow \langle \overline{v}, \overline{v} \rangle + (n+1)\overline{v}_{n-i}^2$
7   Calculate $\langle \overline{u}, \overline{w} \rangle$.
8   **for** $j \leftarrow 0$ **to** n **do**
9          $\langle s, s \rangle \leftarrow \langle \overline{u}, \overline{u} \rangle + (n+1)\overline{u}_{n-j}^2$
10         $\langle s, t \rangle \leftarrow \langle \overline{u}, \overline{w} \rangle + (n+1)\overline{u}_{n-j}\overline{w}_{n-j}$
11         $\langle t, t \rangle \leftarrow \langle \overline{w}, \overline{w} \rangle + (n+1)\overline{w}_{n-j}^2$
12         $m \leftarrow \lfloor \langle s, t \rangle / \langle t, t \rangle \rceil$
13         $norm_{new} \leftarrow \langle s, s \rangle - 2m\langle s, t \rangle + m^2\langle t, t \rangle$
14   **if** $norm_{new} < norm$ **then**
15           $m_{best} \leftarrow m$
16           $norm \leftarrow norm_{new}$

17   **if** $m_{best} \neq 0$ **then**
18        $\overline{u} \leftarrow \overline{u} - m_{best}\overline{w}$
19        $\langle \overline{u}, \overline{u} \rangle \leftarrow \langle \overline{u}, \overline{u} \rangle - 2m_{best}\langle \overline{u}, \overline{w} \rangle + m_{best}^2\langle \overline{w}, \overline{w} \rangle$
20   Mark $\overline{u}$ as reduced.

---

the lazy version of $\overline{u}$, however, may be representing a vector much longer than it should. To prevent reducing with a lazy $\overline{u}$ in later rounds, we need to keep track of the current correct norm of u. This is done on lines 14–16 in Algorithm 4.

Second, because $\overline{u}$ may have changed in the previous round, $\langle \overline{u}, \overline{w} \rangle$ must be recalculated on line 7. To avoid recalculating $\langle \overline{u}, \overline{w} \rangle$ repeatedly, observe that the probability of reducing $\overline{u}$ more than once is not high in the innermost loop. We can keep track of the best m so far, moving the entire if statement on lines 11–14 out and after the for loop.

Applying these two heuristics, we now reach Algorithm 4. This amended algorithm is more efficient because (1) the body of the most inner loop runs in constant time, thus reducing the complexity to calculate all inner products of two vectors from $O(n^3)$ to $O(n^2)$, and (2) the need to rotate $\overline{u}$ is completely eliminated.

4.4 Generalizing Lifting

The regularity of terms in the quotient polynomial $f(x)$ plays a role in the computation of rotations. Consider a cyclotomic polynomial $p(x)$. There might exist another low degree polynomial $r(x)$ such that $p(x)r(x) = x^n \pm 1$. This suggests we promote a vector with dimension $\deg(p(x))$ into dimension $\deg(p(x)) + \deg(r(x))$, thus lowering the computation cost. The same technique of choosing the right pivots can be applied as well.

For example, the next unsolved ideal lattice challenge is dimension 132. One of the ideal lattices in the challenge is generated by the polynomial $f(x) = x^{132} - x^{130} + x^{128} - \cdots + x^4 - x^2 + 1$. Since $(x^2 + 1)f(x) = x^{134} + 1$, we can convert this lattice into a 134-dimensional anti-cyclic lattice with two pivots, one for the $+1$ terms and the other for -1 terms.

5   Parallelization

Let us now look at our parallel variant of Gauss Sieve for a single machine with multiple GPUs. Two layers of parallelization naturally arise in this setting: the workload should first be split (1) across different GPUs, then (2) to different processors within a GPU. These two layers of architectures differ in communication cost. Broadcasting data from the host memory across all the GPUs through PCIe is much more expensive than broadcasting data from the on-chip memory to different processors within a single GPU.

We carefully design these two layers in hope to mitigate communication overhead. Specifically, we view each GPU as an independent sieve (inner layer), and all the GPUs cooperate as a complete parallel sieve (outer layer). In the following subsections, we will see (1) how the problem is divided into independent sub-sieves on different GPUs, so that each sub-sieve acts as a blackbox, ordinary Gauss Sieve, and (2) how the sub-sieve is designed to maximize GPU power.

5.1   Outer Layer

To distribute the work among the GPUs on a single machine, we first recall the work by Bos et al. [BNvdP14], which was originally designed for computer clusters. In their work, each node acts as an independent Gauss Sieve, maintaining its own local list while reducing the same batch of samples broadcast over all the nodes. These nodes communicate only at the end of each iteration, putting any sample that is ever reduced in any of the nodes to the stack. The advantage of this approach is that the long, local lists are never completely moved out of the nodes; only a limited amount of reduced vectors and samples are involved in communication. Communication cost is thus small.

Here, we adapt their method to a machine with multiple GPUs using the following analogy: A cluster is to the machine what a node is to the GPU. As a result, a GPU now works as if it were a node, having its own local list, and communication is done on the host. At each iteration, all the GPUs are given the same batch of samples, either newly generated or from the stack. Each GPU then first reduces its samples mutually with its local list, using the method described in Subsect. 5.2. Next, for each sample, if ever reduced in one GPU, the host compares and chooses the shortest "representative", putting it to the stack.

Reduced vectors from local lists are also put on the stack. Last, the "surviving" samples are appended to the shortest local list. The vectors on the stack become the input for later rounds.

5.2 Inner Layer

The inner layer is a modified version of Ishiguro et al.'s idea. As mentioned in the previous subsection, each GPU can be thought of as an independent sieve, reducing its local list with a batch of samples. First, the local list is reduced with the samples. Then, such samples are mutually reduced with each other. Finally, the samples are reduced with the local list. If any vector is ever reduced during any step, it is marked and later collected on the stack. As showed by Ishiguro et al., any pair of surviving vectors in the local list remains reduced during the process.

Since these three steps share the same pattern — they all reduce one list with another, the same GPU kernel can be used. See Algorithm 1 for general lattices and Algorithm 2 for ideal lattices. To reduce list A with list B, the kernel takes as inputs list A and list B, and in-place outputs the reduced list A. In the kernel, list A is sliced into adequate chunks and distributed to different processors, while list B is broadcast to all processors. The kernel is crafted with care to ensure high throughput, as will be described in the next section.

6 Implementation Details

In this section, we will first see how common performance tuning techniques can be applied to our algorithm. This includes thread- and instruction-level parallelism. Next, we point out more kernel optimization tricks. Finally, we describe two more heuristics that can significantly improve the execution time.

6.1 Vector Layout

On GPUs, each thread has a physical register number limit; depending on how many resources each thread requires, each processor also has a runtime limit for thread numbers. For example, consider the kernel for $n = 100$. On a Maxwell GPU, each thread can use up to 255 registers. If we put both $\bar{u}$ and $\bar{w}$ in one thread, we need $2 \times (100 + 1) = 202$ registers. Although fewer than 255, this is still so much that the processors can only schedule a few threads, limiting thread-level parallelism. At the other extreme, if we spread a vector across too many threads, the overhead of parallel reduction to calculate inner products collectively will take over.

Empirically, we choose to spread a vector across 4 threads. For our target dimension 130, this means each thread takes $\lceil 130/4 \rceil = 33$ elements, with the extra two elements padded with 0. This choice not only reduces register pressure, but also makes the vector length a multiple of 4, which is essential for cache line alignment. To this end, parallel reductions are needed both on line 7 to collectively sum inner products, and before line 17 (after the for loop) to agree on the best m. To make the code more readable, we use the CUB library [Mer] for block load and store in the kernel.

6.2 Instruction-Level Parallelism

Yet another commonly applied trick to increase GPU utilization is to exploit instruction-level parallelism. The idea is to issue independent instructions at once to increase the pipeline usage. However, since the algorithm is very much inherently dependent from line to line, the direct implementation will run very slowly. We do not overlap two independent copies of kernel at the same time, because the register usage is immediately multiplied by two. Instead, we unroll the loop on line 4 with an empirical factor of 8 to facilitate register reuse on line 7. This technique is possible only if the lattice is lifted, since a lattice point is represented in its cyclic form.

Next, we identify two new heuristics due to loop unrolling. First, at the end of each 8th iteration, we choose the best m_{best} from eight possible m_{best}'s. Second, since the prime n is never a multiple of 8, empirically we just omit the remainder of the unrolled loop.

6.3 More Kernel Optimizations

- The rotation on line 5 is tricky because vectors are padded with zero. Therefore, the last thread that contains a vector would have to deal with these zeros. In fact, the padded vector $\bar{v}$ is first stored in the shared memory, then rotated one by one at each iteration. More specifically, at the end of one iteration, the first padded zero is replaced with the next "right" element, and at the start of the next iteration, the vector $\bar{w}$ is read at the "right" offset.
- The vectors in the lists $\bar{U}$ and $\bar{V}$ are loaded in bulks and put in a shared-memory buffer to increase global memory throughput.
- In practice, we choose the first element of a lifted vector as its pivot and rotate reversely. This transforms the index $n - j$ on lines 9–10 to an easier j.
- To ensure high kernel throughput, we empirically tune all the parameters mentioned in the above sections as well as kernel launch parameters, although it is not feasible to try all possible combinations.

6.4 On Faster Convolution

The question naturally arises: why not use FFT or the Karatsuba algorithm to calculate inner products? The reasons are:

- The Karatsuba algorithm reduces the number of multiplications, while adding a lot more additions. On GPUs, however, FMAD is fastest.
- If on line 5, we use them to produce results for all i's simultaneously, there will not be enough register to hold both the results and all the intermediate values during computation.

– The dimension is not a power of 2, which makes the convolution more difficult to be designed efficiently.

There are several techniques to convert non-power-of-2 DFT's into convolutions or FFTs of the same or larger dimensions. The best approach we are aware of is Devil's convolution [Cra96], but this is not easily applicable on GPUs.

6.5   Heuristics

Besides the heuristics for kernel optimization, we also applied two heuristics to speed up in conjunction with the techniques above.

First, as already mentioned in Voulgaris's implementation [Gau], the lists in step 1 are sorted so that only longer vectors (before rotation) are reduced with shorter ones. We also tried to see if this can be applied to steps 2 and 3. Empirically we found that it is not as effective, probably because vectors are shorter during steps 2 and 3; they are less likely to be reduced. We also use the CUB library to sort data on GPUs.

Second, empirically we choose to iterate the innermost loop over only the first 16 values of j (line 8). This is because the rotations of prime cyclotomic vectors have larger norms. The expansion factor for prime cyclotomic lattices is discussed in [Sch13].

7   Experiments

For our experiments, we use a total of eight NVIDIA GeForce GTX TITAN X graphics cards. Four of these cards are installed on the main machine, while the other four are installed on a PCIe extension box.

We use the bases from the Ideal Lattice Challenge [Ide]. Since for dimension n, the prime cyclotomic polynomial has index $n+1$, as an example, we choose the basis for dimension 126 from the file `ideallatticedim126index127seed0.txt`. The input bases for Gauss Sieve are first reduced by BKZ with block size 30 and $\delta = 0.99$.

7.1   Parallel Efficiency

Here we show the parallel efficiency of the outer layer of our parallel framework in Fig. 1. The *(parallel) efficiency* for N GPUs is defined in [BNvdP14] as

$$E = \frac{\text{runtime for } N \text{ GPUs}}{N \cdot \text{runtime for 1 GPU}}.$$

For the dimension 108, the efficiency is 74%, 72%, 55% and 45% for 2, 4, 6 and 8 GPUs, respectively. However, the dimension is so low that the efficiency is quite low as the number of GPUs exceeds 6.

On the other hand, for dimension 112, the efficiency scales better with the number of GPUs. However, we do not yet have the running time for one GPU. If we base the efficiency on 2 GPUs, then the efficiency is the 86%, 81% and 74% for 4, 6 and 8 GPUs, respectively. We believe that in high enough dimensions, the efficiency of 8 GPUs will be more than 70%.

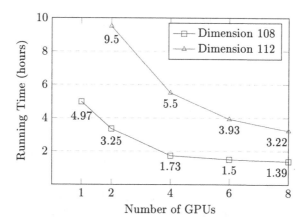

Fig. 1. Parallel efficiency: the numbers of GPUs versus running time and the exact running time is labelled below the node

7.2 Ideal Lattices Versus General Lattices

For general lattices, we use the bases from the SVP Challenge [Lat]. Our single-GPU implementation takes 9.3 h to solve the challenge of dimension 96. In contrast, the implementation from [IKMT14] requires 200 CPU-hour. That is, our single-GPU implementation on general lattices is 21.5 times faster than the CPU version.

For ideal lattices, our 4-GPU implementation requires 5 min to solve the challenge of dimension 96 and our single-GPU implementation requires 8.6 min. In contrast, the implementation from [IKMT14] requires 8 CPU-hours. That is, our 4-GPU (resp. single-GPU) implementation on general lattices is 96 (resp. 55.8) times faster than the CPU version. Note that the polynomial we use is prime-cyclotomic ($x^{96} + x^{95} + \cdots + 1$), which is more complicated than the trinomial polynomial ($x^{96} + x^{48} + 1$) used by [IKMT14].

Combining these two cases, the speed-up from using the property of prime-cyclotomic ideal lattices is $\dfrac{9.3 \text{ hrs}}{8.6 \text{ mins}} = 64.9$ in dimension 96. Applying the complexity estimation from [MV10], we estimate the ratio to be $\dfrac{9.3 \text{ hrs} \cdot 2^{0.52 \cdot 30}}{2734 \text{ hrs}} =$ 169 in dimension 126 and $\dfrac{9.3 \text{ hrs} \cdot 2^{0.52 \cdot 34}}{6583 \text{ hrs}} = 297$ in dimension 130.

In contrast, [IKMT14] shows that the speed-up ratio of using the property of anti-cyclic ideal lattices is around 600 in dimension 128. This gives an evidence that the SVP over prime-cyclotomic ideal lattices is harder than over anti-cyclic ideal lattices by a factor of around 2.

7.3 Chronological Behavior

The chronological behavior of the sieving algorithms is studied intensively in [MV10, Sch11]. We can observe the same behavior in Fig. 2. (a) shows that the

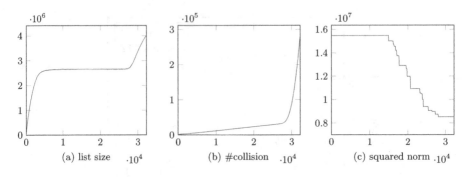

Fig. 2. Behavior of Gauss Sieve for dimension 126 with 8 GPUs: (a) the list size versus iteration; (b) the number of collisions versus iteration; (c) the squared norm of current shortest vector versus iteration

size of the list grows very fast in the beginning, later on it reaches a plateau, and finally when the shortest vector is found, it grows rapidly again. (b) shows the number of collision grows almost linearly but goes up very fast after the shortest vector is found. (c) shows the squared norm of the current shortest vector. The norm starts to drop half-way, and keeps descending until the shortest vector is found. One possible improvement is to reduce the basis by the current founded short vector, as in the work [FK15]. However, since the very first "shorter" vector only shows up half-way, the speed-up ratio by this method is limited by 2.

Our result in Table 1 is the fastest implementation of the Gauss Sieve algorithm so far. A rough space usage estimation is $2 \times 4 \times ListSize \times Dimension$. The factor 4 is due to the data type, 4-byte float, and the 2 is due to an extra buffer for sorting on the device. Therefore, it requires around 0.37, 2.59 and 4.35 GB of memory for dimension 112, 126 and 130, respectively.

Table 1. Results of ideal lattice challenge

Dimension	112	126	130
Number of vectors	444,341	2,759,903	4,490,083
Running time (GPU-hours)	32	2,734	6,583

7.4   Hardness Estimation

Finally, Fig. 3 compares our results with previous works. Obviously, our results are below the estimation of [LP11]. Even more, the slope of ours is flatter than theirs, which means that there is an exponential speed-up. Some of the data from the SVP and Ideal SVP Challenge is computed using an accelerated random sampling algorithm [FK15], but in higher dimensions (say, higher than 136), our

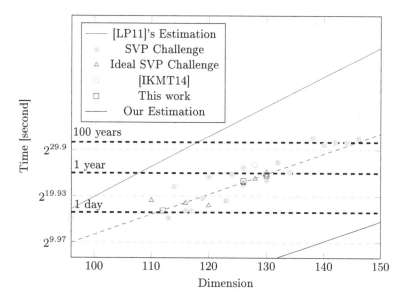

Fig. 3. Comparison of time: the dimension of SVP Challenge versus the running time

fitting curve is also below their results. This might imply that the running time of the Gauss Sieve algorithm grows quite slowly.

Fitting our data using least-square regression, we have $y = 2^{0.435x-31.8}$, as depicted in Fig. 3. To resist attacks using super powerful special-purpose hardware, our conservative model of SVP hardness in ideal lattices, with approximation factor 1.05, is

$$time(SVP_{\alpha=1.05}^{Ideal}) = 2^{0.43n-50} \text{(seconds)}$$

However, we emphasize that the space complexity of the sieve algorithm is exponential, but estimation models of [LP11,CN11] are based on BKZ or BKZ 2.0, which requires only polynomial space. More precisely, our implementation requires $2^{0.19n+7.3}$ bytes of memory.

8   Conclusion

In this work, we propose the lifting technique for prime-cyclotomic ideal lattices, which accelerates the Gauss Sieve algorithm. Moreover, by applying a sequence of transformations described in Sect. 4, the cost to reduce two vectors with all of their rotations decreases from $O(n^3)$ to $O(n^2)$. We also designed and implemented a Gauss Sieve that includes these technique both on a single GPU and on several GPUs. Our implementation is more than 21.5 (resp. 55.8) times faster than the best prior known result on a single CPU core for general (resp. ideal) lattice. Finally, we give a reasonable model to estimate the running time of solving SVP in ideal lattices. Although our model requires an exponential space usage due to the natural property of sieving algorithms, it suggests a bound much lower than the previous model [LP11].

We will release the code to the public domain after we finish up with all the details such as providing a less hostile interface, doing a clean up, and so on.

Acknowledgement. Partially sponsored by MoST projects 105-2923-E-001-003-MY3 and 105-2221-E-001-020-MY3. We would also like to thank Dr. Shinsaku Kiyomoto of KDDI Research for the fruitful discussion.

References

[AD97]   Ajtai, M., Dwork, C.: A public-key cryptosystem with worst-case/average-case equivalence. In: STOC 1997, pp. 284–293. ACM, New York (1997)

[Ajt97]   Ajtai, M.: The shortest vector problem in l_2 is np-hard for randomized reductions. In: Electronic Colloquium on Computational Complexity (ECCC), vol. 4, no. 47 (1997)

[AKS01]   Ajtai, M., Kumar, R., Sivakumar, D.: A sieve algorithm for the shortest lattice vector problem. In: STOC 2001, pp. 601–610. ACM, New York (2001)

[BDGL16] Becker, A., Ducas, L., Gama, N., Laarhoven, T.: New directions in nearest neighbor searching with applications to lattice sieving. In: SODA 2016, pp. 10–24 (2016)

[BL16]   Becker, A., Laarhoven, T.: Efficient (ideal) lattice sieving using cross-polytope LSH. In: Pointcheval, D., Nitaj, A., Rachidi, T. (eds.) AFRICACRYPT 2016. LNCS, vol. 9646, pp. 3–23. Springer, Heidelberg (2016). doi:10.1007/978-3-319-31517-1_1

[BNvdP14] Bos, J.W., Naehrig, M., van de Pol, J.: Sieving for shortest vectors in ideal lattices: a practical perspective. IACR Cryptology ePrint Archive 2014, 880 (2014)

[CN11]   Chen, Y., Nguyen, P.Q.: BKZ 2.0: better lattice security estimates. In: Lee, D.H., Wang, X. (eds.) ASIACRYPT 2011. LNCS, vol. 7073, pp. 1–20. Springer, Heidelberg (2011). doi:10.1007/978-3-642-25385-0_1

[Cra96]   Crandall, R.E.: Topics in Advanced Scientific Computation. Springer-Telos, New York (1996)

[CUD15]   CUDA C programming guide 7.5 (2015). http://docs.nvidia.com/cuda/cuda-c-programming-guide/

[FK15]   Fukase, M., Kashiwabara, K.: An accelerated algorithm for solving SVP based on statistical analysis. JIP **23**(1), 67–80 (2015)

[Gau]   Gauss Sieve implementation by panagiotis voulgaris. https://cseweb.ucsd.edu/~pvoulgar/impl.html

[Gen09]   Gentry, C.: Fully homomorphic encryption using ideal lattices. In: Annual ACM Symposium on Theory of Computing – STOC, pp. 169–178 (2009)

[GGH13]   Garg, S., Gentry, C., Halevi, S.: Candidate multilinear maps from ideal lattices. In: Johansson, T., Nguyen, P.Q. (eds.) EUROCRYPT 2013. LNCS, vol. 7881, pp. 1–17. Springer, Heidelberg (2013). doi:10.1007/978-3-642-38348-9_1

[GNR10]   Gama, N., Nguyen, P.Q., Regev, O.: Lattice enumeration using extreme pruning. In: Gilbert, H. (ed.) EUROCRYPT 2010. LNCS, vol. 6110, pp. 257–278. Springer, Heidelberg (2010). doi:10.1007/978-3-642-13190-5_13

[Ide] Idea Lattice Challenge. http://www.latticechallenge.org/ideallattice-challenge/

[IKMT14] Ishiguro, T., Kiyomoto, S., Miyake, Y., Takagi, T.: Parallel Gauss Sieve algorithm: solving the SVP challenge over a 128-dimensional ideal lattice. In: Krawczyk, H. (ed.) PKC 2014. LNCS, vol. 8383, pp. 411–428. Springer, Heidelberg (2014). doi:10.1007/978-3-642-54631-0_24

[KSD+11] Kuo, P.-C., Schneider, M., Dagdelen, Ö., Reichelt, J., Buchmann, J., Cheng, C.-M., Yang, B.-Y.: Extreme enumeration on GPU and in clouds. In: Preneel, B., Takagi, T. (eds.) CHES 2011. LNCS, vol. 6917, pp. 176–191. Springer, Heidelberg (2011). doi:10.1007/978-3-642-23951-9_12

[Laa15] Laarhoven, T.: Sieving for shortest vectors in lattices using angular locality-sensitive hashing. In: Gennaro, R., Robshaw, M. (eds.) CRYPTO 2015. LNCS, vol. 9215, pp. 3–22. Springer, Heidelberg (2015). doi:10.1007/978-3-662-47989-6_1

[Lat] SVP Challenge. http://www.latticechallenge.org/svp-challenge/

[LP11] Lindner, R., Peikert, C.: Better key sizes (and attacks) for LWE-based encryption. In: Kiayias, A. (ed.) CT-RSA 2011. LNCS, vol. 6558, pp. 319–339. Springer, Heidelberg (2011). doi:10.1007/978-3-642-19074-2_21

[MBL15] Mariano, A., Bischof, C.H., Laarhoven, T.: Parallel (probable) lock-free hash sieve: a practical sieving algorithm for the SVP. ICPP **2015**, 590–599 (2015)

[MDB14] Mariano, A., Dagdelen, Ö., Bischof, C.: A comprehensive empirical comparison of parallel ListSieve and GaussSieve. In: Lopes, L., et al. (eds.) Euro-Par 2014. LNCS, vol. 8805, pp. 48–59. Springer, Heidelberg (2014). doi:10.1007/978-3-319-14325-5_5

[Mer] Duane (Nvidia Coorporation) Merrill. The CUB Library

[MS11] Milde, B., Schneider, M.: A parallel implementation of GaussSieve for the shortest vector problem in lattices. In: Malyshkin, V. (ed.) PaCT 2011. LNCS, vol. 6873, pp. 452–458. Springer, Heidelberg (2011). doi:10.1007/978-3-642-23178-0_40

[MV10] Micciancio, D., Voulgaris, P.: Faster exponential time algorithms for the shortest vector problem. In: SODA 2010, pp. 1468–1480 (2010)

[NV08] Nguyen, P.Q., Vidick, T.: Sieve algorithms for the shortest vector problem are practical. J. Math. Cryptol. **2**(2), 181–207 (2008)

[Sch11] Schneider, M.: Analysis of Gauss-Sieve for solving the shortest vector problem in lattices. In: Katoh, N., Kumar, A. (eds.) WALCOM 2011. LNCS, vol. 6552, pp. 89–97. Springer, Heidelberg (2011). doi:10.1007/978-3-642-19094-0_11

[Sch13] Schneider, M.: Sieving for shortest vectors in ideal lattices. In: Youssef, A., Nitaj, A., Hassanien, A.E. (eds.) AFRICACRYPT 2013. LNCS, vol. 7918, pp. 375–391. Springer, Heidelberg (2013). doi:10.1007/978-3-642-38553-7_22

A Tool Kit for Partial Key Exposure Attacks on RSA

Atsushi Takayasu[1,2(✉)] and Noboru Kunihiro[1]

[1] The University of Tokyo, Tokyo, Japan
a-takayasu@it.k.u-tokyo.ac.jp, kunihiro@k.u-tokyo.ac.jp
[2] National Institute of Advanced Industrial Science and Technology (AIST),
Tokyo, Japan

Abstract. Thus far, *partial key exposure attacks on RSA* have been intensively studied using lattice based Coppersmith's methods. In the context, attackers are given partial information of a *secret exponent* and *prime factors* of *(Multi-Prime) RSA* where the partial information is exposed in various ways. Although these attack scenarios are worth studying, there are several known attacks whose constructions have similar flavor. In this paper, we try to formulate general attack scenarios to capture several existing ones and propose attacks for the scenarios. Our attacks contain all the state-of-the-art partial key exposure attacks, e.g., due to Ernst et al. (Eurocrypt'05) and Takayasu-Kunihiro (SAC'14, ICISC'14), as special cases. As a result, our attacks offer better results than previous best attacks in some special cases, e.g., Sarkar-Maitra's partial key exposure attacks on RSA with the most significant bits of a prime factor (ICISC'08) and Hinek's partial key exposure attacks on Multi-Prime RSA (J. Math. Cryptology '08). We claim that our contribution is not only generalizations or improvements of the existing results. Since our attacks capture general exposure scenarios, the results can be used as a tool kit; the security of some future variants of RSA can be examined without any knowledge of Coppersmith's methods.

Keywords: (Multi-Prime) RSA · Partial key exposure · Lattices · Coppersmith's methods

1 Introduction

Background. Let $N = pq$ be a public RSA modulus where p and q are distinct prime factors with the same bit-size. A public/secret exponent e and d such that $ed = 1 \pmod{\Phi(N)}$ where $\Phi(N)$ is Euler's totient function. There is a variant of RSA called Multi-Prime RSA that have a public modulus $N = \prod_{i=1}^{r} p_i$ where p_i's are all distinct primes with the same bit-size. A public/secret exponent of Multi-Prime RSA satisfies the same equation as the standard RSA. Multi-Prime RSA offers faster decryption/signing by combining with Chinese Remainder Theorem.

From the invention of RSA cryptosystems, hardness of the factorization/RSA problem have been intensively studied. One well known approach in the literature is lattice based Coppersmith's methods [6,7]. The method showed an RSA

© Springer International Publishing AG 2017
H. Handschuh (Ed.): CT-RSA 2017, LNCS 10159, pp. 58–73, 2017.
DOI: 10.1007/978-3-319-52153-4_4

modulus $N = pq$ can be factorized in polynomial time with half the most significant bits of a prime factor. Although Coppersmith's methods requires involved technical analyses, the method has revealed the vulnerability of RSA in many papers. One of the most famous result is Boneh and Durfee's small secret exponent attack on RSA [3] that factorizes an RSA modulus N in polynomial time when $d < N^{1-1/\sqrt{2}} = N^{0.292\cdots}$. Ciet et al. [5] extended the attack for Multi-Prime RSA and their attack works when $d < N^{1-\sqrt{1-1/r}}$.

Boneh et al. [4] proposed several attacks on RSA called *partial key exposure attacks* that make use of the most/least significant bits (MSBs/LSBs) of d. Afterwards, the research becomes a hot topic and numerous papers have been published. Although the original attacks [4] work only for a small e, several improvements [1,12,22,26] have been proposed using Coppersmith's methods [6,7]. In particular, Ernst et al. [12] revealed that RSA becomes vulnerable even for a full size e and Takayasu-Kunihiro's attacks [26] contain Boneh-Durfee's small secret exponent attack [3] as a special case. Besides these results, numerous papers have studied partial key exposure attacks for various attack scenarios; attacks on Multi-Prime RSA with the MSBs/LSBs of d [13], attacks on RSA with the MSBs of a prime factor [21], attacks on RSA with the MSBs/LSBs of d and the MSBs of a prime factor [20], attacks on RSA where the prime factors share the same LSBs [23], attacks on RSA where the prime factors are almost the same sizes [29], attacks on Multi-Prime RSA where all the prime factors are almost the same sizes [25,30,31], and more.

Indeed, there are many papers that study partial key exposure attacks on RSA. However, the situation does not immediately mean that the problem is worth studying in such many papers. Among the above variants of the attack, some papers capture almost the same attack scenarios. Hence, essentially the same algorithms have been proposed in several papers. We do not think the situation is not desirable for the development of the cryptographic research.

Our Contributions. To resolve the situation, we define a general partial key exposure scenario. For the purpose, we classify some existing works with respect to three properties; attackers know partial information of a *secret exponent* and *prime factors* for *Multi-Prime RSA*. Since there are no results that capture the three properties simultaneously, we define a general attack scenario as follows.

Definition 1 $((\alpha, \beta, \gamma, \delta)$-**Partial Key Exposure Attacks on RSA**)**.** *Let* $N = \prod_{i=1}^{r} p_i$ *where all* $p_1, \ldots, p_r$ *are distinct primes of the same bit-size. Let* $e = N^\alpha$ *and* $d = N^\beta$ *such that* $ed = 1 \mod \Phi(N)$*. Given* $(N, e, \tilde{d}, \tilde{\Phi}(N))$ *where* $\tilde{d} \geq N^{\beta-\gamma}$ *is the MSBs/LSBs of* d *and* $|\Phi(N) - \tilde{\Phi}(N)| \leq N^\delta$*, the goal of the problem is to compute* $\Phi(N)$*.*

We parametrize the problem with respect to $(\alpha, \beta, \gamma, \delta)$. Notice that the number of prime factors r is independent of the hardness of the problem. Although partial information of prime factors in previous works are defined in various ways, the above definition captures several exposure scenarios simultaneously. For example, let us focus on an attack on RSA with the most significant bits prime factors and an attack on Multi-Prime RSA. Given $\tilde{p}$ which is the $\delta' \log N$ MSBs of an RSA prime factor p, then we regard $\tilde{\Phi}(N) = N - \tilde{p}N^{1/2-\delta'} - \lfloor N/\tilde{p}N^{1/2-\delta'} \rfloor$ and an attack on RSA with the most significant bits of prime factors is captured by $\delta = 1/2 - \delta'$ since $|\Phi(N) - \tilde{\Phi}(N)|$ is bounded above by $N^{1/2-\delta'}$ within a constant factor [20,21]. Similarly, we regard $\tilde{\Phi}(N) = N$ and an attack on Multi-Prime RSA is captured by $\delta = 1 - 1/r$ since $|\Phi(N) - N|$ is bounded above by $N^{1-1/r}$ within a constant factor [13]. Since we analyze all $0 \le \gamma \le \beta$ and $0 \le \delta \le 1$, our definition covers several existing works simultaneously. Moreover, the definition will cover other unknown variants that will be studied in the future. Then our results can be viewed as a *tool kit* to study partial key exposure attacks as [2]. It means that our results enable even beginners of Coppersmith's methods to examine the security of such future variants without understanding the technical detail of this paper.

We use lattice based Coppersmith's methods to solve integer/modular equations as previous works and obtain the following results.

Theorem 1. *Given the MSBs/LSBs of d, there are polynomial time algorithms to solve $(\alpha, \beta, \gamma, \delta)$-Partial Key Exposure Attacks on RSA when*

$$- \gamma < \frac{3-\delta-2\sqrt{\delta^2+3(\alpha+\beta-1)\delta}}{3}.$$

Theorem 2. *Given the MSBs of d, there are polynomial time algorithms to solve $(1, \beta, \gamma, \delta)$-Partial Key Exposure Attacks on RSA when*

1. $\gamma < 1 - \frac{2}{3}\left(\delta + \sqrt{\delta(4\delta - 3 + 6\beta)}\right)$ *for* $\beta < 1 - \delta - \sqrt{\frac{\delta(1-\delta)}{3}}$,

2. $\gamma < \frac{1+\beta-\delta-\sqrt{4\delta-3(1-\beta)^2}}{2}$ *for* $1 - \delta - \sqrt{\frac{\delta(1-\delta)}{3}} \le \beta < 1 - \delta$ *and* $1/3 \le \delta$, *and for*

   $1 - \delta - \sqrt{\frac{\delta(1-\delta)}{3}} \le \beta < 1 - \sqrt{\frac{\delta}{3}}$ *and* $\delta < 1/3$,

3. $3\lambda\tau - 3(1-\delta)\tau^2 + \tau^3 < \frac{(\delta\tau-\beta+\lambda)^3}{\delta(1+\lambda-2\beta)}$ *where* $\lambda = \max\{\gamma, \beta+\delta-1\}$ *and* $\tau = 1 - \frac{\beta+\delta-1}{\delta-\sqrt{1+\lambda-2\beta}}$ *for* $1 - \delta \le \beta < \frac{3(1-\delta)(1+\delta)}{4}$ *and* $1/3 \le \delta < 2/3$, *and for*

   $1 - \delta \le \beta < \delta - \frac{(2\delta-1)^2}{\delta^2}$ *and* $2/3 \le \delta$,

4. $\gamma \le \frac{3(1-\delta)^2}{4}$ *for* $\frac{3(1-\delta)(1+\delta)}{4} \le \beta < \frac{3(1-\delta)^2+4(1-\delta)}{4}$ *and* $1/3 \le \delta < 2/3$,

5. $\gamma < \frac{2+\beta-2\delta-2\sqrt{(\beta+\delta-1)(\beta+4\delta-1)}}{3}$ *for* $\frac{3(1-\delta)^2+4(1-\delta)}{4} \le \beta$ *and* $1/3 \le \delta$,

6. $\gamma \le 1 - \frac{2\sqrt{3\delta}}{3}$ *for* $1 - \sqrt{\frac{\delta}{3}} \le \beta$ *and* $\delta < 1/3$.

Theorem 3. *Given the LSBs of d, there are polynomial time algorithms to solve $(1, \beta, \gamma, \delta)$-Partial Key Exposure Attacks on RSA when*

1. $\gamma < 1 - \frac{2}{3}\left(\delta + \sqrt{\delta(4\delta - 3 + 6\beta)}\right)$ *for* $\beta < 1 - \delta - \sqrt{\frac{\delta(1-\delta)}{3}}$,

2. $\gamma < \frac{1+\beta-\delta-\sqrt{4\delta-3(1-\beta)^2}}{2}$ *for* $1 - \delta - \sqrt{\frac{\delta(1-\delta)}{3}} \le \beta < 1 - \frac{\delta}{2} - \frac{\sqrt{3\delta(4-\delta)}}{6}$,

3. $\gamma < 1 - \frac{\delta+2\sqrt{\delta(\delta+3\beta)}}{3}$ *for* $1 - \frac{\delta}{2} - \frac{\sqrt{3\delta(4-\delta)}}{6} \le \beta$.

First of all, our results cover all the known best attacks as special cases, e.g., Theorem 1, the conditions 4–6 of Theorem 2, and the condition 3 of Theorem 3 for $\delta = 1/2$ are the same as Ernst et al.'s attack [12]. Extensions of previous works are not trivial at all. In the context of the algorithm construction of Coppersmith's methods, to tackle the equations with the more monomials requires the more involved analyses. Hence, to extend some attacks with more partial information and the extended attacks completely cover the original ones as special cases is challenging in some cases. For example, Ernst et al.'s $(1, \beta, \gamma, 1/2)$-partial key exposure attack [12] for $\gamma = \beta$ do not cover Boneh and Durfee's $(1, \beta, \beta, 1/2)$-partial key exposure attack [3]. It takes about ten years until the desired attacks [26] were proposed. Indeed, in this paper, we have to analyze eight attacks to obtain the best results for all the cases.

Furthermore, our results offer improved attacks in some special cases. More concretely, we improve Sarkar and Maitra's partial key exposure attacks on RSA with partial information of prime factors [20] for small d and Hinek's partial key exposure attacks on Multi-Prime RSA [13]. See Figs. 1 and 2 for detailed comparisons. Indeed, our attacks require smaller portions of partial information of d than their attacks.

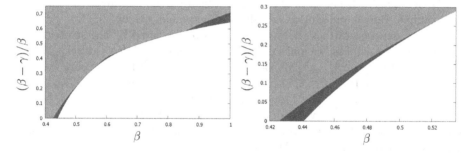

Fig. 1. Comparisons of partial key exposure attacks on RSA with the $\approx \frac{3}{16} \log N$ MSBs of p, i.e., $(1, \beta, \gamma, 5/16)$-partial key exposure attacks. We compare how much portions of d should be exposed for β between Sarkar and Maitra's attack (gray areas) [20] and our Theorems 2 and 3 (red areas). The left (resp. right) figure represents the attack with the MSBs (resp. LSBs). (Color figure online)

Technical Overview. To provide better attacks based on Coppersmith's methods is equivalent to provide better lattice constructions to solve underlying equations. There is a well-known strategy for the construction due to Jochemsz and May [15]. The construction may be simple and easy to understand even for beginners of the research area. Ernst et al. [12] made use of the strategy for their attacks. Sarkar-Maitra [20], Hinek [13], and some other papers extended the attack of Ernst et al. Then, we also follow the strategy and propose extended attacks in Sect. 3; Theorem 1, the conditions 4–6 of Theorem 2, and the condition 3 of Theorem 3. The results based on the strategy are almost naive extensions of

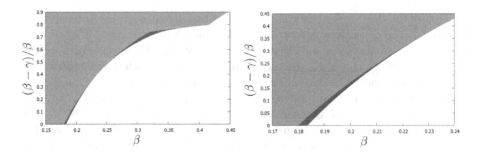

Fig. 2. Comparisons of partial key exposure attacks on Multi-Prime RSA for the number of prime factors $r = 3$, i.e., $(1, \beta, \gamma, 2/3)$-partial key exposure attacks. We compare how much portions of d should be exposed for β between Hinek's attack (gray areas) [13] and our Theorems 2 and 3 (red areas). The left (resp. right) figure represents the attack with the MSBs (resp. LSBs). (Color figure online)

the previous attacks although there are some improved analyses in our results; the condition 6 of Theorem 2 in Sect. 3.3 improves Sarkar-Maitra's attack.

Notice that the Jochemsz-May strategy does not always offer the best attacks and lattice constructions that outperform the strategy require involved analyses. For example, Boneh and Durfee's small secret exponent attack [3]; $(1, \beta, \beta, 1/2)$-partial key exposure attack, does not seem to be captured by the strategy. To construct better attacks, we make use of Takayasu and Kunihiro's attacks [25, 26] where the attack in [25] and [26] solved $(1, \beta, \beta, \delta)$-partial key exposure attacks for $0 \leq \delta \leq 1$ and $(1, \beta, \gamma, 1/2)$-partial key exposure attacks for $0 \leq \gamma \leq \beta$, respectively. Technically, the former and the latter attack constructs a better lattice with respect to the value of δ and γ, respectively. Moreover, they are the only existing partial key exposure attacks that outperform the Jochemsz-May strategy [15] except the Boneh-Durfee attack and its straightforward extension. As we suggested above, these lattice constructions [25, 26] seem to be technically hard to follow. Indeed, there are only a few papers [27, 28] that make use of these results to obtain better results. In this paper, we fully exploit the spirit of the lattice constructions [25, 26] and propose $(1, \beta, \gamma, \delta)$-partial key exposure attacks for arbitrary $0 \leq \gamma \leq \beta$ and $0 \leq \delta \leq 1$. Our attacks cover Takayasu and Kunihiro's attacks [25, 26] for a fixed $\gamma = \beta$ and $\delta = 1/2$, respectively. We study the attacks with the MSBs and LSBs of d in Sects. 4 and 5, respectively.

2   Preliminaries

In this section, we briefly introduce some basic notions of Coppersmith's methods. For more detailed information, see [8, 9, 18, 19].

Let $\boldsymbol{b}_1, \ldots, \boldsymbol{b}_n \in \mathbb{Z}^{n'}$ be linearly independent n'-dimensional vectors. All vectors are row representations. A lattice $L(\boldsymbol{b}_1, \ldots, \boldsymbol{b}_n)$ spanned by the basis vectors $\boldsymbol{b}_1, \ldots, \boldsymbol{b}_n$ is defined as $L(\boldsymbol{b}_1, \ldots, \boldsymbol{b}_n) = \{\sum_{j=1}^{n} c_j \boldsymbol{b}_j : c_j \in \mathbb{Z}\}$. We also use matrix representations $\boldsymbol{B} \in \mathbb{Z}^{n \times n'}$ for the bases where each row corresponds to basis vectors $b_1, \ldots, b_n$. Then, a lattice spanned by the basis matrix B is defined as $L(B) = \{cB : c \in \mathbb{Z}^n\}$. We call n a rank of the lattice, and n' a dimension of the lattice. We call the lattice full-rank when $n = n'$. We define a determinant of a lattice $\det(L(B))$ as $\det(L(B)) = \sqrt{\det(BB^T)}$ where B^T is a transpose of B. By definition, a determinant of a full-rank lattice can be computed as $\det(L(B)) = |\det(B)|$. Moreover, a determinant of a triangular matrix can be easily computed as the product of all diagonals.

For a cryptanalysis, to find short lattice vectors is a very important problem. In 1982, Lenstra et al. [16] proposed a polynomial time algorithm to find short lattice vectors.

Proposition 1 (LLL algorithm [16,17]). *Given a matrix $B \in \mathbb{Z}^{n \times n'}$, the LLL algorithm finds vectors b_1' and b_2' in a lattice $L(B)$. Euclidean norms of the vectors are bounded by*

$$\|b_1'\| \leq 2^{(n-1)/4}(\det(L(B)))^{1/n} \text{ and } \|b_2'\| \leq 2^{n/2}(\det(L(B)))^{1/(n-1)}.$$

The running time is polynomial time in n, n', and input length.

Although the outputs of the LLL algorithm are not the shortest lattice vectors in general, the fact is not the matter in the context of Coppersmith's methods.

Instead of original Coppersmith's methods, we introduce Howgrave-Graham's reformulation to solve modular equations [14] and Coron's reformulation to solve integer equations [10]. Although Coron's method [10] is less efficient than original Coppersmith's method [6] and Coron's other method [11], it is simpler to analyze than the other methods.

For a k-variate polynomial $h(x_1, \ldots, x_k) = \sum h_{i_1, \ldots, i_k} x_1^{i_1} \cdots x_k^{i_k}$, we define a norm of a polynomial $\|h(x_1, \ldots, x_k)\| = \sqrt{\sum h_{i_1, \ldots, i_k}^2}$ and $\|h(x_1, \ldots, x_k)\|_\infty = \max_{i_1, \ldots, i_k} |h_{i_1, \ldots, i_k}|$. At first, we show a modular method since an integer method makes use of the modular method. Coppersmith's method can find solutions $(\tilde{x}_1, \tilde{x}_2)$ of a bivariate modular equation $h(x_1, x_2) = 0 \pmod{e}$ when $|\tilde{x}_1| < X_1, |\tilde{x}_2| < X_2$, and $X_1 X_2$ is reasonably smaller than e. In general, the simpler the Newton polygon of the polynomial is, the larger solutions can be recovered. Let m be a positive integer. We construct n polynomials $h_1(x_1, x_2), \ldots, h_n(x_1, x_2)$ that have the root $(\tilde{x}_1, \tilde{x}_2)$ modulo e^m. Then, we construct a matrix B whose rows consist of coefficients of $h_1(x_1 X_1, x_2 X_2), \ldots, h_n(x_1 X_1, x_2 X_2)$. Applying the LLL algorithm to B and we obtain two short vectors b_1' and b_2', and their corresponding polynomials $h'(x_1, x_2)$ and $h_2'(x_1, x_2)$. If norms of these polynomials are small, they have the root $(\tilde{x}_1, \tilde{x}_2)$ over the integers. The fact comes from the following lemma due to Howgrave-Graham [14].

Lemma 1 ([14]). *Let $h(x_1, \ldots, x_k) \in \mathbb{Z}[x_1, \ldots, x_k]$ be a polynomial over the integers that consists of at most n monomials. Let $X_1, \ldots, X_k$, and R be positive integers. If the polynomial $h(x_1, \ldots, x_k)$ satisfies the following two conditions:*

1. $h(\tilde{x}_1, \ldots, \tilde{x}_k) = 0 \pmod{R}$, *where* $|\tilde{x}_1| < X_1, \ldots, |\tilde{x}_k| < X_k$,
2. $\|h(x_1 X_1, \ldots, x_k X_k)\| < R/\sqrt{n}$.

Then, $h(\tilde{x}_1, \ldots, \tilde{x}_k) = 0$ holds over the integers.

Therefore, if $h'(x_1, x_2)$ and $h'_2(x_1, x_2)$ satisfy Lemma 1, we can compute Gröbner bases or a resultant of them and easily recover $(\tilde{x}_1, \tilde{x}_2)$. By making use of the unravelled linearization, we only analyze triangular matrices in this paper. Better lattice constructions for triangular matrices are well analyzed [18,24] by introducing helpful polynomials. Intuitively, polynomials in lattice bases are called helpful when their diagonals in the triangular basis matrices are smaller than the modulus of the equations e^m. To solve modular equations for larger roots, as many (resp. less) helpful (resp. unhelpful) polynomials as possible should be selected as long as the basis matrices are triangular. We follow the definition from [26] as follows.

Definition 2 (Helpful Polynomials [18,26]). *To solve equations modulo e, consider a basis matrix $\boldsymbol{B}$. We add a new shift-polynomial $h_{[i',j']}(x, y)$ and construct a new basis matrix $\boldsymbol{B}^+$. We call $h_{[i',j']}(x, y)$ a helpful polynomial, provided that $\det(\boldsymbol{B}^+)/\det(\boldsymbol{B}) \leq e^m$. Conversely, if the inequality does not hold, we call $h_{[i',j']}(x, y)$ an unhelpful polynomial.*

Next, we show an integer method. Coppersmith's method can find solutions $(\tilde{x}_1, \tilde{x}_2, \tilde{x}_3)$ of a trivariate integer equation $h(x_1, x_2, x_3) = 0$ when $|\tilde{x}_1| < X_1, |\tilde{x}_2| < X_2, |\tilde{x}_3| < X_3$, and $X_1 X_2 X_3$ is reasonably smaller than $\|h(x_1 X_1, x_2 X_2, x_3 X_3)\|_\infty$. Although we omit details of the method, we set a reasonable integer R and remaining procedures are almost the same as modular case by solving a modular equation $h(x_1, x_2, x_3) = 0 \mod R$. New polynomials $h'(x_1, x_2, x_3)$ and $h'_2(x_1, x_2, x_3)$ obtained by outputs of the LLL algorithm are provably algebraically independent of $h(x_1, x_2, x_3)$. See [10] for the detail. To the best of our knowledge, there are no algorithms to solve integer equations known that outperform the algorithm based on the Jochemsz-May strategy [15]. Hence, we follow the strategy. Let l_j denote the largest exponent of x_j in the polynomial $h(x_1, \ldots, x_k) = \sum h_{i_1, \ldots, i_k} x_1^{i_1} \cdots x_k^{i_k}$. We set an (possibly large) integer W such that $W \leq \|h(x_1, \ldots, x_k)\|_\infty$ and an integer $R := W X_1^{l_1(m-1)+t} \prod_{u=2}^{k} X_j^{l_u(m-1)}$ with some positive integers m and $t = O(m)$ such that $\gcd(R, h_{0, \ldots, 0}) = 1$. We compute $c = h_{0, \ldots, 0}^{-1} \pmod{R}$ and $h'(x_1, \ldots, x_k) := c \cdot h(x_1, \ldots, x_k) \pmod{R}$. We define shift-polynomials g and g' as

$$g : x_1^{i_1} \cdots x_k^{i_k} \cdot h(x_1, \ldots, x_k) \cdot X_1^{l_1(m-1)+t-i_1} \prod_{u=2}^{k} X_j^{l_u(m-1)-i_j} \quad \text{for } x_1^{i_1} \cdots x_k^{i_k} \in S,$$

$$g' : x_1^{i_1} \cdots x_k^{i_k} \cdot R \quad \text{for } x_1^{i_1} \cdots x_k^{i_k} \in M \backslash S,$$

for sets of monomials

$$S := \bigcup_{0 \leq j \leq t} \{x_1^{i_1+j} \cdots x_k^{i_k} | x_1^{i_1} \cdots x_k^{i_k} \text{ is a monomial of } h(x_1, \ldots, x_k)^{m-1}\},$$

$$M := \{\text{monomials of } x_1^{i_1} \cdots x_k^{i_k} \cdot h(x_1, \ldots, x_k) \text{ for } x_1^{i_1} \cdots x_k^{i_k} \in S\}.$$

All these shift-polynomials g and g' modulo R have the root $(\tilde{x}_1, \ldots, \tilde{x}_k)$ that is the same as $h(x_1, \ldots, x_k)$. We construct a lattice with coefficients of

$g(x_1 X_1, \ldots, x_k X_k)$ and $g'(x_1 X_1, \ldots, x_k X_k)$ as the bases. The shift-polynomials generate a triangular basis matrix. Ignoring low order terms of m, LLL outputs short vectors that satisfy Lemma 1 when $\prod_{j=1}^{k} X_j^{s_j} < W^{|S|}$ for $s_j = \sum_{x_1^{i_1} \cdots x_k^{i_k} \in M \setminus S} i_j$. When the condition holds, we can find all the small root. See [15] for the detail.

We should note that these methods require heuristic argument. There are no assurance if new polynomials obtained by outputs of the LLL algorithm are algebraically independent. In this paper, we assume that these polynomials are always algebraically independent and resultants of polynomials will not vanish as previous works.

3   Attacks by Solving Integer Equations

In this section, we solve integer equations and propose three attacks, i.e., Attacks 1–3. The Attack 1, 2, and 3 in Sects. 3.1, 3.2, and 3.3 corresponds to Theorem 1 and the condition 3 of Theorem 3, the conditions 4 and 5 of Theorem 2, and the condition 6 of Theorem 2, respectively. Algorithm constructions in this section are similar to Ernst et al. [12].

3.1   The Attack 1

In this section, we consider $(\alpha, \beta, \gamma, \delta)$-partial key exposure attacks with the MSBs/LSBs of d. When $\tilde{d}$ which is the MSBs/LSBs of d is given, RSA key generation can be written as $e(\tilde{d}\tilde{M} + d'M') = 1 + k\Phi(N)$ with some integer k such that $|k| \leq N^{\alpha+\beta-1}$. When $\tilde{d}$ is the MSBs (resp. LSBs), d' denotes the LSBs (resp. MSBs) of d, and $\tilde{M} = 2^{\lfloor \gamma \log N \rfloor}$ and $M' = 1$ (resp. $\tilde{M} = 1$ and $M' = 2^{\lfloor (\beta-\gamma) \log N \rfloor}$). Then, we find the root of the following polynomial over the integers:

$$f_{i1}(x, y, z) = c + eM'x + y(\tilde{\Phi} + z)$$

where $c = 1 - e\tilde{d}\tilde{M}$. If we can recover the root $(x, y, z) = (-d', k, \Phi(N) - \tilde{\Phi}(N))$, whole secret information can be computed. By definition, the absolute values of the root is bounded above by $X := N^\gamma, Y := N^{\alpha+\beta-1}, Z := N^\delta$. By solving the integer equation based on the Jochemsz-May strategy [15], Theorem 1 and the condition 3 of Theorem 3 can be obtained.

3.2   The Attack 2

In this section, we consider $(1, \beta, \gamma, \delta)$-partial key exposure attacks with the MSBs of d. As in Sect. 3.1, when $\tilde{d}$ which is the MSBs of d is given, RSA key generation can be written as $e(\tilde{d}M + d') = 1 + k\Phi(N)$ with some integer k such that $|k| \leq N^\beta$ and $M = 2^{\lfloor \gamma \log N \rfloor}$. In this section, we use an additional information $\tilde{k} = \lfloor (e\tilde{d} - 1)/\tilde{\Phi}(N) \rfloor$ which is an approximation to k. Indeed, $\tilde{k}$ satisfies the following condition:

$$|\tilde{k} - k| < 2N^\lambda \text{ where } \lambda = \max\{\gamma, \beta + \delta - 1\}. \tag{1}$$

The approximate value enables us to obtain better results for large β. Since Sarkar and Maitra [20] used $\lambda = \max\{\gamma, \beta - 1/2\}$ for $\delta \leq 1/2$, we improve the bound although the following lattice construction is completely the same. We find the root of the following polynomial over the integers:

$$f_{i2}(x, y, z) = c + ex + (\tilde{k} + y)(\tilde{\Phi} + z),$$

where $c = 1 - e\tilde{d}\tilde{M}$ as in Sect. 3.1. If we can recover the root $(x, y, z) = (-d', k - \tilde{k}, \Phi(N) - \tilde{\Phi}(N))$, whole secret information can be computed. The absolute values of the root are bounded above by $X := N^\gamma, Y := N^\lambda, Z := N^\delta$ where $\lambda = \max\{\gamma, \beta + \delta - 1\}$. Although the absolute values of solutions become smaller than those in Sect. 3.1, the result in this section is not always better since the Newton polygon of the polynomial becomes more complex.

We set an (possibly large) integer W such that $W < N^{1+\lambda}$ since $\|f_{i2}(xX, yY, zZ)\|_\infty \geq |\tilde{\Phi}(N)Y| \approx N^{1+\lambda}$. Next, we set an integer $R := WX^{m-1} \cdot Y^{m+r-1+t}Z^{m-1}$ with some integers $m = \omega(r)$ and $t = \tau m$ where $\tau \geq 0$ such that $\gcd(R, c) = 1$. We compute $c' = c^{-1} \mod R$ and $f'_{i2}(x, y, z) := c \cdot f_{i2}(x, y, z) \mod R$. We define shift-polynomials g_{i1} and g'_{i1} as

$$g_{i2} : x^{i_X} y^{i_Y} z^{i_Z} \cdot f'_{i2} \cdot X^{m-1-i_X} Y^{m-1+t-i_Y} Z^{m+r-1-i_Z} \text{ for } x^{i_X} y^{i_Y} z_1^{i_Z} \in S,$$

$$g'_{i2} : x^{i_X} y^{i_Y} z^{i_Z} \cdot R \quad \text{for } x^{i_X} y^{i_Y} z_1^{i_Z} \in M \backslash S,$$

for sets of monomials

$$S := \bigcup_{0 \leq j \leq t} \left\{ x^{i_X} y^{i_Y+j} z^{i_Z} \middle| x^{i_X} y^{i_Y} z^{i_Z} \text{ is a monomial of } f_i(x, y, z_1)^{m-1} \right\},$$

$$M := \left\{ x^{i_X} y^{i_Y} z^{i_Z} \middle| \text{ monomials of } x^{i'_X} y^{i'_Y} z^{i'_Z} \cdot f_i(x, y, z) \text{ for } x^{i'_X} y^{i'_Y} z^{i'_Z} \in S \right\}.$$

By definition of sets of monomials S and M, it follows that

$$x^{i_X} y^{i_y} z^{i_Z} \in S \Leftrightarrow i_X = 0, 1, \ldots, m - 1; i_Y = 0, 1, \ldots, m - 1 + t - i_X;$$

$$i_Z = 0, 1, \ldots, m - 1 - i_X,$$

$$x^{i_X} y^{i_y} z^{i_Z} \in M \Leftrightarrow i_X = 0, 1, \ldots, m; i_Y = 0, 1, \ldots, m + t - i_X;$$

$$i_Z = 0, 1, \ldots, m - i_X.$$

All these shift-polynomials g_{i2} and g'_{i2} modulo R have the root $(x, y, z) = (-d', k - \tilde{k}, \Phi(N) - \tilde{\Phi}(N))$ that is the same as $f_{i2}(x, y, z)$. We build a lattice with these polynomials.

Based on the Jochemsz-May strategy [15], the integer equation $f_{i1}(x, y, z) = 0$ can be solved when $X^{(\frac{1}{3}+\frac{\tau}{2})m^3} Y^{(\frac{1}{2}+\tau+\frac{\tau^2}{2})m^3} Z^{(\frac{1}{2}+\frac{\tau}{2})m^3} < W^{(\frac{1}{3}+\frac{\tau}{2})m^3}$. By substituting $\tau = \frac{1-\gamma-\delta-\lambda}{2\lambda}$, the conditions 4 and 5 of Theorem 2 can be obtained. To follow the definition $\lambda = \max\{\gamma, \beta + \delta - 1\}$, $\lambda = \gamma$ when $\beta < \frac{3(1-\delta)^2+4(1-\delta)}{4}$ and $\lambda = \beta + \delta - 1$ otherwise.

3.3   Attack 3

In this section, we propose a better lattice construction than that in Sect. 3.2. Notice that the Newton polygon of $f_{i2}(x, y, z)$ is symmetric with respect to y and z. Hence, we should add extra shifts for the smaller variable. From the bound of the Attack 2, $Y = N^\lambda = N^{3(1-\delta)^2/4} \geq Z = N^\delta$ when $\delta < 1/3$. Therefore, we add extra shifts for z for such small δ. We construct a lattice that is symmetric with respect to y and z from that in Sect. 3.2 and the integer equation $f_{i2}(x, y, z) = 0$ can be solved when $X^{\left(\frac{1}{3}+\frac{\tau}{2}\right)m^3} Y^{\left(\frac{1}{2}+\frac{\tau}{2}\right)m^3} Z^{\left(\frac{1}{2}+\tau+\frac{\tau^2}{2}\right)m^3} < W^{\left(\frac{1}{2}+\frac{\tau}{2}\right)m^3}$. By substituting $\tau = \frac{1-\lambda-2\delta}{2\delta}$, the condition 6 of Theorem 2 can be obtained. Notice that when $\delta < 1/3$, $\beta + \delta - 1 < \gamma \leq 1 - \frac{2\sqrt{3\delta}}{3}$ always hold for $\beta < 1$.

4   Attacks with the MSBs of d by Solving Modular Equations

In this section, we solve modular equations and propose three attacks, i.e., Attacks 4–6, for $(1, \beta, \gamma, \delta)$-partial key exposure attacks with the MSBs of d. The Attack 4, 5, and 6 in Sects. 4.1, 4.2, and 4.3 correspond to the conditions 2, 3, and 1 of Theorem 2, respectively. Algorithm constructions in Sects. 4.1 and 4.2, that in Sect. 4.3 are similar to Takayasu-Kunihiro's [25,26], respectively.

4.1   The Attack 4

As in Sect. 3.2, when $\tilde{d}$ which is the MSBs of d is given, RSA key generation can be written as $e(\tilde{d}M + d') = 1 + k\Phi(N)$ with some integer k such that $|k| \leq N^\beta$ and $M = 2^{\lfloor \gamma \log N \rfloor}$. Then, we find the root of the following modular polynomial:

$$f_{MSBs,m}(x, y) = 1 + (\tilde{k} + x)(\tilde{\Phi}(N) + y) \pmod{e}$$

where $\tilde{k} = \lfloor (e\tilde{d} - 1)/\tilde{\Phi}(N) \rfloor$ which is an approximation to k as in Sect. 3.2. If we can recover the root $(x, y) = (k - \tilde{k}, \Phi(N) - \tilde{\Phi}(N))$, whole secret information can be computed. To obtain better results than integer equations based method in Sect. 3, we use a linearized variable $z = (\tilde{k} + x)y + 1$. The absolute values of the root are bounded above by $X := N^\lambda, Y := N^\delta, Z := N^{\beta+\delta}$ where $\lambda = \max\{\gamma, \beta + \delta - 1\}$.

To solve the modular equation $f_{MSBs,m}(x, y) = 0$, we use the following shift-polynomials $g_{[u,i]}^{MSBs.m1}(x, y)$ and $g_{[u,i]}^{MSBs.m2}(x, y)$:

$$g_{[u,i]}^{MSBs.m1}(x, y) := x^{u-i} f_{MSBs,m}(x, y)^i e^{m-i} \text{ and}$$

$$g_{[u,j]}^{MSBs.m2}(x, y) := y^j f_{MSBs,m}(x, y)^u e^{m-u}.$$

All these shift-polynomials $g_{[u,i]}^{MSBs.m1}$ and $g_{[u,j]}^{MSBs.m2}$ modulo e^m have the root $(x, y) = (k - \tilde{k}, \Phi(N) - \tilde{\Phi}(N))$ that is the same as $f_{MSBs,m}(x, y)$. We build a lattice with these polynomials. In this section, we show a basic lattice construction to solve the modular equation and the resulting algorithm works when

$1-\delta-\sqrt{\frac{\delta(1-\delta)}{3}} \leq \beta < 1-\delta$ and $1/3 \leq \delta$, and when $1-\delta-\sqrt{\frac{\delta(1-\delta)}{3}} \leq \beta < 1-\sqrt{\frac{\delta}{3}}$ and $\delta < 1/3$. In the lattice construction, we use shift-polynomials $g_{[u,i]}^{MSBs.m1}(x,y)$ and $g_{[u,i]}^{MSBs.m2}(x,y)$ with indices in $\mathcal{I}_x$ and $\mathcal{I}_y$ where

$$\mathcal{I}_x \Leftrightarrow u = 0, 1, \ldots, m; i = 0, 1, \ldots, u \text{ and}$$

$$\mathcal{I}_y \Leftrightarrow u = 0, 1, \ldots, m; j = 1, 2, \ldots, \left\lfloor \frac{\beta-\lambda}{\delta}m + \frac{1+\lambda-\delta-2\beta}{\delta}u \right\rfloor,$$

respectively. Although the selections of shift-polynomials generate non-triangular basis matrices, we partially apply the linearization $z = (\tilde{k}+x)y+1$ and the basis matrices can be transformed into triangular as in [25]. We follow the result and the basis matrices have diagonals

- $X^{u-\lceil l^{MSBs}(i)\rceil} Y^{i-\lceil l^{MSBs}(i)\rceil} Z^{\lceil l^{MSBs}(i)\rceil} e^{m-i}$ for $g_{[u,i]}^{MSBs.m1}(x,y)$ and
- $X^{u-\lceil l^{MSBs}(u+j)\rceil} Y^{u+j-\lceil l^{MSBs}(u+j)\rceil} Z^{\lceil l^{MSBs}(u+j)\rceil} e^{m-u}$ for $g_{[u,j]}^{MSBs.m2}(x,y)$

where $l^{MSBs}(j) := \max\left\{ 0, \frac{\delta j - (\beta-\lambda)m}{1+\lambda-2\beta} \right\}$.

Notice that the result is valid only when $\frac{1+\lambda-\delta-2\beta}{\delta} \leq 1$, i.e., $\beta \geq \frac{1+\lambda-2\delta}{2}$, since unravelled linearization does not work well otherwise in the sense that the diagonals of triangular basis matrices become larger. We define the above polynomial selections for all the $g_{[u,j]}^{MSBs.m2}(x,y)$ to be helpful.

Lemma 2. *Assume there are shift-polynomials* $g_{[u,u'+j']}^{MSBs.m1}(x,y)$ *for* $u = u' + j', \ldots, m$ *and* $g_{[u,u'+j'-u]}^{MSBs.m2}(x,y)$ *for* $u = u' + 1, \ldots, u' + j' - 1$ *in lattice bases. Then, shift-polynomials* $g_{[u',j']}^{MSBs.m2}(x,y)$ *are helpful polynomials when* $u' = 0, 1, \ldots, m; j' = 1, \ldots, \lfloor \frac{\beta-\lambda}{\delta}m + \frac{1+\lambda-\delta-2\beta}{\delta}u \rfloor$, *whereas shift-polynomials* $g_{[u',j']}^{MSBs.m2}(x,y)$ *are unhelpful polynomials when* $u' = 0, 1, \ldots, m; j' > \frac{\beta-\lambda}{\delta}m + \frac{1+\lambda-\delta-2\beta}{\delta}u$.

When $m + \frac{\beta-\lambda}{\delta}m + \frac{1+\lambda-\delta-2\beta}{\delta}m = \frac{1-\beta}{\delta}m \leq 1$, i.e., $\beta \geq 1 - \delta$, shift-polynomials $g_{[u,j]}^{MSBs.m1}(x,y)$ for $u \geq \frac{\beta-\lambda}{2\beta+\delta-\lambda-1}; i \geq \frac{\beta-\lambda}{2\beta+\delta-\lambda-1}$ are unhelpful polynomials and do not contribute for the basis matrices to be triangular. In addition, when $\frac{1+\lambda-\delta-2\beta}{\delta} \leq 0$, i.e., $\beta \geq \frac{1+\lambda-\delta}{2}$, not all the $g_{[u,j]}^{MSBs.m2}(x,y)$ become helpful polynomials. Hence, we use the above collection of shift-polynomials only when $\beta < \min\{1 - \delta, \frac{1+\lambda-\delta}{2}\}$.

The above lattice yields the condition 2 of Theorem 2. Notice that the bound is always larger than $\beta + \delta - 1$. When $\beta \geq 1 - \sqrt{\frac{\delta}{3}}$ and $\delta < 1/3$, the Attack 3 becomes the best.

4.2   The Attack 5

In this section, we propose an attack for larger β, i.e., $\beta \geq 1 - \delta$ for $1/3 \leq \delta$. As discussed above, the polynomial selections in Sect. 4.1 have unhelpful polynomials in this case and we should eliminate them to obtain better results.

For the purpose, in this section, we use shift-polynomials $g_{[u,i]}^{MSBs.m1}(x,y)$ and $g_{[u,j]}^{MSBs.m2}(x,y)$ with indices in $\mathcal{I}_x$ and $\mathcal{I}_y$ where

$$\mathcal{I}_x \Leftrightarrow u = 0, 1, \ldots, m; i = 0, 1, \ldots, \min\{u, t\} \text{ and}$$

$$\mathcal{I}_y \Leftrightarrow u = 0, 1, \ldots, m; j = 1, 2, \ldots, \min\left\{\left\lfloor \frac{\beta - \lambda}{\delta}m + \frac{1 + \lambda - \delta - 2\beta}{\delta}u \right\rfloor, t - u\right\},$$

for some integer t, respectively. The parameter $\tau = t/m$ should be optimized later. The selections of shift-polynomials generate basis matrices that are not triangular. However, we partially apply the linearization $z = (\tilde{k} + x)y + 1$ and the basis matrices can be transformed into triangular as in Sect. 3.3. Moreover, the diagonals of the basis matrices are the same as those in Sect. 3.3. Hence, Lemma 2 also holds. We use the above polynomial selections when $\frac{\beta - \lambda}{\delta}m < t$ and $\frac{1 + \lambda - \delta - 2\beta}{\delta} > 0$ hold, i.e., $\beta < \min\{\delta\tau + \lambda, \frac{1 + \lambda - \delta}{2}\}$, since all the $g_{[u,j]}^{MSBs.m2}(x,y)$ do not become helpful polynomials otherwise.

The above lattice yields the condition 3 of Theorem 2. The attack 2 becomes the best for larger β.

4.3   The Attack 6

In this section, we propose an attack for smaller β, i.e., $\beta < 1 - \delta - \sqrt{\frac{\delta(1-\delta)}{3}}$. As discussed above, the polynomial selections in Sect. 4.1 collect $g_{[u,j]}^{MSBs.m2}(x,y)$ where all the shifts are not helpful. The defect follows from the fact that when $\frac{1 + \lambda - \delta - 2\beta}{\delta} > 1$, the unravelled linearization does not work well and the diagonals of the resulting triangular basis matrices become larger. Hence, in this section, we use shift-polynomials $g_{[u,i]}^{MSBs.m1}(x,y)$ and $g_{[u,j]}^{MSBs.m2}(x,y)$ with indices in $\mathcal{I}_x$ and $\mathcal{I}_y$ where

$$\mathcal{I}_x \Leftrightarrow u = 0, 1, \ldots, m; i = 0, 1, \ldots, u \text{ and}$$

$$\mathcal{I}_y \Leftrightarrow u = 0, 1, \ldots, m; j = 1, 2, \ldots, t + u,$$

for some integer t, respectively. The parameter $\tau = t/m$ should be optimized later. The selections of shift-polynomials generate basis matrices that are not triangular. However, we partially apply the linearization $z = (\tilde{k} + x)y + 1$ and the basis matrices can be transformed into triangular as in Sect. 4.1. Moreover, the diagonals of the basis matrices are the same as those in Sect. 3.3 by modifying $l^{MSBs}(k) := \max\left\{0, \frac{k - \tau m}{2}\right\}$. Hence, Lemma 2 also holds. By substituting $\tau = \frac{1 - 2\delta - \lambda}{2\delta}$, the above lattice yields the condition 1 of Theorem 2.

5   Attacks with the LSBs of d by Solving Modular Equations

In this section, we solve modular equations and propose two attacks, i.e., Attacks 6 and 7, for $(1, \beta, \gamma, \delta)$-partial key exposure attacks with the LSBs of d. The Attack 7 and 8 in Sects. 5.1 and 5.2 corresponds to the conditions 2 and 1 of Theorem 3, respectively. Algorithm constructions in Sect. 5.1 and that in Sect. 5.2 is similar to Takayasu-Kunihiro's [25, 26], respectively.

5.1   The Attack 7

As in Sect. 3.1, when $\tilde{d}$ which is the LSBs of d is given, RSA key generation can be written as $e(\tilde{d} + d'M) = 1 + k\Phi(N)$ with some integer k such that $|k| \leq N^\beta$ and $M = 2^{\lfloor(\beta-\gamma)\log N\rfloor}$. Then, we find the root of the following modular polynomials:

$$f_{LSBs.m1}(x,y) := 1 - e\tilde{d} + x(\tilde{\Phi}(N) + y) \pmod{eM},$$

$$f_{LSBs.m2}(x,y) := 1 + x(\tilde{\Phi}(N) + y) \pmod{e}.$$

If we can recover the root $(x,y) = (k, \Phi(N) - \tilde{\Phi}(N))$, whole secret information can be computed. To obtain better results than integer equations based method in Sect. 3, we use a linearized variable $z = xy + 1$. The absolute values of the root are bounded above by $X := N^\beta, Y := N^\delta, Z := N^{\beta+\delta}$.

To solve the modular equations $f_{LSBs.m1}(x,y) = 0$ and $f_{LSBs.m2}(x,y) = 0$, we use the following shift-polynomials $g_{[u,i]}^{LSBs.m1}(x,y)$ and $g_{[u,j]}^{LSBs.m2}(x,y)$:

$$g_{[u,i]}^{LSBs.m1}(x,y) := x^{u-i} f_{LSBs.m1}(x,y)^i (eM)^{m-i} \text{ and}$$

$$g_{[u,j]}^{LSBs.m2}(x,y) := y^j f_{LSBs.m1}(x,y)^{u-\lceil l^{LSBs}(j)\rceil} f_{LSBs.m2}(x,y)^{\lceil l^{LSBs}(j)\rceil} .$$

$$e^{m-u} M^{m-(u-\lceil l^{LSBs}(j)\rceil)},$$

where $l^{LSBs}(j) = \max\left\{0, \frac{\delta j-(\beta-\gamma)m}{1-2\beta+\gamma-\delta}\right\}$. All these shift-polynomials $g_{[u,i]}^{LSBs.m1}$ and $g_{[u,j]}^{LSBs.m2}$ modulo $(eM)^m$ have the root $(x,y) = (k, \Phi(N) - \tilde{\Phi}(N))$ that is the same as $f_{LSBs.m1}(x,y)$ and $f_{LSBs,m2}(x,y)$. We build a lattice with these polynomials. In this section, we show a basic lattice construction to solve the modular equations and the resulting algorithm works when $1 - \delta - \sqrt{\frac{\delta(1-\delta)}{3}} \leq \beta < 1 - \frac{\delta}{2} - \frac{\sqrt{3\delta(4-\delta)}}{6}$. In the lattice construction, we use shift-polynomials $g_{[u,i]}^{LSBs.m1}(x,y)$ and $g_{[u,j]}^{LSBs.m2}(x,y)$ with indices in $\mathcal{I}_x$ and $\mathcal{I}_y$ where

$$\mathcal{I}_x \Leftrightarrow u = 0, 1, \ldots, m; i = 0, 1, \ldots, u \text{ and}$$

$$\mathcal{I}_y \Leftrightarrow u = 0, 1, \ldots, m; j = 1, 2, \ldots, \left\lfloor \left|\frac{\beta-\lambda}{\delta}m + \frac{1+\lambda-\delta-2\beta}{\delta}u\right| \right\rfloor,$$

respectively. Although the selections of shift-polynomials generate non-triangular basis matrices, we partially apply the linearization $z = xy + 1$ and the basis matrices can be transformed into triangular as in [25]. We follow the result and the basis matrices have diagonals

- $X^u Y^i (eM)^{m-i}$ for $g_{[u,i]}^{LSBs.m1}(x,y)$ and
- $X^{u-\lceil l^{LSBs}(u+j)\rceil} Y^{u+j-\lceil l^{LSBs}(u+j)\rceil} Z^{\lceil l^{LSBs}(u+j)\rceil} e^{m-u} M^{m-(u-\lceil l^{LSBs}(u+j)\rceil)}$
  for $g_{[u,j]}^{LSBs.m2}(x,y)$.

Notice that the result is valid only when $\frac{1+\gamma-\delta-2\beta}{\delta} \leq 1$, i.e., $\beta \geq \frac{1+\gamma-2\delta}{2}$, since unravelled linearization does not work well otherwise. We define the above polynomial selections for all the $g_{[u,j]}^{MSBs.m2}(x,y)$ to be helpful.

Lemma 3. *Assume there are shift-polynomials* $g^{LSBs.m2}_{[u'+i,j'+i]}(x,y)$ *for* $i = 1, 2, \ldots, m - u'$ *in lattice bases. Then, shift-polynomials* $g^{LSBs.m2}_{[u',j']}(x,y)$ *are helpful polynomials when* $u' = 0, 1, \ldots, m; j' = 1, \ldots, \lfloor \frac{\beta-\gamma}{\delta}m + \frac{1+\gamma-\delta-2\beta}{\delta}u' \rfloor$, *whereas shift-polynomials* $g^{LSBs.m2}_{[u',j']}(x,y)$ *are unhelpful polynomials when* $u' = 0, 1, \ldots, m; j' > \frac{\beta-\gamma}{\delta}m + \frac{1+\gamma-\delta-2\beta}{\delta}u'$.

When $\frac{1+\gamma-\delta-2\beta}{\delta} \leq 0$, i.e., $\beta \geq \frac{1+\gamma-\delta}{2}$, all the shift-polynomials $g^{LSBs.m2}_{[u,j]}(x,y)$ in the above selection do not become a helpful polynomial since the assumption in Lemma 3 fails. Hence, we use the above collection of shift-polynomials only when $\beta < \frac{1+\gamma-\delta}{2}$.

The above lattice yields the condition 2 of Theorem 3. When $1 - \frac{\delta}{2} - \frac{\sqrt{3\delta(4-\delta)}}{6} \leq \beta$, Theorem 1 becomes the best.

5.2 The Attack 8

In this section we propose an attack that works when $\beta < 1 - \delta - \sqrt{\frac{\delta(1-\delta)}{3}}$. In the lattice construction, we use the same shift-polynomials $g^{LSBs.m1}_{[u,i]}(x,y)$ and $g^{LSBs.m2}_{[u,j]}(x,y)$ where $l^{LSBs}(j) = \max\{0, j - \tau m\}$ with indices in $\mathcal{I}_x$ and $\mathcal{I}_y$ where

$$\mathcal{I}_x \Leftrightarrow u = 0, 1, \ldots, m; i = 0, 1, \ldots, u \text{ and}$$
$$\mathcal{I}_y \Leftrightarrow u = 0, 1, \ldots, m; j = 1, 2, \ldots, t + u,$$

respectively. The parameter $\tau = t/m$ should be optimized later. Although the selections of shift-polynomials generate non-triangular basis matrices, we partially apply the linearization $z = xy + 1$ and the basis matrices can be transformed into triangular as in Sect. 5.1. The basis matrices have the same diagonals as those in Sect. 5.1 although the function $l^{LSBs}(j)$ is modified.

We set the parameter $\tau = \frac{1-2\delta-\gamma}{2\delta}$ and the above lattice yields the condition 1 of Theorem 2.

Acknowledgement. The first author is supported by a JSPS Fellowship for Young Scientists. This research was supported by CREST, JST, and supported by JSPS Grant-in-Aid for JSPS Fellows 14J08237 and KAKENHI Grant Number 25280001 and 16H02780.

References

1. Blömer, J., May, A.: New partial key exposure attacks on RSA. In: Boneh, D. (ed.) CRYPTO 2003. LNCS, vol. 2729, pp. 27–43. Springer, Heidelberg (2003). doi:10.1007/978-3-540-45146-4_2
2. Blömer, J., May, A.: A tool kit for finding small roots of bivariate polynomials over the integers. In: Cramer, R. (ed.) EUROCRYPT 2005. LNCS, vol. 3494, pp. 251–267. Springer, Heidelberg (2005). doi:10.1007/11426639_15

3. Boneh, D., Durfee, G.: Cryptanalysis of RSA with private key d less than $N^{0.292}$. IEEE Trans. Inf. Theory **46**(4), 1339–1349 (2000)

4. Boneh, D., Durfee, G., Frankel, Y.: An attack on RSA given a small fraction of the private key bits. In: Ohta, K., Pei, D. (eds.) ASIACRYPT 1998. LNCS, vol. 1514, pp. 25–34. Springer, Heidelberg (1998). doi:10.1007/3-540-49649-1_3

5. Ciet, M., Koeune, F., Laguillaumie, F., Quisquater, J.J.: Short private exponent attacks on fast variants of RSA. UCL Crypto Group Technical report series CG-2002/4, University Catholique de Louvain (2002)

6. Coppersmith, D.: Finding a small root of a bivariate integer equation; factoring with high bits known. In: Maurer, U.M. (ed.) EUROCRYPT 1996. LNCS, vol. 1070, pp. 178–189. Springer, Heidelberg (1996). doi:10.1007/3-540-68339-9_16

7. Coppersmith, D.: Finding a small root of a univariate modular equation. In: Maurer, U.M. (ed.) EUROCRYPT 1996. LNCS, vol. 1070, pp. 155–165. Springer, Heidelberg (1996). doi:10.1007/3-540-68339-9_14

8. Coppersmith, D.: Small solutions to polynomial equations, and low exponent RSA vulnerabilities. J. Cryptol. **10**(4), 233–260 (1997)

9. Coppersmith, D.: Finding small solutions to small degree polynomials. In: Silverman, J.H. (ed.) CaLC 2001. LNCS, vol. 2146, pp. 20–31. Springer, Heidelberg (2001). doi:10.1007/3-540-44670-2_3

10. Coron, J.-S.: Finding small roots of bivariate integer polynomial equations revisited. In: Cachin, C., Camenisch, J.L. (eds.) EUROCRYPT 2004. LNCS, vol. 3027, pp. 492–505. Springer, Heidelberg (2004). doi:10.1007/978-3-540-24676-3_29

11. Coron, J.-S.: Finding small roots of bivariate integer polynomial equations: a direct approach. In: Menezes, A. (ed.) CRYPTO 2007. LNCS, vol. 4622, pp. 379–394. Springer, Heidelberg (2007). doi:10.1007/978-3-540-74143-5_21

12. Ernst, M., Jochemsz, E., May, A., de Weger, B.: Partial key exposure attacks on RSA up to full size exponents. In: Cramer, R. (ed.) EUROCRYPT 2005. LNCS, vol. 3494, pp. 371–386. Springer, Heidelberg (2005). doi:10.1007/11426639_22

13. Hinek, M.J.: On the security of multi-prime RSA. J. Math. Cryptol. **2**(2), 117–147 (2008)

14. Howgrave-Graham, N.: Finding small roots of univariate modular equations revisited. In: Darnell, M. (ed.) Cryptography and Coding 1997. LNCS, vol. 1355, pp. 131–142. Springer, Heidelberg (1997). doi:10.1007/BFb0024458

15. Jochemsz, E., May, A.: A strategy for finding roots of multivariate polynomials with new applications in attacking RSA variants. In: Lai, X., Chen, K. (eds.) ASIACRYPT 2006. LNCS, vol. 4284, pp. 267–282. Springer, Heidelberg (2006). doi:10.1007/11935230_18

16. Lenstra, A., Lenstra, H., Lovász, L.: Factoring polynomials with rational coefficients. Math. Ann. **261**, 515–534 (1982)

17. May, A.: New RSA vulnerabilities using lattice reduction methods. Ph.D. thesis, University of Paderborn (2003)

18. May, A.: Using LLL-reduction for solving RSA and factorization problems. In: Nguyen, P.Q., Vallée, B. (eds.) The LLL Algorithm - Survey and Applications. Information Security and Cryptography, pp. 315–348. Springer, Heidelberg (2010). doi:10.1007/978-3-642-02295-1_10

19. Nguyen, P.Q., Stern, J.: The two faces of lattices in cryptology. In: Silverman, J.H. (ed.) CaLC 2001. LNCS, vol. 2146, pp. 146–180. Springer, Heidelberg (2001). doi:10.1007/3-540-44670-2_12

20. Sarkar, S., Maitra, S.: Improved partial key exposure attacks on RSA by guessing a few bits of one of the prime factors. In: Lee, P.J., Cheon, J.H. (eds.) ICISC 2008. LNCS, vol. 5461, pp. 37–51. Springer, Heidelberg (2009). doi:10.1007/978-3-642-00730-9_3

21. Sarkar, S., Maitra, S., Sarkar, S.: RSA cryptanalysis with increased bounds on the secret exponent using less lattice dimension. IACR Cryptology ePrint Archive 2008, 315 (2008)

22. Sarkar, S., Sen Gupta, S., Maitra, S.: Partial key exposure attack on RSA – improvements for limited lattice dimensions. In: Gong, G., Gupta, K.C. (eds.) INDOCRYPT 2010. LNCS, vol. 6498, pp. 2–16. Springer, Heidelberg (2010). doi:10.1007/978-3-642-17401-8_2

23. Sun, H.-M., Wu, M.-E., Steinfeld, R., Guo, J., Wang, H.: Cryptanalysis of short exponent RSA with primes sharing least significant bits. In: Franklin, M.K., Hui, L.C.K., Wong, D.S. (eds.) CANS 2008. LNCS, vol. 5339, pp. 49–63. Springer, Heidelberg (2008). doi:10.1007/978-3-540-89641-8_4

24. Takayasu, A., Kunihiro, N.: Better lattice constructions for solving multivariate linear equations modulo unknown divisors. IEICE Trans. **97-A**(6), 1259–1272 (2014)

25. Takayasu, A., Kunihiro, N.: General bounds for small inverse problems and its applications to multi-prime RSA. In: Lee, J., Kim, J. (eds.) ICISC 2014. LNCS, vol. 8949, pp. 3–17. Springer, Heidelberg (2015). doi:10.1007/978-3-319-15943-0_1

26. Takayasu, A., Kunihiro, N.: Partial key exposure attacks on RSA: achieving the Boneh-Durfee bound. In: Joux, A., Youssef, A. (eds.) SAC 2014. LNCS, vol. 8781, pp. 345–362. Springer, Heidelberg (2014). doi:10.1007/978-3-319-13051-4_21

27. Takayasu, A., Kunihiro, N.: How to generalize RSA cryptanalyses. In: Cheng, C.-M., Chung, K.-M., Persiano, G., Yang, B.-Y. (eds.) PKC 2016. LNCS, vol. 9615, pp. 67–97. Springer, Heidelberg (2016). doi:10.1007/978-3-662-49387-8_4

28. Takayasu, A., Kunihiro, N.: Partial key exposure attacks on RSA with multiple exponent pairs. In: Liu, J.K., Steinfeld, R. (eds.) ACISP 2016. LNCS, vol. 9723, pp. 243–257. Springer, Heidelberg (2016). doi:10.1007/978-3-319-40367-0_15

29. de Weger, B.: Cryptanalysis of RSA with small prime difference. Appl. Algebra Eng. Commun. Comput. **13**(1), 17–28 (2002)

30. Zhang, H., Takagi, T.: Attacks on multi-prime RSA with small prime difference. In: Boyd, C., Simpson, L. (eds.) ACISP 2013. LNCS, vol. 7959, pp. 41–56. Springer, Heidelberg (2013). doi:10.1007/978-3-642-39059-3_4

31. Zhang, H., Takagi, T.: Improved attacks on multi-prime RSA with small prime difference. IEICE Trans. **97-A**(7), 1533–1541 (2014)

Fault and Glitch Resistant
Implementations

Feeding Two Cats with One Bowl: On Designing a Fault and Side-Channel Resistant Software Encoding Scheme

Jakub Breier[1][(✉)] and Xiaolu Hou[2]

[1] Physical Analysis and Cryptographic Engineering,
Temasek Laboratories at Nanyang Technological University, Singapore, Singapore
`jbreier@ntu.edu.sg`
[2] Divison of Mathematical Sciences, Nanyang Technological University,
Singapore, Singapore
`ho0001lu@e.ntu.edu.sg`

Abstract. When it comes to side-channel countermeasures, software encoding schemes are becoming popular and provide a good level of security for general-purpose microcontrollers. However, these schemes are not designed to be fault resistant, and this property is discussed very rarely. Therefore, implementers have to pile up two different countermeasures in order to protect the algorithm against these two popular classes of attacks.

In our paper, we discuss the fault resistance properties of encoding schemes in general. We define theoretical bounds that clearly show the possibilities and limitations of encoding-based countermeasures, together with trade-offs between side-channel and fault resistance. Moreover, we simulate several codes with respect to most popular fault models, using a general-purpose microcontroller assembly implementation. Our algorithm shows how to implement fault resistance to an encoding scheme that currently has the best side-channel resistant capabilities. As a result, we are able to design a code by using automated methods, that can provide the optimal trade-off between side-channel and fault resistance.

Keywords: Software encoding schemes · Side-channel attacks · Fault attacks · Countermeasures

1 Introduction

When it comes to small, constrained devices, such as the ones designed for Internet of Things applications, they are usually easy to access and do not contain comprehensive security measures to protect them. Therefore, even though a strong cryptography is used to protect the communication, hardware attacks pose a serious threat. Side-channel and fault attacks are among the most popular means to breach the device security. When designing a cryptographic implementation, it is necessary to consider countermeasures against these attacks.

© Springer International Publishing AG 2017
H. Handschuh (Ed.): CT-RSA 2017, LNCS 10159, pp. 77–94, 2017.
DOI: 10.1007/978-3-319-52153-4_5

There are two main countermeasure classes to protect implementations against side channel attacks. *Masking* [8] is a software-level countermeasure which tries to "mask" the relationship between the intermediate values and power leakage. *Hiding* [18] tries to reduce the signal and increase noise by utilizing various techniques – it "hides" the operations performed by the device. While masking can make fault attacks more challenging, it does not help to prevent them. On the other hand, some hiding techniques, such as dual-rail precharge logic (DPL), help in preventing fault attacks by detecting faults [16].

In 2011, DPL was extended to software by Hoogvorst et al. [9], by using balanced encoding schemes. Since then, there were several other proposals [5,12,13,17], all of them using various coding techniques to prevent side-channel leakage. However, it was shown, that unlike hardware DPL representation, its software counterpart is not fault resistant by default [2]. Therefore, to prevent both attack techniques, it is necessary to design the coding scheme from the beginning with this goal in mind.

In this paper, we introduce a theoretical background necessary for designing software hiding countermeasures that are resistant to both side-channel and fault attacks. We provide an algorithm for constructing such codes and ranking them according to required properties. We select optimal codes for various distances and number of codewords, and evaluate them – by using detection and correction probabilities and by simulating them in a faulty environment. This simulation is done by using a general-purpose microcontroller implementation and an instruction set simulator that is capable of injecting different fault models into any instruction of the code. Our results show that the codes generated by our algorithm provide a high security level with respect to both side-channel and fault attacks.

The rest of the paper is organized as follows. Section 2 provides an overview of the related work in this field, together with necessary background on coding theory. Section 3 defines the properties of codes with respect to fault attacks. Section 4 details our algorithm, and provides estimated and simulated results on chosen codes. These results are further discussed in Sect. 5. Finally, Sect. 6 concludes this paper and provides a motivation for further work.

2   General Background

In this section we provide a necessary background on software encoding-based side-channel countermeasures and coding theory for developing a combined countermeasure. Subsect. 2.1 overviews the related work in the field. Subsect. 2.2 provides basic definitions that are used later in this paper.

2.1   Related Work

After the paper by Hoogvorst et al. [9] presented a method to extend the DPL to software implementations, several works were published in the area of software hiding schemes.

Rauzy et al. [13] developed a scheme that encodes the data by using bit-slicing, where only one bit of information is processed at a time. They claim this kind of protection is 250 times more resistant to power analysis attacks compared to the unprotected implementation, while being 3 times slower. For testing, they used PRESENT cipher, running on an 8-bit microcontroller.

Chen et al. [5] proposed an encoding scheme that adds a complementary bit to each bit of the processed data, resulting in a constant Hamming weight code. Their countermeasure was implemented on a PRINCE cipher, using an 8-bit microcontroller.

Servant et al. [17] introduced a constant weight implementation for AES, by using a (3,6)-code. To improve the performance, they split 8-bit variables into two 4 bit words and encode them separately. This implementation was also capable of detecting faults with 93.75% probability. Their implementation used a 16-bit microcontroller.

Maghrebi et al. [12] proposed an encoding scheme that differs from the previous proposals. For their case, they did not assume the Hamming weight leakage model for register bits, therefore they concluded that balanced codes might not be the optimal ones to use. In their method, they first obtain the profile of a device to get a vector of register bit leakages. Then they estimate leakage values for each codeword and build a code by using codewords with the lowest leakage. Their algorithm selects the optimal code by ranking the codes based on the difference in power consumption between the codewords and on the power consumption variance. Our algorithm extends this idea by adding the variance of register bits in order to achieve better leakage characteristics and by adding conditions for error detection and correction.

In general, none of the previous schemes have been designed for fault resistance. Schemes proposed in [5,13] have been analyzed with respect to fault attacks by Breier et al. [2], concluding that without additional modifications to assembly code, the probability of a successful fault attack is non-negligible. Therefore, to improve the current state-of-the-art, we focus on designing fault tolerant and side-channel resistant coding schemes.

When it comes to combined countermeasures, in [15], Schneider et al. proposed a hardware countermeasure based on combining threshold implementation with linear codes. As stated in the paper, their proposal is not considered for software targets. In the execution process, there are multiple checking steps that protect the implementation against faults. However, in software, it would be easy to overcome such checks by multiple fault injections [19]. Also, it would be possible to inject faults that are impossible with hardware implementations, such as instruction skips [3].

Our contributions in this work are:

- We define theoretical bounds for encoding schemes with respect to fault attacks that are necessary to take into account when designing a fault resistant scheme.

- We show how to design a code that is capable of protecting the implementation against side-channel and fault attacks and we show trade-offs between these two resistances.
- We improve the ranking algorithm proposed in [12] (current state-of-the-art) for constructing side-channel resistant codes with better properties – by ranking the codes according to the codeword with the highest leakage, and by calculating the register bit variance. Furthermore, we add the conditions for selecting the codes with the desired error detection/correction capabilities in an automated way.
- We analyze the codes constructed by our algorithm – we calculate leakages, fault detection and correction probabilities, and we simulate the assembly code implementing the codes on a general-purpose microcontroller.

2.2   Coding Theory Background

A *binary code*, denoted by $\mathcal{C}$, is a subset of the n-dimensional vector space over $\mathbb{F}_2 - \mathbb{F}_2^n$, where n is called the *length* of the code $\mathcal{C}$. Each element $c \in \mathcal{C}$ is called a *codeword* in $\mathcal{C}$ and each element $x \in \mathbb{F}_2^n$ is called a *word* [10, p. 6]. Take two codewords $c, c' \in \mathcal{C}$, the *Hamming distance* between c and c', denoted by dis (c, c'), is defined to be the number of places at which c and c' differ [10, p. 9]. More precisely, if $c = c_1 c_2 \ldots c_n$ and $c' = c_1' c_2' \ldots c_n'$, then

$$\text{dis}(c, c') = \sum_{i=1}^{n} \text{dis}(c_i, c_i'),$$

where c_i and c_i' are treated as binary words of length 1 and hence

$$\text{dis}(c_i, c_i') = \begin{cases} 1 & \text{if } c_i \neq c_i' \\ 0 & \text{if } c_i = c_i' \end{cases}.$$

Furthermore, for a binary code $\mathcal{C}$, the *(minimum) distance* of $\mathcal{C}$, denoted by dis $(\mathcal{C})$, is [10, p. 11]

$$\text{dis}(\mathcal{C}) = \min\{\text{dis}(c, c') : c, c' \in \mathcal{C}, c \neq c'\}.$$

Definition 1 [6, p. 75]. *For a binary code C of length n with* dis $(\mathcal{C}) = d$, *let* $M = |\mathcal{C}|$ *denote the number of codewords in C. Then C is called an (n, M, d)-binary code.*

This minimum distance of a binary code is closely related to the error-detection and error-correction capabilities of $\mathcal{C}$.

Definition 2 [10, p. 12]. *Let u be a positive integer. C is said to be u-error-detecting if, whenever there is at least one but at most u errors that occur in a codeword in C, the resulting word is not in C.*

From the definition, it is easy to prove that C is u-error-detecting if and only if $\text{dis}(C) \geq u + 1$ [10, p. 12]. A common decoding method that is used is *nearest neighbor decoding*, which decodes a word $x \in \mathbb{F}_2^n$ to the codeword c_x such that

$$\text{dis}(x, c_x) = \min_{c \in C} \text{dis}(x, c). \tag{1}$$

When there are more codewords c_x that satisfy (1), the *incomplete decoding rule* requires a retransmission [10, p. 10].

Definition 3 [10, p. 13]. *Let v be a positive integer. C is v−error-correcting if minimum distance decoding with incomplete decoding rule is applied, v or fewer errors can be corrected.*

Remark 1. C is v-error correcting if and only if $\text{dis}(C) \geq 2v + 1$ [10, p. 13].

Definition 4 [7]. *An (n, M, d)-binary code C is called an equidistant code if $\forall c, c' \in C, \text{dis}(c, c') = d$.*

For our purpose, we will use binary code for protecting the underlying implementation.

We propose two choices of lookup tables:

1. Correction Table: This table will treat a word $x \in \mathbb{F}_2^n$ the same as the codeword $c_x \in C$ which satisfies $\text{dis}(c_x, x) \leq \lfloor \frac{d-1}{2} \rfloor$, where d is the distance of C. Note that this is equivalent to using *bounded distance decoding* [11, p. 36] and taking the bounded distance to be $\lfloor \frac{d-1}{2} \rfloor$. To use this table we require that $\text{dis}(C) \geq 3$.
2. Detection Table: This is a normal lookup table that returns a null value when $x \notin C$ is accessed.

We will give a theoretical criterion to measure the bit flip fault resistant capability of a binary code when it is used as an encoding countermeasure against fault injection attacks in Sect. 3. Afterwards we propose two coding schemes. The encoding schemes will be simulated (and implemented) and evaluated in Sect. 4.

Let m be a positive integer such that $1 \leq m \leq n$, where n is the code length.

Definition 5. *An m-bit fault is a fault injected in the codeword that flips exactly m bits. We assume each bit has an equal probability to be flipped.*

Definition 6. *When the fault is analyzed, we adopt the following terminologies:*

- Corrected: *fault is detected and corrected.*
- Null: *fault is detected and results into zero output.*
- Invalid: *fault is detected and results into an output that is not a codeword.*
- Valid: *fault is not detected and fault injection is successful, i.e. it results in the output of a valid but incorrect codeword.*

3   Theoretical Analysis

In this section we will first give the theoretical analysis for the fault resistant capabilities of binary code in general. Then we propose two different coding schemes and analyze the fault resistant properties of codes used under those two schemes.

3.1   Correction Table

Definition 7. *For an (n, M, d)-binary code $\mathcal{C}$ such that $d \geq 3$, let*

$$F_{c,m} := \left\{ \boldsymbol{x} \in \mathbb{F}_2^n : \mathrm{dis}\,(\boldsymbol{c}, \boldsymbol{x}) = m \text{ and } \exists \boldsymbol{c}' \in \mathcal{C} \text{ such that } \mathrm{dis}\,(\boldsymbol{x}, \boldsymbol{c}') \leq \left\lfloor \frac{d-1}{2} \right\rfloor \right\}.$$

Then

$$p_{m,(e)} := \begin{cases} 1 & m \leq \lfloor \frac{d-1}{2} \rfloor \\ 1 - \frac{1}{M\binom{n}{m}} \sum_{c \in \mathcal{C}} |F_{c,m}| & m > \lfloor \frac{d-1}{2} \rfloor \end{cases} \tag{2}$$

is called the m-bit fault resistance probability with error correction for $\mathcal{C}$.

As mentioned earlier, when a Correction Table is used, it is equivalent to using bounded distance decoding. When $m \leq \lfloor \frac{d-1}{2} \rfloor$ bits are flipped, by Remark 1, the error will be corrected and hence $p_{m,(e)} = 1$. When $m > \lfloor \frac{d-1}{2} \rfloor$ bits are flipped, the fault will be valid if the resulting word is at distance at most $\lfloor \frac{d-1}{2} \rfloor$ from any codeword. Thus by Definition 6, $1 - p_{m,(e)}$ gives the theoretical probability of a *Valid* fault and the bigger $p_{m,(e)}$ is, the more resistant the binary code to m-bit fault. Furthermore, when $m = 1$, the fault will be corrected and most of the cases are expected to return *Corrected*.

Another interesting fault model is random fault, i.e. assuming there is an equal probability for m-bits fault to occur $\forall 1 \leq m \leq n$. Taking this into account, we define

Definition 8. *For an (n, M, d)-binary code $\mathcal{C}$ such that $d \geq 3$, let $p_{m,(e)}$ be its m-bit fault resistance probability with error for $1 \leq m \leq n$, then*

$$p_{\mathrm{rand},(e)} := \sum_{m=1}^{n} \frac{1}{n} p_{m,(e)}$$

is called the overall resistance index with error correction for $\mathcal{C}$.

As suggested by the name, the bigger $p_{\mathrm{rand},(e)}$ is, the more resistant the code $\mathcal{C}$ is to random faults.

3.2   Detection Table

Now we consider Detection Table.

Definition 9. *For an (n, M, d)-binary code C such that $d \geq 2$, let*

$$S_m := \sum_{c \in C} |\{c' \in C : \text{dis}(c', c) = m\}|.$$

Then

$$p_m := 1 - \frac{S_m}{M \binom{n}{m}} \tag{3}$$

is called the m-bit fault resistance probability for C.

When an m-bit fault is injected in the codeword, if the resulting word is not a codeword then the value will be set to *Null*. The only case when the fault is valid is when after m bits are flipped, the resulting word is still a codeword. Thus by Definition 6, $1 - p_m$ gives the theoretical probability of a *Valid* fault. Hence the bigger p_m is, the better the binary code is m-fault resistant.

Remark 2. When $m < d$, no codeword is at distance m from each other and hence $p_m = 1$.

Note that if $S_n = M$, i.e. for each codeword $c \in C$, there exists a $c' \in C$ such that $\text{dis}(c, c') = n$, then we have

$$p_n = 1 - \frac{M}{M \binom{n}{n}} = 1 - 1 = 0.$$

That means, for this code, n-bit fault will always be injected successfully. In view of this, we exclude these kind of codes from our selection (see Algorithm 1). In practice, n and M are fixed known values, from Eq. (3), to get bigger p_m the goal of choosing the code C is to make S_m small. There are several ways of achieving this depending on the preference of the user:

1. For small values of m, make $p_m = 0$: choose code with a bigger minimum distance d, then p_m will be 1 for more values of m. Of course, there is a limit for the minimum distance that can be achieved (see Table 1). This particular scheme will be discussed in Sect. 3.3, where it is called Detection Scheme.
2. A certain m_0-bit fault resistance is desired: choose code such that $S_{m_0} = 0$.
3. Sacrificing one m_0-bit fault resistance to achieve m-bit fault resistance for all other values of $m \neq m_0$: this is possible by using equidistant codes. That is, take code such that $|S_{m_0}| = M$. This particular scheme will be discussed in Sect. 3.3, where it is called Equidistant Detection Scheme.
4. Making all p_m almost equally large: choose C such that S_m are similar for all $m > d$. Note that

$$\sum_{m=d+1}^{n} S_m = M$$

is always true.

Similar to last subsection, considering random fault, we define

Definition 10. *For an* (n, M, d)-*binary code* C *such that* $d \geq 2$, *let* p_m *be its* m-*bit fault resistance probability for* $1 \leq m \leq n$, *then*

$$p_{\text{rand}} := \sum_{m=1}^{n} \frac{1}{n} p_m$$

is called the overall resistance index *for* C.

Note that the bigger p_{rand} is, the more resistant the code C is to random faults.

Lemma 1. *For an* (n, M, d)-*binary code* C, *if it is equidistant, then*

$$p_m = \begin{cases} 1 & m \neq d \\ 1 - \frac{M-1}{\binom{n}{d}} & m = d \end{cases}, \text{ and } p_{\text{rand}} = 1 - \frac{M-1}{\binom{n}{d} n}.$$

3.3 Coding Schemes

Here we propose two different coding schemes:

1. Detection Scheme: using binary code which has minimum distance at least 2.
2. Correction Scheme: using binary code which has minimum distance at least 3 with error correction enabled lookup table.

Furthermore, as will be seen from the rest of this paper, equidistant codes have different behaviors than codes that are not equidistant. Hence when equidistant codes are used, we emphasize the usage by referring to the schemes as "Equidistant detection scheme" and "Equidistant correction scheme" respectively.

We will analyze the m-bit fault resistant probability (with error) as well as overall resistance index (with error) for each of them using (n, M, d) binary codes for $n = 8, 9, 10$ and $M = 4, 16$. We chose $M = 4$ because it is easy to analyze and explain, and $M = 16$ because it can encode one nibble of the data, therefore it is usable in a practical scenario[1].

Firstly, we discuss the possible values of the minimum distance d. As is well known in coding theory, fixing the length of the code n and minimum distance d, M is upper bounded by certain value. This upper bound is tight for small values n and d and still open for a lot of other values [6, p. 247]. In particular, for $n = 8, 9, 10$ and different values of d we know the exact possible values of M. In return, the possible values of d are known when n, M are fixed. In Table 1 we list the possible minimum distances that can be achieved for $n = 8, 9, 10$ and $M = 4$ or 16. Note that the values are taken from [6, pp. 247, 248] and [4].

---

[1] To illustrate the usage of the schemes we refer the readers to the extended version of this paper (ia.cr/2016/931).

Table 1. Possible (n, M, d)-binary codes for $n = 8, 9, 10$, $M = 16$ and $n = 8, M = 4$.

n	M	d
8	4	$2, 3, 4, 5$
8	16	$2, 3, 4$
9	16	$2, 3, 4$
10	16	$2, 3, 4$

For equidistant binary code, we have the following constraint on d:

Lemma 2. *Let C be an (n, M, d) equidistant binary code such that $M \geq 3$, then d is even.*

Proof. Recall the *Hamming weight* of a word $x \in \mathbb{F}_2^n$, denoted by $\mathrm{wt}(x)$ is defined to be the number of nonzero coordinates in x [10, p. 46]. And we have the following relation (see [10, Corollary 4.3.4 and Lemma 4.3.5])

$$wt(x) + wt(y) \equiv \mathrm{dis}(x, y) \mod 2.$$

Take an (n, M, d) equidistant binary code C and any three distinct codewords $x, y, z \in C$, we have

$$\mathrm{dis}(x, y) + \mathrm{dis}(y, z) + \mathrm{dis}(z, x) \equiv 2wt(x) + 2wt(y) + 2wt(z) \equiv 0 \mod 2.$$

Hence, d cannot be odd.

Furthermore we have $M \leq n + 1$ [7]. Thus we will only consider $(8, 4, 2)$ and $(8, 4, 4)$ equidistant binary codes. The fact that such codes exist can be derived from [7].

4 Evaluation Methodology and Results

In this section, we will utilize the findings stated in Sect. 3 to build the codes with the optimal side-channel and fault detection properties. First, we construct an algorithm that finds the codes based on searching criteria in Sect. 4.1. Then we show properties of the codes that were produced by the algorithm in Sect. 4.2. To verify our theoretical results, we simulate fault injections into these codes, by using the fault simulator which will be explained in Sect. 4.3. Finally, we present and discuss the simulation results in Sect. 4.4.

4.1 Code Generation and Ranking Algorithm

When it comes to device leakage, it normally depends on the processed intermediate values. In [12], they proposed the first encoding scheme that assumed a stochastic leakage model over the Hamming weight model. In such model, leakage is formulated as follows:

$$T(x) = L(x) + \epsilon, \tag{4}$$

where L is the leakage function mapping the deterministic intermediate value (x) processed in the register to its side-channel leakage, and ϵ is the (assumed) mean-free Gaussian noise. For 8-bit microcontroller case, we can specify this function as $L(x) = \alpha_0 + \alpha_1 x_1 + \ldots \alpha_8 x_8$, where x_i is the i-th bit of the intermediate value, and α_i is the i-th bit weight leakage for specific register [14]. The α_i values can be obtained by using the following equation:

$$\alpha = (\mathbf{A}^T \mathbf{A})^{-1} \mathbf{A}^T \mathbf{T}, \tag{5}$$

where $\mathbf{A}$ is a matrix of intermediate values and $\mathbf{T}$ is a set of traces. After the device profiling which obtains the α, we can use our ranking algorithm for selecting the optimal code (Algorithm 1). Note that one can still use the Hamming weight model – for that case, α has to be defined as unity. In the following, we will explain how the algorithm works.

First, the inputs have to be specified – length (n), number of the codewords (M), minimum distance (d) and leakages of the register bits (α_i). Depending on these values, the algorithm analyzes every possible set of M codewords that can be a potential code candidate. Lines 2–3 iterate over every combination of two codewords. Lines 4–6 test if the minimum distance condition is fulfilled. Then, lines 7–10 check, whether for each codeword there exists another codeword which is at distance n from it – if yes, we skip this set. This condition is necessary in order to get a code resistant against n-bit flip (we will detail such case in Sect. 5). Lines 11–13 compute the 3 values that are used in order to calculate the values for the whole code in the later phase: estimated power consumption for the codeword, stored in table A, estimated variance for bit leakages in the codeword, stored in table B, and the highest bit leakage value, stored in table C. Next, the codeword value is stored in the index table I.

Lines 14–16 use the values from tables A, B, C to compute the register leakage variance ($\mu_{S[x]}$ denotes the mean leakage for a word $S[x]$), highest variance for bit leakages within registers, and value of the highest bit leakage within registers for the set S. These values are stored in tables D, E, F, respectively, and are used in the final evaluation.

The final evaluation is the last phase of the algorithm. First, it takes a subset of D with the best register leakage variance (μ_S denotes the mean leakage for codewords in S). It narrows this subset to candidate codes with the lowest value of the highest bit leakage according to set E. From these, it chooses the code with the lowest bit leakage variance using table F.

4.2   Estimated Values for Chosen Codes

Codes with the best side-channel and fault resistance properties according to Algorithm 1 with 4 and 16 codewords and various lengths can be found in Table 2. Their detailed properties are stated in Table 3[2].

---

[2] For more codes with cardinality 16 and various distances, please refer to the extended version of this paper (ia.cr/2016/931).

Algorithm 1. Ranking algorithm that chooses the code with the optimal leakage properties.

Input   : n: the codeword bit-length, M: number of codewords, d: minimum distance of the code, α_i: the leakage bit weights of the register, where i in $[\![1, n]\!]$

Output: An (n, M, d) binary code

1 **for** *Every set S of M words* **do**
2   **for** $x == 0;\ x < |S|;\ x{+}{+}$ **do**
3   **for** $y == x + 1;\ y < |S|;\ y{+}{+}$ **do**
4   Calculate the distance dis $(S[x], S[y])$;
5   **if** dis $(S[x], S[y]) < d$ *(or* dis $(S[x], S[y])! = d,\ depends\ on equidistance\ condition)$ **then**
6   **continue with a different set** S;
7   **if** dis $(S[x], S[y]) == n$ **then**
8                  $n_{distance}{+}{+}$

9   **if** $n_{distance} == n$ **then**
10   **continue with a different set** S;
11   Compute the estimated power consumption for codeword $S[x]$ and store the result in table A: $A[S[x]] = \Sigma_{i=1}^{n} \alpha_i S[x][i]$;
12   Compute the estimated variance for bit leakages in $S[x]$ and store the result in table B: $B[S[x]] = \Sigma_{i=1}^{n}((\alpha_i S[x][i]) - \mu_{S[x]})^2$;
13   Compute the bit with the highest bit leakage in $S[x]$ and store the result in table C: $C[S[x]] = max(\alpha_i S[x][i])$;

14   Compute the register leakage variance for codewords in S and store the result in table D: $D[S] = \Sigma_{S[x]=1}^{|S|}(A[S[x]] - \mu_S)^2$;
15   Choose the highest variance for register bit leakages for codewords in S and store the result in table E: $E[S] = max(B)$;
16   Choose the value of the highest register bit leakage among the codewords in S and store the result in table F: $F[S] = max(C)$;

17 Get the optimal candidate using the following criteria:
    1. Choose the candidates with the lowest register variances from $D[S]$;
    2. From this set, choose the candidates with the lowest value of the highest leakage according to $F[S]$;
    3. Finally, choose from the previous set, take the candidate with the lowest bit leakage variance according to $E[S]$;

return M codewords in case all the conditions are met, or an empty set otherwise

---

For calculating the register variance, we follow the similar methodology as used in [12], together with their generated α values, but we improved their ranking algorithm by calculating the bit variances inside registers and by selecting the code which has the lowest leakage value for the highest leaking codeword. First part of Table 3 shows these three values, with the order of preference according to our ranking algorithm. Second part of the table shows bit fault resistance

Table 2. Codes used in evaluation.

Code	Distance	Denoted by
0x21, 0x22, 0x24, 0x30	$= 2$	$C_{8,4,eq2}$
0x17, 0x41, 0x44, 0x94	$>= 2$	$C_{8,4,min2}$
0x35, 0x4A, 0x8D, 0x9A	$>= 3$	$C_{8,4,min3}$
0x3B, 0x52, 0x68, 0xA2	$= 4$	$C_{8,4,eq4}$
0x37, 0x4B, 0x70, 0x9E	$>= 4$	$C_{8,4,min4}$
0x4E, 0x61, 0x9B, 0xB4	$>= 5$	$C_{8,4,min5}$
0x2E, 0xFB, 0xFC, 0x76, 0xB7, 0xE1, 0xCE, 0x5F, 0xD2, 0xD5, 0x6D, 0x43, 0xA2, 0x8B, 0x58, 0x44	$>= 3$	$C_{8,16,min3}$
0xBC, 0x1FA, 0x1FD, 0xD7, 0x1E1, 0x1B7, 0x167, 0x1CB, 0xEE, 0x15F, 0x1C6, 0x174, 0x7B, 0x1D0, 0xCD, 0x19E	$>= 3$	$C_{9,16,min3}$
0x12F, 0x3F7, 0x3F8, 0xFB, 0x1DE, 0x3CD, 0x2EE, 0x1E2, 0x35B, 0x27D, 0x2E1, 0x1D1, 0xF4, 0xC7, 0x2D2, 0x364	$>= 4$	$C_{10,16,min4}$

probabilities, denoted by p_m for m-bit flips in the codeword, as well as overall resistance index, denoted by p_{rand} for the code. The last part of the table shows the fault resistance probabilities with error correction, denoted by $p_{m,(e)}$, as well as overall resistance index with error correction, which is denoted by $p_{\text{rand},(e)}$. We do not consider codes with distance 1 because such codes do not provide protection against 1-bit flips and therefore the fault protection would be very low. However, such codes can still be used for minimizing the side-channel leakage.

In general, if we aim for higher distance values, we get better detection and correction capabilities, but the side-channel leakage is higher as well. That is because if the distance is higher, it is more likely that the variance of leakage among the codewords is bigger. Also, we can see that equidistant codes have a constant detection probability of 1 except the case when number of bit flips is the same as the code distance. Moreover, if we sum up the probabilities of all the bit flip faults for non-equidistant codes, the overall detection probability is lower. However, the side-channel leakage of equidistant codes is more than 10 times higher compared to non-equidistant codes.

4.3   Fault Simulation

The fault simulator we used was customized for the purpose of evaluating a microcontroller assembly table look-up implementation of the encoding schemes presented in this paper. More details on this simulator are provided in [1]. This simulator helps us to extend the theoretical results to real-world results, where one has to use capabilities of microprocessors for computing the results.

A high-level overview is given in Fig. 1. There are three instructions in total – the first two LDI load the two operands into registers r0 and r1. Both of the

Table 3. Side-channel and fault properties of the codes.

$\alpha = [0.613331, 0.644584, 0.602531, 0.190986, 0.586268, 0.890951, 1.838814, 1.257943, 0.899922, 0.614699]$

Code	$C_{8,4,eq2}$	$C_{8,4,min2}$	$C_{8,4,min3}$	$C_{8,4,eq4}$	$C_{8,4,min4}$	$C_{8,4,min5}$	$C_{8,16,min3}$	$C_{9,16,min3}$	$C_{10,16,min4}$
Codeword variance	4.537×10^{-4}	1.460×10^{-5}	1.045×10^{-4}	9.555×10^{-3}	4.997×10^{-4}	8.032×10^{-4}	0.1190	0.0091	0.0134
Highest leakage	1.4772	2.4413	2.6648	2.7935	3.2823	3.2769	2.0515	2.0515	3.7101
Bit variance	0.0232	0.3821	0.4830	0.4560	0.3768	0.3779	0.4560	0.4657	0.3430
P_1	1	1	1	1	1	1	1	1	1
P_2	0.8929	0.9821	1	1	1	1	1	1	1
P_3	1	0.9821	0.9911	1	1	1	0.9040	0.9464	1
P_4	1	0.9857	0.9857	0.9571	0.9857	1	0.9161	0.9563	0.9512
P_5	1	0.9911	0.9911	1	0.9732	0.9643	0.9687	0.9772	0.9950
P_6	1	1	0.9821	1	0.9821	0.9643	0.9598	0.9821	0.9899
P_7	1	1	0.9375	1	1	1	0.8906	0.9826	0.9958
P_8	1	1	1	1	1	1	1	1	0.9833
P_9	-	-	-	-	-	-	-	1	0.9875
P_{10}	-	-	-	-	-	-	-	-	1
P_{rand}	0.9866	0.9926	0.9859	0.9946	0.9926	0.9911	0.9549	0.9827	0.9903
$P_{1,(e)}$	-	-	1	1	1	1	1	1	1
$P_{2,(e)}$	-	-	0.9464	1	1	1	0.4241	0.6250	1
$P_{3,(e)}$	-	-	0.9196	0.7857	0.9286	1	0.4844	0.6845	0.6583
$P_{4,(e)}$	-	-	0.9143	0.9571	0.8786	0.8571	0.4071	0.6280	0.9214
$P_{5,(e)}$	-	-	0.8661	0.7857	0.8482	0.8571	0.4286	0.6875	0.7004
$P_{6,(e)}$	-	-	0.8036	1	0.8214	0.75	0.5536	0.7932	0.9435
$P_{7,(e)}$	-	-	0.8125	1	0.875	0.75	0.6094	0.8576	0.8750
$P_{8,(e)}$	-	-	0.5	1	1	1	0.1250	0.86111	0.9250
$P_{9,(e)}$	-	-	-	-	-	-	-	1	0.8375
$P_{10,(e)}$	-	-	-	-	-	-	-	-	0.8750
$P_{rand,(e)}$	-	-	0.8453	0.9410	0.9190	0.9018	0.5040	0.7930	0.8736

operands are already encoded according to one of the coding schemes. The LPM instruction loads the data from the look-up table stored in the memory by using the values in r0 and r1, and the result is stored to register r2. This part works as a standard instruction set simulator. During each execution, a fault is injected into the code. For each type of fault, we test all the possible combinations of codewords, and we disturbed all the instructions in our code. We have tested the following fault models:

- **Bit faults:** in this fault model, one to n bits in the destination register change its value to a complementary one.
- **Random byte faults:** The *random byte fault* model changes random number of bits in the destination register.
- **Instruction skip:** instruction skip is a very powerful model that is capable of removing some countermeasures completely. We have tested a single instruction skip on all three instructions in the code.
- **Stuck-at fault:** in this fault model, the value of the destination register changes to a certain value, usually to all zeroes. Therefore, we have tested this value in our simulator.

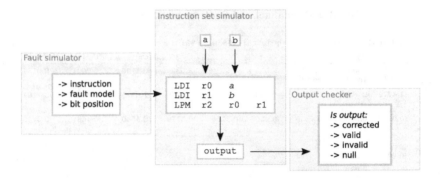

Fig. 1. Fault simulator operation overview.

After the output is produced under a faulty condition, it is analyzed by the output checker, which decides on its classification. Outputs can be of four types (*Corrected, Valid, Invalid, Null*), and these types are described in detail in Sect. 2.2.

4.4   Simulated Results

Figure 2 shows plots for $\mathcal{C}_{8,4,min4}$ and $\mathcal{C}_{8,4,eq4}$, with and without the error correction. Instruction skip faults and stuck-at faults show zero success when attacking any of the generated codes. When it comes to bit flips, we can see that for better fault tolerance, one should not use the error correction capabilities, since the properties of such codes allow changing the faulty codeword into another codeword, depending on the number of bit flips and minimum distance of the code. When deciding whether to choose an equidistant code or not, situation is the same as in Table 3 – equidistant codes have slightly better fault detection properties, but worse side-channel leakage protection. Therefore, it depends on the implementer to choose a compromise between those two.

5   Discussion

First, we would like to explain the difference between the calculated results in Table 3 and simulated results in Fig. 2 in equidistant code $\mathcal{C}_{8,4,min4}$. Table 3 shows theoretical results assuming that error happens before using the lookup table. However, in a real-world setting, fault can be injected at any point of the execution, including the table look-up, or even after obtaining the result from the table. That is also why there are *Invalid* faults, despite the table always outputs *Null* in case of being addressed by a word that does not correspond to any codeword. Because there are three instructions in the assembly code, faulting the destination register of the last one after returning the value from the table results into 1/3 of *Invalid* faults in all the cases except instruction skips.

To explain the condition on lines 7–8 of the Algorithm 1, we can take the code with $n = 8$, $M = 16$, and $d = 4$ as an example. The simulation result for this code

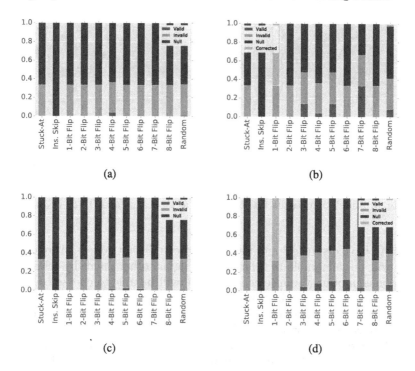

Fig. 2. Simulation results for $\mathcal{C}_{8,4,eq4}$ with equidistant detection scheme in (a) and with equidistant correction scheme in (b); $\mathcal{C}_{8,4,min4}$ with detection scheme in (c) and with correction scheme in (d).

is stated in Fig. 3(a). There are no codes with these parameters that could satisfy the above mentioned condition – all 480 codes that can be constructed, have the property that if any codeword is faulted by n bit flip, it will change to other codeword. Therefore, such codes are not suitable for protecting implementations against fault attacks. For this reason, it is more suitable to use the $\mathcal{C}_{8,16,min3}$ code, stated in Fig. 3(b), that does not suffer from such property.

To summarize the results, we point out the following findings:

- Correction Scheme is not suitable for fault tolerant implementations – while it can be helpful in non-adversary environments, where it can be statistically verified, how many bits are usually faulted, and therefore, a proper error correction function can be specified, in adversary-based settings, one cannot estimate the attacker capabilities. In case of correcting 1 bit error for example, attacker who can flip multiple bits will have a higher probability of producing *Valid* faults, compared to using detection scheme with the same code.
- We can design optimal code either from the fault tolerance perspective, or from side-channel tolerance perspective – if we consider both, a compromise has to be made, depending on which attack is more likely to happen or how powerful an attacker can be in either setting. If we sacrifice the fault tolerance, we will normally get a code with distance 2 (e.g. side-channel resistant codes

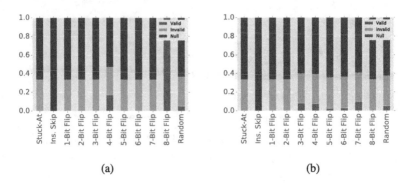

Fig. 3. Simulation results for the codes: (a) $\mathcal{C}_{8,16,min4}$ and (b) $\mathcal{C}_{8,16,min3}$.

in [12] all have distance 2 and they are not equidistant codes), therefore such codes will be vulnerable to 2-bit faults. On the other hand, by relaxing the power consumption variance condition, we will be able to choose codes with bigger distance, being able to resist higher number of bit faults.

- Both types of resistances can be improved if we sacrifice the memory and choose codes with greater lengths.
- Equidistant detection schemes is a good option in case the implementation can be protected against certain number of bit flips – because all the *Valid* faults are achieved only if the attacker flips the same number of bits as is the distance. However, this condition does not hold in case of equidistant correction schemes.

6   Conclusions

In this paper, we provided a necessary background for constructing side-channel and fault attack resistant software encoding schemes. Current encoding schemes only cover side-channel resistance, and either do not discuss fault resistance, or only state it as a side product of the construction, such as [17]. Our work defines theoretical bounds for fault detection and correction and provides a way to construct efficient codes that are capable of protecting the underlying computation against both physical attack classes.

To support our result with a practical case study, we simulated the table look-up under faulty conditions, by using a microcontroller assembly code. As expected, the codes constructed by using our algorithm provide noticeably better fault resistance properties compared to state-of-the-art, while keeping the side-channel leakage at the minimum.

For the future work, we would like to use our scheme to implement all the operations in a symmetric cryptographic algorithm and test both the side-channel leakage and fault tolerance in a real world setting. Also, we would like to examine the timing leakage implications of the table look-ups with respect to processed data.

Acknowledgments. The authors would like to thank Dr. Punarbasu Purkayastha for the useful discussions and the anonymous reviewers for their valuable suggestions. The research of X. Hou is supported by Nanyang President Graduate Scholarship.

References

1. Breier, J.: On analyzing program behavior under fault injection attacks. In: 2016 11th International Conference on Availability, Reliability and Security (ARES), pp. 1–5. IEEE, August 2016 (to appear)
2. Breier, J., Jap, D., Bhasin, S.: The other side of the coin: analyzing software encoding schemes against fault injection attacks. In: 2016 IEEE International Symposium on Hardware Oriented Security and Trust (HOST), pp. 209–216. IEEE (2016)
3. Breier, J., Jap, D., Chen, C.N.: Laser profiling for the back-side fault attacks: with a practical laser skip instruction attack on AES. In: Proceedings of 1st ACM Workshop on Cyber-Physical System Security, CPSS 2015, pp. 99–103. ACM, New York (2015). http://doi.acm.org/10.1145/2732198.2732206
4. Brouwer, A.E., Shearer, L.B., Sloane, N., et al.: A new table of constant weight codes. In: IEEE Trans Inform Theory. Citeseer (1990)
5. Chen, C., Eisenbarth, T., Shahverdi, A., Ye, X.: Balanced encoding to mitigate power analysis: a case study. In: Joye, M., Moradi, A. (eds.) CARDIS 2014. LNCS, vol. 8968, pp. 49–63. Springer, Heidelberg (2015). doi:10.1007/978-3-319-16763-3_4
6. Conway, J.H., Sloane, N.J.A.: Sphere Packings, Lattices and Groups, vol. 290. Springer Science & Business Media, Berlin (2013)
7. Fu, F.W., Kløve, T., Luo, Y., Wei, V.K.: On equidistant constant weight codes. Discret. Appl. Math. **128**(1), 157–164 (2003)
8. Goubin, L., Patarin, J.: DES and differential power analysis the "Duplication" method. In: Koç, Ç.K., Paar, C. (eds.) CHES 1999. LNCS, vol. 1717, pp. 158–172. Springer, Heidelberg (1999). doi:10.1007/3-540-48059-5_15
9. Hoogvorst, P., Danger, J.L., Duc, G.: Software implementation of dual-rail representation. In: COSADE, Darmstadt, Germany (2011)
10. Ling, S., Xing, C.: Coding Theory: A First Course. Cambridge University Press, Cambridge (2004)
11. MacWilliams, F.J., Sloane, N.J.A.: The Theory of Error Correcting Codes. Elsevier, Amsterdam (1977)
12. Maghrebi, H., Servant, V., Bringer, J.: There is wisdom in harnessing the strengths of your enemy: customized encoding to thwart side-channel attacks - extended version. Cryptology ePrint Archive, Report 2016/183 (2016). http://eprint.iacr.org/
13. Rauzy, P., Guilley, S., Najm, Z.: Formally proved security of assembly code against leakage. IACR Cryptology ePrint Archive 2013, 554 (2013)
14. Schindler, W., Lemke, K., Paar, C.: A stochastic model for differential side channel cryptanalysis. In: Rao, J.R., Sunar, B. (eds.) CHES 2005. LNCS, vol. 3659, pp. 30–46. Springer, Heidelberg (2005). doi:10.1007/11545262_3
15. Schneider, T., Moradi, A., Güneysu, T.: ParTI - towards combined hardware countermeasures against side-channel and fault-injection attacks. Cryptology ePrint Archive, Report 2016/648 (2016). http://eprint.iacr.org/2016/648
16. Selmane, N., Bhasin, S., Guilley, S., Graba, T., Danger, J.L.: WDDL is protected against setup time violation attacks. In: FDTC, pp. 73–83 (2009)

17. Servant, V., Debande, N., Maghrebi, H., Bringer, J.: Study of a novel software constant weight implementation. In: Joye, M., Moradi, A. (eds.) CARDIS 2014. LNCS, vol. 8968, pp. 35–48. Springer, Heidelberg (2015). doi:10.1007/978-3-319-16763-3_3
18. Tiri, K., Verbauwhede, I.: A logic level design methodology for a secure DPA resistant ASIC or FPGA implementation. In: DATE 2004, Paris, France, pp. 246–251 (2004)
19. Trichina, E., Korkikyan, R.: Multi fault laser attacks on protected CRT-RSA. In: 2010 Workshop on Fault Diagnosis and Tolerance in Cryptography (FDTC), pp. 75–86, August 2010

An Efficient Side-Channel Protected AES Implementation with Arbitrary Protection Order

Hannes Gross[(⊠)], Stefan Mangard, and Thomas Korak

Institute for Applied Information Processing and Communications (IAIK),
Graz University of Technology, Inffeldgasse 16a, 8010 Graz, Austria
{hannes.gross,stefan.mangard,thomas.korak}@iaik.tugraz.at

Abstract. Passive physical attacks, like power analysis, pose a serious threat to the security of digital circuits. In this work, we introduce an efficient side-channel protected Advanced Encryption Standard (AES) hardware design that is completely scalable in terms of protection order. Therefore, we revisit the private circuits scheme of Ishai *et al.* [13] which is known to be vulnerable to glitches. We demonstrate how to achieve resistance against multivariate higher-order attacks in the presence of glitches for the same randomness cost as the private circuits scheme. Although our AES design is scalable, it is smaller, faster, and less randomness demanding than other side-channel protected AES implementations. Our first-order secure AES design, for example, requires only 18 bits of randomness per S-box operation and 6 kGE of chip area. We demonstrate the flexibility of our AES implementation by synthesizing it up to the 15th protection order.

Keywords: Domain-Oriented Masking · Private circuits · Threshold implementations · ISW · Side-channel analysis · DPA · Hardware security · AES

1 Introduction

The increasing number of interconnected devices demand security not only on a cryptographic level but also on a physical level. Without countermeasures against physical attacks, devices are defenseless against attackers which have physical access. An attacker can easily extract device internal secrets by measuring the power consumption [14] or the electromagnetic emanation [19] of the device during security critical operations.

The most promising approach to achieve resistance against passive physical attacks is to make sensitive computations independent from the processed data by using so-called masking schemes. There exist many masking schemes, the scheme of Goubin and Patarin [10], or Ishai *et al.*'s private circuits [13], and the Trichina gate [22]. However, the aforementioned schemes have been shown to be

© Springer International Publishing AG 2017
H. Handschuh (Ed.): CT-RSA 2017, LNCS 10159, pp. 95–112, 2017.
DOI: 10.1007/978-3-319-52153-4_6

vulnerable against glitches and thus rigorous care has to be taken during the implementation to avoid leakage caused by glitches.

There exist masking schemes that are inherently immune against glitches. The most popular scheme is the threshold implementation (TI) masking scheme introduced by Nikova et al. [18]. It has been extensively researched and extended by Bilgin et al. [1,4] during the last years. There exist many protected hardware implementations that are based on TI [2,3,17].

Recently, Reparaz et al. introduced the Consolidated Masking Scheme [20] (CMS). One interesting aspect of the CMS scheme is the possibility to reduce the number of required input shares of TI from $td + 1$ to $d + 1$, where d corresponds to the attack order and t is the algebraic degree of the function that should be protected. At CHES 2016, De Cnudde et al. [7] demonstrated the suitability of using only $d + 1$ shares on an AES hardware design. The design requires less chip area than related work, but at the cost of an increased randomness demand compared to $td + 1$ TI. More specifically, the CMS scheme requires $(d + 1)^2$ random bits for protecting one $GF(2^n)$ multiplication as required multiple times for the AES S-box.

Producing a high amount of random numbers in hardware, however, is not trivial and goes hand in hand with an increased chip area usage, a higher energy consumption, and has also a negative influence on the throughput of a design. Therefore, for the efficiency of masked implementations the randomness demand is crucial.

Our Contribution. In this work[1], we demonstrate how the randomness requirements for $d + 1$ masking can be lowered from $(d + 1)^2$ to only $d(d + 1)/2$. In order to achieve this, we revisit the private circuits scheme [13] which is known to be vulnerable to glitches. We perform a similar approach under the premise of glitches, and demonstrate how to achieve d^{th}-order protection in the presence of glitches for the same randomness cost and without losing genericity. We show the suitability of our approach by implementing a dth-order protected AES-128 encryption-only hardware design. Our first-order AES implementation requires only 18 fresh random bits per S-box calculation, which is a third of the random bits of the CMS implementation of De Cnudde et al. [7]. Our AES design is also very compact in terms of chip area and requires only 6 kGE of chip area and 246 clock cycles per encryption. Furthermore, our approach is generic in terms of protection order, allowing our AES design to be synthesized for any desired protection order. The number of required clock cycles per encryption, however, is independent of the protection order. We demonstrate the genericity of our design by stating post-synthesis hardware results up to the 15th protection order. The VHDL source code of the generic AES design is published online [11], which we hope will help future research and make comparisons easier.

---

[1] An earlier version of this work has been published online [12] under the title "Domain-Oriented Masking: Compact Masked Hardware Implementations with Arbitrary Protection Order".

2   Private Circuits and the ISW Transformation

The original idea of Ishai *et al.* [13] was to build a so-called private circuit compiler that can transform arbitrary circuits into circuits that resist passive physical attacks, like chip probing and side-channel analysis, up to a protection order d. For this purpose, the circuit's data signals are first split into a number of shares, which when recombined through addition over $GF(2)$ result in the original value. The sharing is done based on uniformly distributed random numbers. A sharing of a signal x can be written as shown in Eq. 1, where the shares are denoted by capital letters with the name of the shared signal in the subscript index.

$$x = \underbrace{A_x + B_x + C_x + \ldots}_{d+1 \text{ shares}} \tag{1}$$

The security of masking schemes is typically shown in the so-called d-probing model. A masking scheme provides security of order d in this model, if each combination of up to d signals is independent of all unshared intermediate signals. It was demonstrated by Faust *et al.* [8] and Rivain and Prouff [21] that there indeed exists a relation between the number of probed wires in the d-probing model and the attack order for a differential power analysis (DPA) attack.

The security of the sharing of x in Eq. 1 against a d-probing attacker follows from the fact that the attacker only gets access to the $d + 1$ shares of x $(A_x, B_x, \ldots)$ but not to x itself. The circuit is secure against d probes, as long as no signal in the circuit contains a combination of more than one share of x. To keep this share independence, also all gates of the circuit are required to fulfill this requirement.

The basic idea of the ISW transformation in order to achieve this, is therefore to transform the original circuit in a way that it only consists of protected NOT and AND gates. While the protected implementation of the NOT gate is straightforward and only requires the negation of one share (see Eq. 2), the protected implementation of the AND gate is more difficult and requires the introduction of fresh randomness to fulfill the independence requirement.

$$\neg x = \neg A_x + B_x + C_x + \ldots \tag{2}$$

AND Gate. For the correct and secure realization of the AND gate in the ISW scheme (with x and y as the input and q as the output), Eq. 3 needs to be expanded, securely evaluated, and compressed again to $d + 1$ output shares.

$$q = xy = (A_x + B_x + C_x + \ldots)(A_y + B_y + C_y + \ldots) \tag{3}$$

To achieve independence during the compression, some terms need to be first remasked by using fresh randomness denoted by Z shares in the following. Equation 4 shows an example for an ISW implementation of an AND gate for $d = 2$ in our notation. For a general description of the compression algorithm see [13], for details. The correctness of the AND gate in Eq. 4 is given because all random Z shares appear exactly twice in additive manner, and the rest of the terms are the one of the expanded Eq. 3.

$$\begin{aligned}
A_q &= A_x A_y + Z_0 + Z_1 \\
B_q &= B_x B_y + (Z_0 + A_x B_y + B_x A_y) + Z_2 \\
C_q &= C_x C_y + (Z_1 + A_x C_y + C_x A_y) + (Z_2 + B_x C_y + C_x B_y)
\end{aligned} \tag{4}$$

For the security of Eq. 4, the order in which the terms are summed up is critical. While the calculation order can be easily controlled for software implementations, the order in which the terms are summed up cannot so easily be controlled in the combinatorial logic of hardware implementations.

Like many other masking schemes, the private circuits approach is therefore considered to be vulnerable to so-called glitches [15]. Glitches are caused in the combinatorial path of hardware circuits because the electric signals do not propagate with unlimited speed. Instead signal arrival times and delays at the logic gates can cause several changes at the output of a gate before the gate output reaches its final state (for more details see, *e.g.*, [16]). Digital designers also have only marginal influence on the exact placement of the logic gates, the signal timings, and the order in which the signals are combined. A secure masking scheme thus needs to be inherently immune against glitches without relying on correct placement of the gates and signal timings.

Since there exist secure ISW implementations in software, a straightforward approach of its implementation in hardware would be to emulate the behavior of a processor running the ISW transformed software. As a result, the output of each AND and each XOR operation would be first stored in a register before any further processing is performed. However, this approach is neither very resource friendly nor efficient in terms of throughput.

In the next section, we thus introduce a secure construction of a masked AND gate in hardware and argue its security in the d-probing model for the case that glitches are taken into account. Our masked AND gate uses the same multiplication terms as the ISW AND gate, and has the same randomness requirements and a generic structure. However, in contrast to ISW, the introduced masked AND gate is resistant to glitches and has a balanced gate distribution which is desirable in order to minimize the delay of a hardware implementation. We start our construction and security argumentation for a first-order secure masked AND gate before we generalize the concept to arbitrary protection orders.

3   A Glitch-Resistant Masked AND Gate

The basic idea behind our glitch-resistant masked AND gate, is to split the calculation of Eq. 3 into independent share *domains*. Each share of a signal is associated with one specific domain. This is also reflected in the notation that is used in this paper. The shares A_x and B_x of a data signal x, for example, are associated with the domains labeled A and B, respectively.

The AND gate uses $d+1$ shares per signal in order to achieve d^{th}-order security and there are $d+1$ domains in this case. The intuition behind this approach is to keep the shares of all domains independent from shares of other domains. This independence ensures d^{th}-order security according to the d-probing model when considering glitches.

The critical parts of the circuit, are the parts that need to process inputs from multiple domains. In this case dedicated measures need to be taken before the terms can be securely served as inputs of a domain. By adding a fresh random share Z to these terms, the terms can be reassociated to a targeted domain. Furthermore, the usage of a register in this case prevents that glitches propagate from one domain to the another domain.

We first start with the introduction of the glitch-resistant AND gate for first-order security before this approach is extended to arbitrary protection orders.

3.1 1st-Order Secure AND Gate

A first-order secure AND gate (see Fig. 1) consists of two domains labeled A and B. The inputs x and y are provided to the AND gate by the shares A_x and B_x, and A_y and B_y, respectively. The sharings for x and y are required to be uniformly random and independent of each other. The AND gate returns the shares A_q and B_q of the output q. The calculations are performed in three steps in order to map the input shares to the output shares. We refer to these steps as *calculation*, *resharing* and *integration*.

Calculation: In the first step, the actual calculation of the logic function (expanded Eq. 3) is performed and the terms A_xA_y, A_xB_y, B_xA_y and B_xB_y are calculated. The terms that can be directly associated with one domain (inner-domain terms) are the terms A_xA_y and B_xB_y, respectively. These terms are not critical from a security point of view. Any computation on inner-domain terms associated with one specific domain, only lead to outputs that again depend only on shares associated with this domain.

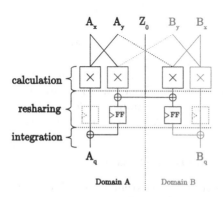

Fig. 1. First-order secure AND gate (Color figure online)

In case of terms that contain different domain labels (cross-domain terms), there is less freedom. In fact these calculations are only secure for independently shared input signals. If shares of the same signal would be combined for example, the independence would be trivially broken. For example, the term $A_x B_x$ would leak information about x. However, shares associated with different domains that correspond to different signals of the unprotected circuit can be combined without violating the requirement for d^{th}-order security. In fact, there is no leakage about x or y when calculating $A_x B_y$. This results from the requirement that x and y are independently shared. There is also no leakage caused by $B_x A_y$ for an independent sharing of x and y. Cross-domain terms of the AND gate that can not directly be associated with one domain are plotted red in Fig. 1.

Resharing: The integration of the cross-domain terms into a specific domain is prepared in the resharing step. By adding a fresh random Z share to these terms, the term becomes statistically independent from all other shares and can therefore be associated with any arbitrary domain in the next step. In case of the 1st-order secure AND gate, the same fresh share Z_0 is used for the resharing of the product terms $A_x B_y$ and $B_x A_y$. This does not lead to any first-order leakage, because a probing attacker restricted to one probing needle cannot find a single signal in the AND gate that correlates to the unshared inputs x and y or the output q.

In order to prevent that any glitch propagates through the resharing step, a register is included as last part of the resharing step. The two registers in grey dotted lines are optional registers and are only required for pipelining purposes but not for the security of the AND gate.

Integration: During the integration phase, the reshared cross-domain terms are added to the inner-domain terms, which concludes the calculation of the AND gate. Please note that this addition leads to glitches at the XOR gate at the output of the domain. However, as the resharing step finishes with a register no glitches can occur that depend on x or y. In terms of correctness, it is important to point out that the fresh share Z_0 becomes part of both domains. Hence, it holds that $q = A_q + B_q$.

In summary, the security against a first-order probing attacker is given because each domain contains either inner-domain terms that contain only shares that are already associated with one specific domain, or cross-domain terms that are reshared with a fresh random Z share which is only used once in each domain. An attacker thus always needs to combine at least two signals to get one signal that depends on one of the independently shared inputs x or y.

3.2   Higher-Order Secure AND Gate

The first-order AND gate can be extended to arbitrary protection orders. The generalization requires to first extend the *calculation* step to produce a correct sharing with $d+1$ shares for any given protection order d. In the *resharing* phase it needs to be ensured that the fresh random Z shares are distributed over the domains in a way that (1) each cross-domain term is reshared with a Z share that is unique inside the targeted domain, and (2) none of the signal combinations created in the *integration* phase reveals more than the inner-domain terms or shares of the respective domain.

Calculation: The same rules as for the first-order AND gate apply for the higher-order generalization. Again, any combination of shares can be safely used inside their associated domain without any restrictions. Cross-domain terms, however, require independently shared signals to prevent the case that two shares of the same sharing are combined. This ensures that by probing a cross-domain term, the attacker does not learn more about the inputs x and y then when probing a share of x and y directly.

The calculation step can be generalized for $d + 1$ input shares as shown in Eq. 5. Each row of this formula stands for one domain with a dedicated label calculating one share of the output q. The terms in the diagonal (bold) are the inner-domain terms containing only shares from one specific domain and hence only leak about shares of this domain. The cross-domain terms do not leak more information on the inputs x and y then when probing one share of x and one share of y directly. Hence, with this formula the sharing for the *calculation* step for the AND gate resists a d-probing attacker for an arbitrary numbers of shares. An example for a second-order AND gate is given in Fig. 2.

$$
\begin{aligned}
A_q &= \underbrace{\mathbf{A_x A_y}}_{t_{0,0}} + \underbrace{(A_x B_y + Z_0)}_{t_{0,1}} + \underbrace{(A_x C_y + Z_1)}_{t_{0,2}} + \underbrace{(A_x D_y + Z_3)}_{t_{0,3}} + \underbrace{(A_x E_y + Z_6)}_{t_{0,4}} + \cdots \\
B_q &= \underbrace{(B_x A_y + Z_0)}_{t_{1,0}} + \underbrace{\mathbf{B_x B_y}}_{t_{1,1}} + \underbrace{(B_x C_y + Z_2)}_{t_{1,2}} + \underbrace{(B_x D_y + Z_4)}_{t_{1,3}} + \underbrace{(B_x E_y + Z_7)}_{t_{1,4}} + \cdots \\
C_q &= \underbrace{(C_x A_y + Z_1)}_{t_{2,0}} + \underbrace{(C_x B_y + Z_2)}_{t_{2,1}} + \underbrace{\mathbf{C_x C_y}}_{t_{2,2}} + \underbrace{(C_x D_y + Z_5)}_{t_{2,3}} + \underbrace{(C_x E_y + Z_8)}_{t_{2,4}} + \cdots \\
D_q &= \underbrace{(D_x A_y + Z_3)}_{t_{3,0}} + \underbrace{(D_x B_y + Z_4)}_{t_{3,1}} + \underbrace{(D_x C_y + Z_5)}_{t_{3,2}} + \underbrace{\mathbf{D_x D_y}}_{t_{3,3}} + \underbrace{(D_x E_y + Z_9)}_{t_{3,4}} + \cdots \\
E_q &= \underbrace{(E_x A_y + Z_6)}_{t_{4,0}} + \underbrace{(E_x B_y + Z_7)}_{t_{4,1}} + \underbrace{(E_x C_y + Z_8)}_{t_{4,2}} + \underbrace{(E_x D_y + Z_9)}_{t_{4,3}} + \underbrace{\mathbf{E_x E_y}}_{t_{4,4}} + \cdots
\end{aligned}
\tag{5}
$$

with rows labeled Q_0, Q_1, Q_2, Q_3, Q_4 on the left.

Resharing: A core property for the generalization of this AND gate implementation is how the required fresh random Z shares can be efficiently distributed among the cross-domain terms in a correct manner. From Eq. 5 it can be seen that there are exactly $d(d + 1)$ cross-domain terms which need to be reshared. It is also important to note that there are exactly two cross-domain terms that combine shares from the same two domains. For example shares from domain A and B are only combined in the terms $A_x B_y$ and $B_x A_y$. We use the same fresh Z share for cross-domain terms that combine shares from the same two domains. Hence, we use $d(d + 1)/2$ fresh shares for a dth-order AND gate, which is the same amount as in the ISW scheme.

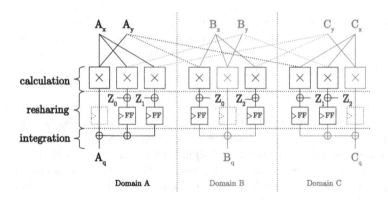

Fig. 2. Second-order secure AND gate

Since no probing of any intermediate signal created in the *calculation* phase contains more than one share of each input x or y, and in the *resharing* phase we add fresh random shares to the cross-domain terms, no advantage to a d-probing attacker is given during these phases.

Integration: In the integration phase, the terms associated with each domain are added up at the output of the AND gate. Because a digital designer has no influence on the sequence in which these terms are added up (without forcing it through registers), the higher-order secure AND gate needs to provide probing security for each possible partial sum of these terms. In particular, it has to be taken care of that each of these possible partial sums an attacker could probe reveals only the shares of the domains she is probing in. This is ensured by the resharing shown in Eq. 5, where each Z share is only reused for cross-domain terms with the same domain association.

In order to exploit the reuse of Z shares, it would be necessary to probe in the two domains that use the cross-domain terms with the reused Z share. However, the two cross-domain terms that use the same Z share contain only the shares of the same domains. Hence, there is no advantage for the attacker due the reuse.

For example, the share Z_0 in Fig. 2 is used on the terms $A_x B_y$ and $B_x A_y$ and these two terms only occur in the domains A and B. An attacker that probes any partial sum of the terms in A learns only about shares in domain A. When probing any partial sum of the terms in B, there is only information about shares associated with B. A second-order attacker that learns about partial sums in A and B learns about shares from the domains A and B in any case. The fact that the cross-domain terms $A_x B_y$ and $B_x A_y$ reuse Z_0 does not provide any advantage to an attacker.

Based on Eq. 5, the fact that the AND gate fulfills d^{th}-order security can also be verified visually. In this matrix the diagonal terms are formed by the inner-domain terms. These inner-domain terms also divide the matrix into an upper and lower triangular matrix in which each of the fresh random Z shares is used exactly once. The triangle formed by the Z shares is mirrored along the diagonal. The mirroring of the Z shares ensures that each possible combination of partial sums from any two domains removes at most one fresh random share, and reveals only the shares associated with both domains. Because this applies for all combinations of partial sums of all different domains, an attacker restricted to d probes obtains at most d shares per signal. However, for this security argumentation to hold it needs to be always ensured that the sharings of the inputs x and y are independent.

The domain equations of the matrix in Eq. 5 can also be written in closed form as shown in Eq. 6.

$$Q_i = t_{i,i} + \sum_{j>i}^{d} (t_{i,j} + Z_{(i+j*(j-1)/2)}) + \sum_{j<i}^{d} (t_{i,j} + Z_{(j+i*(i-1)/2)}) \tag{6}$$

This equation is also the basis for the scalable AES design in the next section. Furthermore, we note that the approach can be easily extended to arbitrary finite fields. Consequently, our glitch-resistant masked AND gate, which equals a multiplication in $GF(2)$, can be extended to arbitrary large $GF(2^n)$ multiplications by replacing the AND gates in the calculation step by GF multipliers. Operations that are linear over $GF(2^n)$ like XOR or logic negation, on the other hand, can be applied to the shares without domain crossings. We use this property for an efficient implementation of the AES S-box in the next section.

4 d^{th}-Order Secure AES Implementation

To compare the efficiency of our approach with existing masked implementations, we implemented the AES-128 encryption-only design suggested by Moradi et al. [17]. Moradi's design was also used and modified by Bilgin et al. [2,3,6] and recently by De Cnudde et al. [7] for a $d + 1$ share CMS TI.

The control path of our modified AES design consists of a linear-feedback shift register (LFSR), the round constant generation module (RCON), and some additional logic gates to generate the control signals (see [17] for more details). Our LFSR module has a cycle length of 23. In each round, the first 16 cycles are spent on *AddRoundKey* and *SubBytes*. Then there are four cycles used for *MixColumns* and to calculate the first four bytes of the next round key. Then there are two dummy rounds inserted to bring the state register in correct position for further processing before in the final cycle the ShiftRows transformation is performed. The datapath mainly consists of the S-box, the key and state registers which are implemented as shift registers, the *MixColumns* module, and some multiplexers.

4.1 AES S-Box

The by far most complex and most security critical part of the AES implementation is the S-box. Figure 3 shows our design of a 1th-order protected variant of

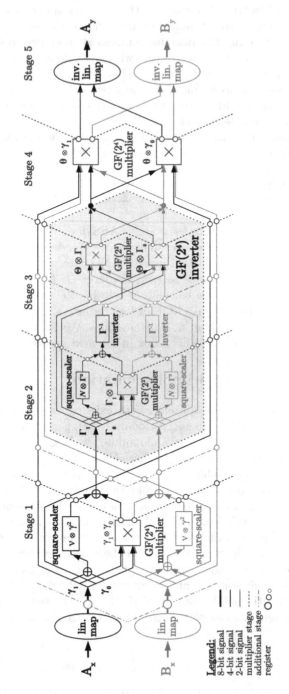

Fig. 3. First-order masked AES S-box with seven pipeline stages (Color figure online)

Canright's [5] AES S-box design. The S-box consists of many linear operations like the linear mappings at the input and the output, the square scalers, the sub-field inverters, and the adders. These are the parts that can be implemented share-wise for both domains in a straightforward way. The Galois field multipliers with different field order form the non-linear parts of the S-box. Canright's S-box makes repeated use of a finite field isomorphism to express $GF(2^8)$ elements as multiple elements in lower subfields—down to eight elements in $GF(2)$. These $GF(2^n)$ multipliers are replaced by the generalization of the masked AND gate of Sect. 3 for GF multipliers. Therefore, the standard-cell library AND cells used for the calculation step in the masked AND gate are simply replace by the according GF multipliers.

To maximize the efficiency of the implementation, seven pipelining stages are added to the S-box. The pipelining registers are marked with circles and appear along the red and green dotted lines in Fig. 3. Red dotted lines indicate multiplier related stages which are also labeled Stage 1-5 in order to refer to them more easily. The green marked registers are required to ensure independence in the presence of glitches for the inputs of the adjacent GF gates. To make the S-box secure and efficient at the same time, it is necessary to pinpoint all GF gates that have related input sharings. These gates need to be treated more carefully than the one with independent inputs. We now discuss the security of each multiplication stage individually which reveals that the additional pipeline stages (plotted in green) are required at multiplication stages 1, 2, and 3, but not at 4 and 5.

Stage 1. The $GF(2^4)$ gate in Stage 1 receives its inputs from the linear mapping at the S-box input. The linear mapping takes the 8-bit input shares A_x and B_x and linearly combines these eight bits inside their respective domain (see [5] for more details). Because of the different signal transition times and gate delays, it is therefore possible that the output of the linear mapping temporarily consists of bits with related sharing. Applying these bits directly to the GF gate from Fig. 1—while the linear mapping has not yet settled—would thus violate the independence in the cross-domain terms associated GF multipliers. To avoid these glitches, registers are inserted after the linear maps to ensure the signals are settled before the bits are applied to the GF gate.

Stage 2 and 3. The situation is similar at Stage 2 and Stage 3. At these stages, glitches can occur from the combination of the square scaler outputs with the outputs of the GF gate. Again these glitches can be avoided by inserting pipelining stages at the marked positions in Fig. 2.

Stage 4. For the GF gates in Stage 4, the inputs are the pipelined S-box inputs and the output of the GF gates of the previous stage. The output of the GF gate of Stage 3 originate from the inputs of the $GF(2^4)$ inverter which is remasked in Stage 1 (the masking is effective at latest at Stage 2). Therefore, the inputs of the Stage 4 GF gates are clearly independent and so no registers are required here.

Stage 5. The output mapping in this stage is again a linear transformation and uncritical as long as it is not followed by a nonlinear transformation that is unprepared for related sharing of its inputs. However, in our design of the AES core the output of the S-box is either stored in the key or state registers before it is used again, or fed into the S-box which is also uncritical because the input multiplier of either S-box variant is already prepared to process related input sharings.

The rest of the S-box is implemented according to the original Canright design but without some of its optimizations that would not be beneficial for our implementation. Canright's design, for example, reuses some temporary results in other parts of the S-box. Storing temporary results would lead to many additional pipelining registers for our design of the S-box and is therefore not suitable. For the generalization of the S-box to higher protection orders, the black (or blue) parts in Fig. 3 are basically duplicated and the secure GF gates are generated as described in Sect. 3.

5   Implementation Results

All stated numbers are post-synthesis results for a 90 nm UMC Low-K process with 1.0 V power supply and 0.1 MHz clock frequency (in accordance with related work). Our designs are compiled with the Cadence Encounter RTL compiler version v08.10-s28_1 and routed with Cadence NanoRoute v08.10-s155. Please note that in general hardware result for different technologies, compiled and synthesized with different tool chains are difficult to compare. Furthermore, the functionality implemented by different modules is not always consistent with other implementations. The comparison of chip area results with related work should therefore be seen under this premise. To make comparison with our generic AES design easier for future work, we therefore decided on publishing the source code online [11].

Anyway, for a masked hardware design the number required fresh random bits is even more crucial for the efficiency of an implementation than the stated chip area of the designs. The generation of fresh random bits with high entropy requires additional hardware and involves, e.g., complex analog circuitry or pseudo random number generators based on symmetric primitives. Both options have a critical influence on the chip area requirements, the energy budget, and on the delay or throughput.

First-Order Secure AES. Table 1 compares our first-order secure AES hardware implementation with existing related work. The $d + 1$ share designs of [7] with 6.7 kGE and our design with 6 kGE are smaller than the $td + 1$ TI designs. The size difference mainly comes from the fact that $td + 1$ TI requires at least three shares for securely calculating non-linear functions while the first-order $d + 1$ share designs require only two shares.

In comparison with $d + 1$ TI design [7] which requires 54 random bits per S-box calculation, our design requires with 18 bits only a third of its random bits. Nevertheless, our design achieves the same throughput as the $td + 1$ TI design of Bilgin *et al.* with 52 Kbps for a 100 kHz clock and requires 14 bits less fresh randomness.

Table 1. First-order secure AES-128 implementation results

Design/module	Chip area		Randomness	Cycles	Throughput @0.1 MHz
	[%]	[kGE]	[Bits/S-box]		[Kbps.]
Our implementation (90 nm)					
This work	**100.0**	**6.0**	**18**	**246**	**52**
S-box	37.3	2.2			
State registers	34.0	2.0			
Key registers	21.0	1.3			
Control, et cetera	7.7	0.5			
td + 1 threshold implementations (180 nm)					
Moradi et al. [17]	11.0/10.8[a]		48	266	48
Bilgin et al. [2]	9.1/8.2[a]		44	246	52
Bilgin et al. [3]	8.1/7.3[a]		32	246	52
d + 1 threshold implementations (45 nm)					
De Cnudde et al. [7]	6.7/6.3[a]		54	276	46

[a]This variant uses the *compile_ultra* flag which is not available in our tool chain.

Second-Order Secure AES. In Table 2, a comparison of our second-order AES design with other second-order secure designs is given. In case of the $td + 1$ TI design the chip area was estimated by De Cnudde et al. [7]. Again, there is a noticeable gap between the $td + 1$ share design with about 14.9 kGE and the $d + 1$ share designs with about 10 kGE in terms of chip area resulting from the increased amount of shares (five shares versus three shares). Considering the randomness demand of the designs, our design requires 54 bits which is more than two times less than the $td + 1$ design with 126 fresh random bits, and three times less than the $d + 1$ TI design with 162 bits. In terms of throughput, our AES design requires 246 cycles instead of 276 cycles per encryption.

5.1 d^{th}-Order AES Implementation Results

The generic construction of our AES implementation not only allows the calculation of the number of required fresh random bits of $9d(d+1)$, but furthermore it is possible to synthesize the AES implementation for arbitrary protection orders by just changing one input parameter of our hardware design.

Figure 4 shows the post-synthesis area results for the different components in relation to the protection order. It can be observed that the state key and control logic requirements grow linearly with the protection order. The S-box and the contained GF gates grow quadratically. For the S-box, the size increases from 37.4% for the first-order implementation to about 78.5% for the 15th-order. The relative size of the state and key register decrease from 34% and 21% to around 12.2% and 7.5%, respectively. The smallest amount of chip area is spent on the control logic which stays almost constant.

Table 2. Second-order secure AES-128 implementation results

Design/module	Chip area		Randomness	Cycles	Throughput @0.1 MHz
	[%]	[kGE]	[Bits/S-box]		[Kbps]
Our implementation (90 nm)					
This work	**100.0**	**10.0**	**54**	**246**	**52**
S-box	45.1	4.5			
State registers	30.3	3.0			
Key registers	18.7	1.9			
Control, et cetera	5.9	0.6			
td + 1 threshold implementation (estimated [7], 45 nm)					
De Cnudde *et al.* [6]	*18.6/14.9*[a]		126	276	46
d + 1 Threshold Implementation (45 nm)					
De Cnudde *et al.* [7]	10.5/10.3[a]		162	276	46

[a]This variant uses the *compile_ultra* flag which is not available in our tool chain.

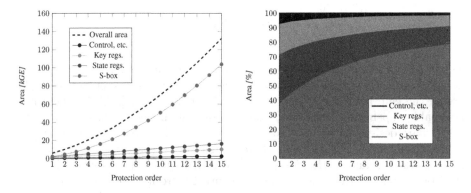

Fig. 4. Area requirements absolute (left) and in percent (right) per protection order

6 Side-Channel Evaluation

To show the resistance of our AES design against side-channel analysis attacks, different instances of the Welch's t-test are used (see Goodwill *et al.* [9] for details). The intention of this test is that for a side-channel secure implementation, a set of randomly picked (unshared) inputs should not show any statistically differences in the power traces for a set with constant inputs. For these two sets the so-called t value is calculated. If the t value is outside the confidence interval of ± 4.5 the null-hypothesis is rejected with confidence greater than 99.999% for large sizes of N.

Our evaluation approach is quite similar to what is checked in the d-probing model. Instead of using power trace values of, e.g., an FPGA implementation of our design, the t values of each individual signal are recorded for a post-synthesis netlist of our AES design during simulation. In comparison to an FPGA based validation this approach has three advantages: (1) the signals are completely noise free, (2) if any statistical differences are found, the violating signals can be directly pin-pointed, (3) if ASIC implementations are targeted, the synthesized netlist is closer to the final ASIC implementation than an FPGA implementation.

First-Order AES Design. The results of the first-order t-test for our first-order secure design are shown in Fig. 5 (left) for up to one million traces. The t-value stays below the ±4.5 border as required by the t-test to succeed. To demonstrate the soundness of our evaluation setup we also performed a second-order t-test. However, for the second-order t-test in a bivariate attack setting, performing individual t-tests for each signal separately is no longer feasible. The evaluation of each signal combined with every other signal for different points in time would take too long. Therefore, one single trace is calculated that sums up all signal transitions together. We then combine in each case two trace points over centered product pre-processing for all points in time within an eight clock cycles period (the delay of the S-box). As expected the t-tests fail with

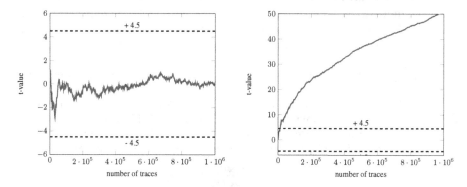

Fig. 5. First-order t-test (left) and second-order t-test (right) for first-order secure AES design

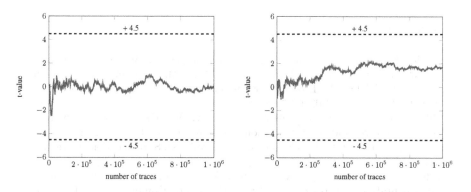

Fig. 6. First-order t-test (left) and second-order t-test (right) for second-order secure AES design

great confidence with t values clearly above the ± 4.5 border even for just a hundred traces.

Second-Order AES Design. The t-tests for the second-order AES design are illustrated in Fig. 6. In both cases the t-tests do not indicate any leakage. We thus conclude that our implementation seems to be correct and secure in a bivariate second-order attack scenario.

7   Conclusions

In this work we introduced a generic hardware design of the AES. In contrast to existing implementations, our design is freely scalable in terms of resistance to side-channel analysis attacks. Because of its $d + 1$ share design principle it is also very efficient. With only 6 kGE of chip area, our design is the smallest published first-order (and beyond) masked AES implementation to this date.

Since the generation of random numbers with high entropy is a very demanding task for hardware implementations, we consider the randomness requirements to be even more decisive for the efficiency of a masked hardware implementation. In comparison with the recently published $d + 1$ share AES design [7], our design requires just $d(d + 1)/2$ fresh random shares instead of $(d + 1)^2$.

Acknowledgements. This work has been supported by the Austrian Research Promotion Agency (FFG) under grant number 845589 (SCALAS). The HECTOR project has received funding from the European Unions Horizon 2020 research and innovation programme under grant agreement No. 644052. This project has received funding from the European Research Council (ERC) under the European Union's Horizon 2020 research and innovation programme (grant agreement No. 681402).

References

1. Bilgin, B., Daemen, J., Nikov, V., Nikova, S., Rijmen, V., Assche, G.: Efficient and first-order dpa resistant implementations of KECCAK. In: Francillon, A., Rohatgi, P. (eds.) CARDIS 2013. LNCS, vol. 8419, pp. 187–199. Springer, Heidelberg (2014). doi:10.1007/978-3-319-08302-5_13
2. Bilgin, B., Gierlichs, B., Nikova, S., Nikov, V., Rijmen, V.: A more efficient AES threshold implementation. In: Pointcheval, D., Vergnaud, D. (eds.) AFRICACRYPT 2014. LNCS, vol. 8469, pp. 267–284. Springer, Heidelberg (2014). doi:10.1007/978-3-319-06734-6_17
3. Bilgin, B., Gierlichs, B., Nikova, S., Nikov, V., Rijmen, V.: Trade-offs for threshold implementations illustrated on AES. IEEE Trans. Comput. Aided Des. Integr. Circuits Syst. **34**(7), 1188–1200 (2015)

4. Bilgin, B., Nikova, S., Nikov, V., Rijmen, V., Stütz, G.: Threshold implementations of all 3×3 and 4×4 S-boxes. In: Prouff, E., Schaumont, P. (eds.) CHES 2012. LNCS, vol. 7428, pp. 76–91. Springer, Heidelberg (2012). doi:10.1007/ 978-3-642-33027-8_5
5. Canright, D.: A very compact S-box for AES. In: Rao, J.R., Sunar, B. (eds.) CHES 2005. LNCS, vol. 3659, pp. 441–455. Springer, Heidelberg (2005). doi:10. 1007/11545262_32
6. Cnudde, T., Bilgin, B., Reparaz, O., Nikov, V., Nikova, S.: Higher-order threshold implementation of the AES S-box. In: Homma, N., Medwed, M. (eds.) CARDIS 2015. LNCS, vol. 9514, pp. 259–272. Springer, Heidelberg (2016). doi:10.1007/ 978-3-319-31271-2_16
7. De Cnudde, T., Reparaz, O., Bilgin, B., Nikova, S., Nikov, V., Rijmen, V.: Masking AES with $d + 1$ shares in hardware. In: Gierlichs, B., Poschmann, A.Y. (eds.) CHES 2016. LNCS, vol. 9813, pp. 194–212. Springer, Heidelberg (2016). doi:10. 1007/978-3-662-53140-2_10
8. Faust, S., Rabin, T., Reyzin, L., Tromer, E., Vaikuntanathan, V.: Protecting circuits from leakage: the computationally-bounded and noisy cases. In: Gilbert, H. (ed.) EUROCRYPT 2010. LNCS, vol. 6110, pp. 135–156. Springer, Heidelberg (2010). doi:10.1007/978-3-642-13190-5_7
9. Goodwill, G., Jun, B., Jaffe, J., Rohatgi, P.: A testing methodology for side-channel resistance validation. In: NIST Non-Invasive Attack Testing Workshop (2011)
10. Goubin, L., Patarin, J.: DES and differential power analysis the "Duplication" method. In: Koç, Ç.K., Paar, C. (eds.) CHES 1999. LNCS, vol. 1717, pp. 158–172. Springer, Heidelberg (1999). doi:10.1007/3-540-48059-5_15
11. Gross, H.: DOM Protected Hardware Implementation of AES. https://github.com/ hgrosz/aes-dom (2016)
12. Gross, H., Mangard, S., Korak, T.: Domain-oriented masking: compact masked hardware implementations with arbitrary protection order. Cryptology ePrint Archive, Report 2016/486 (2016). http://eprint.iacr.org/2016/486
13. Ishai, Y., Sahai, A., Wagner, D.: Private circuits: securing hardware against probing attacks. In: Boneh, D. (ed.) CRYPTO 2003. LNCS, vol. 2729, pp. 463–481. Springer, Heidelberg (2003). doi:10.1007/978-3-540-45146-4_27
14. Kocher, P., Jaffe, J., Jun, B.: Differential power analysis. In: Wiener, M. (ed.) CRYPTO 1999. LNCS, vol. 1666, pp. 388–397. Springer, Heidelberg (1999). doi:10. 1007/3-540-48405-1_25
15. Mangard, S., Popp, T., Gammel, B.M.: Side-channel leakage of masked CMOS gates. In: Menezes, A. (ed.) CT-RSA 2005. LNCS, vol. 3376, pp. 351–365. Springer, Heidelberg (2005). doi:10.1007/978-3-540-30574-3_24
16. Mangard, S., Schramm, K.: Pinpointing the side-channel leakage of masked AES hardware implementations. In: Goubin, L., Matsui, M. (eds.) CHES 2006. LNCS, vol. 4249, pp. 76–90. Springer, Heidelberg (2006)
17. Moradi, A., Poschmann, A., Ling, S., Paar, C., Wang, H.: Pushing the limits: a very compact and a threshold implementation of AES. In: Paterson, K.G. (ed.) EUROCRYPT 2011. LNCS, vol. 6632, pp. 69–88. Springer, Heidelberg (2011). doi:10.1007/978-3-642-20465-4_6
18. Nikova, S., Rechberger, C., Rijmen, V.: Threshold implementations against side-channel attacks and glitches. In: Ning, P., Qing, S., Li, N. (eds.) ICICS 2006. LNCS, vol. 4307, pp. 529–545. Springer, Heidelberg (2006). doi:10.1007/11935308_38

19. Quisquater, J.-J., Samyde, D.: Electromagnetic analysis (EMA): measures and counter-measures for smart cards. In: Attali, I., Jensen, T. (eds.) E-smart 2001. LNCS, vol. 2140, pp. 200–210. Springer, Heidelberg (2001). doi:10.1007/3-540-45418-7_17
20. Reparaz, O., Bilgin, B., Nikova, S., Gierlichs, B., Verbauwhede, I.: Consolidating masking schemes. In: Gennaro, R., Robshaw, M. (eds.) CRYPTO 2015. LNCS, vol. 9215, pp. 764–783. Springer, Heidelberg (2015)
21. Rivain, M., Prouff, E.: Provably secure higher-order masking of AES. In: Mangard, S., Standaert, F.-X. (eds.) CHES 2010. LNCS, vol. 6225, pp. 413–427. Springer, Heidelberg (2010). doi:10.1007/978-3-642-15031-9_28
22. Trichina, E.: Combinational logic design for AES subbyte transformation on masked data. IACR Cryptology ePrint Archive, 2003 (2003)

Side-channel Resistant Implementations

Time-Memory Trade-Offs for Side-Channel Resistant Implementations of Block Ciphers

Praveen Kumar Vadnala[(✉)]

Riscure, Delft, The Netherlands
vadnala@riscure.com

Abstract. Currently, the most efficient first-order masked implementations use the classical randomized table countermeasure, which induces a penalty factor of around 2–3 in execution time compared to an unmasked implementation. However, an S-box with n-bit input and m-bit output requires $2^n m$ bit memory; for example, AES requires 256 bytes of RAM. Conversely, generic S-box computation method due to Rivain-Prouff requires almost no memory, but the penalty factor to achieve first-order resistance is roughly 30–35. Therefore, we suggest studying time-memory trade-offs for block-cipher implementations based on an adaptation of a table compression technique proposed by IBM. We use the similar approach to study time-memory trade-offs for second-order masked implementations as well. We show that for the case of AES, reasonably efficient implementations can be obtained with just 40 bytes of RAM in both the cases and hence they can be used in highly memory constrained devices.

Keywords: Side-channel attacks · Masking · S-box compression · Time-memory trade-off · AES-128

1 Introduction

Side-Channel Attacks. Implementations of cryptographic algorithms leak information about the secret key, which could potentially be exploited by an attacker. Examples of such leakages include timing [Koc96], power consumption [KJJ99], and electromagnetic emission [AARR03]. These so-called *side-channel attacks* are very powerful in the sense that one can completely break the security of cryptographic devices with a very inexpensive setup. In particular, Differential Power Analysis (DPA) attacks have been subjected to extensive research as they require little knowledge about the implementation details.

Masking. To counteract DPA attacks, several countermeasures have been proposed in the literature. Masking, besides hiding, is one of the most widely used countermeasures to prevent side-channel attacks [CJRR99]. The basic idea behind masking is that each sensitive variable (a function of a known variable and the secret key) used in the algorithm is split into two shares where one is generated randomly (called *mask*) and the second share is computed using the

© Springer International Publishing AG 2017
H. Handschuh (Ed.): CT-RSA 2017, LNCS 10159, pp. 115–130, 2017.
DOI: 10.1007/978-3-319-52153-4_7

mask and the sensitive variable. All subsequent operations in the algorithm are applied separately on the shares, which are subsequently combined at the end to produce the desired ciphertext (or plaintext).

Higher-Order Masking. The masking scheme described above (called *first-order masking*) can be attacked by combining the leakages from both the shares (called *second-order attack*) [Mes00,OMHT06]. To thwart that, we divide the sensitive variable into three shares, among which two are randomly generated, while the third share is computed from the two random shares and the sensitive variable (called *second-order masking*) [RDP08]. In fact, this approach can be generalized to any number of shares. Namely, a d-th order (also called higher-order) masking scheme [RP10,Cor14] involving d shares can be attacked by combining the leakages from $d + 1$ shares of the sensitive variable.

In general, block ciphers consist of several round transformations, where each transformation is a combination of linear and non-linear layers. Applying masking to a linear function is easy since we can evaluate the function on the shares independently. However, it is not straightforward in the case of non-linear functions such as S-boxes. The non-linear layer of the cipher is often implemented as a lookup table for performance reasons, which is typically stored in ROM. One approach widely used to obtain masked implementations here is to randomize the lookup table for every execution of the cipher. This requires creating a new table in RAM, which is of equal size as the original lookup table.

1.1   Classical Randomized Table Countermeasure

We recall the classical randomized table countermeasure, which is secure against first-order attacks only, as suggested in [CJRR99]. An (n, m) S-box table $S(u)$ is first randomized in RAM by letting

$$T(u) = S(u \oplus r) \oplus s$$

for all $u \in \{0, 1\}^n$, where $r \in \{0, 1\}^n$ is the input mask and $s \in \{0, 1\}^m$ is the output mask.

To evaluate $S(x)$ from the masked value $x_1 = x \oplus r$, it suffices to compute $y_1 = T(x_1)$, as we get $y_1 = T(x_1) = S(x_1 \oplus r) \oplus s = S(x) \oplus s$; this shows that y_1 is indeed a masked value of $S(x)$. In other words, the randomized table countermeasure consists of first re-computing in RAM a temporary table with inputs shifted by r and with masked outputs, so that later it can be evaluated on a masked value $x_1 = x \oplus r$ to obtain a masked output. In the case of AES, this method requires a table of 256 bytes in RAM.

1.2   Compression of Lookup Table

In [RRST02] a compression scheme was proposed for lookup tables, which reduces the RAM requirement. Namely, masking using randomized lookup table can be implemented for AES using 128 bytes of RAM only. We recall the original scheme from [RRST02] below.

For simplicity, let us consider the AES S-box, though the original scheme can be applied to any S-box. AES S-box ($S(u)$ for all $u \in \{0,1\}^8$) is an 8-bit to 8-bit S-box. Let us rewrite $S(u)$ as $S(u) = S_0(u) \| S_1(u)$ for all $u \in \{0,1\}^8$, where $S_0(u)$ and $S_1(u)$ are 4-bit values. Let the two random input masks be $r_1, r_2 \in \{0,1\}^8$ and the output mask be $s \in \{0,1\}^4$. The new randomized table is defined as:

$$T(u) = S_0(u \oplus r_1) \oplus S_1(u \oplus r_2) \oplus s \tag{1}$$

This table has 2^8 entries, each requiring 4 bits of memory thus totaling to 128 bytes. Now let us assume that the sensitive variable x is represented by two shares ($x_1 = x \oplus r$, r), where $r = r_1 \oplus r_2$. From (1) we can write:

$$S_0(x) = S_0(x_1 \oplus r_1 \oplus r_2) = T(x_1 \oplus r_2) \oplus S_1(x_1) \oplus s$$
$$S_1(x) = S_1(x_1 \oplus r_1 \oplus r_2) = T(x_1 \oplus r_1) \oplus S_0(x_1) \oplus s$$

which gives:

$$S_0(x) \oplus s = T(x_1 \oplus r_2) \oplus S_1(x_1) \tag{2}$$
$$S_1(x) \oplus s = T(x_1 \oplus r_1) \oplus S_0(x_1) \tag{3}$$

By accessing the table T at ($x_1 \oplus r_1$) and ($x_1 \oplus r_2$), and the original S-box look up table at x_1 (which gives $S_0(x_1)$ and $S_1(x_1)$), we can compute the masked values of $S_0(x)$ and $S_1(x)$. The masked value of $S(x)$ can then be obtained as $S(x) \oplus (s\|s) = S_0(x) \oplus s \| S_1(x) \oplus s$. We recall below the algorithms for creating the compressed table (Algorithm 1) and performing the table lookup operation (Algorithm 2).

Algorithm 1. Table T creation: first-order compression

Require: Two random numbers $r_1, r_2 \in \{0,1\}^n$, output mask: $s \in \{0,1\}^{m/2}$, an (n,m)
    S-box lookup function where $S(u) = S_0(u)\|S_1(u)$ for all $u \in \{0,1\}^n$
Ensure: Table T
1: **for** $u := 0$ to $2^n - 1$ **do**
2:    $T(u) \leftarrow S_0(u \oplus r_1) \oplus S_1(u \oplus r_2) \oplus s$
3: **end for**

Algorithm 2. TableLookup

Require: The masked input $x_1 = x \oplus r$, table T from Algorithm 1, two random
    numbers $r_1, r_2 \in \{0,1\}^n$, output mask $s \in \{0,1\}^{\frac{m}{2}}$, and an (n,m) S-box lookup
    function where $S(u) = S_0(u)\|S_1(u)$ for all $u \in \{0,1\}^n$
Ensure: Two shares of $S(x)$
1: $t \leftarrow \mathsf{rand}(\frac{m}{2})$
2: $a \leftarrow T(x_1 \oplus r_2) \oplus S_1(x_1)$
3: $b \leftarrow T(x_1 \oplus r_1) \oplus S_0(x_1) \oplus t$
4: **return** $(a\|b, s\|(s \oplus t))$

In step 3 of Algorithm 2, we further randomize the second lookup operation so as to make it independent of the first lookup in Step 2. We then concatenate the results from the two table lookups to obtain the first share of the S-box output: $S(x)$. The second share can then be obtained by concatenating ($s \oplus t$) with s. It is easy to see that this scheme is secure against first-order DPA attacks as all the intermediate variables computed here are uniformly random and hence are independent of x and $S(x)$. Secondly, this compression scheme can be generalised to any S-box input and output sizes. Moreover, one can obtain a better compression factor by splitting $S(x)$ into more shares; for example, a 8-bit S-box could be split into 8 tables (one for each output bit) and the resulting randomised table T would be 8 times smaller, at the cost of increasing the running time for every table look-up.

Our Contribution. The classical randomized table method requires $2^n m$ bit memory for an (n, m) S-box and induces a penalty of 2–3 in execution time compared to an unmasked implementation. On another hand, generic S-box computation method due to Prouff and Rivain (recalled in Algorithm 5) needs almost no memory; however, it requires execution time in the order of 30–35 times more compared an unmasked implementation. In this paper, we study the time-memory trade-offs for implementing the randomized lookup table by using the compression scheme recalled above.

We first generalize the compression scheme so that the size of table T can be reduced further. Here, the table size is determined based on a parameter, which we call *compression level* (denoted by l). Then, by applying the Rivain-Prouff countermeasure to the generic compression scheme, we obtain time-memory trade-offs for S-box lookup table implementations that are secure against first-order DPA attacks. Next, we propose a similar compression scheme as well as time-memory trade-offs for the second-order secure S-box computation scheme proposed by Rivain *et al.* [RDP08]. We apply all these schemes to the case of AES-128 and provide the performance results on a 32-bit ARM Cortex-M3 microcontroller. Our results show that we can obtain relatively efficient implementations just under 40 bytes of RAM in both cases.

Outline. We first give our generic compression scheme as well as the time-memory trade-offs for first-order secure lookup table implementations in Sect. 2. The corresponding schemes for the second-order secure lookup table implementations are presented in Sect. 3. We provide the implementation results of all our schemes for the case of AES-128 in Sect. 4. Finally, we conclude the paper in Sect. 5.

2   First-Order Secure Compression Scheme

The original compression scheme requires working with 4-bit nibbles, which might be inefficient to implement, as we need to perform bit manipulations.

To avoid that, we can use a similar approach as in [RDP08] and keep the table entry size to a byte.

2.1 A Variant of the Compression Scheme

We consider again the case of 8-bit to 8-bit S-box S: $(\{0,1\}^8 \rightarrow \{0,1\}^8)$. We define functions S_0 and S_1 $(\{0,1\}^7 \rightarrow \{0,1\}^8)$ for all $u \in \{0,1\}^7$ as follows:

$$S_0(u) = S(u\|0), \ S_1(u) = S(u\|1) \ \forall \ u \in \{0,1\}^7$$

Let $r_1, r_2 \in \{0,1\}^7$ and $s \in \{0,1\}^8$ be random masks, and define the randomized table as:

$$T_1(u) = S_0(u \oplus r_1) \oplus S_1(u \oplus r_2) \oplus s \tag{4}$$

which is a 7-bit to 8-bit table. Hence it requires only 128 bytes in memory.

Let $x_1 = x \oplus (r^{(1)}\|r^{(2)})$ be a masked data, where $r^{(1)}$ contains the most significant 7 bits of the mask and $r^{(2)}$, the least significant bit. We write

$$x = x^{(1)}\|x^{(2)}$$
$$x_1 = x_1^{(1)}\|x_1^{(2)}$$

which gives

$$x_1^{(1)} = x^{(1)} \oplus r^{(1)}$$

and

$$x_1^{(2)} = x^{(2)} \oplus r^{(2)}$$

We then re-share $r^{(1)}$ into two shares $r_1, r_2 \in \{0,1\}^7$ so that $r_1 \oplus r_2 = r^{(1)}$. Hence $x_1^{(1)}$ is given as:

$$x_1^{(1)} = x^{(1)} \oplus r_1 \oplus r_2$$

We have from (4):

$$S_0(x^{(1)}) = S_0(x_1^{(1)} \oplus r_1 \oplus r_2) = T_1(x_1^{(1)} \oplus r_2) \oplus S_1(x_1^{(1)}) \oplus s$$
$$S_1(x^{(1)}) = S_1(x_1^{(1)} \oplus r_1 \oplus r_2) = T_1(x_1^{(1)} \oplus r_1) \oplus S_0(x_1^{(1)}) \oplus s$$

which gives:

$$S_0(x^{(1)}) \oplus s = T_1(x_1^{(1)} \oplus r_2) \oplus S_1(x_1^{(1)}) \tag{5}$$
$$S_1(x^{(1)}) \oplus s = T_1(x_1^{(1)} \oplus r_1) \oplus S_0(x_1^{(1)}) \tag{6}$$

In the second step we define a 1-bit to 8-bit table:

$$U = (S_0(x^{(1)}) \oplus s, S_1(x^{(1)}) \oplus s)$$

If $x^{(2)} = 0$ then we have $S(x) = S_0(x^{(1)})$, while if $x^{(2)} = 1$ we have $S(x) = S_1(x^{(1)})$, therefore:

$$U(x^{(2)}) = S(x) \oplus s$$

Hence we must evaluate the table U at $x^{(2)}$. However, the bit $x^{(2)}$ cannot be accessed directly, as it leaks information about the sensitive variable x. We use the standard randomized table technique to prevent this. We define this randomized table as:

$$T_2(i) = U(i \oplus r^{(2)})$$

for $i \in \{0,1\}$. We then retrieve the value stored at $x_1^{(2)}$ from table T_2: $T_2(x_1^{(2)}) = U(x^{(2)}) = S(x) \oplus s$, which gives the masked value of $S(x)$.

This variant can easily be generalized to smaller table size in RAM. It suffices to pack 2^ℓ S-box values at the beginning instead of only 2. We describe the generalized method in the next subsection.

2.2   Generic Compression Scheme

We now present a generic compression scheme for any S-box $S : \{0,1\}^n \to \{0,1\}^m$. With the classical randomized table method such an S-box requires a table of 2^n entries, where each entry is of size m bits. If we want to pack 2^l S-box values, we need two tables T_1 and T_2 of size 2^{n-l} and 2^l entries respectively. Let $L = 2^l, N = 2^n$ and $P = 2^{n-l}$.

Algorithm 3. Table T_1 creation: generic first-order compression

---

Require: An (n,m) S-box lookup function S where $S_i(u)$ for all $0 \le u \le P - 1$ is defined as $S_i(u) = S(u||i)$ for $0 \le i \le L - 1$
Ensure: Table T_1, L random numbers $r_i \in \{0,1\}^{n-l}$ for $0 \le i \le L - 1$, and output mask: $s \in \{0,1\}^m$
1: **for** $i := 0$ to $L - 1$ **do**
2:     $r_i \leftarrow \mathsf{rand}\,(n-l)$
3: **end for**
4: $s \leftarrow \mathsf{rand}\,(m)$
5: **for** $u := 0$ to $P - 1$ **do**
6:     $T_1(u) \leftarrow \left(\left(\bigoplus_{0 \le i \le L-1} S_i(u \oplus r_i) \right) \oplus s \right)$
7: **end for**

We define the function S_i $(\{0,1\}^{n-l} \to \{0,1\}^m)$ for $0 \le i \le L - 1$ as follows:

$$S_i(u) = S(u||i) \; \forall \, u \, \in \{0,1\}^{n-l}$$

We generate random numbers for the two tables T_1 and T_2: r_i (for $0 \le i \le L - 1$) and t as follows:

$$r_i \leftarrow \{0,1\}^{n-l}$$
$$t \leftarrow \{0,1\}^l$$

Now the table T_1 which is of $n - l$-bit to m-bit is defined as:

$$T_1(u) = \left(\bigoplus_{0 \le i \le L-1} S_i(u \oplus r_i) \right) \oplus s \qquad (7)$$

We give the algorithm to generate table T_1 in Algorithm 3. Next we describe a method to compute $S(x) \oplus s$ from table T_1. We divide the variables into two parts of size $n - l$ and l bits respectively. For example, x is now written as: $x = x^{(1)} \| x^{(2)}$ and r as $(r^{(1)} \| r^{(2)})$ etc.

From (7) we can get $S_i(x^{(1)})$ for $0 \leq i \leq L - 1$ as follows:

$$S_i(x^{(1)}) \oplus s = T_1((x_1^{(1)} \oplus r_i) \oplus r^{(1)}) \oplus \bigoplus_{j \in \{\{0:L-1\} - \{i\}\}} (S_j((x_1^{(1)} \oplus r_i \oplus r_j) \oplus r^{(1)})) \quad (8)$$

Let us now define table $U : \{0,1\}^l \to \{0,1\}^m$ for $i \in \{0, L - 1\}$ as follows:

$$U(i) = S_i(x^{(1)}) \oplus s = S(x^{(1)} \| i) \oplus s$$

Using the standard randomized table technique, we describe table $T_2 : \{0,1\}^l \to \{0,1\}^m$ as:

$$T_2(i) = U(i \oplus t)$$

for $i \in \{0, L - 1\}$. We then compute

$$T_2((x_1^{(2)} \oplus t) \oplus r^{(2)}) = U(x^{(2)})$$
$$= S(x^{(1)} \| x^{(2)}) \oplus s$$
$$= S(x) \oplus s$$

Algorithm 4. Generic compression scheme for first-order secure S-box computation

Require: Two input shares: $(x_1 = x \oplus r, r) \in \{0,1\}^n$, an (n, m) S-box lookup function S where $S_i(u)$ for all $u \in \{0, P - 1\}$ is defined as $S_i(u) = S(u\|i)$ for $0 \leq i \leq L - 1$; table T_1, L random numbers $r_i \in \{0,1\}^{n-l}$ for $0 \leq i \leq L - 1$, and an output mask: $s \in \{0,1\}^m$ from Algorithm 3

Ensure: $S(x) \oplus s$

1: Let $x_1^{(1)} \| x_1^{(2)} \leftarrow x_1$ where $x_1^{(1)}$ and $x_1^{(2)}$ are of size $n - l$ bits and l bits respectively
2: $t \leftarrow \mathsf{rand}\ (l)$
3: $t_1 \leftarrow (x_1^{(2)} \oplus t) \oplus r^{(2)}$ ▷ Change the mask to t
4: **for** $i := 0$ to $L - 1$ **do**   ▷ **Create table T_2**
5:    $k \leftarrow i \oplus t$
6:    $ind_1 \leftarrow x_1^{(1)} \oplus r_k \oplus r^{(1)}$ ▷ compute $x^{(1)} \oplus r_k$
7:    $ssum \leftarrow 0$
8:    **for** $j := 0$ to $L - 1$ **do**   ▷ Evaluate (8)
9:       **if** $k \neq j$ **then**
10:          $ind_2 \leftarrow ind_1 \oplus r_j$
11:          $ssum \leftarrow ssum \oplus S_j(ind_2)$
12:       **end if**
13:    **end for**
14:    $T_2(i) \leftarrow T_1(ind_1) \oplus ssum$ ▷ Store the entry in table T_2
15: **end for**
16: **return** $T_2(t_1)$ ▷ Return masked S-box output

We give the full algorithm to compute masked S-box output from masked inputs using our generic compression scheme in Algorithm 4. For each possible value in $\{0, L - 1\}$ we compute the corresponding table entry using (8). Note that the table T_2 is shifted by t so as to avoid the leakage of the sensitive variable x. We finally retrieve the value stored at $T_2(t_1)$, since $i = t_1$ implies:

$$
\begin{aligned}
k &= i \oplus t \\
&= (x_1^{(2)} \oplus t) \oplus r^{(2)} \oplus t \\
&= (x_1^{(2)} \oplus r^{(2)}) \\
&= x^{(2)}
\end{aligned}
$$

which gives the value $S(x) \oplus x$, as required.

Theorem 1. *Algorithm 4 is secure against first-order DPA attacks.*

Proof. The intermediate variables $r_1, r_2, \cdots, r_{L-1}$ are uniformly distributed in $\{0, 1\}^{n-l}$ and hence are independent of the sensitive variables x and $S(x)$. Similarly, the intermediate variables t, t_1, ind_1, and ind_2 are uniformly distributed in $\{0, 1\}^l$ and are independent of the sensitive variables x and $S(x)$. As ind_2 is uniformly distributed in $\{0, 1\}^l$ and $j \in \{\{0 : L - 1\} - \{k\}\}$ so that $ind_2 \neq x^{(1)}$, $S(ind_2 \| j)$ and $ssum$ are also independent of x and $S(x)$. Finally, as ind_1 is uniformly distributed in $\{0, 1\}^{n-l}$ and $ssum$ is independent of the sensitive variables, $T_1(ind_1) \oplus ssum$ (and as a result $T_2(t_1)$) is also independent of x and $S(x)$. As all the intermediate variables present in Algorithm 4 are independent of the sensitive variables, we can conclude that it is secure against first-order DPA attacks.   □

Application to AES. It is known that a straightforward implementation of randomized lookup table for AES takes 256 bytes ($n = 8$, $l = 0$). When we apply our generic compression method, we get implementations with varying memory requirements depending on the value of l as shown in Table 1. It can be seen that the memory required for table T_1 reduces exponentially as l increases. On another hand, the memory required for table T_2 increases at the same rate. The best case scenario occurs when $l = 4$, since we need only 32 bytes overall.

2.3   Time-Memory Trade-Offs for First-Order Masking

We now present a method to obtain time-memory trade-offs for implementing block ciphers secure against first-order DPA attacks. Our method essentially is a combination of the generic compression scheme presented in Sect. 2.2 and the generic secure S-box computation method proposed by Prouff and Rivain [PR07]. We recall in Algorithm 5, the first-order secure method to compute masked S-box output from masked input due to Prouff and Rivain.

Table 1. Memory requirement for tables T_1, T_2 (in bytes) and number of calls to the random number generator for masked implementation of AES for different values of l

l	T_1	T_2	Rand
1	128	2	3
2	64	4	5
3	32	8	9
4	16	16	17
5	8	32	33
6	4	64	65
7	2	128	129

Algorithm 5. Sec1O-masking

Require: Two input shares: $(x_1 = x \oplus r, r) \in \{0,1\}^n$, output mask: $s \in \{0,1\}^m$, and an (n, m) S-box lookup function S
Ensure: Masked S-box output: $S(x) \oplus s$
1: **for** $a := 0$ to $2^n - 1$ **do**
2:    $cmp \leftarrow \mathsf{compare}(a, r)$
3:    $R_{cmp} \leftarrow (S(x_1 \oplus a) \oplus s)$
4: **end for**
5: **return** R_1

Algorithm 5 takes two input shares of x ($x_1 = x \oplus r$, r), an output mask s, the (n, m) lookup table S and computes $S(x) \oplus s$ without any first-order leakage corresponding to the sensitive variable x. For all the possible values of $a \in \{0,1\}^n$, it computes $S(x_1 \oplus a) \oplus s$ and stores it in one of the two registers R_0 and R_1 based on the result from the comparison. Namely, if $a = r$, the comparison returns true and the result is stored in R_1; otherwise it is stored in R_0. When $a = r$, the value stored at R_1 is $S(x_1 \oplus a) \oplus s = S(x) \oplus s$, which is returned at the end.

This technique clearly requires $\mathcal{O}(1)$ memory, whereas the time complexity is $\mathcal{O}(2^n)$. By applying this technique to table T_2 in Algorithm 4 for different values of l, we can obtain time-memory trade-offs for first-order secure S-box implementations. Namely, we do not store the table T_2 anymore. Instead, we apply the similar technique as in Algorithm 5 for computing the entries in table T_2 and hence require only two registers irrespective of the size l. In this case, we need to replace Step 14 of Algorithm 4 with two steps:

$$cmp \leftarrow \mathsf{compare}(i, t_1)$$
$$R_{cmp} \leftarrow T_1(ind_1) \oplus ssum$$

Then we return the value stored at R_1, which is $S(x) \oplus s$, as required. The memory requirements for table T_1 with this technique are similar to that of

Algorithm 4 (given in Table 1). However, we do not need any memory for lookup table T_2 as we make use of two registers R_0 and R_1 to achieve the same.

3   Second-Order Secure Compression Scheme

In this section we show that we can apply the compression mechanism to second-order secure implementations in [RDP08] as well. For simplicity, let us take the example of AES S-box S: $(\{0,1\}^8 \rightarrow \{0,1\}^8)$ and define functions S_0 and S_1 $(\{0,1\}^7 \rightarrow \{0,1\}^8)$ as earlier:

$$S_0(u) = S(u||0), \ S_1(y) = S(u||1) \ \forall \ u \in \{0,1\}^7$$

We are given three input shares of x: $x_1, x_2, x' = x \oplus x_1 \oplus x_2$ and we need to compute three output shares of $S(x)$. Let us divide the shares into two parts: $x = y||b$, $x_1 = y_1||b_1$, $x_2 = y_2||b_2$ and $x' = y'||b'$. Let $r_0, r_1 \in \{0,1\}^7$ and $s_1, s_2 \in \{0,1\}^8$ be random masks, and let us define the randomized table:

$$T_1(a') = S_0(y' \oplus a \oplus r_0) \oplus S_1(y' \oplus a \oplus r_1) \oplus s_1 \oplus s_2 \tag{9}$$

where $y' = y \oplus y_1 \oplus y_2$, $a' = a \oplus ((y_1 \oplus y_3) \oplus y_2)$ for every $a \in \{0,1\}^7$ and random $y_3 \in \{0,1\}^7$. When $a' = y_3 \oplus r_0$, we have:

$$a = y_3 \oplus r_0 \oplus y_1 \oplus y_3 \oplus y_2 = y_1 \oplus y_2 \oplus r_0$$
$$T_1(y_3 \oplus r_0) = S_0(y) \oplus S_1(y \oplus r_0 \oplus r_1) \oplus s_1 \oplus s_2$$
$$S_0(y) \oplus s_1 \oplus s_2 = T_1(y_3 \oplus r_0) \oplus S_1(\tilde{y})$$

where $\tilde{y} = y \oplus r_0 \oplus r_1$. Similarly when $a' = y_3 \oplus r_1$ we have:

$$a = y_3 \oplus r_1 \oplus y_1 \oplus y_3 \oplus y_2 = y_1 \oplus y_2 \oplus r_1$$
$$T_1(y_3 \oplus r_1) = S_0(y \oplus r_0 \oplus r_1) \oplus S_1(y) \oplus s_1 \oplus s_2$$
$$S_1(y) \oplus s_1 \oplus s_2 = T_1(y_3 \oplus r_1) \oplus S_0(\tilde{y})$$

Once we have masked values of $S_0(y)$ and $S_1(y)$, we can find $S(x) = S(y||b)$ using a 1-bit to 8-bit table U, which is defined as:

$$U(b) = S(y||b) = S_{(b)}(y) = T_1(y_3 \oplus r_{(b)}) \oplus S_{(b \oplus 1)}(\tilde{y}) \oplus s_1 \oplus s_2$$

If we store the table U directly, this could leak information about the bit b. Hence we use a randomized table T_2 created using the generic second-order scheme from [RDP08]. Let $x' = y'||b'$ and $b' = b \oplus b_1 \oplus b_2$. For random $b_3 \in \{0,1\}$ we define:

$$T_2(a') = S_{(b' \oplus a)}(y) \oplus s_1 \oplus s_2 = T_1(y_3 \oplus r_{(b' \oplus a)}) \oplus S_{(b' \oplus a \oplus 1)}(\tilde{y})$$

where $a' = a \oplus ((b_3 \oplus b_1) \oplus b_2)$ for $a \in \{0,1\}$. When $a' = b_3$ we have:

$$a = b_3 \oplus (b_3 \oplus b_1 \oplus b_2) = b_1 \oplus b_2$$
$$T_2(b_3) = T_1(y_3 \oplus r_{(b)}) \oplus S_{(b \oplus 1)}(\tilde{y})$$

which gives the masked value of $S(y||b)$: $S(x) \oplus s_1 \oplus s_2$.

We now give the generic compression scheme for second-order secure look-up table. Let the S-box be $S : \{0,1\}^n \to \{0,1\}^m$. Assume that we want to pack 2^l values in T_1. Hence, the size of tables T_1 and T_2 become 2^{n-l} and 2^l respectively. Let $L = 2^l, N = 2^n, M = 2^m, P = 2^{n-l}$ and let S_i ($\{0,1\}^{n-l} \to \{0,1\}^m$) be:

$$S_i(y) = S(y||i) \; \forall \; y \; \in \{0,1\}^{n-l}$$

We define the first randomized table T_1 for all $a \in \{0,1\}^{n-l}$ as:

$$T_1(a') = \left(\left(\bigoplus_{0 \leq i \leq L-1} S_i(y' \oplus a \oplus r_i) \right) \oplus s_1 \right) \oplus s_2 \tag{10}$$

where $a' = a \oplus ((y_1 \oplus y_3) \oplus y_2)$ for a random y_3. We give the algorithm to compute the table entries of T_1 in Algorithm 6.

Algorithm 6. Table T_1 creation: generic second-order compression

---

Require: Three input shares: $(y' = y \oplus y_1 \oplus y_2, y_1, y_2) \in \{0,1\}^{n-l}$, output masks: $s_1, s_2 \in \{0,1\}^m$, $r_i \in \{0,1\}^{n-l}$ for $0 \leq i \leq L-1$, an (n, m) S-box lookup function S and S_i for $0 \leq i \leq L-1$ where $S_i(y) = S(y||i)$
Ensure: Table T_1, y_3
1: $y_3 \leftarrow \mathsf{rand}(n-l)$
2: $y'' \leftarrow (y_1 \oplus y_3) \oplus y_2$
3: **for** $a := 0$ to $P - 1$ **do**
4:      $a' \leftarrow a \oplus y''$
5:      $T_1(a') \leftarrow \left(\left(\bigoplus_{0 \leq i \leq L-1} S_i(y' \oplus a \oplus r_i) \right) \oplus s_1 \right) \oplus s_2$
6: **end for**

Next we describe a method to compute table T_2. When $a' = y_3 \oplus r_0$, we have:

$$a = y_3 \oplus r_0 \oplus y_1 \oplus y_3 \oplus y_2 = y_1 \oplus y_2 \oplus r_0$$

$$T_1(y_3 \oplus r_0) = S_0(y) \oplus \left(\bigoplus_{1 \leq i \leq L-1} S_i(y \oplus r_0 \oplus r_i) \right) \oplus s_1 \oplus s_2$$

$$S_0(y) \oplus s_1 \oplus s_2 = T_1(y_3 \oplus r_0) \oplus \bigoplus_{1 \leq i \leq L-1} S_i(y \oplus r_0 \oplus r_i)$$

In general, for $0 \leq j \leq L-1$ and $a' = y_3 \oplus r_j$, we have:

$$S_j(y) \oplus s_1 \oplus s_2 = T_1(y_3 \oplus r_j) \oplus \bigoplus_{i \in \{0:L-1\}-\{j\}} S_i(y \oplus r_j \oplus r_i)$$

Algorithm 7. Table T_2 creation: generic second-order compression
***
Require: Three input shares: $(b' = b \oplus b_1 \oplus b_2, b_1, b_2) \in \{0,1\}^l$, $r_i \in \{0,1\}^{n-l}$ for $0 \leq i \leq L - 1$, an (n,m) S-box lookup function S and S_i for $0 \leq i \leq L - 1$ where $S_i(y) = S(y\|i)$, table T_1, y_3, y', y_1, y_2
Ensure: Table T_2, b_3
1: $b_3 \leftarrow \mathrm{rand}(l)$; $b'' \leftarrow (b_1 \oplus b_3) \oplus b_2$
2: **for** $a := 0$ to $L - 1$ **do**
3:     $a' \leftarrow a \oplus b''$
4:     $t_1 \leftarrow T_1(y_3 \oplus r_{(b' \oplus a)})$
5:     $T_2(a') \leftarrow t_1 \oplus \left(\bigoplus_{i \in \{\{0:L-1\}-\{a\}\}} (S_{(b' \oplus i)}(((y' \oplus r_{(b' \oplus a)} \oplus r_{(b' \oplus i)}) \oplus y_1) \oplus y_2)) \right)$
6: **end for**
***

We subsequently store each of these values in table T_2 without any second-order leakage from the sensitive variable x. We formally describe our technique to compute the entries of table T_2 in Algorithm 7. Finally, Algorithm 8 gives our generic compression scheme for second-order secure S-box computation.

Algorithm 8. Generic compression scheme for second-order secure S-box computation
***
Require: Three input shares: $(x' = x \oplus x_1 \oplus x_2, x_1, x_2) \in \{0,1\}^n$, output masks: $s_1, s_2 \in \{0,1\}^m$, an (n,m) S-box lookup function S and S_i for $0 \leq i \leq L-1$ where $S_i(y) = S(y\|i)$
Ensure: $S(x) \oplus s_1 \oplus s_2$
1: Let $y'\|b' \leftarrow x'$ where y' is of size $n - l$ bits and b' l-bits
2: **for** $i := 0$ to $L - 1$ **do**
3:     $r_i \leftarrow \mathrm{rand}\,(n - l)$
4: **end for**
5: Create table T_1 using Algorithm 6
6: Create table T_2 using Algorithm 7
7: **return** $T_2(b_3)$
***

Security Analysis. It is easy to prove the security of our generic compression scheme against second-order DPA attacks. The security of Algorithms 6 and 7 follow directly from the proofs given in [RDP08]. Let us denote the sets of intermediate variables from Algorithms 6 and 7 by I_1 and I_2 respectively. We already know that $I_1 \times I_1$ and $I_2 \times I_2$ are secure against second-order attacks. We can use the similar arguments as in [RDP08] to show that $I_1 \times I_2$ is also independent of the sensitive variables, which essentially proves the security of Algorithm 8.

Second-Order Time-Memory Trade-Offs. Rivain *et al.* also presented a technique to perform second-order secure S-box computation using only two registers and 2^n bits of memory for an (n,m) S-box (Algorithm 3 in [RDP08]).

Hence, we can obtain time-memory trade-offs for second-order secure S-box computation similar to the first-order case given in Sect. 2.3.

Initially, we proceed as in the case of generic compression scheme and create table T_1 of P entries. However, instead of creating table T_2, we work with two registers R_0 and R_1. The correct output is stored in one of these two registers, which is based on a random bit c. We iterate over all possible values of $a \in \{0,1\}^l$ as in the case of Algorithm 7. Instead of storing all the entries in table T_2, we store the value of $(S_{(b'\oplus a)}(y) \oplus s_1) \oplus s_2$ in one of the two registers based on the comparison between $(b_1 \oplus a, b_2)$. When $b_1 \oplus a = b_2$:

$$a = b_1 \oplus b_2$$
$$R_c = (S_{b'\oplus b_1 \oplus b_2}(y) \oplus s_1) \oplus s_2$$
$$= (S_b(y) \oplus s_1) \oplus s_2$$

Hence the correct value is stored in R_c and the wrong value in $R_{\bar{c}}$. Finally, we return the value stored in R_c which is $(S(y||b)\oplus s_1)\oplus s_2 = (S(x)\oplus s_1)\oplus s_2$, as required. Here, we need to ensure that the comparison function does not leak any information about x, for which we use the first-order secure compare function described in Appendix A of [RDP08][1].

4   Implementation Results

We applied both our proposed schemes (compression as well as time-memory trade-off) to AES-128 so as to obtain implementations that are secure against first and second-order DPA attacks. For simplicity, we only considered the case of using the same mask for all the 16 S-boxes in our first-order masking. However, our schemes can also be easily applied to multi-mask implementations [OS05]. We implemented these masking schemes for all possible values of compression levels l i.e., for $l = (0,1,2,3,4,5,6,7)$. We give here the performance results of our implementations on NXP-LPC1769, a 32-bit ARM Cortex-M3 based microcontroller. We only consider the classical randomized lookup table method and second-order masking scheme from Rivain et al. for comparison as rest of the schemes incur significantly higher penalty. For example, penalty factor for first and second-order secure implementations using Rivain-Prouff higher-order masking scheme [RP10] are 50 and 96 respectively (from [Cor14]).

Our implementation results for first-order masking are given in Table 2. On the left, we can see the results for AES-128 for different values of l when we apply our generic compression scheme. The columns in the table denote the compression level l, execution time in milliseconds, penalty factor compared to an unmasked implementation, and the required number of bytes in RAM respectively. On the right we show our results for time-memory trade-offs for different values of l. We see that in both the cases, for $l > 4$ the penalty factor is significantly high. This is due to the fact that the number of required randoms

---

[1] Note that this particular scheme has a flaw if the device leaks in the Hamming distance model [CGP+12].

and size of the table T_2 (or the number of comparisons in case of time-memory trade-offs) increase exponentially with the compression level l. Secondly, the penalty factor in both the cases for $l > 3$ is actually higher than that of no memory case $i.e.$, for $l = 0$ on the right side of the table. Hence, for AES-128, we can conclude that compression level greater than 3 is not useful in practice. Another interesting observation here is that the penalty factor for $l = 2$, which requires almost four times less memory compared to the classical randomized table method ($l = 0$) is around three times only.

We give the results for our implementation of second-order masking in Table 3. We can see that the results are identical to that of first-order masking. Namely, for $l > 4$ the penalty is very high and for $l > 3$ the penalty is higher than the no memory case. Note also that the penalty factor for $l = 3$ is only twice that of $l = 0$, while requiring only one fourth of the memory.

Table 2. Running time in milliseconds and penalty factor for first-order generic compression scheme (left) and time-memory trade-offs (right).

ℓ	Time	PF	Mem
0	54	1.8	256
1	81	2.7	130
2	168	5.6	68
3	380	12.6	40
4	1100	36.6	32
5	3800	126.6	40
6	13900	463.3	68
7	51900	1730	130

ℓ	Time	PF	Mem
0	1054	35.1	0
1	94	3.1	128
2	182	6.1	64
3	418	13.9	32
4	1300	43.3	16
5	4400	146.6	8
6	15800	526.6	4
7	57600	1920	2

Table 3. Running time in milliseconds and penalty factor for second-order generic compression scheme (left) and time-memory trade-offs (right).

ℓ	Time	PF	Mem
0	914	30.4	256
1	1006	33.5	130
2	1560	52	68
3	1870	62.4	40
4	3560	118.6	32
5	10340	344.6	40
6	36990	1233	68
7	141210	4707	130

ℓ	Time	PF	Mem
0	2214	73.8	32
1	1177	39.2	129
2	1730	57.6	65
3	2030	67.6	33
4	3680	122.6	18
5	10310	343.6	12
6	36140	1204.6	12
7	136730	4556.6	18

5 Conclusion

In this paper, we studied the time-memory trade-offs for implementations of block ciphers secure against first and second-order DPA attacks. We first generalized the compression scheme for lookup tables so that it works for any compression level. We then applied Rivain-Prouff countermeasure to the generic compression scheme and obtained time-memory trade-offs for first-order secure implementations. Similarly, we also obtained generic compression scheme and time-memory trade-offs for second-order secure implementations as well. We implemented AES-128 on a 32-bit ARM based microcntroller using our proposed schemes. Our results show that one can obtain relatively efficient implementations for only 40 bytes of RAM, which can be useful for highly memory constrained devices. Moreover, reasonably efficient implementations that use multi-mask method (*i.e.* 16 different tables used for 16 S-boxes) can also be obtained with 620 bytes of RAM compared to 4KB RAM required in straightforward implementations. This improvement allows one to implement multi-mask method efficiently even on microcontrollers with 8KB RAM (*e.g.* Cortex-M0).

We describe two directions for future work. Firstly, it would be interesting to perform similar analysis on higher-order lookup table method proposed by Coron [Cor14] to obtain time-memory trade-offs there as well. Secondly, it is possible to improve the implementation results for second-order masking on devices with large sized registers (for *e.g.* AES on 32-bit devices) using improved algorithm from [RDP08]. However, note that such a reduction would be applicable to all the cases and hence it will not change the relative performance of the results given here.

Acknowledgments. I would like to thank Jean-Sébastien Coron for introducing me to this problem and Debdeep Mukhopadhyay for hosting me at IIT Kharagpur, India during this work. I would also like to thank Srinivas Vivek and Sikhar Patranabis for helpful discussions.

References

[AARR03] Agrawal, D., Archambeault, B., Rao, J.R., Rohatgi, P.: The EM side—channel(s). In: Kaliski, B.S., Koç, K., Paar, C. (eds.) CHES 2002. LNCS, vol. 2523, pp. 29–45. Springer, Heidelberg (2003). doi:10.1007/3-540-36400-5_4

[CGP+12] Coron, J.-S., Giraud, C., Prouff, E., Renner, S., Rivain, M., Vadnala, P.K.: Conversion of security proofs from one leakage model to another: a new issue. In: Schindler, W., Huss, S.A. (eds.) COSADE 2012. LNCS, vol. 7275, pp. 69–81. Springer, Heidelberg (2012). doi:10.1007/978-3-642-29912-4_6

[CJRR99] Chari, S., Jutla, C.S., Rao, J.R., Rohatgi, P.: Towards sound approaches to counteract power-analysis attacks. In: Wiener, M. (ed.) CRYPTO 1999. LNCS, vol. 1666, pp. 398–412. Springer, Heidelberg (1999). doi:10.1007/3-540-48405-1_26

[Cor14] Coron, J.-S.: Higher order masking of look-up tables. In: Nguyen, P.Q., Oswald, E. (eds.) EUROCRYPT 2014. LNCS, vol. 8441, pp. 441–458. Springer, Heidelberg (2014). doi:10.1007/978-3-642-55220-5_25

[KJJ99]   Kocher, P., Jaffe, J., Jun, B.: Differential power analysis. In: Wiener, M. (ed.) CRYPTO 1999. LNCS, vol. 1666, pp. 388–397. Springer, Heidelberg (1999). doi:10.1007/3-540-48405-1_25

[Koc96]   Kocher, P.C.: Timing attacks on implementations of Diffie-Hellman, RSA, DSS, and other systems. In: Koblitz, N. (ed.) CRYPTO 1996. LNCS, vol. 1109, pp. 104–113. Springer, Heidelberg (1996). doi:10.1007/3-540-68697-5_9

[Mes00]   Messerges, T.S.: Using second-order power analysis to attack DPA resistant software. In: Koç, Ç.K., Paar, C. (eds.) CHES 2000. LNCS, vol. 1965, pp. 238–251. Springer, Heidelberg (2000). doi:10.1007/3-540-44499-8_19

[OMHT06] Oswald, E., Mangard, S., Herbst, C., Tillich, S.: Practical second-order DPA attacks for masked smart card implementations of block ciphers. In: Pointcheval, D. (ed.) CT-RSA 2006. LNCS, vol. 3860, pp. 192–207. Springer, Heidelberg (2006). doi:10.1007/11605805_13

[OS05]   Oswald, E., Schramm, K.: An efficient masking scheme for AES software implementations. In: Song, J.-S., Kwon, T., Yung, M. (eds.) WISA 2005. LNCS, vol. 3786, pp. 292–305. Springer, Heidelberg (2006). doi:10.1007/11604938_23

[PR07]   Prouff, E., Rivain, M.: A generic method for secure SBox implementation. In: Kim, S., Yung, M., Lee, H.-W. (eds.) WISA 2007. LNCS, vol. 4867, pp. 227–244. Springer, Heidelberg (2007). doi:10.1007/978-3-540-77535-5_17

[RDP08]   Rivain, M., Dottax, E., Prouff, E.: Block ciphers implementations provably secure against second order side channel analysis. In: Nyberg, K. (ed.) FSE 2008. LNCS, vol. 5086, pp. 127–143. Springer, Heidelberg (2008). doi:10.1007/978-3-540-71039-4_8

[RP10]   Rivain, M., Prouff, E.: Provably secure higher-order masking of AES. In: Mangard, S., Standaert, F.-X. (eds.) CHES 2010. LNCS, vol. 6225, pp. 413–427. Springer, Heidelberg (2010). doi:10.1007/978-3-642-15031-9_28

[RRST02] Rao, J.R., Rohatgi, P., Scherzer, H., Tinguely, S.: Partitioning attacks: or how to rapidly clone some GSM cards. In: IEEE Symposium on Security and Privacy, pp. 31–41. IEEE Computer Society (2002)

Hiding Higher-Order Side-Channel Leakage

Randomizing Cryptographic Implementations in Reconfigurable Hardware

Pascal Sasdrich[1(✉)], Amir Moradi[1], and Tim Güneysu[2]

[1] Horst Görtz Institute for IT-Security, Ruhr-Universität Bochum,
Bochum, Germany
{pascal.sasdrich,amir.moradi}@rub.de
[2] University of Bremen & DFKI, Bremen, Germany
tim.gueneysu@uni-bremen.de

Abstract. First-order secure Threshold Implementations (TI) of symmetric cryptosystems provide provable security at a moderate overhead; yet attacks using higher-order statistical moments are still feasible. Cryptographic instances compliant to Higher-Order Threshold Implementation (HO-TI) can prevent such attacks, however, usually at unacceptable implementation costs. As an alternative concept we investigate in this work the idea of *dynamic hardware modification*, i.e., random changes and transformations of cryptographic implementations in order to render higher-order attacks on first-order TI impractical. In a first step, we present a generic methodology which can be applied to (almost) every cryptographic implementation. In order to investigate the effectiveness of our proposed strategy, we use an instantiation of our methodology that adapts ideas from White-Box Cryptography and applies this construction to a first-order secure TI. Further, we show that dynamically updating cryptographic implementations during operation provides the ability to avoid higher-order leakages to be practically exploitable.

1 Introduction

Side-channel analysis (SCA) uses information leakage by measuring physical device internals, e.g., timing [9], power consumption [10] or electromagnetic emanations [2], to extract cryptographic secrets. Modern side-channel countermeasures are classified either as *hiding* or *masking* [11]. While *hiding* countermeasures aim to decrease the Signal-to-Noise ratio (SNR) in order to hide information leakage in random noise, *masking* countermeasures tackle information leakage using secret sharing and multi-party computation techniques. The idea of Threshold implementation (TI) has been developed based on Boolean masking in particular to target hardware implementations [15]. However, the initial concept of TI was only suitable to counteract first-order side-channel leakages, still allowing attacks using higher-order statistical moments to successfully recover cryptographic secrets. Naturally, higher-order Threshold Implementations (HO-TI) have been proposed to solve this problem [4]. Despite, HO-TI

© Springer International Publishing AG 2017
H. Handschuh (Ed.): CT-RSA 2017, LNCS 10159, pp. 131–146, 2017.
DOI: 10.1007/978-3-319-52153-4_8

might be limited to uni-variate scenarios [18] as well as they come with increased time overhead and area demands due to the ever increasing number of minimum shares for higher-order protection. Therefore, combining first-order secure TI with *hiding* countermeasures to achieve (practical) higher-order protection might be an alternative solution.

Previous Work: Although the threat of side-channel attacks is well known, many cryptographic devices are vulnerable to side-channel analysis due to their static design and behavior which allows attacks based on statistical and differential analysis. Introducing dynamic behavior in terms of ever-changing and morphing implementations and circuits could help to overcome these problems. However, this is not a trivial task and poses big challenges to designers of cryptographic implementations in particular using static hardware devices. In recent years, several research into this direction has been performed and published but still existing solutions are at an early stage and have to face many difficulties. In 2008, Mentens et *al.* [12] introduced a first work for using dynamic reconfiguration of modern FPGAs as countermeasure against power and fault attacks. However, their solution had to struggle with slow reconfiguration times as well as too small complexity which still allowed efficient analysis and attacks. Besides, their solution specifically targeted reconfigurable hardware which provides dynamic reconfiguration features. Moradi and Mischke [13] examined the opportunities of using dual ciphers as alternative representations (in particular for AES) in order to achieve protection against side-channel attacks. Though, dual ciphers maintain structural properties of the original representation which again could be exploited using statistical analysis. Recently, Sasdrich et *al.* [19] proposed the application of affine equivalence representations of cryptographic S-boxes to change the cipher implementation dynamically during runtime. Although the complexity of this approach is quite high, it exploits very specific properties of the cryptographic components, so that this approach cannot be generically applied to cryptographic implementations.

Contribution: Our contribution in this work is twofold: First, we present a generic approach to change the representations of cryptographic implementations dynamically in order to introduce non-static behavior. Our methodology can be applied to (almost) every cryptographic implementation and circuit independent of the cryptographic algorithm or scheme. Our approach uses random substitution of basic elements along with random encoding of intermediate connections and offers high flexibility and scalability of attack complexities depending of the used level of abstraction and granularity of the underlying circuit. Second, we investigate and analyze a specific instantiation of our approach to randomize a TI. In particular, we are going to examine a first-order PRESENT TI as a case study which is implemented in reconfigurable hardware. The randomization of intermediate signals, in terms of random non-linear 4-bit encodings, is chosen dynamically during runtime and injected into each implemented look-up table in order to substitute them by different representations. In particular, this approach adapts ideas and techniques from the area of White-Box Cryptography (WBC), although we want to emphasize that we do not aim to achieve resistance against attacks in the white-box adversary model. Eventually, we conduct practical side-channel measurements for our case study. Using a leakage assessment methodology, we focus on effects of our countermeasure on higher-order statistical properties and moments and show that our approach can increase the protection against higher-order side-channel attacks from a practical point of view.

Outline: The remainder of this article is organized as follows: Sect. 2 summarizes and provides important background information on directed graphs, Threshold Implementations and White-Box Cryptography. In Sect. 3 we present a description of our generic approach to dynamically update and randomize cryptographic implementations which is applied in a case study in Sect. 4 using a specific instantiation based on a PRESENT TI and 4-bit non-linear encodings (as proposed for WBC). Section 5 provides side-channel evaluation results using power measurements and state-of-the-art leakage assessment methodologies. Eventually, our work concludes in Sect. 6.

2   Background

This section briefly introduces the theory of directed graphs before we recapitulate the background of Threshold Implementations (TI) and White-Box Cryptography (WBC).

2.1   Notations

We denote single-bit random variables using lower-case characters, bold ones for multi-bit vectors, bars for shared representations, lowering indices for elements within a vector and raising indices for elements of a shared vector.

Furthermore, let us denote any element $x \in \mathbb{GF}(2^m)$ as vector of m single bit elements $\langle x_1, \ldots, x_m \rangle$. The shared representation $\bar{x}$ of a vector x using Boolean masking with s shares is given as $\bar{x} = (x^1, \ldots, x^s)$, where

$$x = \bigoplus_{i=1}^{s} \bar{x} = \bigoplus_{i=1}^{s} x^i = \bigoplus_{i=1}^{s} \langle x_1^i, \ldots, x_m^i \rangle.$$

Eventually, we denote functions using sans serif fonts and sets using calligraphic ones.

2.2   Directed Graphs

Directed Graphs (or digraphs) are use for many applications in order to abstractly model a certain problem and find according solutions. In general, a graph is a set of nodes that are connected by some edges. For a directed graph, the edge are provided with a certain direction.

Definition 1. *A **directed graph** or **digraph** is an ordered pair of sets $\mathcal{G} = (\mathcal{V}, \mathcal{A})$ where $\mathcal{V}$ is a set of vertices and $\mathcal{A}$ is a set of ordered pairs $a_{ij} = \langle v_i, v_j \rangle$ (called arrows or directed edges) with $v_i, v_j \in \mathcal{V}$.*

In particular, each vertex has a certain number of connected edges. Due to the direction of the edges, we can distinguish between edges that arrive at a vertex and edges that leave a vertex. The number of arriving edges is given by the in-degree of a node, whereas the number of leaving edges is given by the out-degree.

Definition 2. *In a directed graph, the **in-degree** $deg^+(v)$ and the **out-degree** $deg^-(v)$ of a vertex $v \in \mathcal{V}$ count the number of directed edges connecting to and from a vertex respectively. It holds, that $\sum_{v \in \mathcal{V}} deg^+(v) = \sum_{v \in \mathcal{V}} deg^-(v) = |\mathcal{A}|$.*

Eventually, every node (connected to a digraph) has to have at least one arriving or leaving edge. In case the node has an in-degree of zero, it is called source, since it only serves as starting point for several edges. Similarly, a node without any leaving edges is called sink, since it is only an ending point for some edges. In the following, we consider the source nodes as starting points of our directed graph, whereas the sink will be the final points.

2.3 Threshold Implementation

Threshold Implementation (TI) is a widely used technique to protect hardware devices against physical attacks. In particular, TI is based on Boolean masking and multi-party computation and provides provable security, even in the presence of glitches[1]. In general, any Threshold Implementation has to provide the following three properties:

Correctness: Given a vector $\bar{x} = (x^i, \ldots, x^s)$ in its shared representation, we can compute any function $\mathsf{F}(\bar{x}) = \bar{y}$ on it but have to ensure *correctness*, i.e., the result $\bar{y} = (y^i, \ldots, y^t)$ has to be shared representation of $y = \mathsf{F}(x)$ with $t \geq s$. But for this purpose, we can use according component functions f^i to evaluate F for each share individually. However, finding *correct* component functions is not trivial, in particular if F is a non-linear function [3]. In addition, each component function has to ensure further properties such as *non-completeness* and *uniformity*.

Non-completeness: As mentioned before, each resulting share $(y^i, \ldots, y^t)$ is given by an individual evaluation of a component function $\mathsf{f}^{i \in \{1, \ldots, t\}}(\cdot)$ over the input shares. However, in order to achieve security in sense of first-order statistical moments, each component function has to provide *non-completeness*. This means that each component function $\mathsf{f}^{i \in \{1, \ldots, t\}}(\cdot)$ must be *non-complete*, i.e., independent of at least one input share.

---

[1] For a more detailed description, please refer to the original articles [4,15,16].

Uniformity: The security of Threshold Implementations as masking schemes is based on the *uniform* distribution of the mask respectively the shared representation which serve as input for a function evaluation. However, since results of a function, e.g., an S-box, are used as input to another function, the outputs of the functions again have to be *uniformly* distributed. This means, given a set of all possible input sharings $\mathcal{X} = \{\bar{x} | \bigoplus_{i=1}^{s} \bar{x} = x\}$ the set of all possible output sharings, i.e., $\{(f^1(\cdot), \ldots, f^t(\cdot) | \bar{x} \in \mathcal{X}\}$ should be drawn *uniformly* from the set of $\mathcal{Y} = \{\bar{y} | \bigoplus_{i=1}^{t} \bar{y} = y\}$ as all possible sharings of $y = F(x)$.

2.4 White-Box Cryptography

The concept of *White-Box Cryptography* is concerned with the protection of implementations of cryptographic algorithms in the presence of white-box adversaries that have virtually unlimited capabilities and access to an implementation as well as full control of the execution environment. Implementations are assumes secure against white-box adversaries if they provide an adversary with not more information than given by a black-box access, in other words, the white-box implementation should behave as virtual black box.

As a matter of fact, an ideal white-box implementation of a cryptographic algorithm would consist of a single look-up table which maps every possible plaintext to an according ciphertext (for a given and fixed secret key). However, for modern ciphers that provide security levels and key sizes of 128 bits and more, this approach is obviously impractical. Consequently, alternative approaches which can be realized in practice are necessary. In 2002, Chow et al. proposed practical white-box implementations for DES [6] and AES [7] using divide-and-conquer strategies to build white-box implementations using *networks of randomized look-up tables*.

In general, the proposed strategy can be applied for any key-alternating, round-based, symmetric block cipher E_K to derive its white-box implementation E'_K and it can be described as:

$$E'_K = \underbrace{(f^{r+1})^{-1} \circ E^r_{k_r} \circ f^r}_{table(s)} \circ \cdots \circ \underbrace{(f^3)^{-1} \circ E^2_{k_2} \circ f^2}_{table(s)} \circ \underbrace{(f^2)^{-1} \circ E^1_{k_1} \circ f^1}_{table(s)}$$

$$= (f^{r+1})^{-1} \circ E^r_{k_r} \circ \cdots \circ E^2_{k_2} \circ E^1_{k_1} \circ f^1 = (f^{r+1})^{-1} \circ E_K \circ f^1,$$

In this context, $E^{i \in \{1, \ldots, r\}}$ is a single round of E_K, and $f^{i \in \{1, \ldots, (r+1)\}}$ are encoding functions which are chosen randomly in order to randomize and hide any key material inside the look-up tables. Besides, in order to ensure full protection of the first and last round of a block cipher and to prevent so called *Code Lifting* attacks [8], white-box implementations usually use external encodings (here given as f^1 respectively $(f^{r+1})^{-1}$).

3   Methodology

In this section, we introduce our methodology to dynamically update and randomize cryptographic implementations using a generic approach. We first state some important observations that directly lead to a generic representation of the problem. This is followed by an algorithmic solution to achieve dynamic updates of cryptographic implementations.

3.1   Generic Approach

In general, our generic approach can be applied to any cryptographic implementation. However, the provided physical platform has to allow some changes of the implementation during runtime. Since we want to focus on hardware implementations throughout this work, we particularly target reconfigurable hardware in terms of Field-Programmable Gate Arrays (FPGA). Eventually, we present a solution that achieves on-the-fly dynamic randomization of cryptographic implementations.

> **Observation 1.** *Any cryptographic implementation can be represented as network or sequence of modular or atomic functions subsequently applied on an internal state.*

Consequently, we can model any cryptographic implementation as a directed graph. Depending on the level of abstraction and the desired granularity (e.g., system level, gate level, etc.), each node of the graph represents a single or multiple modular and atomic functions of the algorithm. Besides, the edges which connecting the nodes in a certain direction represent the data flow of the internal state.

> **Observation 2.** *Any cryptographic implementation can be modeled by different but equivalent directed graphs.*

In general, the numbers of nodes and edges required to model a cryptographic implementation is not determined and particularly not limited by an upper bound. Principally, we can add new nodes and edges arbitrarily to the graph to find new representations (with sufficient complexity). However, we still have to maintain and ensure correctness of the overall implementation.

3.2   Morphing Algorithm for Cryptographic Implementations

Based on this observations, we developed a generic algorithm to morph a digraph of a cryptographic implementation into an equivalent but encoded digraph while still maintaining correctness of the implementation.

According to Algorithm 1, each arrow $\langle v_i, v_j \rangle$ of a digraph is replaced by an encoded directed edge. For this purpose, both adjacent vertices have to be replaced as well. The starting vertex v_i is replace such that it not only performs

---

Algorithm 1. Morphing algorithm for cryptographic implementations

---

Input : $\mathcal{G} = (\mathcal{V}, \mathcal{A})$: digraph representing a cryptographic implementation.
Output: $\mathcal{G}^* = (\mathcal{V}^*, \mathcal{A}^*)$: digraph representing an encoded cryptographic
          implementation.

$\mathcal{G}^* = (\mathcal{V}^*, \mathcal{A}^*)$: $\mathcal{V}^* \leftarrow \mathcal{V}$, $\mathcal{A}^* \leftarrow \mathcal{A}$

for $\forall v_i \in \mathcal{V}^*$ **do**

   $\mathcal{D} \leftarrow \emptyset$

   $s \leftarrow \mathsf{f}(v_i)$, $\mathcal{V}^* \leftarrow \mathcal{V}^* \setminus \{v_i\}$

   for $\forall v_j \in \mathcal{V}^*$ **do**

      if $a_{ij} \in \mathcal{A}^*$ **then**

         $\mathcal{D} \leftarrow \mathcal{D} \cup \mathsf{f}^{-1}(v_j)$

         $\mathcal{V}^* \leftarrow \mathcal{V}^* \setminus \{v_j\}$, $\mathcal{A}^* \leftarrow \mathcal{A}^* \setminus \{a_{ij}\}$,

      end

   end

   for $\forall d_i \in \mathcal{D}$ **do**

      $\mathcal{V}^* \leftarrow \mathcal{V}^* \cup \{s, d_i\}$, $\mathcal{A}^* \leftarrow \mathcal{A}^* \cup \langle s, d_i \rangle$,

   end

end

return $\mathcal{G}^*$

---

its originally provided function but in addition performs an encoding function f to the state. In order to maintain correctness of the implementation, the ending vertex v_j has to cancel the applied encoding using the inverse (decoding) function f^{-1} before performing its original function to the state.

3.3   Applicable Encoding Functions

In this section, we will briefly discuss properties and requirements on encoding functions that are applicable within our algorithm. First of all, the encoding function should be a randomly drawn function in order to perform a randomization of the implementation during the update. However, each encoding function has to fulfill a few minimal requirements and has to provide some properties to be compatible with our methodology. Obviously, the encoding function has to be injective, i.e., it has to be information preserving in order to allow a correct operation of the original implementation. Apart from that, input and output sizes of the encoding functions will depend on the desired granularity of the algorithm and can differ as long as the output size is at least the input size. Besides, the chosen encoding function can have any complexity (but still should be reasonable efficient). Possible realizations of encoding functions could be: linear functions [23], non-linear bijections (like S-boxes) [7], or any other instance which meets the requirements.

3.4 Verification and Semantic Equivalence Checking

Since our methodology should not affect the correctness of the final result of the original implementation, we have to ensure semantic equivalence of the randomized implementations. Therefore, our approach has to include checking and verification steps. As mentioned before, the randomly drawn encoding functions have to meet minimal requirements which has to be checked and verified continuously during the operation. Correctness of the final result, i.e., semantic equivalence of the randomized implementation, is ensured by only encoding single edges (or small paths[2]) and including the inverse decoding function at the same time.

4   Case Study: PRESENT Threshold Implementation

Throughout this section, we present a practical realization of our proposed countermeasure using an encoded PRESENT TI as case study. Before investigating the feasibility of our approach in terms of hiding higher-order side-channel leakage, we give a detailed description of our practical architecture realized on a modern Xilinx FPGA and elaborate our design strategy.

4.1   Adversary Model

Although our practical instantiation employes certain ideas and concepts of White-Box Cryptography in terms of using encoded look-up tables to hide secret key material, we want to emphasize that we still do not consider adversaries of the white-box model. It is obvious that every adversary who has full access and control of the execution environment can circumvent our proposed countermeasures in order to extract secret keys from the implementation using more powerful attacks, e.g., an algebraic analysis of the look-up tables. However, we therefore only consider adversaries of the gray-box model, i.e., adversaries that still can access the implementation but can only gain helpful information through side-channel leakage.

4.2   Design Considerations

PRESENT [5] is a lightweight symmetric block cipher based on a block size of 64 bits. In particular, it is a Substitution-Permutation network (SPN) with 31 rounds. It provides two different key sizes (80 bit or 128 bit) and derives 32 different 64-bit round-keys based on the initial key. Since nowadays, it is advised against using 80-bit keys, we opted to implement and focus on PRESENT-128.

---

[2] Given for instance a linear operation within a cryptographic implementation (e.g., MixColumns of the AES algorithm) and the application of linear encoding functions would allow to keep encoded intermediate values. However, the decoding function then has to consider the inversion of the linear operation as well.

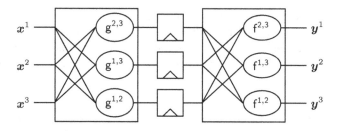

Fig. 1. First-order TI of the PRESENT S-box

Threshold Implementation of PRESENT: Our implementation is based on the first TI that was presented in [17]. In particular, we apply the decomposition of the S-box into two quadratic functions g and f that was proposed by Poschmann et al. in order to benefit from the minimal number of shares (i.e., $m = n = 3$). Since the permutation of the PRESENT cipher is a linear function, it can be applied to each share individually and without modification. Due to the decomposition of the S-box, additional register stages have to be placed in between g and f in order to prevent side-channel leakage caused by glitches. The final structure of the first-order TI of the PRESENT S-box is shown in Fig. 1.

Encoding the TI: Before instantiating and implementing our proposed algorithm taking the example of a first-order secure PRESENT TI, we have to find an architecture which supports dynamic updates of sub-functions or components and can be implemented on an FPGA. Given the basic structure of a TI of the PRESENT S-box as shown in Fig. 1 we chose the component functions as basic building blocks that have to be updated on-the-fly. Besides, we opted to implement each function as look-up tables because it is a natural choice for FPGAs but also allows fast updates.

Starting from this, the PRESENT TI can be implemented as network of look-up tables, each operating on 4-bit nibbles of the internal state. In a next step, the output of each look-up table is encoded using a non-linear 4-bit bijection. In order to maintain correctness, all subsequent look-up tables have to apply the according decoding function before being evaluated, i.e., the original table has to be combined with the according inverse bijection. In general, this approach reflects basic ideas and concepts of White-Box Cryptography as initially proposed by Chow et al. in [6,7].

However, this strategy has some important implications that effect the final hardware architecture. First, the secret key has to be known during design time since it is included within the look-up tables. Hence, the (shared) key is fixed and combined with the look-up tables of the first layer of the TI S-box. Second, since the permutation layer is a linear functions which operates on single bits, we cannot perform the permutation on 4-bit encoded values. Instead, we have to implement the permutation layer as sequence of look-up tables that decode and re-encode the nibbles while performing the original permutation.

Eventually, our encoded TI is implemented using different look-up tables for each sub-function and all rounds. However, this complicates the task of implementing our design efficiently using an round-based approach. None the less, modern FPGAs provide useful features that allow an efficient implementation (as presented in Sect. 4.3).

Dynamic Update of Encodings: So far, our TI is encoded statically using arbitrary non-linear functions applied during design time. However, in order to perform dynamic randomization during operation time, we want to modify these initial encodings. Therefore, in general, we have to find solutions for the following two issues:

1. How to find or compute random non-linear functions on-the-fly?
2. How to inject random non-linear functions into our hardware implementation during runtime?

Random 4-bit non-linear functions, i.e., a random permutations of the sequence $\{0, 1, \ldots, 15\}$ can be generated using a linear-time algorithm using swapping operations and sampling uniform random numbers [22]. Although the permutation generation is slightly biased, this effect can be neglected in the context of side-channel analysis.

Since our encoded TI is implemented as network of look-up tables, injecting random non-linear functions can be realized as table re-computation and re-ordering. In particular, we can apply arbitrary functions to the output of a table by replacing each table entry by the according encoded value. The decoding function can be applied to the input of a table by re-odering the table entries according to the decoded address value. Fortunately, this procedure is independent of the previous injection of random functions, i.e., if we first apply a random function n_1 follow by a second function n_2 this is the same as applying another function n_3 with $n_3 = n_2 \circ n_1$. Hence, we can continually update our implementation using random non-linear functions without increasing the size of our implementation by just performing table re-computations and re-orderings.

Eventually, for the given PRESENT implementation, we have to update 5904 4-bit encodings per encryption in order to perform a full *dynamic hardware modification* process. Since there are 16! different 4-bit encodings, the final randomization complexity of our methodology (for the given case study) is about 2^{56}.

4.3 Practical Implementation on Reconfigurable Hardware

The deliberate application of modern reconfigurable hardware in terms of a Xilinx Kintex-7 FPGA provides several interesting advantages and allows a practical evaluation and implementation in order to confirm the feasibility of the proposed approach. In particular, the selected Kintex-7 XC7K160T FPGA implements roughly 12 Mb of block RAM (BRAM) in the form of 325 individual memory instances, each providing 32-Kb of general purpose memory as well as a true dual-port feature. Note that the dual-port option is of particular importance for the dynamic update of our implementation since we can use one port solely to perform the cryptographic operations whereas the second port is used to perform the dynamic table re-ordering and re-computation.

Besides, since all look-up tables of our architecture are 256×4-bit tables, each BRAM primitive could store up to 32 different look-up tables. Fortunately, PRESENT has only 31 different rounds, so we can arrange tables of the same operation but different rounds in the same BRAM instance. This strategy yields a round-based hardware architecture as presented in Fig. 2. Moreover, since each BRAM still provides enough memory to store another table we can use this free table entry to store an updated table. Hence, after performing the table re-ordering and re-computation and storing the updated table in the free segment, a context switch is performed, i.e., the storage of the old table is released and the updated table is applied during operation. But since the update is performed through the second port while the first port is continuously used for operation, our strategy does not affect the overall performance.

Table 1 provides the implementation numbers of our design, including control logic and a reconfiguration unit that generates new random 4-bit encodings on-the-fly. Obviously, a lion's share of the used resources is necessary to implement the encoding generation. Basically, the round function can be implemented in 192 BRAM instances – the remaining logic in terms of LUTs is necessary to control and operate the table update using the second port of the BRAM. Eventually, the control logic implements a small finite state machine (FSM) that controls both the round function and the reconfiguration engine and provides an interface for external access and control.

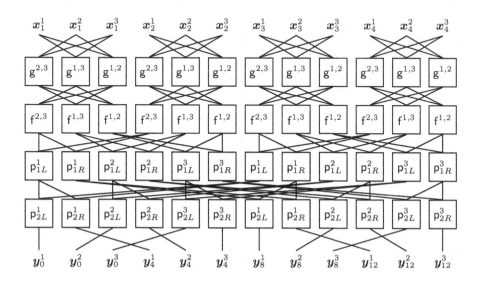

Fig. 2. Quarter round of encoded PRESENT TI

Table 1. Area consumption of our hardware architecture

Module/component	Resource utilization				
	Logic	Memory			Area
	(LUT)	(FF)	(LUTRAM)	(BRAM)	(Slices)
Control logic	11	24	0	0	13
Round function	96	0	0	192	87
g-Layer	0	0	0	48	0
f-Layer	0	0	0	48	0
p_1-Layer	0	0	0	48	0
p_2-Layer	0	0	0	48	0
Reconfiguration	3129	3222	1952	0	2373
Context engine	22	44	32	0	18
Encoding engine	2880	2880	1920	0	2258
Randomness generator	136	256	0	0	40
Total	3236	3246	1952	192	2473

4.4 Comparison

In Table 2, we provide a comparison of different approaches to achieve higher-order side-channel resistance (except for the *1st-order TI*) by the example of PRESENT. Obviously, our approach offers competitive results, both in terms of performance and area utilization, although it has an increased demand for BRAM instances. Still, the security of our proposed countermeasure may not only be limited to second-order attacks but it may also affect higher-order leakages, hence providing better security than a *2n-order TI* (at least from a practical point of view).

Table 2. Comparison of different PRESENT Hardware Architectures

Scheme/ implementation	Logic (LUT)	Memory (FF)	(LUTRAM)	(BRAM)	Latency (cycles)	Freq. (MHz)	Throughput (MBit/s)	Ref.
1st-order TI	808	384	-	-	64	207	413	[14]
2nd-order TI	2245	1680	-	-	128	204	406	[14]
Affine equivalence	1834	742	-	1	64	112	224	[19]
Glitch-free duplication	5442	12672	-	-	704	459	458	[14]
Dynamic hardware mod.	3236	3246	1952	192	124	153	315	**New**

5 Practical Side-Channel Evaluation

We evaluated side-channel information of our design implemented on a physical device using a SAKURA-X FPGA platform [1] which provides a Xilinx Kintex-7 XC7K160T FPGA for practical side-channel evaluations using the power consumption of the device. Measuring the voltage drop over a 1Ω resistor in the V_{dd} path of the FPGA using a digital oscilloscope with a sampling rate of 500 MS/s, 20 MHz bandwidth limitation, and a stable, jitter-free clock frequency of 24 MHz, we could practically examine vulnerabilities of our proposed design.

5.1 Non-specific t-Test

A common technique to investigate the resistance and vulnerabilities of physical cryptographic implementations against side-channel attacks is the *Test Vector Leakage Assessment* (TVLA) methodology. The evaluation is based on Welsh's (two-tailed) t-test, sometimes also referred to as *fix vs. random* or *non-specific* t-test, and can be extended naturally to higher-order statistical moments [20,21].

5.2 Results

In this section we provide practical evaluation results using the non-specific t-test on the first, second and third statistical order. Besides, we include the evolution of the absolute maximum of the t-test over the number of used traces. In total, we performed measurements and evaluations for two different evaluation profiles: first, reference measurements without sharing (i.e., all-zero masks) and omitted dynamic update, and second, measurements using shared values and including our proposed countermeasure in terms of dynamically updating and randomizing the implementation.

Profile 1: Before evaluating the feasibility and effectiveness of our proposed approach, we have to ensure the correctness of our implemented first-order TI using reference measurements. In order to provide such a reference, we measured one million power traces while the PRNG that generates the random masks for sharing and random encodings was disabled, i.e., all masks were set to zero and the dynamic update was omitted. We expect to detect and observer leakage on all considered statistical orders which is confirmed by our evaluation results shown in Fig. 3. One the left-hand side, we provide the results of the non-specific t-test for the first, second and third order after measuring and evaluating the total number of one million traces while on the right-hand side, the development of the absolute maximum for the t-test on each statistical order over the number of evaluated traces is shown.

Profile 2: Eventually, we extend the previous measurement profile by applying our proposed approach in order to hide higher-order side-channel leakages by

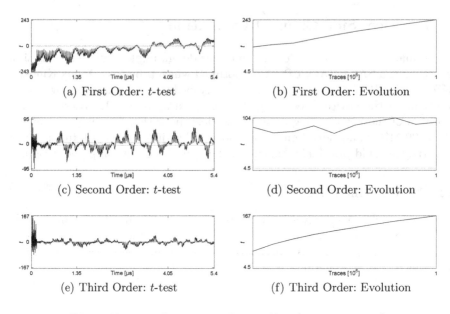

Fig. 3. Non-specific t-test results: profile 1 (1 000 000 traces)

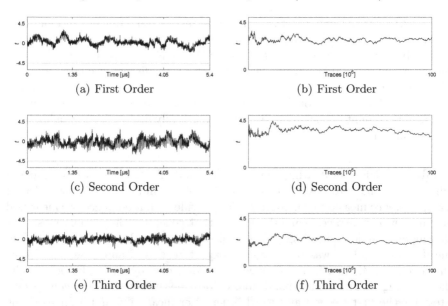

Fig. 4. Non-specific t-test results: profile 2 (100 000 000 traces)

continuously performing dynamic updates of the look-up tables of our implementation. Again, we do not expect to detect any first-order leakage due to the application of a first-order TI but moreover the leakage detectable at higher statistical orders should be prevented as well. The evaluation results shown in

Fig. 4 confirm the correctness of these assumptions since we could not detect any leakage after measuring 100 million power traces – neither at the first, second nor third statistical order – which hence also confirms the effectiveness of our proposed approach.

6 Conclusion

In this work we have presented a generic strategy and methodology in order to apply dynamic and random updates to cryptographic implementations and circuits in order to hide higher-order side-channel leakages. Using a case study based on a first-order PRESENT TI and a random updates based on non-linear encodings, we have shown the feasibility and practicability of proposed concept using side-channel power measurements and applying the state-of-the-art leakage assessment methodologies. Eventually, we can conclude that our methodology presents a viable alternative to building higher-order Threshold Implementations and convinces by its generality and scalability.

Acknowledgment. This work has been co-funded by the Commission of the European Communities through the Horizon 2020 program under project number 645622 PQCRYPTO.

References

1. Side-Channel AttacK User Reference Architecture. http://satoh.cs.uec.ac.jp/SAKURA/index.html
2. Agrawal, D., Archambeault, B., Rao, J.R., Rohatgi, P.: The EM side—channel(s). In: Kaliski, B.S., Koç, K., Paar, C. (eds.) CHES 2002. LNCS, vol. 2523, pp. 29–45. Springer, Heidelberg (2003). doi:10.1007/3-540-36400-5_4
3. Beyne, T., Bilgin, B.: Uniform First-Order Threshold Implementations. Cryptology ePrint Archive, Report 2016/715 (2016). http://eprint.iacr.org/2016/715
4. Bilgin, B., Gierlichs, B., Nikova, S., Nikov, V., Rijmen, V.: Higher-order threshold implementations. In: Sarkar, P., Iwata, T. (eds.) ASIACRYPT 2014. LNCS, vol. 8874, pp. 326–343. Springer, Heidelberg (2014). doi:10.1007/978-3-662-45608-8_18
5. Bogdanov, A., Knudsen, L.R., Leander, G., Paar, C., Poschmann, A., Robshaw, M.J.B., Seurin, Y., Vikkelsoe, C.: PRESENT: an ultra-lightweight block cipher. In: Paillier, P., Verbauwhede, I. (eds.) CHES 2007. LNCS, vol. 4727, pp. 450–466. Springer, Heidelberg (2007). doi:10.1007/978-3-540-74735-2_31
6. Chow, S., Eisen, P., Johnson, H., Oorschot, P.C.: A white-box DES implementation for DRM applications. In: Feigenbaum, J. (ed.) DRM 2002. LNCS, vol. 2696, pp. 1–15. Springer, Heidelberg (2003). doi:10.1007/978-3-540-44993-5_1
7. Chow, S., Eisen, P., Johnson, H., Oorschot, P.C.: White-box cryptography and an AES implementation. In: Nyberg, K., Heys, H. (eds.) SAC 2002. LNCS, vol. 2595, pp. 250–270. Springer, Heidelberg (2003). doi:10.1007/3-540-36492-7_17
8. Delerablée, C., Lepoint, T., Paillier, P., Rivain, M.: White-box security notions for symmetric encryption schemes. In: Lange, T., Lauter, K., Lisoněk, P. (eds.) SAC 2013. LNCS, vol. 8282, pp. 247–264. Springer, Heidelberg (2014). doi:10.1007/978-3-662-43414-7_13

9. Kocher, P.C.: Timing attacks on implementations of Diffie-Hellman, RSA, DSS, and other systems. In: Koblitz, N. (ed.) CRYPTO 1996. LNCS, vol. 1109, pp. 104–113. Springer, Heidelberg (1996). doi:10.1007/3-540-68697-5_9

10. Kocher, P., Jaffe, J., Jun, B.: Differential power analysis. In: Wiener, M. (ed.) CRYPTO 1999. LNCS, vol. 1666, pp. 388–397. Springer, Heidelberg (1999). doi:10.1007/3-540-48405-1_25

11. Mangard, S., Oswald, E., Popp, T.: Power Analysis Attacks - Revealing the Secrets of Smart Cards. Springer, Heidelberg (2007)

12. Mentens, N., Gierlichs, B., Verbauwhede, I.: Power and fault analysis resistance in hardware through dynamic reconfiguration. In: Oswald, E., Rohatgi, P. (eds.) CHES 2008. LNCS, vol. 5154, pp. 346–362. Springer, Heidelberg (2008). doi:10.1007/978-3-540-85053-3_22

13. Moradi, A., Mischke, O.: Comprehensive evaluation of AES dual ciphers as a side-channel countermeasure. In: Qing, S., Zhou, J., Liu, D. (eds.) ICICS 2013. LNCS, vol. 8233, pp. 245–258. Springer, Heidelberg (2013). doi:10.1007/978-3-319-02726-5_18

14. Moradi, A., Wild, A.: Assessment of hiding the higher-order leakages in hardware. In: Güneysu, T., Handschuh, H. (eds.) CHES 2015. LNCS, vol. 9293, pp. 453–474. Springer, Heidelberg (2015). doi:10.1007/978-3-662-48324-4_23

15. Nikova, S., Rechberger, C., Rijmen, V.: Threshold implementations against side-channel attacks and glitches. In: Ning, P., Qing, S., Li, N. (eds.) ICICS 2006. LNCS, vol. 4307, pp. 529–545. Springer, Heidelberg (2006). doi:10.1007/11935308_38

16. Nikova, S., Rijmen, V., Schläffer, M.: Secure hardware implementation of nonlinear functions in the presence of glitches. J. Cryptol. 24(2), 292–321 (2011)

17. Poschmann, A., Moradi, A., Khoo, K., Lim, C., Wang, H., Ling, S.: Side-channel resistant crypto for less than 2,300 GE. J. Cryptol. 24(2), 322–345 (2011)

18. Reparaz, O., Bilgin, B., Nikova, S., Gierlichs, B., Verbauwhede, I.: Consolidating masking schemes. In: Gennaro, R., Robshaw, M. (eds.) CRYPTO 2015. LNCS, vol. 9215, pp. 764–783. Springer, Heidelberg (2015). doi:10.1007/978-3-662-47989-6_37

19. Sasdrich, P., Moradi, A., Güneysu, T.: Affine equivalence and its application to tightening threshold implementations. In: Dunkelman, O., Keliher, L. (eds.) SAC 2015. LNCS, vol. 9566, pp. 263–276. Springer, Heidelberg (2016). doi:10.1007/978-3-319-31301-6_16

20. Schneider, T., Moradi, A.: Leakage assessment methodology. In: Güneysu, T., Handschuh, H. (eds.) CHES 2015. LNCS, vol. 9293, pp. 495–513. Springer, Heidelberg (2015). doi:10.1007/978-3-662-48324-4_25

21. Schneider, T., Moradi, A.: Leakage assessment methodology - extended version. J. Cryptogr. Eng. 6(2), 85–99 (2016)

22. Veyrat-Charvillon, N., Medwed, M., Kerckhof, S., Standaert, F.-X.: Shuffling against side-channel attacks: a comprehensive study with cautionary note. In: Wang, X., Sako, K. (eds.) ASIACRYPT 2012. LNCS, vol. 7658, pp. 740–757. Springer, Heidelberg (2012). doi:10.1007/978-3-642-34961-4_44

23. Xiao, Y., Lai, X.: A secure implementation of white-box AES. In: 2nd International Conference on Computer Science and its Applications (2009)

Digital Signatures
and Random Numbers

Surnaming Schemes, Fast Verification, and Applications to SGX Technology

Dan Boneh[1] and Shay Gueron[2,3(✉)]

[1] Department of Computer Science, Stanford University, Stanford, USA
dabo@cs.stanford.edu
[2] Department of Mathematics, University of Haifa, Haifa, Israel
[3] Intel Development Center, Intel Corporation, Haifa, Israel
shay@math.haifa.ac.il

Abstract. We introduce a new cryptographic primitive that we call *surnaming*, which is closely related to digital signatures, but has different syntax and security requirements. While surnaming can be constructed from a digital signature, we show that a direct construction can be somewhat simpler.

We explain how surnaming plays a central role in Intel's new Software Guard Extensions (SGX) technology, and present its specific surnaming implementation as a special case. These results explain why SGX does not require a PKI or pinned keys for authorizing enclaves.

SGX motivates an interesting question in digital signature design: for reasons explained in the paper, it requires a digital signature scheme where verification must be as fast as possible, the public key must be short, but signature size is less important. We review the RSA-based method currently used in SGX and evaluate its performance.

Finally, we propose a new hash-based signature scheme where verification time is much faster than the RSA scheme used in SGX. Our scheme can be scaled to provide post-quantum security, thus offering a viable alternative to the current SGX surnaming system, for a time when post-quantum security becomes necessary.

Keywords: Digital signatures · Fast verification · Software Guard Extensions (SGX) technology · Post-quantum secure signatures

1 Introduction

Intel has recently introduced a powerful security architecture called Software Guard Extensions (SGX for short). It is available on the 6th Generation Intel® Core^TM processor (microarchitecture codename Skylake). This technology enables applications to operate on secret data without fear of compromise. The basic element in SGX is called a secure "enclave", which contains some application code (and data). SGX allows one to run an enclave on the processor, and enjoy complete isolation: nothing else running on the processor can access its memory. The system memory is considered untrusted, so memory reads/writes

© Springer International Publishing AG 2017
H. Handschuh (Ed.): CT-RSA 2017, LNCS 10159, pp. 149–164, 2017.
DOI: 10.1007/978-3-319-52153-4_9

are encrypted with integrity and replay protection. An enclave is initialized by loading executable code into a segment of memory, using special SGX instructions, and then calling the instruction EINIT to initialize the enclave. EINIT verifies a digital signature, presumably signed by the enclave's author. After initialization, the enclave code can invoke another instruction, EGETKEY, which generates a secret key that is unique to all enclaves written by this author, and running on this platform.

Surprisingly, SGX does not use a PKI or pinned keys in the processor when verifying the signature: the inputs to EINIT include both the signature *and* the associated public key, but no certificates. Without a PKI or pinned keys, signature verification is meaningless, and yet SGX uses neither. This is not a vulnerability because the process of initializing the enclave does not use the digital signature to authorize the enclave. Instead, it uses the digital signature in an unusual way, which we call a *surnaming scheme*. We define surnaming schemes formally in Sect. 3 and explain why they are sufficient for SGX. We show that surnaming can be constructed from a digital signature scheme, and vice versa, but surprisingly, surnaming can be implemented more efficiently. Specifically, a surnaming scheme can be implemented without conditional statements, whereas digital signature verification always requires conditional statements, at the very least to make an accept/reject decision.

The Need for Fast Verification in SGX. EINIT is a *processor instruction* that needs to verify a digital signature atomically. It needs to complete execution within the allowed latency for processor instructions, which is dictated by the no-interrupt latency that the OS tolerates. This hard limit applies even if EINIT is used infrequently. As a result, SGX must use a digital signature with the fastest possible verification time. In our settings, the signature size is unimportant, because the input to EINIT is a pointer to a large data structure containing the signature. This introduces an unusual design requirement: construct a secure signature scheme with the fastest possible verification time, irrespective of signature size. For a technical reason that we explain below, the signature scheme must also generate short public-keys.

With this in mind, in Sect. 4 we review the signature scheme currently used in SGX. It is based on RSA with 3072 bit keys, coupled with an optimization called QVRSA. The optimization makes signature verification significantly faster than in the standard RSA and ECDSA signature schemes.

In Sect. 5 we propose an alternative mechanism based on hash-based signatures, optimized for fast verification. After much optimization we obtain a scheme where signature verification is faster than QVRSA. Moreover, scaling the parameters of our hash-based scheme, makes it a viable candidate for post-quantum security. This gives a post-quantum secure alternative for SGX enclave verification.

Our Contributions.

- We provide an explanation and justification for the unorthodox way that SGX uses digital signatures.

- We formalize the concept of surnaming and show secure methods to construct a surnaming scheme from a signature scheme and vice versa. We also show how to obtain a surnaming scheme that is somewhat simpler than the signature scheme from which it is derived.
- We analyze the security of the surnaming mechanism used in SGX, and explain why it meets the performance requirements.
- We show a hash-based signature scheme whose verification time is even faster than that of QVRSA. Moreover, this variant can be extended to give a post-quantum secure scheme.

2   SGX and its Surnaming Mechanism

SGX technology is designed to allow a general purpose computer platform to run application software in a trustworthy manner, and to handle secrets that are inaccessible to anyone outside the defined trust boundaries. These trust boundaries encompass only the CPU internals, implying, in particular, that the system memory is untrusted. SGX is a complex technology that involves many details (see [2, 3, 15, 16, 18]). We provide here *only* an outlined description of the SGX elements needed for this paper.

Enclaves. The basic primitive in SGX is the "enclave". An enclave consists of code, data, and metadata (CDM hereafter) that realize some application that the enclave's author (A) prepares. The enclave is organized as a collection of 4KB "pages". The identity of an enclave consists of the following information: (a) the CDM inside the enclave (before initialization)[1]; (b) the order of loading the pages into memory (and their linear addresses); and (c) the security attributes (Read/Write/eXecute) of each page. The SHA256 digest of this information is called MRENCLAVE, and represents the cryptographic identity of the enclave. Two enclaves with the same identity are considered equivalent.

To prepare an enclave after writing the application code, A is expected to: (a) compute MRENCLAVE by hashing the appropriate data; (b) generate a private-public key pair (pk, sk_{sign}); (c) sign MRENCLAVE using sk_{sign}, to obtain signature σ; (d) ship σ, pk, and the expected MRENCLAVE, together with the enclave.

SGX ships with a Software Development Kit (SDK) that automates the execution of some of the above steps. A is responsible for generating the private-public key pair securely, and to sign MRENCLAVE (that the SDK computes). The SDK processes the information, produces the required outputs, and wraps them in the required format. In our context, the SHA256 digest of pk is called MRSIGNER. These are different identifiers: MRENCLAVE identifies the enclave's contents (its CDM), thus reflects its intended functionality; MRSIGNER identifies A.

The enclave code and data are available in the clear before instantiation. That is, the CDM is visible, and – more importantly – auditable (some pieces of

---

[1] The author (A) may decide which parts of the CDM should be baked into the enclave's identity, by specifying the pages to be measured. For example, non-initialized data, or SSA pages, can be skipped.

the CDM may be encrypted, but the decryption key should not be pre-installed).
Consequently, the entity (`Service-Provider`) that hands secrets to the enclave
can work with A to pre-approve the enclave (i.e., its intended functionality).
Secrets such as keys, passwords, and other sensitive data, need to be handed to
the enclave from a 3rd party (either from another enclave or from outside the
platform), *after* it is loaded and instantiated on the platform. To this end, the
enclave must convince a remote service provider (`Service-Provider`) who owns
a secret, that it is trustworthy, and can be provisioned with secrets. Furthermore,
after an enclave is provisioned with secrets, it should be able to securely store
them outside of the enclave for subsequent use.

Instantiating an Enclave[2]. SGX includes special CPU instructions that are
used to "build" an enclave: `ECREATE` (sets up and records the configuration information), `EADD` (records the offset of a page inside the enclave, and its security
attributes, and copies the CDM page from non trusted memory to trusted (protected) memory, `EEXTEND` (records the pointer and the data stored in a 128 byte
chunk of the enclave page), and `EINIT` (described below). The enclave is built
by invoking `ECREATE`, and then, for each page of its CDM, invoking `EADD`, followed
by 32 invocations of `EEXTEND`. This flow copies the CDM, incrementally, from general purpose (unprotected) memory, and locks it in a protected memory region,
while (incrementally) measuring `MRENCLAVE` and logging the size of its CDM. The
build process ends by invoking the `EINIT` instruction. `EINIT` has several roles,
and we describe only those that are relevant to our discussion: (a) "finalize" the
SHA256 computation of `MRENCLAVE` (i.e., add the padding block that reflects the
recorded enclave size); (b) "verify", using the input `pk`, that the input σ is a
valid signature on the (measured) `MRENCLAVE`; and (c) compute `MRSIGNER` and
store it in the protected memory region (only after completing a successful verification). After the build process terminates successfully, the initialized enclave
is considered "instantiated," and ready to run (in a protected enclave mode).

Isolation During Run Time. An instantiated enclave runs in a special "secure
enclave" mode where a hardware based access control mechanism isolates it
from all other processes (at all privilege levels) that run on the platform, and
from external hardware devices that are attached to the system. Furthermore,
the enclave operates from a memory region that is cryptographically protected
by a dedicated hardware unit (Memory Encryption Engine [11]), that protects
privacy, integrity and freshness (anti-replay). In other words, the enclave can
protect its secrets during run-time.

Acquiring Secrets. Secrets need to be delivered to the enclave after its origin,
identity, and execution environment are verified. `Service-Provider` is expected
to vet the enclave's Trusted Computing Base (TCB) before it trusts it and
provisions it with secrets (in particular, to guarantee that the enclave would
perform some pre-approved intended functionality). To this end, SGX offers the
means for an enclave to prove to an off-platform party its `MRENCLAVE` value, its
`MRSIGNER` value, and its execution environment (enclave mode, the CPU security

---

[2] This is a conceptual flow, but actual software might implement a different one.

level, and the Security Version Number (SVN)). The details of the tools, the protocols, and the Provisioning and Attestation services are outside the scope of this paper (details appear in [3,16]).

Handling Secrets. The enclave needs the ability to store its secrets to non-volatile memory, in order to use them in subsequent runs. For this purpose, SGX includes the EGETKEY instruction that the (instantiated) enclave software can invoke. EGETKEY can be used to obtain a *Sealing key* which is unique to the specific platform, to the enclave (either its identity or its author), and to the SGX SVN. The enclave software can use the Sealing key to encrypt its secret information and store it on untrusted media. In subsequent runs, the enclave can retrieve the Sealing key by invoking EGETKEY, and decrypt the sealed information.

EGETKEY computes the Sealing key by applying a PRF (Pseudo Random Function) whose key is derived from a secret key (*PlatformKey*) that is unique to the platform (and the SVN). The PRF runs over several non-secret fields, including either MRENCLAVE or MRSIGNER, as determined by the parameters that the enclave software feeds to EGETKEY.

We discuss here only the Sealing keys that are produced by using MRSIGNER (and ignore Sealing keys that are produced by using MRENCLAVE). They have the following desirable property: enclaves running on the same platform with the same SVN, and written by the same author A, will obtain the same Sealing key when calling EGETKEY. This lets two enclaves written by the same developer (running on the same platform) share secret state. This greatly simplifies the process of updating the software running in an enclave — all versions of a single enclave share the same Sealing key.

Remark 1. The value of MRSIGNER represents A, the enclave's software developer. Service-Provider, who owns the secrets that need to be delivered to the instantiated enclave, is not necessarily A. However, Service-Provider can communicate, offline, with A and establish trust in A's MRSIGNER identity. This implies that Service-Provider would trust all enclave software that A produces (e.g., different versions of the same application), and allow these applications to share secrets (on a given platform) through the common Sealing key. The "ISV SVN" field, that A can embed in the enclave, is also a component in the Sealing key derivation. It enables A to control which of its application share a common Sealing key. For this reason, A is sometimes called the *Sealing Authority*.

The Strict Performance Constraints. SGX raises special constraints. Since EINIT is a processor instruction, its allowed latency is strictly limited, regardless of how (in)frequently it is invoked. A too long instruction can affect the stability of the OS. Thus, it is essential for SGX to specify a signature scheme that has a very fast verification. In contrast, the time to generate a signature is irrelevant, as it occurs offline. Similarly, signature size is not a significant concern — the argument to EINIT is a pointer to a memory location that stores inputs in a data structure (SIGSTRUCT), that included the signature, and this structure can be of arbitrary size. Note that EINIT computes MRSIGNER by hashing the input public key. This step adds to the latency of EINIT, adding the requirement for a short public key so that hashing is fast.

3 Surnaming Schemes

We now define the cryptographic primitive needed to authorize enclaves. We call this primitive a *surnmaing scheme.*

Recall that a digital signature scheme is a triple of algorithms (G, S, V). Algorithm G generates a private-public key pair, $(\mathrm{pk}, \mathrm{sk_{sign}})$. Algorithm $S(\mathrm{sk_{sign}}, m)$ signs the message m and outputs a signature σ. Algorithm $V(\mathrm{pk}, m, \sigma)$ verifies the signature, and outputs true or false. An example application is a software/firmware update mechanism, where a software vendor signs an update using $\mathrm{sk_{sign}}$, and every device verifies the signature before installing the update. This assumes every device has or can acquire, a trusted copy of pk.

A surnaming scheme has a different syntax and security requirements. The purpose of surnaming is to allow an author (A) to use its public-private key pair $(\mathrm{pk}, \mathrm{sk_{sign}})$ to sign multiple messages $m_1, \ldots, m_k$ and distribute triples $(\mathrm{pk}, m_1, \sigma_1)$, $\ldots$, $(\mathrm{pk}, m_k, \sigma_k)$ so that A is assured of the following properties:

- if a verifier (V) is presented with a triplet $(\mathrm{pk}, m_j, \sigma_j)$ for $1 \leq j \leq k$, it can apply some pre-agreed algorithm, Surname, to the given values to generate a constant id that depends only on pk but not on m or σ.
- if V is presented with a triplet $(\mathrm{pk}', m', \sigma')$ where $m' \notin \{m_1, \ldots, m_k\}$, then Surname outputs a constant $\mathrm{id}' \neq \mathrm{id}$, even when σ' and pk' are chosen adversarially.

We define this concept formally below.

On the Term "Surnaming". The Surname algorithm assigns the same "family name" (surname) to all the members of a "family" of messages (that were produced by A), and only to members of that family. This surname (denoted by id) is subsequently used by the verifier V as input to a PRF with a secret key owned by V, in order to generate a secret key that can be computed for any message in the family, and only for those messages. In other words, the "family secrets" are shared among all the family members, but with no others. This motivates the term "surnaming" for describing the scheme.

Note the significant difference from signatures: in a surnaming scheme, V does not need to trust m_j or pk, hence no PKI or pinning is needed for verifying a triple (pk, m, σ). Moreover, verification does not make an accept/reject decision. It just outputs some constant. The time it takes V to produce id is hereafter referred to as the *verification time.*

3.1 Surnaming Schemes: Definition

Definition 1. *A surnaming scheme operates over a message space $\mathcal{M}$. The scheme is a triple of algorithms* (Setup, Authorize, Surname) *where*

- *Setup: outputs sk.*

- **Authorize**(sk, m) (m ∈ M) outputs σ.
- **Surname**(m, σ) (m ∈ M): outputs id or ⊥.

Correctness requirement: for all **sk** *output by* **Setup**, *and all* $m, m' \in \mathcal{M}$:

$$\text{if } \sigma \leftarrow \textbf{Authorize}(\textbf{sk}, m) \text{ and } \sigma' \leftarrow \textbf{Authorize}(\textbf{sk}, m')$$
$$\text{then } \textbf{Surname}(m, \sigma) = \textbf{Surname}(m', \sigma').$$

That is, if m and m' are authorized, then **Surname** on either one produces the same surname (id).

Mapping to SGX. To understand the relation to SGX, it is helpful to think of the following mapping. The surnaming scheme is used by two parties: A (enclave author) and V (verifier), which is the processor that executes EINIT. The authorization key **sk** is A's private key and m is the enclave's CDM. The author runs Authorize(sk, CDM) to obtain the enclave authorization token σ which is packed into the enclave metadata. On the SGX machine, the EINIT instruction runs Surname(m, σ) and the output is some constant id. The "consumer" of the output id is the EGETKEY instruction: it uses id to generate a local secret by computing Sealing Key = PRF(*PlatformKey*, id). The correctness property of the surnaming scheme ensures that all enclaves that are authorized by a single developer will lead to the generation of the same identifier id, and therefore obtain the same Sealing Key when running on the same platform.

Security Definition. Define the following challenger-adversary game:

1. Challenger generates random **sk** $\xleftarrow{\text{R}}$ Setup.
2. Adversary adaptively submits messages (at least one) $m_1, m_2, \ldots$, and gets back $\sigma_i := \textbf{Authorize}(\textbf{sk}, m_i)$ for $i = 1, 2, \ldots$.
3. Eventually, adversary outputs (m, σ) where m is not in $\{m_1, m_2, \ldots\}$.

The adversary wins if $\textbf{Surname}(m, \sigma) = \textbf{Surname}(m_1, \sigma_1)$.

Definition 2. *A surnaming scheme (***Setup, Authorize, Surname***) is secure if no efficient adversary can win the game with non-negligible probability.*

The security definition captures the intuition that an adversary who obtains authorization tokens for arbitrary enclaves of its choice, cannot construct a useful authorization token σ for some other enclave m. That is, $\textbf{Surname}(m, \sigma)$ will be different from the output of **Surname** for the valid enclaves $m_1, m_2, \ldots$.

Secure Surnaming from a Secure Signature Scheme. It is not difficult to see that a secure surnaming scheme can be constructed from a secure digital signature. Let (G, S, V) be a signature scheme, where V outputs 0 or 1. The derived surnaming scheme is defined as follows:

- **Setup**: run G to get sk_{sign} and vk. Set the secret Surnaming key to be $\text{sk} := (\text{sk}_{\text{sign}}, \text{pk})$.
- **Authorize**(sk, m): run $\text{sig} \leftarrow S(\text{sk}_{\text{sign}}, m)$ and output $\sigma \leftarrow (\text{sig}, \text{pk})$.
- **Surname**(m, σ): output $\big(\text{pk}, V(\text{pk}, m, \text{sig})\big)$.

The scheme is correct: $\texttt{Surname}(m, \sigma)$ outputs $(\texttt{pk}, 1)$ whenever σ is a valid authorization for m. Security follows from the following simple theorem.

Theorem 1. *The derived surnaming scheme is secure assuming (G, S, V) is a signature scheme that is existentially unforgeable under a chosen message attack.*

Proof (Sketch). An adversary that defeats the derived surnaming scheme queries the challenger on a sequence of messages $m_1, m_2, \ldots$ and finally produces a pair $(m, \ \sigma = (\texttt{sig}, \texttt{pk}))$ such that $\texttt{Surname}(m, \sigma)$ outputs $(\texttt{pk}, 1)$ and m is new. But then $V(\texttt{pk}, m, \texttt{sig}) = 1$, which is an existential forgery for the underlying signature scheme.   □

Surnaming Implies Signatures. Next we show that every secure surnaming scheme implies a secure signature scheme. Let $(\texttt{Setup}, \texttt{Authorize}, \texttt{Surname})$ be a surnaming scheme. Define the following signature scheme (G, S, V):

- Algorithm G works as follows:
  - run $\texttt{Setup}$ to get $\texttt{sk}$,
  - run $\texttt{Authorize}(\texttt{sk}, 0)$ to get σ,
  - run $\texttt{Surname}(m, \sigma)$ to get $\texttt{id}$.

  Set $\texttt{pk} := \texttt{id}$ and output the key pair $(\texttt{pk}, \texttt{sk})$.
- Algorithm $S(\texttt{sk}, m)$: output $\sigma \leftarrow \texttt{Authorize}(\texttt{sk}, m)$.
- Algorithm $V(\texttt{pk}, m, \sigma)$: accept if $\texttt{Surname}(m, \sigma) = \texttt{pk}$.

The following theorem shows that the constructed signature scheme is secure.

Theorem 2. *If $(\texttt{Setup}, \texttt{Authorize}, \texttt{Surname})$ is a secure surnaming scheme then (G, S, V) is a signature scheme secure against existential forgery under a chosen message attack.*

Proof. Suppose there is an attacker $\mathcal{A}$ on the signature scheme. We use it to build an attacker $\mathcal{B}$ on the underlying Surnaming scheme. $\mathcal{B}$ begins by choosing a random message m' and asking its challenger to authorize m', thereby receiving $\sigma' \xleftarrow{\text{R}} \texttt{Authorize}(\texttt{sk}, m')$. Then $\texttt{Surname}(m', \sigma') = \texttt{pk}$. Next, $\mathcal{B}$ runs the signature attacker $\mathcal{A}$, giving it $\texttt{pk}$. It responds to $\mathcal{A}$'s signature queries by asking $\mathcal{B}$'s challenger to authorize the messages output by $\mathcal{A}$. Eventually $\mathcal{A}$ outputs an existential forgery (m, σ). Since (m, σ) is a valid signature, we know that $\texttt{Surname}(m, \sigma) = \texttt{pk}$, even though m was never authorized. This breaks the underlying surnaming scheme because $\texttt{Surname}(m, \sigma) = \texttt{Surname}(m', \sigma')$. We assume that $\mathcal{M}$ is sufficiently large so that $m \neq m'$ with high probability.   □

3.2   Surnaming with Conditional-Free Verification

Signature verification algorithms necessarily require conditional statements to decide if a given signature is valid. The $\texttt{Surname}$ algorithm in a surnaming scheme can be implemented with no conditional statements. Nothing needs to be checked. This is a significant advantage of surnaming schemes over traditional signatures, primarily because signature verification checks have often been implemented incorrectly in practice. Bleichenbacher's attack on low-exponent RSA signatures [1] is a famous example of faulty signature verification, where the error was a result of an incorrect PKCS1 padding check. Another example is the large subgroup attack on some discrete-log based signature schemes where the verifier forgets to check if the given signature components are in the prescribed subgroup [19]. Even the original DSA specification from NIST contained a security error in signature verification where the verifier did not properly verify that the size of the two signature elements are in the required range [23]. We demonstrate how to implement a surnaming scheme with no conditional statements. An example, we use RSA signatures with PKCS1 padding[3].

RSA-PKCS1 Surnaming Scheme:

– Setup: Run the RSA key generation algorithm to obtain $\mathbf{pk} = (N, e)$ and $\mathbf{sk}_{\text{sign}} = (N, d)$. Here N is the RSA modulus and e is the RSA public exponent, and d is the RSA private exponent. Output $\mathbf{sk} = (\mathbf{pk}, \mathbf{sk}_{\text{sign}})$.
– Authorize($\mathbf{sk}, m$): RSA sign m, that is set

$$m' := \text{PKCS1PAD} \parallel \text{SHA256}(m)$$

as the PKCS1 padded message and treat m' as an integer. Then use $\mathbf{sk}$ to compute $s := (m')^d \pmod{N}$ and output $\sigma := (\mathbf{pk}, s)$.
– Surname(m, σ): let $\sigma = (\mathbf{pk}, s)$ and do:
  • compute $s' := s^e \pmod{N}$,
  • remove the message hash, namely set

$$s'' := s' - \text{SHA256}(m) \pmod{N},$$

  when σ is valid, this zeroes out the 256 least-significant bits of s'',
  • output $\mathbf{id} := (\mathbf{pk}, s'')$

Note that no conditional statements are used in Surname. The point is that instead of checking the pad, as required during RSA signature verification, we simply output the pad as part of the id. This eliminates the risk of incorrectly validating the PKCS1 pad.

In the next section we show that the SGX EINIT instruction essentially uses this RSA surnaming mechanism to derive the constant id, and EGETKEY uses the result in order to provide a Sealing Key to an enclave that invokes it.

The RSA surnaming scheme is a correct surnaming scheme: when σ is a valid authorization for m then Surname(m, σ) produces an id containing the public key and the PKCS1 pad. The same id is obtained for every properly authorized message. The following theorem captures the security of the RSA surnaming scheme.

---

[3] For RSA3072, using SHA256 hash, the PKCS1 pad is (see [17]):

PKCS1PAD = 00 || 01 || FF[330B] || 00 || 3031300D060960864801650304020105000420.

Theorem 3. *The RSA surnaming scheme is a secure surnaming scheme assuming RSA-PKCS1 is existentially unforgeable under a chosen message attack.*

Proof. Suppose there is an attacker $\mathcal{A}$ on the RSA surnaming scheme. We use it to build an attacker $\mathcal{B}$ on the RSA signature scheme. $\mathcal{B}$ runs the surnaming attacker $\mathcal{A}$. It responds to $\mathcal{A}$'s authorization queries by asking $\mathcal{B}$'s challenger to sign the messages output by $\mathcal{A}$ using RSA-PKCS1. Eventually $\mathcal{A}$ outputs a valid forgery (m, σ). We know that $\mathtt{Surname}(m, \sigma)$ outputs $(\mathrm{pk}, \ \mathrm{PKCS1PAD} \| 0^{256})$. But this means that σ is a valid RSA-PKCS1 signature for m and therefore (m, σ) is a valid existential forgery for RSA-PKCS1.   □

More generally, any signature scheme with message recovery can be converted into a surnaming scheme with conditional-free $\mathtt{Surname}$. Examples include RSA, Rabin [4], and Nyberg-Reupell [21] signatures. Even pairing-based BLS [8] signatures give a surnaming scheme with a conditional-free $\mathtt{Surname}$. Recall that BLS signatures are verified by testing that $e(g, \mathrm{pk}) = e(H(m), \sigma)$ where e is a paring function, H is a hash function, and g is a fixed group generator. When used in $\mathtt{Surname}(m, \sigma)$ one can instead output $e(g, \mathrm{pk})/e(H(m), \sigma)$ so that $\mathtt{Surname}$ contains no conditional statements. If the signature is valid the ratio will be 1. Otherwise, it will be some other value. Since the output of a pairing function is never zero, we need not worry about division by zero.

4   The SGX Surnaming Scheme

The performance of the *processor instruction* $\mathtt{EINIT}$ limits the possible choice of signature primitives that SGX can use. To understand the available options, consider Table 1 which shows the verification performance of some standard 128-bit security signature schemes. The measurements were done on an Intel Skylake processor, which is the (first) processor that supports SGX. The performance of ECDSA and RSA3072 with public exponent $e = 2^{16} + 1$ is prohibitive. RSA3072 with a short public exponent $e = 3$ is $\approx 2.76x$ faster than with $e = 2^{16} + 1$. Finally, note that RSA3072 with $e = 3$ using a QVRSA verification method [10] is the fastest option, by a wide margin. Indeed, SGX's surnaming scheme is based on these primitives (see full version [7]). We describe it here.

Table 1. Signature verification performance for several signature schemes.

Scheme	Cycles per verification	Comments
ECDSA (P256)	264,609	OpenSSL 1.0.2
ECDSA (P256) optimized	226,986	OpenSSL patched [12]
RSA3072 ($e = 2^{16} + 1$)	122,928	OpenSSL
RSA3072 ($e = 3$)	44,500	50,400 with padding check
RSA3072 with QVRSA ($e = 3$)	12,000	Optimized implementation

The QVRSA Optimization. Because SGX needs a signature with fast verification, it uses a variant of RSA verification called *Quick Verification RSA*, or QVRSA, as proposed by Gueron [10]. This optimization is a way to speed up RSA verification for any public exponent, in particular, of the form $2^k + 1$. QVRSA is based on handing the verifier some pre-computed constants with which the verifier can compute $T = (\sigma)^e \pmod{N}$ using only integer arithmetic instead of modular arithmetic. Computing these (public) constants does not require knowledge of d, and can even be done by post processing a signature that a secure platform (e.g., an HSM) generates. QVRSA is especially effective with $e = 3$, which is our case. Here, only two constants $q1, q2$ are needed:

$$q1 = \left\lfloor \sigma^2/N \right\rfloor, \quad q2 = \left\lfloor (\sigma^3 - q1 \cdot S \cdot N)/N \right\rfloor \tag{1}$$

The verifier is given m, σ, N, $q1$, $q2$, and applies the following algorithm to compute $\sigma^3 \pmod{N}$.

Algorithm 1. The QVRSA algorithm for computing σ^3 mod N
Input: m, $\sigma, N, q1, q2$ (s.t., $0 < \sigma, q1, q2 < N < 2^{3072}$)
  (1)   if $\neg(\sigma < N)$ then verification = FAILURE
  (2) $T1 = \sigma^2 - q1 \cdot N$
  (3)   if $\neg(0 < T1 < N)$ then verification = FAILURE
  (4) $T2 = \sigma \cdot T1 - q2 \cdot N$
  (5)   if $\neg(0 < T2 < N)$ then verification = FAILURE
Output: if (verification = FAILURE) output FAILURE; else output $T2$.

Theorem 4. *Algorithm 1 returns T2 if and only if q1 and q2 satisfy Condition 1, and in that case, $T2 = \sigma^3 \pmod{N}$.*

The proof, and some additional comments are given in the full version [7].

The SGX Conditional-Free Verification. In Sect. 3.2 we explained that RSA surnaming can be implemented with no conditional statements. This technique is employed in SGX, in the EINIT instruction, and makes it possible to validate the given RSA signature without ever checking the PKCS1 pad[4]. We refer the readers to [2], and the use of the 352 bytes strings "PKCS Padding Buffer" and the "HARDCODED_PKCS1_5_PADDING" in EINIT and EGETKEY. See also the full version [7].

5   Alternative Signatures with Fast Verification

The previous sections motivate two fundamental questions: is there an alternative signature scheme where verification is faster than QVRSA? Moreover, is there a post-quantum secure signature scheme with fast verification? We answer both questions positively by designing a hash-based signature scheme for our settings.

First, consider the Merkle tree-based signature scheme [20]. Optimizing it for fast verification, irrespective of signature size, is an unusual point in the

---

[4] EINIT executes the correct padding check anyway, but security does not depend on the padding check.

design space. Previous schemes, such as SPHINCS [5] and others [6,9], focus on minimizing signature size to reduce network traffic. Our goal is to minimize verification speed, which calls for a very different set of optimizations.

We begin by constructing a hash-based *one-time* signature, which generally belong to one of two families: Winternitz [9,20] and HORS [22] (both are extensions of Lamport's signature). HORS is designed to produce short signatures, but verification is slower than with Winternitz. We therefore opt to optimize a Winternitz-like construction. Our scheme is built from three primitives:

- a one-way function $f : X \to X$,
- a collision resistant hash $h : X^v \to Y$ for some v, and
- an enhanced target collision resistant (eTCR) hash $\hat{h} : \{0,1\}^* \times R \to X$, as defined in [14].

The eTCR hash $\hat{h}$ is used to hash the input message using a random nonce chosen by the signature algorithm. We use a construction due to Halevi and Krawczyk [14] that shows how to build an eTCR hash function from a Merkle-Damgård function such as SHA256. The reason to use an eTCR hash is that it lets us shrink the size of the message hash without affecting security. The smaller message hash greatly speeds up verification.

Concretely, for 128-bit security we set $X := \{0,1\}^{128}$ and $Y := \{0,1\}^{256}$, and for post-quantum settings we set $X := \{0,1\}^{256}$ and $Y := \{0,1\}^{384}$. We emphasize that, thanks to our use of an eTCR, for 128-bit classical security, the message hash $\hat{h}$ can output only 128 bits and still provide 128-bit classical security.

Our scheme is parametrized by a small constant d, called the chain depth. We explore constructions with $d = 2, 4$ or 8, because larger values are not helpful for fast verification (on current processors). The constant d determines the length of a hash chain based on the function f, as explained below. We use the notation $f^{(v)}(x)$ to denote the composition of f with itself v times, e.g., $f^{(2)}(x) = f(f(x))$.

To describe the scheme we need the following two quantities:

$$n := \lceil \log(|Y|)/\log(d) \rceil \quad \text{and} \quad \ell := \lceil \log(n(d-1)+1)/\log(d) \rceil. \quad (2)$$

The total number of hash chains constructed during key generation is $n + \ell$ and a (one-time) signature consists of $n + \ell$ values in X. For example, when $Y = \{0,1\}^{256}$ and $d = 2$, we have $n = 256$ and $\ell = 9$, i.e., we construct $256 + 9 = 265$ chains and the signature contains 265 values in X. Setting d to 4 reduces the number of chains to $128 + 5 = 133$.

The signature scheme works as follows (its security is discussed in the full version [7]):

- Algorithm G:
  (1) choose random $x_0, \ldots, x_{n+\ell-1}$ in X,
  (2) for $i = 0, \ldots, n+\ell-1$ compute $y_i := f^{(d-1)}(x_i)$, that is, construct $n+\ell$ hash chains,
  (3) output $\text{sk} := (x_0, \ldots, x_{n+\ell-1}) \in X^{n+\ell}$ and $\text{pk} := h(y_0, \ldots, y_{n+\ell-1}) \in Y$.

- Algorithm $S(\mathrm{sk}, m)$:
  (1) choose a random $r \in R$ and compute $\hat{h}(m, r) \in Y$, treat it as a positive integer written in base d with digits $0 \leq m_0, \ldots, m_{n-1} < d$. Then $\hat{h}(m, r) = m_0 + m_1 d + \ldots + m_{n-1} d^{n-1}$.
  (2) set $w := n(d-1) - (m_0 + \ldots + m_{n-1})$ and write w in base d with digits $0 \leq m_n, \ldots, m_{n+\ell-1} < d$,
  (3) for $i = 0, \ldots, n + \ell - 1$ set $s_i := h^{(m_i)}(x_i)$,
  (4) output the signature $\sigma := (r, s_0, \ldots, s_{n+\ell-1})$.
- Algorithm $V(\mathrm{pk}, m, \sigma)$: with $\sigma = (r, s_0, \ldots, s_{n+\ell-1})$ do:
  (1) compute $\hat{h}(m, r)$ and write it base d, namely $0 \leq m_0, \ldots, m_{n+\ell-1} < d$,
  (2) for $i = 0, \ldots, n + \ell - 1$ let $y_i := f^{(d-1-m_i)}(s_i)$,
  (3) accept the signature if $\mathrm{pk} = h(y_0, \ldots, y_{n+\ell-1})$ and reject otherwise.

Verification time is dominated by steps (2) and (3): evaluating f at $(d-1)(n+\ell)$ points in the worst case (half that on average), and computing h given $n + \ell$ quantities in X as input. These steps can be implemented to take advantage of pipelining available in modern processors. The free parameter to play with is the chain depth d that offers a different balance between the number of evaluations of f and the length of the string that needs to be hashed using h. Our optimization chooses the optimal d for different choices of h and f.

In the context of SGX, the enclave author uses surnaming to authorize a small number of enclaves, mostly needed for software updates. Therefore, a single signing-key is used only a small number of times. Our experiments are geared for supporting at most 1,000 enclave/versions per signing key.

We can extend a one-time signature to a 1000-times (stateful) signatures by generating a thousand one-time public-keys and publishing a Merkle tree root of these public-keys as the global public-key. Each leaf of the tree can sign one enclave, and a signature will include a Merkle proof of inclusion of the relevant public-key in the Merkle tree. The verification algorithm should check this proof, and this adds another $\lceil \log_2(1000) \rceil = 10$ hash computations (a small overhead compared to the amount of hashing needed to verify the one-time signature).

We used our general signature construction to derive several concrete signature schemes with fast verification. Table 2 lists six candidates for the one-way function f ($f1, \ldots, f6$) and four candidates for the collision resistant function h ($h1, \ldots, h4$). The functions $f1, f2, h1, h2$ target classical 128-bit security, whereas the other functions target 128-bit post-quantum security. Some of our experiments use the Simpira permutations [13], a recently proposed family of cryptographic permutations (see also [7]).

Performance Results for Classical 128-bit Security[5]. Table 3 (left side) gives the cycle count for signature verification using different d, f and h selections (see full version [7]) that target classical 128-bit security. Comparison to Table 1 shows that our hash-based signatures have much faster verification than

---

[5] Buchmann et al. [9] show that when using a 128-bit function f, Winternitz security for a chain of depth 4 and 8 is slightly less than 2^{128}. This is because the composition of random functions is slightly easier to invert than inverting the base function f.

Table 2. Different options for one way functions $(f : X \to X)$ and collision resistant functions $(h : \{0,1\}^* \to Y)$ for different choices of X and Y.

n (bits)	X	$f : X \to X$
256	$\{0,1\}^{128}$	$f1(x) = AES128_{K0}(x) \oplus x$
256	$\{0,1\}^{128}$	$f2(x) = AES128_x(0)$
256	$\{0,1\}^{256}$	$f3(x) = Simpira_2(x) \oplus x$
256	$\{0,1\}^{256}$	$f4(x) = a\|b$ s.t., $a = AES256_x(0)$ $b = AES256_a(0)$
256	$\{0,1\}^{256}$	$f5(x) = Rijndael256_{K0}(x) \oplus x$
256	$\{0,1\}^{256}$	$f6(x) = Rijndael256_x(0)$

	Y	$h : \{0,1\}^* \to Y$
256	$\{0,1\}^{256}$	$h1(x) = \text{SHA256}(x)$
256	$\{0,1\}^{256}$	$h2(x) = SimpiraHash(x)$
384	$\{0,1\}^{384}$	$h3(x) = \text{SHA384}(x)$
384	$\{0,1\}^{384}$	$h4(x) = SimpiraHash(x)$

RSA and ECDSA. As expected, QVRSA which SGX uses has a very competitive verification speed. However, our best hash-based signatures option have faster verification, even if we add ~2400 cycles measured to be the overhead of computing the 10 additional hashes to support 1000-time signatures.

Performance Results for 128-bit Post-quantum Security. Table 3 (right side) provides the cycle counts for signature verification for 128-bit quantum security parameters, where RSA and ECDSA are not competitors due to their insecurity against quantum attacks. Lattice based signatures may provide a viable alternative, but they require large public keys and hashing those keys during verification may dominate verification time. Moreover, lattice based signatures are based on specific algebraic assumptions which may or may not hold in a post-quantum world. In contrast, hash-based signatures are unlikely to be affected by quantum machines. The optimal chain depth is $d = 4$. The functions $(f3, h4)$ give the best performance, but also require the strongest security assumptions. The functions $(f4, h3)$ are the most conservative, but are slower than the fastest Rijndael-based construction. These results may renew the interest in the Rijndael cipher with a 256-bit block. The combination $(f2, h1)$ is the most conservative in terms of the assumptions needed for security.

Table 3. Signature verification performance. The reported numbers measure processor cycle counts for signature verification (lower is better). Left table: 128-bit classical security parameters, with $d = 2, 4, 8$ and $h \in \{h1, h2\}$ and $f \in \{f1, f2\}$. Right table: 128-bit quantum security parameters, with $d = 2, 4, 8$ and $h \in \{h3, h4\}$ and $f \in \{f3, f4, f5, f6\}$.

d		f1	f2
2	h2	2,363	4,805
2	h1	14,720	17,472
4	h2	**2,002**	6,247
4	h1	9,903	14,211
8	h2	3,001	8,478
8	h1	9,790	15,643

d		f3	f4	f5	f6
2	h4	18,157	34,018	21,161	80,708
2	h3	48,599	64,125	51,549	111,049
4	h4	**13,759**	36,813	18,048	102,729
4	h3	30,051	53,108	34,360	118,982
8	h4	16,240	52,073	22,892	152,566
8	h3	28,668	64,487	35,180	165,060

6   Conclusions

We formalized the concept of a surnaming mechanism, as needed for Intel's SGX technology. Although the concept is closely related to digital signatures, there are important subtle differences. Most notably, surnaming schemes do not require a PKI or pinned keys. Our abstraction explains why SGX is able to verify enclaves without relying on a PKI.

SGX gives rise to an unusual design requirement: the need for a signature scheme with the fastest possible verification and a short public-key, where the signature size is immaterial. We reviewed the QVRSA method used in SGX, and explained its benefits over other options. We then presented an alternative approach based on hash-based signatures that, after considerable optimization work, yields a signature scheme that has faster verification than QVRSA. To make SGX post-quantum secure, it would need to move away from QVRSA (in addition to other necessary changes). Our experiments with quantum-secure hash-based signatures show that they are a practical option for replacing RSA3072 in SGX.

Finally, we mention that surnaming is of fundamental importance to SGX, and the surname generated by EINIT is used for multiple purposes beyond just generating a Sealing key. Examples include authorizing a Launch Enclave and generating a Provisioning key (see full version [7]). This paper sheds light on this mechanism and explains its cryptographic underpinnings.

Acknowledgments. The first author is supported by NSF, DARPA, the Simons foundation, and a grant from ONR. Opinions, findings and conclusions or recommendations expressed in this material are those of the author(s) and do not necessarily reflect the views of DARPA. The second author is supported by the PQCRYPTO project, which is partially funded by the European Commission Horizon 2020 research Programme, grant #645622, by the Blavatnik Interdisciplinary Cyber Research Center (ICRC) at the Tel Aviv University, and by the ISRAEL SCIENCE FOUNDATION (grant No. 1018/16).

References

1. An attack on RSA digital signature. A NIST document (2006). http://csrc.nist. gov/groups/ST/toolkit/documents/dss/RSAstatement_10-12-06.pdf
2. Intel® Software Guard Extensions Programming Reference (2014). https:// software.intel.com/en-us/isa-extensions/intel-sgx
3. Anati, I., Gueron, S., Johnson, S., Scarlata, V.: Innovative technology for CPU based attestation and sealing. In: Proceedings of the 2nd International Workshop on Hardware and Architectural Support for Security and Privacy, vol. 13 (2013)
4. Bellare, M., Rogaway, P.: The exact security of digital signatures-how to sign with RSA and Rabin. In: Maurer, U. (ed.) EUROCRYPT 1996. LNCS, vol. 1070, pp. 399–416. Springer, Heidelberg (1996). doi:10.1007/3-540-68339-9_34
5. Bernstein, D.J., Hopwood, D., Hülsing, A., Lange, T., Niederhagen, R., Papachristodoulou, L., Schneider, M., Schwabe, P., Wilcox-O'Hearn, Z.: SPHINCS: practical stateless hash-based signatures. In: Oswald, E., Fischlin, M. (eds.) EUROCRYPT 2015. LNCS, vol. 9056, pp. 368–397. Springer, Heidelberg (2015). doi:10. 1007/978-3-662-46800-5_15

6. Bleichenbacher, D., Maurer, U.: On the efficiency of one-time digital signatures. In: Kim, K., Matsumoto, T. (eds.) ASIACRYPT 1996. LNCS, vol. 1163, pp. 145–158. Springer, Heidelberg (1996). doi:10.1007/BFb0034843

7. Boneh, D., Gueron, S.: Surnaming schemes, fast verification, and applications to SGX technology (2016). http://crypto.stanford.edu/~dabo/pubs/abstracts/surnaming.html

8. Boneh, D., Lynn, B., Shacham, H.: Short signatures from the Weil pairing. J. Cryptol. **17**(4), 297–319 (2004)

9. Buchmann, J., Dahmen, E., Ereth, S., Hülsing, A., Rückert, M.: On the security of the Winternitz one-time signature scheme. In: Nitaj, A., Pointcheval, D. (eds.) AFRICACRYPT 2011. LNCS, vol. 6737, pp. 363–378. Springer, Heidelberg (2011). doi:10.1007/978-3-642-21969-6_23

10. Gueron, S.: Quick verification of RSA signatures. In: 2011 Eighth International Conference on Information Technology: New Generations (ITNG), pp. 382–386, April 2011

11. Gueron, S.: A memory encryption engine suitable for general purpose processors. Cryptology ePrint Archive, Report 2016/204 (2016). http://eprint.iacr.org/

12. Gueron, S., Krasnov, V.: Improved P256 ECC performance by means of a dedicated function for modular inversion modulo the P256 group order. OpenSSL patch (2015). https://mta.openssl.org/pipermail/openssl-dev/2015-December/003821.html

13. Gueron, S., Mouha, N.: Simpira v2: a family of efficient permutations using the AES round function. Cryptology ePrint Archive, Report 2016/122 (2016)

14. Halevi, S., Krawczyk, H.: Strengthening digital signatures via randomized hashing. In: Dwork, C. (ed.) CRYPTO 2006. LNCS, vol. 4117, pp. 41–59. Springer, Heidelberg (2006). doi:10.1007/11818175_3

15. Hoekstra, M., Lal, R., Pappachan, P., Phegade, V., Del Cuvillo, J.: Using innovative instructions to create trustworthy software solutions. In: Proceedings of the 2nd International Workshop on Hardware and Architectural Support for Security and Privacy, HASP 2013, p. 11:1. ACM, New York (2013)

16. Johnson, S., Scarlata, V., Rozas, C., Brickell, E., Mckeen, F.: Extensions, Intel® Software Guard: EPID provisioning and attestation services. White Paper (2016)

17. Kaliski, B.S.: Public-Key Cryptography Standards (PKCS) #1: RSA CryptographySpecifications Version 2.1. RFC 3447, October 2015

18. McKeen, F., Alexandrovich, I., Berenzon, A., Rozas, C.V., Shafi, H., Shanbhogue, V., Savagaonkar, U.R.: Innovative instructions and software model for isolated execution. In: Proceedings of the 2nd International Workshop on Hardware and Architectural Support for Security and Privacy, HASP 2013, p. 10:1. ACM, New York (2013)

19. Menezes, A.: Another look at HMQV. Cryptology ePrint Archive, Report 2005/205 (2005). http://eprint.iacr.org/

20. Merkle, R.C.: A certified digital signature. In: Brassard, G. (ed.) CRYPTO 1989. LNCS, vol. 435, pp. 218–238. Springer, Heidelberg (1990). doi:10.1007/0-387-34805-0_21

21. Nyberg, K., Rueppel, A.: A new signature scheme based on the DSA giving message recovery. In: Proceedings of the 1st ACM Conference on Computer and Communications Security, CCS 1993 (1993)

22. Reyzin, L., Reyzin, N.: Better than BiBa: short one-time signatures with fast signing and verifying. In: Batten, L., Seberry, J. (eds.) ACISP 2002. LNCS, vol. 2384, pp. 144–153. Springer, Heidelberg (2002). doi:10.1007/3-540-45450-0_11

23. Rivest, R.L., Hellman, M.E., Anderson, J.C., Lyons, J.W.: Responses to NIST's proposal. Commun. ACM **35**(7), 41–54 (1992)

On the Entropy of Oscillator-Based True Random Number Generators

Yuan Ma[1,2(✉)], Jingqiang Lin[1,2,3], and Jiwu Jing[1,2,3]

[1] Data Assurance and Communication Security Research Center,
Chinese Academy of Sciences, Beijing, China
{yma,linjq,jing}@is.ac.cn
[2] State Key Laboratory of Information Security,
Institute of Information Engineering, Chinese Academy of Sciences, Beijing, China
[3] University of Chinese Academy of Sciences, Beijing, China

Abstract. True random number generators (TRNGs) are essential for cryptographic systems, and they are usually evaluated by the concept of entropy. In general, the entropy of a TRNG is estimated from its stochastic model, and reflected in the statistical results of the generated raw bits. Oscillator-based TRNGs are widely used in practical cryptographic systems due to its elegant structure, and its stochastic model has been studied in different aspects. In this paper, we investigate the applicability of the different entropy estimation methods for oscillator-based TRNGs, including the *bit-rate entropy*, the *lower bound* and the *approximate entropy*. Particularly, we firstly analyze the two existing stochastic models (one of which is phase-based and the other is time-based), and deduce consistent bit-rate entropy results from these two models. Then, we design an approximate entropy calculation method on the output raw bits of a simulated oscillator-based TRNG, and this statistical calculation result well matches the bit-rate entropy from stochastic models. In addition, we discuss the extreme case of tiny randomness where some methods are inapplicable, and provide the recommendations for these entropy evaluation methods. Finally, we design a hardware verification method in a real oscillator-based TRNG, and validate these estimation methods in the hardware platform.

Keywords: Oscillators · True random number generators · Entropy estimation · Stochastic model

1 Introduction

Random number generators (RNGs) are widely used in cryptographic systems to generate sensitive parameters, such as keys, seeds of pseudo-random number generators, and initialization vectors. The security of many cryptographic schemes and protocols is built on the randomness of RNGs. The output of a RNG is expected to be a bit sequence with the properties of *unbiasedness*, *independence* and *unpredictability*. Statistical tests (such as NIST SP 800-22 [13] and

© Springer International Publishing AG 2017
H. Handschuh (Ed.): CT-RSA 2017, LNCS 10159, pp. 165–180, 2017.
DOI: 10.1007/978-3-319-52153-4_10

Diehard [11]) cannot evaluate the unpredictability of the sequence, as deterministic sequences with good statistical properties are able to pass the statistical tests.

The concept of *entropy*, which measures the uncertainty in bits (e.g., bit-rate entropy), is used to evaluate the unpredictability of a RNG. For a true RNG (TRNG), the predictability comes from the randomness of physical noises. The international standard ISO/IEC 18031 [7] and Germany standard AIS 31 [8] recommend to establish the entropy estimator with a stochastic model for TRNG evaluation. The stochastic model describes the extraction process from physical random noises to digitized random bits based on reasonable physical assumptions.

Oscillator-based sampling is a typical structure adopted by many TRNG designs, and the stochastic models of oscillator-based TRNGs have been well studied in recent years. To figure out the entropy of oscillator-based TRNGs, Killmann and Schindler [9] established a common stochastic model by a *time-based* approach, and gave a tight lower bound of the entropy; using the similar approach, Ma *et al.* [10] presented a calculation method to obtain the precise entropy. In addition, Amaki *et al.* [1] calculated the probabilities of certain bit patterns by using a Markov state transition matrix, but they evaluated the security using the Poker test [6] rather than entropy estimation. Baudet *et al.* [2] proposed a *phase-based* approach and provided a concise analytical formula for the entropy calculation (including the n-bit entropy and the lower bound). The entropy can be rapidly figured out by substituting the TRNG design parameters, including the jitter ratio and the frequencies of the sampling and sampled signals. This formula is then employed to estimate the entropy for a sufficient-entropy TRNG design [4].

While the entropy is estimated with these stochastic models based on the TRNG design parameters, the approximate entropy (ApEn) is obtained *statistically* based on the output bit sequence of a TRNG. ApEn is calculated by comparing the distributions of m-bit and $(m+1)$-bit blocks in the bit sequence. However, the parameter m in ApEn shall be chosen carefully to trade off between the accuracy of entropy estimation and the computation complexity.

Although various entropy estimation methods have been proposed in literature, a comprehensive and systematical study for their accuracy and applicability (e.g., the consistency of different methods, the estimation error between theory and experiment, the extreme cases of design parameters) is still lacking. In this paper, we investigate the applicability of different entropy calculation methods for oscillator-based TRNGs, including the bit-rate entropy, the lower bound and the approximate entropy. Particularly, we make the following contributions.

- We present two bit-rate entropy calculation methods based on the time-based and phase-based n-bit entropy stochastic models [2,10], respectively. The results are analyzed, and we deduce consistent bit-rate entropy results from these two models by expanding the original analytical expression.
- We propose an approximate entropy calculation method for the output bit sequence of oscillator-based TRNGs, where the parameter m is obtained from

the autocorrelation coefficient of the bit sequence. The ApEn calculation result of a simulated oscillator-based TRNG well matches the bit-rate entropy from stochastic models, which confirms the correctness of the theoretical results.

- We investigate the applicability of these entropy estimation methods in the extreme case with tiny randomness (i.e., the accumulated jitter is very small within the sampling interval). As it is possible to make an overestimation of the entropy in such case, we provide an alternative method to acquire a conservative estimation for the entropy.
- We design a hardware verification method in a real oscillator-based TRNG. In the experiment, we calculate the randomness factor under the white noise, and validate these estimation methods in the hardware platform.

The rest of the paper is organized as follows. In Sect. 2, we introduce the preliminary about the principle and existing entropy estimation methods for oscillator-based TRNGs. We propose our evaluation method on the different types of entropy in Sect. 3. In Sect. 4, we present the evaluation results and investigate the case of tiny randomness. In Sect. 5, we investigate the effectiveness of the estimation methods in the hardware platform. In Sect. 6, we conclude the paper.

2   Preliminary

In this section, we first introduce the principle of oscillator-based TRNGs. Then we summarize the methods of entropy estimation. The types of entropy include n-bit entropy, lower bound of entropy and approximate entropy.

2.1   Oscillator-Based TRNGs

The basic structure of oscillator-based TRNGs contains an unstable oscillator generating a fast oscillating signal with jitter, and a sampling reference clock that is assumed without jitter, as shown in Fig. 1. The randomness comes from jitter in the fast signal periods that is caused by noises. In general, the noises that affect jitter are assumed to be independent and identically distributed (i.i.d.) for the simplicity of the model. As an exception, the model of [9] partially allows short-term dependency in the half-periods of the fast oscillating signal.

We firstly present some definitions of the parameters in the aspect of time evolution. The half-periods of fast oscillating signal are assumed to be i.i.d. with mean $m_X = E(X_k)$ and variance $s_X^2 = V(X_k)$, where X_k is the k-th half-period. The fixed sampling interval is denoted as Δt.

As the tiny jitter ($s_X/m_X \ll 1$) accumulates within the sampling interval, the probability of guessing the sampling point lying in the high or low voltage is decreasing. Hence, the jitter ratio and the frequency ratio jointly determine the quality of this type of TRNG, and the integrated factor is often called as *quality factor* [2,10]. Another considerable factor is the divisibility of the half-period m_X to the sampling interval Δt, which is measured using variable

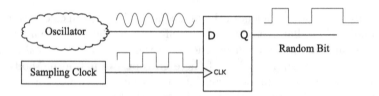

Fig. 1. The basic structure of oscillator-based TRNGs

$r = \Delta t / m_X \mod 1$. The divisibility increases when r approaches to 0.5 from either 0 or 1 and reaches its maximum at $r = 0.5$. The cases of $r = 0$ and $r = 0.5$ represent the worst and the best case of the TRNG output quality, respectively. This property has been discovered in [1,2,10].

2.2 n-bit entropy

The n-bit entropy represents the amount of entropy for n-bit output random sequences. In general, there are two methods to get the n-bit entropy, time-based and phase-based. The basic idea is to calculate the probability of n-bit pattern, which is denoted as $p(\boldsymbol{b})$, from the stochastic model, and then iterate all the patterns to get n-bit entropy via Eq. (1).

$$H_n = \sum_{\boldsymbol{b} \in \{0,1\}^n} -p(\boldsymbol{b}) \log p(\boldsymbol{b}). \tag{1}$$

Time-Based Method. Ma *et al.* [10] use the classic model of [9] in the aspect of time evolution. They utilize the waiting time W_i to represent the relationship between the adjacent sampling bits, where W_i is the distance of the i-th sampling position to the closest following edge of fast oscillating signal. They use a set of conditional probability functions to calculate the n-bit pattern probability by iterating, and eliminate W_i from the final expression by probability integration for the uniform distribution of W_i. Here we do not list the detailed computing process. Furthermore, they gave several curves from the worst to the best case to demonstrate the entropy variation using numerical computation, but an analytical probability or entropy expression was not given in their work.

Phase-Based Method. Baudet *et al.* [2] use the phase-oriented approach to model the stochastic behavior of the oscillating signal. The phase evolution of an oscillation is modeled by a Wiener stochastic process $\varphi(t)$ with drift $\mu > 0$ and volatility $\sigma^2 > 0$. The parameters are equivalent to the time-based definitions following the equations: $\mu = \frac{1}{2m_X}$ and $\sigma^2 = \frac{s_X^2}{4m_X^3}$.

Another quality factor is denoted as $Q = \sigma^2 \Delta t = \frac{s_X^2 \Delta t}{4m_X^3}$. The frequency ratio of the fast signal to the slow one is denoted as $\nu = \mu \Delta t = \frac{\Delta t}{2m_X}$, so $r = 2\nu \mod 1$. Note that, as the investigated target is the same as the time-based method, two sets of parameters can be converted to each other. The quality fact Q in the phase-based method equals $4q^2$, where $q = \sqrt{\frac{\Delta t}{m_X} \cdot \frac{s_X}{m_X}}$ is the parameter defined in the time-based model [10]. For convenience, we use Q and r to compute the entropy for either time-based or phase-based method in the subsequent.

The following two formulas computing the probability and n-bit entropy are provided in their work, where $B = e^{-2\pi^2 Q}$.

1. The probability to output a vector $\mathbf{b} = (b_1, \ldots, b_n) \in \{0, 1\}^n$ satisfies

$$p(\mathbf{b}) = \frac{1}{2^n} + \frac{8}{2^n \pi^2} (\sum_{j=1}^{n-1} (-1)^{b_j + b_{j+1}}) \cos(2\pi\nu) B + O(B^2). \qquad (2)$$

2. The entropy of such an output is

$$H_n = \sum_{\mathbf{b} \in \{0,1\}^n} -p(\mathbf{b}) \log p(\mathbf{b}) = n - \frac{32(n-1)}{\pi^4 \ln(2)} \cos^2(2\pi\nu) B^2 + O(B^3). \qquad (3)$$

2.3 Lower Bound of Entropy

Min-entropy or lower bound of entropy is the most conservative measurement of entropy, and is useful in determining the worst-case entropy of a TRNG. In the aspect of entropy calculating complexity, min-entropy or a lower bound has considerable advantages for dependent stochastic process, as only the probability in the worst case is involved. The methods for calculating a lower bound of entropy for oscillator-based TRNG are presented in [2,9], and the worst case is also investigated in [1].

The calculation expression of the lower bound [9], which is denoted as H_{lo}, was presented in Eq. (4):

$$H(B_i|B_{i-1}, \ldots, B_1) \geq H_{lo} = H(B_i|W_{i-1}) \approx \int_0^s H(R^{(s-u)} \bmod 2) P_W(du), \quad (4)$$

where B_i is the ith sampling bit and $R^{(s-u)}$ represents the number of crossing edges in the duration of $(s - u)$. The idea is inspired by the fact that W_i tells more information about B_{i+1} than all the previous bits. Following the similar idea, [2] also provides an analytical expression for H_{lo}, as shown in Eq. (5).

$$H_{lo} = 1 - \frac{4}{\pi^2 \ln(2)} e^{-4\pi^2 Q} + O(e^{-6\pi^2 Q}) \qquad (5)$$

2.4 Approximate Entropy

Approximate entropy (ApEn) is originally proposed to quantify the unpredictability of fluctuations in a time series. ApEn is a statistical value derived from the tested sequences. Note that, although the entropy of an TRNG shall be estimated from the stochastic model of a TRNG, but ApEn of the raw bits of a TRNG can also reflect the contained randomness. ApEn randomness test is also adopted in the NIST statistical test suite [13], which compares the frequency of overlapping blocks of two consecutive/adjacent lengths (m and $m+1$) against the expected result for a random sequence. The calculation process of ApEn for $\mathbf{b} = (b_1,\ldots,b_n)$ is presented in Algorithm 1. The block length m in Algorithm 1 has an important impact on ApEn calculation, which is treated as a trade-off. The larger value of m improves the accuracy of entropy estimation, but meanwhile significantly increase the computation complexity and the required length of the tested bit sequence.

Algorithm 1. Approximate entropy calculation [13]

Input: block length m, bit sequence $\mathbf{b} = (b_1,\ldots,b_n) \in \{0,1\}^n$
Output: $ApEn$
1: Augment the n-bit sequence to create n overlapping m-bit sequences by appending $m-1$ bits from the beginning of the sequence to the end of the sequence.
2: Make a frequency count of the n overlapping blocks. The count is represented as $\#i$, where i is the m-bit value.
3: Compute $C_i^m = \#i/n$ for each value of i.
4: Compute $\delta_m = \Sigma_{i=0}^{2^m-1} C_i^m \log_2 C_i^m$.
5: Replace m by $m+1$ and repeat Steps 1-4.
6: Compute $ApEn = \delta_{m+1} - \delta_m$.
7: **return** $ApEn$

---

3   Our Evaluation Method

In this section, we provide three estimation methods for the entropy: phase-based, time-based and ApEn. The former two utilize the jitter parameters to perform the estimation in theory, while the latter analyzes the output sequences.

3.1   Bit-Rate Entropy Calculation

In practice, the concept of entropy per bit is preferred for entropy evaluation, rather than the n-bit entropy. As the unit of the lower bound and ApEn is one bit, it is necessary to transfer n-bit entropy to entropy per bit, which is called bit-rate entropy. The bit-rate entropy is closely related to the expected workload that is necessary to guess (sufficiently long) sequences of random bits [7]. In addition, a precise Shannon entropy expression, which contains more parameters, allows the TRNG designers to optimize their structures and specifically adjust the parameters to get more entropy.

The bit-rate entropy H should be calculated from infinitely long random sequences, as Eq. (6) shows. As n shall be infinity, the calculation of H is nearly infeasible in either statistical or iterative computation. One way is to figure out reliable H is to deduce the precise expression of H_n in terms of n. Another possible case is that n actually can be a finite value, rather than being asymptotically infinite.

$$H = \lim_{n \to \infty} \frac{H_n}{n} = \lim_{n \to \infty} H(B_n | B_{n-1}, ..., B_1) \tag{6}$$

Time-Based Method. In the aspect of time evolution, it is observed that the correlation between two adjacent sampling bits is decreasing with the increase of the sampling interval. When the sampling interval is sufficient long, the sampling bits can be treated as independent. Here we provide a method to determine the required sampling interval for independent sampling bits.

The correlation coefficient of adjacent bits B_i and B_{i+1} is represented as:

$$\mathrm{cor}(B_i, B_{i+1}) = \frac{\mathrm{COV}(B_i, B_{i+1})}{\sqrt{\mathrm{Var}(B_i)\mathrm{Var}(B_{i+1})}},$$

where $\mathrm{COV}(\cdot)$ is the covariance function, and $\mathrm{Var}(\cdot)$ represents the variance. Then, using the stationary property [9] that $\mathrm{Prob}(B_i = 1) = \mathrm{Prob}(B_{i+1} = 1)$, the correlation coefficient is deduced as:

$$\mathrm{cor}(B_i, B_{i+1}) = \frac{\mathrm{Prob}(B_i = 1, B_{i+1} = 1) - \mathrm{Prob}(B_i = 1)^2}{\mathrm{Prob}(B_i = 1) \cdot \mathrm{Prob}(B_i = 0)}.$$

For different values of Q, we compute the correlation coefficients, as shown in Fig. 2. We observe that the dependence oscillatingly decreases with the increasing of Q. The absolute value of the coefficient drops below 10^{-3} when Q is larger than 0.16, where we consider that the correlation can be ignored and the adjacent sampling bits are treated as independent.

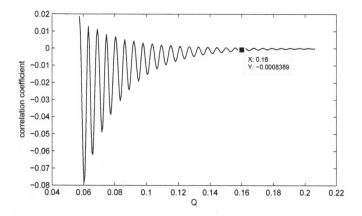

Fig. 2. The correlation coefficient in terms of Q

Using this conclusion, we further determine the longest timing distance, within which all the sampling behaviors are dependent. That is to say, when the distance between two sampling bits is longer than that distance, the two bits are treated to be independent. We denote l as the correlation length, which means the $(i + l)$th sampling bit B_{i+l} is only dependent on the previous l bits. Given a value of Q for a oscillator-based TRNG, the correlation length is deduced as $l = \lfloor (\frac{Q_{ind}}{Q}) \rfloor$, where Q_{ind} is the required Q value for the independence, and Q_{ind} is set to 0.16 in this paper. Then, combining with the additional conclusion that the sampling process is stationary [9], H can be derived as

$$H = \lim_{n \to \infty} H(B_n|B_{n-1}, ..., B_1) = H(B_{l+1}|B_l, ..., B_1). \tag{7}$$

A lower threshold of the coefficient certainly is helpful for getting a more reliable result, but the derived correlation length may be too large to complete the iterating computation of entropy within an acceptable time. The maximum l in our computation is limited below 15. For $Q_{ind} = 0.16$, setting $l = 15$ means reliable values can be acquired for $Q > 0.0107$.

Phase-Based Method. In [2], since an analytical expression of n-bit entropy exists, for n is approaching infinity, the approximated bit-rate entropy is expressed as Eq. (8).

$$H \approx 1 - \frac{32}{\pi^4 \ln(2)} \cos^2(2\pi\nu)B^2 \tag{8}$$

Note that using this equation to calculate bit-rate entropy is tentative, since H_n in Eq. (3) is not provably uniform in n [2]. The problem of non-uniformness in n dose not exist in the time-based method, because the parameter l has been chosen before calculating Eq. (7). In the following sections, we will learn that Eq. (8) is applicable under some parameters, but has non-ignorable errors under other parameters. Hence, in the next section, we improve the equation by performing further expansion of original expression, and validate the effectiveness of the improvement by comparing with the time-based method and the ApEn of simulated sequences.

3.2   Approximate Entropy for Short-Term Dependent Bits

ApEn is a statistical result to estimate the entropy of the tested sequence. An important parameter in the algorithm is the block length m, which partially determine the estimation accuracy of the algorithm. The ideal case is that the tested bits are independent beyond the bit interval of m, which means the estimation algorithm can have a comprehensive overlook on the tested sequence. Fortunately, for the output of oscillator-based TRNGs, the correlation lags are limited due to the independence condition, hence the sampling bits only have short-term dependence.

In the statistical method, we first use the autocorrelation test to find out the correlation length in the sequence, and set m as the length to calculate ApEn. The autocorrelation test is based on the autocorrelation plot [3], which is a commonly-used tool for checking randomness in a data set. Here, we do not adopt the autocorrelation test in [12] for random bits, because the basis of that test is the uniformity of the tested sequence. Otherwise (the uniformity is not satisfied), a higher correlation value will be acquired and autocorrelation test is failed. Hence, we return to the original test approach that only focuses on the correlation. The autocorrelation coefficient is represented as $R_h = C_h/C_0$, and C_h is the autocovariance function: $C_h = \frac{1}{n}\sum_{t=1}^{n-h}(b_t - \bar{b})(b_{t+h} - \bar{b})$, where $\bar{b}$ is the mean of $b_1, ..., b_n$, and C_0 is the variance function: $C_0 = \frac{1}{n}\sum_{t=1}^{n}(b_t - \bar{b})^2$.

For randomness tests, it is recommended to use 99 % confidence band to justify whether the test is passed or not. In this case, the test is passed when C_h lies in the interval $[-z_{1-\alpha/2}/\sqrt{n}, z_{1-\alpha/2}/\sqrt{n}]$, where the significance level $\alpha = 0.01$ and z is the cumulative distribution function of the standard normal distribution. Therefore, for the calculation of approximate entropy for short-term dependent bits, we provide the following statistical method on the oscillator-based TRNG output, as shown in Algorithm 2. Note that, due to the Type-I error in the hypothesis test, the intrinsic independent sequences still has the probability of α to fail the test. However, this fact, which increases the correlation length m, would not lead to estimation error of the entropy as long as the sequence length is satisfied, since larger m is preferred for estimation.

Algorithm 2. Approximate entropy calculation for short-term dependent bits
---
Input: $h = 1$, bit sequence $\mathbf{b} = (b_1, \ldots, b_n) \in \{0, 1\}^n$
Output: $ApEn$
1: **while** $|C_h| > z_{0.995}/\sqrt{n}$ **do**
2:     $h = h + 1$
3: **end while**
4: Compute $ApEn$ using Algorithm 1 with the parameter $m = h$
5: **return** $ApEn$

---

4   Entropy Evaluation

In this section, by comparing the results of different entropy calculation methods, we evaluate the applicability and accuracy of these methods for oscillator-based TRNGs. Particularly, as the original analytical formula has biases on the bit-rata entropy estimation for some TRNG parameters, we present a more accurate formula by performing further deducing, and the correctness is verified with other entropy results. Finally, we investigate the limitations of these methods in the case of tiny randomness, i.e., very small Q.

4.1   Bit-Rate Entropy Calculation Results

We use the proposed time-based and phase-based methods to calculate bit-rate entropy, and the results in terms of Q and r are shown in Fig. 3. However, we find that the approximated bit-rate entropy derived from Eq. (8) is not consistent with that calculated by Eq. (7). The inconsistency has been preliminarily pointed out in [10]. Note that the entropy at $r = 1 - x$ is identical to that at $r = x$, where $x \in (0, 0.5]$, thus we only present the cases for r ranging from 0 to 0.5. More precisely, the difference between the two results is maximized with the parameter r approaching 0.5, as shown in Fig. 3. Their results are almost identical in the worse cases ($r \in [0, 0.2]$), but in the other cases of r with a modest Q value the deviation occurs, especially at $r = 0.5$.

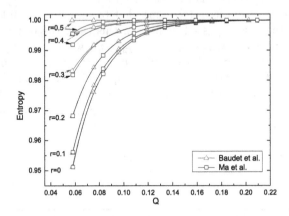

Fig. 3. The bit-rate entropy calculated from Ma *et al.*'s (time-based) and Baudet *et al.*'s (phase-based) methods

From the physical perspective, the r value is related to the fractional part of the ratio of sampling interval to the mean of half-periods ($\Delta t / m_X$). From the theoretical result of [10], to achieve a sufficient bit-rate entropy (such as 0.9999), the required sampling frequency in the best case is about two times faster than that in the worst case under the same quality factor. Hence, in the condition of fixed Q, the value of r has a non-negligible impact on the entropy, as shown in Fig. 3. Also, from the perspective of the designer, adjusting r can significantly improve the entropy without the degradation of the sampling frequency.

4.2 Improved Bit-Rate Entropy Expression Formula

We expand the original approximated expression formula of n-bit entropy by performing further deducing. The improved results are presented in Theorem 1.

Theorem 1. *For* $r = \frac{\Delta t}{m_X}$ mod 1 *and* $Q = \frac{s_X^2 \Delta t}{4 m_X^3}$, *the n-bit entropy is:*

$$
H_n \doteq n - \frac{32}{\ln(2)\pi^4} \cos^2(\pi r)(n - 1)e^{-4\pi^2 Q}
$$
$$
- \frac{32}{\ln(2)\pi^4} \Big[\cos^4(\pi r)(1.524n - 0.092) - 2.379 \cos^2(\pi r)(n - 2) + (n - 2) \Big] e^{-8\pi^2 Q}
$$
$$
+ O(e^{-10\pi^2 Q}). \tag{9}
$$

The (trial) approximated bit-rate entropy is expressed as:

$$H \approx 1 - \frac{32B^2}{\ln(2)\pi^4}\cos^2(\pi r) - \frac{32B^4}{\ln(2)\pi^4}\Big[1.524\cos^4(\pi r) - 2.379\cos^2(\pi r) + 1\Big].$$

(10)

In the improved expression Eq. (9), the first two terms are derived from the original one. Our work focuses on the deduction of the third term, the higher-order term. We strictly follow the same assumptions used in [2], but perform the further deduction on the entropy calculation process based on series expansion. The proof details are presented in the full version of this paper.

4.3   Bit-Rate Entropy Comparison: Time-Based Vs. Phase-Based

In order to validate the reliability of the improved result, we first compare it with the bit-rate entropy derived from the time-based method. The comparison result is shown in Fig. 4. We can see that after our improvement the two results become very close in all six cases from $r = 0$ to $r = 0.5$. Note that as the expression is analytical, the derived entropy is not only the data lying in the six curves, but the values for all the possible cases of Q and r. We remark that it is no surprise that the two entropy results are identical, because the focusing target and the physical assumption of small jitter are both the same. The equivalence between the two models has been presented in [2].

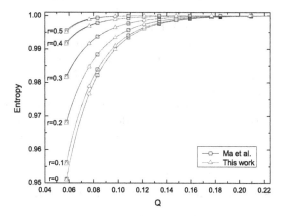

Fig. 4. The comparison of bit-rate entropy with the improved formula

Furthermore, from Eq. (10) we also explain why the original expression has a significant estimation error when r is large. When the coefficients of B^2, B^4,... $(0 < B = e^{-2\pi^2 Q} < 1)$ are comparable, the subsequent terms after B^2 can be ignored for large Q and the estimation error is acceptable. However, with r increasing from 0 to 0.5, the coefficient of B^2 decreases from maximum to 0, while the impact of B^4 increases. Especially, when r approaches 0.5, the coefficient of B^2 approaches 0, while the B^4 term does not become zero due to the existence of the constant 1 in the coefficient of B^4. Therefore, in this case the impact of B^4 term cannot be ignored and only using B^2 term to estimate the entropy is not enough.

Another important observation is that expanding the Taylor series to B^4 is enough to reliably estimate the entropy for such $Q \in (0.06, +\infty)$, where Q is not a very small value. The improved expression might have a bias when Q becomes a much smaller value, as the impact of the higher-order term of B (such as B^6) exists. But we must admit that getting the higher-order term of B seems infeasible, as the series in Eq. (3) after further expanding are too complex.

4.4 Bit-Rate Entropy Vs. Approximate Entropy

After the improvement, the bit-rate entropy values derived from the two methods are consistent, but it is necessary to confirm that the theoretical results is consistent with the experimental. For this purpose, we use the approximate entropy, which is a statistical measurement from the output bit sequence, to verify the applicability of the entropy evaluation method. Note that, the statistical entropy values are also random for random sequences, so directly using ApEn to do entropy estimation would lead to measurement errors. However, it is valuable to compare the trends of ApEn and bit rate entropy, which can be treated as experimental and theoretical results, respectively.

Following the assumptions in the aspect of time evolution, we perform a simulation experiment to calculate ApEn. In the experiment, the fast signal periods are independent and identically distributed, and the distribution is set as the normal distribution $N(1, 0.01^2)$. Each ApEn is computed from 10^5 sampling bits for each sampling interval which corresponds the values of Q and r. As the two bit-rate entropy results are almost the same, we use the improved phase-based result as the reference to compare with ApEn. The comparison results from $r = 0$ to $r = 0.5$ are shown in Fig. 5. We find that the two sets of results are well-matched for all r values. Therefore, Algorithm 2 is suitable to estimate the bit-rate entropy for this type of short-term dependent sequences. A more precise results can be acquired by averaging the estimated values of many statistical experiments.

4.5 Entropy Estimation for Smaller Quality Factor

In the previous entropy estimation results, the investigated values of Q are not very small, which are available for the entropy evaluation of most practical TRNGs. However, for very samll Q values, the presented entropy calculation methods are not applicable. The reasons are explained as follows.

- For the time-based bit-rate entropy calculation method, a very small Q means that the dependent length l is very large. For example, when $Q = 0.005$, l equals to 32, meaning that the traversal space should be 2^{32}, which is infeasible for computation. In this case, the estimation would be larger than the real entropy value, i.e., the overestimation occurs.

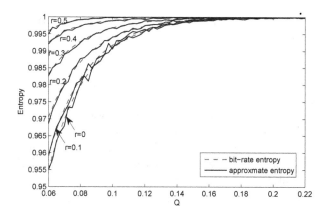

Fig. 5. The comparison between the bit-rate entropy and the approximate entropy

- For the phase-based bit-rate entropy calculation method, when Q decreases, the estimation error increases with no limitation, as the H_n expression is not uniform in n. Therefore, the approximated formula is not applicable to estimate the bit-rate entropy in this case, though our improvement has extended the applicable range of the formula.
- For the presented approximate entropy estimation method, a very small Q makes the statistical correlation lasts very long lags, which causes that the parameter m in Algorithm 2 is too large to complete the computation. For example, in our experiment, when $Q = 0.01$ the statistical m of Algorithm 2 is about 30, thus the workload for the traversal loops and the requirement for the sequence length are unacceptable in this case. The problem also leads to an overestimation for the entropy of the tested sequence.

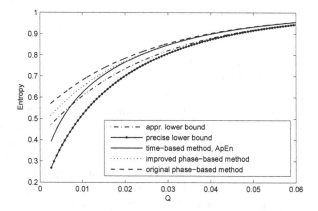

Fig. 6. The comparison of entropy values with small Q at $r = 0$

Actually, as we mentioned, the lower bound expression formula has be presented in [2]. As Eq. (5) shows, the expression formula of the lower bound also contains a term of O. Using this approximated expression also causes overestimation of the entropy when Q is smaller than 0.01. As shown in Fig. 6, the approximated lower bound becomes larger than the bit-rate entropy derived from time-based methods with the worst case of $r = 0$, though the bit-rate entropy might have been overestimated. However, we emphasize that the computation of this O term in Eq. (5) is feasible since the traverse of 2^n states is nonexistent. Therefore, we present the calculation results for the precise lower bound of entropy for smaller Q values, as labeled in Fig. 6. The comparison result indicates that the precise expression of the lower bound eliminates the overestimation of entropy for very small Q values.

5   On the Relationship with Physical RNGs

The existing models [2,10] assume that the oscillating period or phase increment is independently distributed due to the influence of white noise. This is a common assumption in literature, which allows to guarantee the simplicity of the model. However, in real TRNG circuits, the jitter or phase is also influenced by colored noises (such as $1/f$ noise) more than white noise, and the phenomenon has been demonstrated in recent works [5,10,14]. Under these colored noises, the period jitter has long-term dependence, and the dependence is also inherited by the sampling bit sequence [10]. In practical TRNGs, it is infeasible to perform similar confirmatory experiment as our simulation where the entropy is calculated via the output sequence, as the randomness amount is inevitably increased by colored noises and the offset r is hard to be precisely measured.

Fortunately, the white noise is independent from colored noises in principle, so the existing model and corresponding entropy estimation methods can still work for estimating the contribution of the white noise. When the estimated contribution (i.e., entropy) derived from *independent jitter* is sufficient, we can also claim that the entropy of the TRNG is satisfied. In practical entropy evaluation, the independent jitter can be acquired by employing an inner measurement method that excludes the dependent component of the jitter in the measurement (such as [5]). This evaluation approach neatly sidesteps the impact of colored noises.

We perform the hardware experiment on an FPGA (Field Programmable Gate Array) platform (Xilinx XC5LX110T), where two ring oscillators are implemented using Look-Up Tables (LUTs) with the close frequency of 280.5 MHz. The sampling interval is set as the period number of one oscillating signal, and the counting period number of the other signal is treated as the random variable, thus the random bit is the LSB of the counting number. Here, we do not use the number of half-periods to eliminate the impact of the imbalance of the duty cycle, and the change is compatible with the above models. The period number of the sampling signal is set to $256 \times i$, where $i \in \{20, 21, ..., 40\}$. For each sampling interval, we collect the random number sequence with length 2^{20}, and calculate the ApEn of the bit sequence. Particularly, the quality factor that is

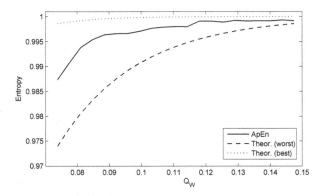

Fig. 7. The comparison between ApEn and theoretical entropy in the physical RNG

influenced by white noises Q_W is computed by employing the method of [5]. The comparison between the ApEn and the theoretical entropy (worst case and best case) is depicted in Fig. 7. It is observed that ApEn increases between the worst and the best case of theoretical entropy as expected. As the bit sequence has been affected by colored noises, its statistical randomness is much better than the worst case of the theoretical entropy. From Fig. 7, we can conclude that our improved theoretical entropy is suitable to estimate the lower and upper bounds of the output bit sequence.

6   Conclusion

Entropy estimation is essential for TRNG security testing, and a reliable result of entropy estimation is preferred for both designers and verifiers. In this paper, we investigate the applicability the different entropy calculation methods for oscillator-based TRNGs, including bit-rate entropy, the lower bound and approximate entropy. In the evaluation, we present two effective methods for bit-rate entropy calculation in theory, and design a specific method for the approximate entropy. The evaluation results indicate that the theoretical estimation results are consistent with the experimental measurements, thus the presented methods are reliable for not small Q values. The mutual verifications among these estimation methods make us believe that the calculated results are reliable. Furthermore, for the case with very small quality factor, the existing entropy estimation methods are inapplicable, thus we recommend to use the precise lower bound as a conservative estimation. In the hardware experiment, we validate that the ApEn still lies in the interval between the worst and best case of the theoretical entropy, though the bit sequence is effected by colored noises.

Acknowledgments.. This work was partially supported by National Basic Research Program of China (973 Program No. 2013CB338001), National Natural Science Foundation of China (No. 61602476, No. 61402470) and Strategy Pilot Project of Chinese Academy of Sciences (No. XDA06010702).

References

1. Amaki, T., Hashimoto, M., Mitsuyama, Y., Onoye, T.: A worst-case-aware design methodology for noise-tolerant oscillator-based true random number generator with stochastic behavior modeling. IEEE Trans. Inf. Forensics Secur. **8**(8), 1331–1342 (2013)
2. Baudet, M., Lubicz, D., Micolod, J., Tassiaux, A.: On the security of oscillator-based random number generators. J. Cryptol. **24**(2), 398–425 (2011)
3. Box, G.E.P., Jenkins, G.: Time Series Analysis: Forecasting and Control, pp. 28–32. Holden-Day, San Francisco (1976)
4. Fischer, V., Lubicz, D.: Embedded evaluation of randomness in oscillator based elementary TRNG. In: Batina, L., Robshaw, M. (eds.) CHES 2014. LNCS, vol. 8731, pp. 527–543. Springer, Heidelberg (2014). doi:10.1007/978-3-662-44709-3_29
5. Haddad, P., Teglia, Y., Bernard, F., Fischer, V.: On the assumption of mutual independence of jitter realizations in P-TRNG stochastic models. In: IEEE Design, Automation and Test in Europe Conference and Exhibition (DATE), pp. 1–6 (2014)
6. Information Technology Laboratory: FIPS 140-2: Security Requirement For Cryptographic Modules (2011)
7. ISO/IEC 18031: Information Technology - Security Techniques - Random bit generation (2011)
8. Killmann, W., Schindler, W.: A proposal for functionality classes for random number generators (2011). http://www.bsi.bund.de/SharedDocs/Downloads/DE/BSI/Zertifizierung/Interpretationen/AIS_31_Functionality_classes_for_random_number_generators_e.pdf?__blob=publicationFile
9. Killmann, W., Schindler, W.: A design for a physical RNG with robust entropy estimators. In: Oswald, E., Rohatgi, P. (eds.) CHES 2008. LNCS, vol. 5154, pp. 146–163. Springer, Heidelberg (2008). doi:10.1007/978-3-540-85053-3_10
10. Ma, Y., Lin, J., Chen, T., Xu, C., Liu, Z., Jing, J.: Entropy evaluation for oscillator-based true random number generators. In: Batina, L., Robshaw, M. (eds.) CHES 2014. LNCS, vol. 8731, pp. 544–561. Springer, Heidelberg (2014). doi:10.1007/978-3-662-44709-3_30
11. Marsaglia, G.: Diehard Battery of Tests of Randomness. http://www.stat.fsu.edu/pub/diehard/
12. Menezes, A., Oorschot, P.V., Vanstone, S.: Handbook of Applied Cryptography. CRC Press, Boca Raton (1997)
13. Rukhin, A., et al.: A statistical test suite for random and pseudorandom number generators for cryptographic applications. NIST Special Publication 800–22. http://csrc.nist.gov/publications/nistpubs/800-22-rev1a/Spp800-22rev1a.pdf
14. Valtchanov, B., Fischer, V., Aubert, A., Bernard, F.: Characterization of randomness sources in ring oscillator-based true random number generators in FPGAs. In: DDECS, pp. 48–53 (2010)

Post-quantum Cryptography

Provably Secure Password Authenticated Key Exchange Based on RLWE for the Post-Quantum World

Jintai Ding[1][✉], Saed Alsayigh[1], Jean Lancrenon[2], Saraswathy RV[1], and Michael Snook[1]

[1] University of Cincinnati, Cincinnati, USA
jintai.ding@gmail.com
[2] University of Luxembourg, Luxembourg City, Luxembourg

Abstract. Authenticated Key Exchange (AKE) is a cryptographic scheme with the aim to establish a high-entropy and secret session key over a insecure communications network. *Password*-Authenticated Key Exchange (PAKE) assumes that the parties in play share a simple password, which is cheap and human-memorable and is used to achieve the authentication. PAKEs are practically relevant as these features are extremely appealing in an age where most people access sensitive personal data remotely from more-and-more pervasive hand-held devices. Theoretically, PAKEs allow the secure computation and authentication of a high-entropy piece of data using a low-entropy string as a starting point. In this paper, we apply the recently proposed technique introduced in [19] to construct two lattice-based PAKE protocols enjoying a very simple and elegant design that is an parallel extension of the class of Random Oracle Model (ROM)-based protocols PAK and PPK [13,41], but in the lattice-based setting. The new protocol resembling PAK is three-pass, and provides *mutual explicit authentication*, while the protocol following the structure of PPK is two-pass, and provides *implicit authentication*. Our protocols rely on the Ring-Learning-with-Errors (RLWE) assumption, and exploit the additive structure of the underlying ring. They have a comparable level of efficiency to PAK and PPK, which makes them highly attractive. We present a preliminary implementation of our protocols to demonstrate that they are both efficient and practical. We believe they are suitable quantum safe replacements for PAK and PPK.

Keywords: Diffie-Hellman · Key Exchange · Authenticated · PAKE · RLWE

1 Introduction

Password-Authenticated Key Exchange and Dictionary Attacks
Authenticated Key Exchange (AKE) is a cryptographic service with the aim of allowing several entities to jointly establish a high-entropy and secret session key over a completely insecure communications network. That the protocol includes

© Springer International Publishing AG 2017
H. Handschuh (Ed.): CT-RSA 2017, LNCS 10159, pp. 183–204, 2017.
DOI: 10.1007/978-3-319-52153-4_11

authentication of the purported peers is essential to prevent man-in-the-middle attacks. In order to achieve this, it is required that some form of long-term authentication material already be in place prior to the exchange occurring. For instance, the entities could each have their own public-key/secret-key pair (e.g. for STS [18], or HMQV [35]), certified by a trusted authority, or they can all share a single symmetric key specifically dedicated to running an AKE with which to establish other session keys (e.g. the protocols in [7]).

In *Password*-Authenticated Key Exchange (PAKE), it is assumed that the parties in play share a simple password. This differs from the shared-symmetric-key case in that the password is not necessarily a cryptographically strong piece of data. Indeed, most passwords have very low entropy so that they can retain their main advantage over strong keying material: they are cheap and human-memorable. Moreover, these features are extremely appealing in an age where most people access sensitive personal data remotely from more-and-more pervasive hand-held devices. Thus, PAKEs are practically relevant. From a theoretical standpoint, they are quite unique in that they allow the secure computation and authentication of a high-entropy piece of data using a low-entropy string as a starting point.

From a security modeling perspective, the use of passwords as authentication material presents specific challenges. A password's low entropy makes it easy to discover by brute force, assuming an attacker can get its hands on a piece of password-dependent data that a guess can be checked against. Such attacks are known as *dictionary attacks*. There are two types: In an *offline* attack, the adversary observes protocol runs - possibly also interacting with the entities involved - and then goes offline to do password testing privately. To avoid this, the protocol messages and session keys must look computationally independent from the password. In an *online attack*, the attacker needs to be *actively involved* in a protocol interaction in order to determine the verdict on its guess(es). The most natural online attack available is to simply run the protocol with an arbitrary password guess as input, and observe whether the protocol run succeeds or fails. It is clear that this attack is unavoidable; thus a PAKE must be designed such that the adversary can test at most a *constant* (ideally, *one*) number of passwords per online interaction.

PAKEs and the Post-Quantum World. Based on the above reasons, PAKEs have been very heavily studied in the past three decades. Adequate formal security models have appeared [6,13], and a plethora of protocols have been designed and analyzed (e.g., [1,14,31,34,41]). The current pool of practical protocols[1] can essentially be classified into two categories: the first we shall call the class of *Random Oracle Model* (ROM)-based PAKEs (such as [3,6,9,14,15,30,41]), and the second, the class of *Common Reference String* (CRS)-based PAKEs (such as [16,23,27,32,34]). Roughly, the protocols in the first category have very simple and elegant designs, but rely crucially on the ROM [8] for their proofs of security,

---

[1] Some impractical yet complexity-theoretically efficient protocols have been studied, for theoretical reasons. See e.g. [25,26,43].

while the protocols in the second category use sophisticated cryptographic tools[2] to achieve standard-model security (assuming a CRS is in place[3]). The bottom line is that the simplicity and efficiency of ROM-based protocols (and the fact that if carefully instantiated, they are not known to have been broken) makes them much more attractive for concrete deployment than CRS-based ones.

Searching for tools that can resist against adversaries attacking using a quantum computer is currently one of the fundamental issues in cryptographic research. Indeed, the security of all public-key algorithms based on classical hard problems will no longer be assured as soon as a quantum computer of satisfactory size exists. In the US, the National Security Agency (NSA) [44] published a webpage announcing preliminary plans for transitioning from its suite B cryptographic tools to quantum resistant algorithms, which are specified by the National Institute of Standards and Technology (NIST) and are used by the NSA's Information Assurance Directorate in solutions approved for protecting classified and unclassified National Security Systems (NSS). It is clear that the effort to develop quantum-resistant technologies is intensifying and, in the immediate future, NIST, which has the authority to establish the security standards of the US government, shall have a call to select post-quantum cryptosystem standards. Regardless of which of the aforementioned categories they belong to, most known PAKEs rest their security on either group-type or factoring-type complexity assumptions, making them unsuitable in a possibly upcoming post-quantum world. Therefore, searching for PAKEs that can be based on provably secure *lattice* assumptions is natural. In the current literature, as far as we know one single PAKE stands out precisely for this reason: the Katz–Vaikuntanathan protocol [33] relies instead on lattice-type assumptions for its security. Unfortunately, it is CRS-based, and therefore not very efficient.

1.1   Our Contributions

In this paper, we propose two lattice-based PAKE protocols enjoying a very simple and elegant design that is extremely similar to that of the class of ROM-based protocols. More specifically, our protocols can be viewed as direct analogues of the PAK and PPK [13,41] protocols in the lattice-based setting. The protocol resembling PAK is three-pass, and provides *mutual explicit authentication*, while the protocol following the structure of PPK is two-pass, and provides *implicit authentication*. Most importantly, our protocols have a comparable level of efficiency to PAK and PPK, which makes them highly attractive.

The starting point for our construction is the recently proposed technique introduced in [19], and used in [50] to design a lattice-based variant of the HMQV protocol. As in the latter paper, our protocols rely on the Ring-Learning-with-Errors (RLWE) assumption, and exploit the additive structure of the underlying

---

[2] In particular, they use *universal hash proof systems* [17] over complex languages.

[3] A CRS is essentially a publicly available string to which a secret trapdoor is theoretically associated, but never used by protocol participants. During a proof of security, the simulator gets access to this trapdoor.

RLWE ring. We therefore obtain two protocols which are suitable quantum-safe replacements for PAK and PPK.

It is indeed true that we can build the PAKE protocol using RLWE in a rather straightforward manner. Though the general structure of the proofs for our protocols is very similar to that of the original PAK protocol's security proof, where the security of PAK relies on an adversary being unable to solve the Diffie–Hellman Problem, the techniques used in our paper are intricate and completely different.

We manage to establish a full proof of security for our RLWE-PAK protocol, in the classic security model proposed by Bellare et al. [6]. To simplify the proof, first we define the Pairing with Errors problem, which can be reduced to the RLWE problem. This new problem is used multiple times in the proof, and allows us to build intermediate steps that did not appear in the original proofs for PAK and PPK.

The complete replacement of the Diffie–Hellman core of PAK with the new lattice-based core means that the distinguishers used in the PAK proof have to be completely replaced with lattice-handling analogues. The distinguishers have to compensate for the presence of the password in the protocol without being able to directly remove its influence, as they have no access to the value of the password itself. In the proof, there are three places where we have to build distinguishers to solve the PWE problem. Since such distinguishers are completely new and subtle, we need to use novel methods to construct them. Only by applying these new distinguishers are we able to link the security directly to the PWE problem.

From the construction in [19], we can use the same idea to build in a completely parallel way a PAKE using the LWE problem instead of the RLWE problem. Here we need to use matrix multiplications, and need to make sure that the order of multiplications is correct.

Finally, we created a proof-of-concept implementation of our new PAKE (and the "implicit" version) to demonstrate its efficiency and practicality. This part is moved to Appendix E of the full version due the lack of room.

1.2   Related Work

AKE protocol research is far too vast to describe in full, hence we only survey those portions of it most relevant to this work. These are PAKE, and AKE based on lattice-type assumptions. We also only consider protocols in the two-party setting.

PAKE Protocols and Security Models. PAKE was essentially invented by Bellovin and Merritt in [9]. The authors raised the problem of dictionary attacks in this particular setting, proposed some protocols - most notably the Encrypted Key Exchange (EKE) protocol - and offered an informal security analysis of their designs. Jablon [30] later proposed another protocol - Simple Password Exponential Key Exchange (SPEKE) - avoiding some of the pitfalls of EKE, but again with only an informal analysis.

The search for good security models began with the work of Lucks [38] and Halevi and Krawczyk [28]. Laying down adequate foundations for the provable security of PAKE was a particularly subtle task, since one cannot prevent the adversary from guessing and trying out candidate passwords in on-line impersonation attempts, and the small size of the dictionary means that the adversary's natural advantage in succeeding at this is *non-negligible*. Good models capturing this phenomenon were finally exhibited by Bellare et al. [6] and Boyko et al. [13], building respectively on the AKE models proposed by Bellare et al. in [7] and Shoup in [46]. The model in [6] was further refined by Abdalla et al. in [4]. The notion of universally composable PAKE has also since been introduced by Canetti et al. [16].

A great deal of protocols have been proposed and analyzed, especially since the apparition of adequate security models. Some extremely efficient examples include the protocols in [2,3,6,13–15,29,30,36,40,41]. On one hand, these are mostly two-or-three-pass protocols, with very few group operations. For instance, the explicitly authenticated PAK protocol in [41] is three-pass, and sends 2 group elements (and 2 confirmation bitstrings of length linear in the security parameter) in total over the network. It also requires a total of only 4 exponentiations, and 2 group multiplications. On the other hand, these protocols' security is very heavily reliant on idealized assumptions[4]. In 2001, a milestone was reached with the work of Katz et al. [32], which showed that it is possible to provably realize PAKE in a *practical way without idealized assumptions*, but at the expense of having a CRS in place. Many works generalizing and optimizing this result followed, such as [22,23,27,31,34], all using a CRS. It was further shown in [16] that without idealized assumptions, universally composable PAKE is possible only if some other trusted setup - e.g. a CRS - is in place. However, all of these CRS-using protocols are generally much less practical than the ROM-using ones mentioned before. While it is possible to achieve a low number of passes using a CRS - e.g. [34] is a two-pass protocol - the number of group computations and elements sent is typically high. To our knowledge, the latest techniques [2] discovered to reduce this still do not beat ROM-based PAKEs in efficiency. For instance, Abdalla et al. [2] report on being able to bring the total group element and exponentiation counts of the Groce-Katz protocol [27] down to 6 and 18 respectively, and those of [34] down to 10 and 28 respectively.

Finally, some work has been devoted to determining if PAKE can be efficiently realized in a reasonable security model with *neither idealized assumptions nor any form of trusted setup*. Goldreich et al. [25] were the first to answer in the affirmative, but assuming non-concurrent protocol executions. Their work was followed up by Nguyen et al. [43], who found a more efficient construction, but in a weaker model. Later, Jain et al. [26] were able to further lift the restriction on concurrent executions. These works are viewed as being mainly of theoretical interest, as the protocols, although *theoretically* efficient, are far less practical then even the CRS-based protocols.

---

[4] The ROM is one of them; another is the *ideal cipher* model, see [6].

AKE from Lattices. Some work was done to address the problem of finding AKE protocols based on lattice-type assumptions. The protocols in [20,21,37] are essentially lattice-based instantiations of generic constructions that use key-encapsulation mechanisms to construct AKEs. In 2012, Ding et al. [19] first proposed simple LWE and RLWE analogues of the unauthenticated Diffie-Hellman protocol. Later there appeared a few variants of Ding's key exchange [5,11,12,45], with the slight modification that the new rounding technique from [19] for least significant bits was adjusted to work for most significant bits. A true LWE-based AKE was proposed in Zhang et al. [50], where the protocol proposed by Ding et al. [19] was leveraged to build a RLWE version of the HMQV protocol.

In all of these works, the authentication mechanism used is reliant on the deployment of a *public-key infrastructure*. In the case of password authentication, the only known protocol to this day appears to be that of Katz et al. [33]. It too can be viewed as a lattice-based instantiation of a generic construction. This is because most known CRS-based frameworks for PAKE make use of an encryption scheme that is both *secure against adaptive chosen-ciphertext attacks* and *equipped with a universal hash proof system* [17], and the heart of [33] is essentially a lattice-based instantiation of such a scheme.

2 Preliminaries

2.1 Security Model

Here, we review the security model from [6]. It basically models the communications between a fixed number of users - which are clients and servers - over a network that is fully controlled by a probabilistic, polynomial-time adversary $\mathcal{A}$. Users are expected to both establish and use session keys over this network. Therefore, $\mathcal{A}$ is given access to a certain number of queries which reflect this usage. It may initialize protocol communications between user instances of its choice, deliver any message it wants to these instances, and observe their reaction according to protocol specification. It may also reveal session keys established by instances, thereby modeling loss of keys through higher-level protocol use. Finally, we even allow the adversary to obtain user passwords, in order to capture forward secrecy. We describe this formally now.

Let P be a PAKE protocol.

Security Game. An algorithmic game initialized with a security parameter k is played between a challenger $\mathcal{CH}$ and a probabilistic polynomial time adversary $\mathcal{A}$. $\mathcal{CH}$ will essentially run P on behalf of honest users, thereby simulating network traffic for $\mathcal{A}$.

Users and Passwords. We assume a fixed set $\mathfrak{U}$ of *users*, partitioned into two non-empty sets $\mathfrak{C}$ of *clients* and $\mathfrak{S}$ of *servers*. We also assume some fixed, non-empty dictionary D of size L. Before the game starts, for each $\mathcal{C} \in \mathfrak{C}$ a password $pw_{\mathcal{C}}$ is drawn uniformly at random from D and assigned to $\mathcal{C}$ outside of $\mathcal{A}$'s view. For each server $\mathcal{S} \in \mathfrak{S}$, we set $pw_{\mathcal{S}} := \big(f(pw_{\mathcal{C}})\big)_{\mathcal{C}}$, where $\mathcal{C}$ runs through all of $\mathfrak{C}$, and f is some efficiently computable one-way function specified by P. (In our case, f will be essentially a hash of the password.) $\mathcal{CH}$ also generates P's public parameters on input 1^k, and gives these to $\mathcal{A}$. We assume that $\mathcal{A}$ is polynomial-time in k as well. The game can then begin.

User Instances. During the game, to any user $\mathcal{U} \in \mathfrak{U}$ is associated an unlimited number of user instances $\varPi_{\mathcal{U}}^i$, where i is a positive integer. The adversary may activate any of these instances using the queries listed below, causing them to initiate and run the protocol.

At any point in time, an instance $\varPi_{\mathcal{U}}^i$ may *accept*. When this happens, it holds a *Partner IDentity* (PID) $pid_{\mathcal{U}}^i$, a *Session IDentity* (SID) $sid_{\mathcal{U}}^i$, and a *Session Key* (SK) $sk_{\mathcal{U}}^i$. The PID is the identity of the user that instance believes it is talking to. The SK is what $\varPi_{\mathcal{U}}^i$ is aiming to ultimately compute. The SID is a string which uniquely identifies the protocol run and ensuing session in which the SK is to be used in. Often the SID is defined as the ordered concatenation of messages sent and received by an instance, except possibly the last message. (In our case, we will need to modify this a bit.)

Queries. The queries $\mathcal{A}$ may make to any given instance $\varPi_{\mathcal{U}}^i$ during the game are as follows:

- Send($\mathcal{U}, i, M$): Causes message M to be sent to instance $\varPi_{\mathcal{U}}^i$. The instance computes what the protocol P says, updates its state, and gives the output to $\mathcal{A}$. We also assume that $\mathcal{A}$ sees if the query causes $\varPi_{\mathcal{U}}^i$ to accept or terminate.
- Execute($\mathcal{C}, i, \mathcal{S}, j$): Causes P to be executed to completion between $\varPi_{\mathcal{C}}^i$ (where $\mathcal{C} \in \mathfrak{C}$) and $\varPi_{\mathcal{S}}^j$ (where $\mathcal{S} \in \mathfrak{S}$) and hands $\mathcal{A}$ the execution's transcript.
- Reveal($\mathcal{U}, i$): Returns the SK $sk_{\mathcal{U}}^i$ held by $\varPi_{\mathcal{U}}^i$ to $\mathcal{A}$.
- Test($\mathcal{U}, i$): For this query to be valid, instance $\varPi_{\mathcal{U}}^i$ must be *fresh*, as defined below. If this is the case, the query causes a bit b to be flipped. If $b = 1$, the actual SK $sk_i^{\mathcal{U}}$ is returned to $\mathcal{A}$; otherwise a string is drawn uniformly from the SK space and returned to $\mathcal{A}$. Note that this query can be asked only once during the game.
- Corrupt($\mathcal{U}$): Returns $\bigl(f(pw_{\mathcal{C}})\bigr)_{\mathcal{C}}$ to $\mathcal{A}$ if $\mathcal{U} \in \mathfrak{S}$ else returns $pw_{\mathcal{U}}$ to $\mathcal{A}$.

Ending the Game. Eventually, $\mathcal{A}$ ends the game, and outputs a single bit b'. We return to the use of this bit in the definition of security below.

Partnering and Freshness. In order to have a meaningful definition of security, we need to introduce the notions of instance *partnering* and instance *freshness*. Essentially, an instance $\varPi_{\mathcal{U}}^i$ is *fresh* if the adversary does not already know that instance's SK through trivial means provided by the security model's queries, for instance by using a Reveal query on the instance in question. Furthermore, since instances are supposed to be sharing keys under normal circumstances, it also makes sense to consider freshness destroyed if an instance's proper communicant has been revealed as well. Thus, we need to formally define what this proper communicant is:

Definition 1. *Let $\Pi_{\mathcal{U}}^i$ and $\Pi_{\mathcal{V}}^j$ be two instances. We shall say that $\Pi_{\mathcal{U}}^i$ and $\Pi_{\mathcal{V}}^j$ are partnered if (i) one is in $\mathfrak{C}$ and one is in $\mathfrak{S}$, (ii) both have accepted, (iii) $pid_{\mathcal{U}}^i = \mathcal{V}$ and $pid_{\mathcal{V}}^j = \mathcal{U}$, (iv) $sid_{\mathcal{U}}^i = sid_{\mathcal{V}}^j =: sid$ and this value is not null, and (v) no other instance accepts with a SID of sid.*

Capturing the notion of *forward secrecy* requires freshness to be carefully defined around the *corrupt* query. Intuitively, if a corruption occurs after an instance has had a correct exchange with a proper partner, then those instances' shared session key should still remain secure. However, we cannot guarantee anything for an instance that has interacted with the adversary after a corruption. More formally:

Definition 2. *An instance $\Pi_{\mathcal{U}}^i$ is fresh if none of the following events occur: (i) Reveal$(\mathcal{U}, (i)$ was queried, (ii) a Reveal$(\mathcal{V}, j)$ was queried, where $\Pi_{\mathcal{V}}^j$ is $\Pi_{\mathcal{U}}^i$'s partner, if it has one, or (iii) Corrupt$(\mathcal{V})$ was queried for some $\mathcal{V}$ before the Test query and a Send$(\mathcal{U}, i, M)$ query occurs for some M.*

Definition of Security. We now turn to actually measuring the adversary's success rate in breaking P. $\mathcal{A}$'s objective is to tell apart a random string from a true SK belonging to a fresh instance. This is the whole purpose of the Test query. Let $Succ_{\mathsf{P}}^{ake}(\mathcal{A})$ be the event:

"$\mathcal{A}$ makes a Test$(\mathcal{U}, i)$ query where $\Pi_{\mathcal{U}}^i$ has terminated and is fresh and $b' = b$, where b is the bit selected when Test$(\mathcal{U}, i)$ was made, and b' is the bit $\mathcal{A}$ output at the end of the game."

$\mathcal{A}$'s *advantage* is then defined as:

$$Adv_{\mathsf{P}}^{ake}(\mathcal{A}) = 2Pr[Succ_{\mathsf{P}}^{ake}(\mathcal{A})] - 1$$

It is easy to see that if we have two protocols P and P' then for any adversary $\mathcal{A}$ we have $Pr[Succ_{\mathsf{P}}^{ake}(\mathcal{A})] = Pr[Succ_{\mathsf{P'}}^{ake}(\mathcal{A})] + \epsilon$ if and only if $Adv_{\mathsf{P}}^{ake}(\mathcal{A}) = Adv_{\mathsf{P'}}^{ake}(\mathcal{A}) + 2\epsilon$.

2.2   Ring Learning with Errors

Ring Learning with Errors. Here, we introduce some notation and recall informally the *Ring Learning with Errors* assumption, introduced in [39]. For our purpose, it will be more convenient to use an assumption we call the Pairing with Errors PWE, which we state formally at the end of the section, and which can easily be shown holds under RLWE.

We denote the security parameter k. Recall that a function f is negligible in k if for every $c > 0$, there exists a N such that $f(k) < \frac{1}{k^c}$ for all $k > N$. The ring of polynomials over $\mathbb{Z}$ (respectively, $\mathbb{Z}_q = \mathbb{Z}/q\mathbb{Z}$) we denote by $\mathbb{Z}[x]$ (resp., $\mathbb{Z}_q[x]$). Let $n \in \mathbb{Z}$ be a power of 2. We consider the ring $R = \mathbb{Z}[x]/(x^n + 1)$. For any positive $q \in \mathbb{Z}$, we set $R_q = \mathbb{Z}_q[x]/(x^n + 1)$. For any polynomial y in R or R_q, we identify y with its coefficient vector in $\mathbb{Z}^n$ or $\mathbb{Z}_q^n$, respectively. Recall that for a fixed $\beta > 0$, the *discrete Gaussian distribution* over R_q (parametrized by β)

is naturally induced by that over $\mathbb{Z}^n$ (centered at 0, with standard deviation β). We denote this distribution over R_q by χ_β. More details can be found in [50].

For a fixed $s \in R_q$, let A_{s,χ_β} be the distribution over pairs $(a, as + 2x) \in R_q \times R_q$, where $a \leftarrow R_q$ is chosen uniformly at random and $x \leftarrow \chi_\beta$ is independent of a. The *Ring Learning with Errors* assumption is the assumption that for a fixed s sampled from χ_β, the distribution A_{s,χ_β} is computationally indistinguishable from the uniform distribution on R_q^2, given polynomially many samples.

We define the norm of a polynomial to be the norm of its coefficient vector. Then we have the following useful facts:

Lemma 1. *Let R be defined as above. Then, for any $s, t \in R$, we have $\|s \cdot t\| \leq \sqrt{n} \cdot \|s\| \cdot \|t\|$ and $\|s \cdot t\|_\infty \leq n \cdot \|s\|_\infty \cdot \|t\|_\infty$.*

Lemma 2 ([24,42]). *For any real number $\alpha = \omega(\sqrt{\log n})$, we have $\Pr_{\mathbf{x} \leftarrow \chi_\alpha}[\|\mathbf{x}\| > \alpha\sqrt{n}] \leq 2^{-n+1}$.*

We now recall the Cha and Mod_2 functions defined in [50]. We denote $\mathbb{Z}_q = \{-\frac{q-1}{2}, \ldots, \frac{q-1}{2}\}$ and consider the set $E := \{-\lfloor\frac{q}{4}\rfloor, \ldots, \lfloor\frac{q}{4}\rceil\}$, i.e. the "middle" of $\mathbb{Z}_q$. Recall that Cha is the characteristic function of *the complement of E*, which returns 0 if the input is in E and 1 if it is not in E, and that $\mathsf{Mod}_2 \colon \mathbb{Z}_q \times \{0,1\} \to \{0,1\}$ is defined as:

$$\mathsf{Mod}_2(v, b) = ((v + b \cdot \frac{q-1}{2}) \bmod q) \bmod 2.$$

These two functions have fundamental features which can be seen in the following two lemmas.

Lemma 3 ([50]). *Let n be the security parameter, and let $q = 2^{\omega(\log n)} + 1$ be an odd prime. Let $v \in \mathbb{Z}_q$ be chosen uniformly at random. For any $b \in \{0,1\}$ and any $v' \in \mathbb{Z}_q$, the output distribution of $\mathsf{Mod}_2(v+v', b)$ given $\mathsf{Cha}(v)$ is statistically close to uniform on $\{0,1\}$.*

Lemma 4 ([50]). *Let q be an odd prime, $v \in \mathbb{Z}_q$ and $e \in \mathbb{Z}_q$ such that $|e| < q/8$. Then, for $w = v + 2e$, we have $\mathsf{Mod}_2(v, \mathsf{Cha}(v)) = \mathsf{Mod}_2(w, \mathsf{Cha}(v))$.*

They also can be extended to R_q by applying them coefficient-wise to the coefficients in $\mathbb{Z}_q$ that define the ring elements. In other words, for any ring element $v = (v_0, \ldots, v_{n-1}) \in R_q$ and binary-vector $\mathbf{b} = (b_0, \ldots, b_{n-1}) \in \{0,1\}^n$, we set $\mathsf{Cha}(v) = (\mathsf{Cha}(v_0), \ldots, \mathsf{Cha}(v_{n-1}))$ and $\mathsf{Mod}_2(v, \mathbf{b}) = (\mathsf{Mod}_2(v_0, b_0), \ldots, \mathsf{Mod}_2(v_{n-1}, b_{n-1}))$.

The PWE Assumption. We now state the Pairing with Errors (PWE) assumption, under which we prove that our protocols are secure. We return to the general notations of Sect. 2.2, but using the Gaussian distribution χ_β for a fixed $\beta \in \mathbb{R}_+^*$. For any $(X, s) \in R_q^2$, we set $\tau(X, s) = \mathsf{Mod}_2(Xs, \mathsf{Cha}(Xs))$. Let $\mathcal{A}$ be probabilistic, polynomial-time algorithm taking inputs of the form (a, X, Y, W), where $(a, X, Y) \in R_q^3$ and $W \in \{0,1\}^n$, and outputting a list of values in $\{0,1\}^n$. $\mathcal{A}$'s objective will be for the string $\tau(X, s)$ to be in its output, where s is randomly chosen from R_q, Y is a "small additive perturbation" of as, and W is $\mathsf{Cha}(Xs)$ itself. Formally, let

$$Adv_{R_q}^{PWE}(\mathcal{A}) \overset{def}{=} Pr\Big[a \leftarrow R_q; s \leftarrow \chi_\beta; X \leftarrow R_q; e \leftarrow \chi_\beta;$$

$$Y \leftarrow as + 2e; W \leftarrow \mathsf{Cha}(Xs) : \tau(X, s) \in \mathcal{A}(a, X, Y, W)\Big]$$

Let $Adv_{R_q}^{PWE}(t, N) = \max_{\mathcal{A}} \Big\{ Adv_{R_q}^{PWE}(\mathcal{A}) \Big\}$, where the maximum is taken over all adversaries of time complexity at most t that output a list containing at most N elements of $\{0, 1\}^n$. The PWE assumption states that for t and N polynomial in k, $Adv_{R_q}^{PWE}(t, N)$ is negligible in k.

We also have decision version of PWE problem that can be defined as follows. Clearly, if DPWE is hard, so is PWE.

Definition 3. *(DPWE) Given* $(a, X, Y, w, \sigma) \in R_q \times R_q \times R_q \times \{0, 1\}^n \times \{0, 1\}^n$ *where* $w = \mathsf{Cha}\, K$ *for some* $K \in R_q$ *and* $\sigma = \mathsf{Mod}_2(K, w)$. *The Decision Pairing with Errors problem (DPWE) is to decide whether* $K = Xs + 2g$ *and* $Y = as + 2e$ *for some* s, g, *and* e *drawn from* χ_β, *or* (K, Y) *is uniformly random in* R_q^2.

Before we show the reduction of the DPWE problem to the RLWE problem, we would like to give a definition to what we called the RLWE-DH problem which can be reduced to RLWE problem.

Definition 4. *(RLWE-DH) Let* R_q *and* χ_β *be defined as above. Given as input ring elements* $a, X, Y,$ *and* K, *where* (a, X) *is uniformly random in* R_q^2, *the RLWE-DH problem is to tell if* K *is* $X \cdot s_y + 2g_y$ *for some* $g_y \leftarrow \chi_\beta$ *and* $Y = a \cdot s_y + 2e_y$ *for some* $s_y, e_y \leftarrow \chi_\beta$, *or* (K, Y) *is uniformly random in* R_q^2.

Now we state the reduction theorems without proof due to the lack of the space. Look at Appendix A of the full version for the proof details.

Theorem 1. *Let* R_q *and* χ_β *be defined as above. The RLWE-DH problem is hard to solve if RLWE problem is hard.*

Now we show the reduction of the DPWE problem to the RLWE-DH problem by the following theorem.

Theorem 2. *Let* R_q *and* χ_β *be defined as above. The DPWE problem is hard if the RLWE-DH problem is hard.*

As a result from Theorems 1 and 2, we can say that if RLWE is a hard problem then DPWE is also hard, and thus so is PWE.

3   Protocol Description

We turn to studying the protocols RLWE-PAK and RLWE-PPK, and their security.

3.1   Password-Authenticated RLWE Key Exchange

Let n be a power of 2, and $f(x) = x^n + 1$. Let $q = 2^{\omega(\log n)} + 1$ be an odd prime such that $q \bmod 2n = 1$. Let $H_1 \colon \{0,1\}^* \to R_q$ be a hash function, $H_l \colon \{0,1\}^* \to \{0,1\}^\kappa$ for $l \in \{2,3\}$ be hash functions for verification of communications, and $H_4 \colon \{0,1\}^* \to \{0,1\}^\kappa$ be a Key Derivation Function (KDF), where κ is the bit-length of the final shared key. We model the hash functions and KDF as random oracles. Let a be a fixed element chosen uniformly at random from R_q and given to all users. Let χ_β be a discrete Gaussian distribution with parameter $\beta \in \mathbb{R}_+^*$. We will make use of the Cha and Mod_2 functions defined in [50] and recalled above. The function f used to compute client passwords' verifiers is set as $f = -H_1(\cdot)$. Our protocol consists of the following steps:

Initiation. Client $\mathcal{C}$ randomly samples $s_\mathcal{C}, e_\mathcal{C} \leftarrow \chi_\beta$, computes $\alpha = a s_\mathcal{C} + 2e_\mathcal{C}$, $\gamma = H_1(pw_\mathcal{C})$ and $m = \alpha + \gamma$ and sends $< \mathcal{C}, m >$ to party $\mathcal{S}$.

Response. Server $\mathcal{S}$ receives $< \mathcal{C}, m >$ from party $\mathcal{C}$ and checks that $m \in R_q$; if not, it aborts. Otherwise it computes $\alpha = m + \gamma'$ where $\gamma' = -H_1(pw_\mathcal{C})$. Server $\mathcal{S}$ then randomly samples $s_\mathcal{S}, e_\mathcal{S} \leftarrow \chi_\beta$ and computes $\mu = a s_\mathcal{S} + 2e_\mathcal{S}$ and $k_\mathcal{S} = \alpha \cdot s_\mathcal{S}$.

Next, Server $\mathcal{S}$ computes $w = \mathsf{Cha}(k_\mathcal{S}) \in \{0,1\}^n$ and $\sigma = \mathsf{Mod}_2(k_\mathcal{S}, w)$. Server $\mathcal{S}$ sends μ, w, and $k = H_2(\mathcal{C}, \mathcal{S}, m, \mu, \sigma, \gamma')$ to party $\mathcal{C}$ and computes the value $k'' = H_3(\mathcal{C}, \mathcal{S}, m, \mu, \sigma, \gamma')$.

Initiator finish. Client $\mathcal{C}$ checks that $\mu \in R_q$, and computes $k_\mathcal{C} = s_\mathcal{C} \cdot \mu$ and $\sigma = \mathsf{Mod}_2(k_\mathcal{C}, w)$. Client $\mathcal{C}$ verifies that $H_2(\mathcal{C}, \mathcal{S}, m, \mu, \sigma_\mathcal{C}, \gamma')$ matches the value of k received from Server $\mathcal{S}$ where $\gamma' = -\gamma$. If it does not, Client $\mathcal{C}$ ends the communication.

If it does, Client $\mathcal{C}$ computes $k' = H_3(\mathcal{C}, \mathcal{S}, m, \mu, \sigma, \gamma')$ and derives the session key $sk_\mathcal{C} = H_4(\mathcal{C}, \mathcal{S}, m, \mu, \sigma, \gamma')$. It then sends k' back to Server $\mathcal{S}$, and sets $sid_\mathcal{C} = (\mathcal{C}, \mathcal{S}, m, \mu)$.

Responder finish. Finally, Server $\mathcal{S}$ verifies that $k' = H_3(\mathcal{C}, \mathcal{S}, m, \mu, \sigma, \gamma')$ the same way Client $\mathcal{C}$ verified k. If this is correct, Server $\mathcal{S}$ then derives the session key by computing $sk_\mathcal{S} = H_4(\mathcal{C}, \mathcal{S}, m, \mu, \sigma, \gamma')$. It sets $sid_\mathcal{S} = (\mathcal{C}, \mathcal{S}, m, \mu)^5$. Otherwise, $\mathcal{S}$ refuses to compute a session key.

Theorem 3 (Correctness). *Let q be an odd prime such that $q > 16\beta^2 n^{3/2}$. Let two parties, $\mathcal{C}$ and $\mathcal{S}$, honestly follow the protocol described above. Then, the two will end with the same key with overwhelming probability.*

Proof. To show the correctness of RLWE-PAK, it is sufficient to show that the key material derived at each end verifies $\mathsf{Mod}_2(k_\mathcal{C}, \mathsf{Cha}(k_\mathcal{S})) = \mathsf{Mod}_2(k_\mathcal{S}, \mathsf{Cha}(k_\mathcal{S}))$. By Lemma 4, if $k_\mathcal{C}$ and $k_\mathcal{S}$ are sufficiently close then we are done. Specifically, if $|k_\mathcal{C} - k_\mathcal{S}| < q/4$ then both sides have the same value, σ. If we compare the two, we find that $k_\mathcal{C} - k_\mathcal{S} = 2[e_\mathcal{S} s_\mathcal{C} - e_\mathcal{C} s_\mathcal{S}]$. By Lemma 2, each individual $e_\mathcal{S}, s_\mathcal{C}, e_\mathcal{C}, s_\mathcal{S}$

---

[5] We purposefully excluded the hint w from the session identifier in order to avoid a trivial bit-flipping attack that makes the proof fail in theory, but otherwise leaves the protocol security unaffected.

term has norm less than $\beta\sqrt{n}$ with overwhelming probability. Applying Lemma 1 and the triangle inequality, we have that $\|k_\mathcal{C} - k_\mathcal{S}\| \leq 4\beta^2 n^{3/2} < q/4$ with overwhelming probability. Hence $\mathsf{Mod}_2(k_\mathcal{C}, \mathsf{Cha}(k_\mathcal{S})) = \mathsf{Mod}_2(k_\mathcal{S}, \mathsf{Cha}(k_\mathcal{S}))$. $\square$

4   Proof of Security for **RLWE-PAK**

Our proof of security follows the one in the PAK suite paper by MacKenzie [41]. We essentially adapt it to our PWE instantiation. The objective is to show that an adversary $\mathcal{A}$ attacking the system is unable to gain any information on the SK of a fresh instance with a greater advantage than through an online dictionary attack. In what follows, we distinguish Client Action (CA) queries and Server Action (SA) queries. The adversary makes a:

- **CA0** query if it instructs some unused $\Pi_\mathcal{C}^i$ to send the first message to some $\mathcal{S}$;
- **SA1** query if it sends some message to a previously unused $\Pi_\mathcal{S}^j$;
- **CA1** query if it sends a message to some $\Pi_\mathcal{C}^i$ expecting the second protocol message;
- **SA2** query if it sends some message to a $\Pi_\mathcal{S}^j$ expecting the last protocol message.

For the convenience of the reader, certain events corresponding to $\mathcal{A}$ making password guesses - against a client instance, against a server instance, and against a client instance and server instance that are partnered - are defined:

- $testpw(\mathcal{C}, i, \mathcal{S}, pw, l)$: for some m, μ, γ', w and k, $\mathcal{A}$ makes an $H_l(< \mathcal{C}, \mathcal{S}, m, \mu, \sigma, \gamma' >)$ query, a CA0 query to $\Pi_\mathcal{C}^i$ with input $\mathcal{S}$ and output $< \mathcal{C}, m >$, a CA1 query to $\Pi_\mathcal{C}^i$ with input $< \mu, k, w >$ and an $H_1(pw)$ query returning $-\gamma' = as_h + 2e_h \in R_q$, where the latest query is either the $H_l(.)$ query or the CA1 query. $\sigma = \mathsf{Mod}_2(k_\mathcal{S}, w) = \mathsf{Mod}_2(k_\mathcal{C}, w)$, $k_\mathcal{S} = \alpha s_\mathcal{S}, k_\mathcal{C} = \mu s_\mathcal{C}$ and $m = \alpha - \gamma'$. The associated value of this event is output of $H_l(.), l \in \{2, 3, 4\}$.
- $testpw!(\mathcal{C}, i, \mathcal{S}, pw)$: for some w and k a CA1 query with input $< \mu, k, w >$ causes a $testpw(\mathcal{C}, i, \mathcal{S}, pw, 2)$ event to occur, with associated value k.
- $testpw(\mathcal{S}, j, \mathcal{C}, pw, l)$: for some m, μ, γ', w and k $\mathcal{A}$ makes an $H_l(< \mathcal{C}, \mathcal{S}, m, \mu, \sigma, \gamma' >)$ query and previously made SA1 query to $\Pi_\mathcal{S}^j$ with input $< \mathcal{C}, m >$ and output $< \mu, k, w >$, and an $H_1(pw)$ query returning $-\gamma'$, where $\sigma = \mathsf{Mod}_2(k_\mathcal{S}, w) = \mathsf{Mod}_2(k_\mathcal{C}, w)$, $k_\mathcal{S} = \alpha s_\mathcal{S}, k_\mathcal{C} = \mu s_\mathcal{C}$ and $m = \alpha - \gamma'$. The associated value of this event is output of $H_l(.), l \in \{2, 3, 4\}$ generated by $\Pi_\mathcal{S}^j$.
- $testpw!(\mathcal{S}, j, \mathcal{C}, pw)$: a SA2 query to $\Pi_\mathcal{S}^j$ is made with k', where a $testpw(\mathcal{S}, j, \mathcal{C}, pw, 3)$ event previously occured with associated value k'.
- $testpw^*(\mathcal{S}, j, \mathcal{C}, pw)$: $testpw(\mathcal{S}, j, \mathcal{C}, pw, l)$ occurs for some $l \in \{2, 3, 4\}$.
- $testpw(\mathcal{C}, i, \mathcal{S}, j, pw)$: for some $l \in \{2, 3, 4\}$, both a $testpw(\mathcal{C}, i, \mathcal{S}, pw, l)$ event and a $testpw(\mathcal{S}, j, \mathcal{C}, pw, l)$ event occur, where $\Pi_\mathcal{C}^i$ is paired with $\Pi_\mathcal{S}^j$ and $\Pi_\mathcal{S}^j$ is paired with $\Pi_\mathcal{C}^i$ after its SA1 query.

- $testexecpw(\mathcal{C}, i, \mathcal{S}, j, pw)$: for some m, μ, γ', w, $\mathcal{A}$ makes an $H_l(< \mathcal{C}, \mathcal{S}, m,$ $\mu, \sigma, \gamma' >)$ query for $l \in \{2, 3, 4\}$, and previously made an $\mathsf{Execute}(\mathcal{C}, i, \mathcal{S}, j)$ query that generates m and μ and an $H_1(pw)$ query returning $-\gamma' = as_h + 2e_h \in R_q$, where $\sigma = \mathsf{Mod}_2(k_\mathcal{S}, w) = \mathsf{Mod}_2(k_\mathcal{C}, w)$, $k_\mathcal{S} = \alpha s_\mathcal{S}$, $k_\mathcal{C} = \mu s_\mathcal{C}$ and $m = \alpha - \gamma'$.
- $correctpw$: before any $\mathsf{Corrupt}$ query, either a $testpw!(\mathcal{C}, i, \mathcal{S}, pw)$ event occurs for some $\mathcal{C}, i$ and $\mathcal{S}$, or a $testpw^*(\mathcal{S}, j, \mathcal{C}, pw_\mathcal{C})$ event occurs for some $\mathcal{S}, j$, and $\mathcal{C}$.
- $correctpwexec$: a $testexecpw(\mathcal{C}, i, \mathcal{S}, j, pw_\mathcal{C})$ event occurs for some $\mathcal{C}, i, \mathcal{S},$ and j.
- $doublepwserver$: before any $\mathsf{Corrupt}$ query happens, both a $testpw^*(\mathcal{S}, j, \mathcal{C}, pw)$ event and $testpw^*(\mathcal{S}, j, \mathcal{C}, pw')$ occur for some $\mathcal{S}, j, \mathcal{C}, pw$ and pw', with $pw \neq pw'$.
- $pairedpwguess$: a $testpw(\mathcal{C}, i, \mathcal{S}, j, pw_\mathcal{C})$ event occurs, for some $\mathcal{C}, i, \mathcal{S},$ and j.

Theorem 4. *Let* P:=RLWE-PAK, *using group* R_q, *and with a password dictionary of size* L. *Fix an adversary* $\mathcal{A}$ *that runs in time* t, *and makes* $n_{se}, n_{ex}, n_{re}, n_{co}$ *queries of type* $\mathsf{Send}, \mathsf{Execute}, \mathsf{Reveal}, \mathsf{Corrupt}$, *respectively, and* n_{ro} *queries to the random oracles. Then for* $t' = O(t + (n_{ro} + n_{se} + n_{ex})t_{exp})$:

$$Adv_\mathsf{P}^{ake}(\mathcal{A}) = \frac{n_{se}}{L} + O\left(n_{se}Adv_{R_q}^{PWE}(t', n_{ro}^2) + Adv_{R_q}^{DRLWE}(t', n_{ro})\right.$$
$$\left. + \frac{(n_{se} + n_{ex})(n_{ro} + n_{se} + n_{ex})}{q^n} + \frac{n_{se}}{2^\kappa}\right)$$

Proof. We study a sequence of protocols - $\mathsf{P}_0, \mathsf{P}_1, \cdots, \mathsf{P}_7$ - with the following properties. First $\mathsf{P}_0 = \mathsf{P}$ and P_7 is by design only possible to attack using natural online guessing. Secondly, we have

$$Adv_{\mathsf{P}_0}^{ake}(\mathcal{A}) \leq Adv_{\mathsf{P}_1}^{ake}(\mathcal{A}) + \epsilon_1 \leq \cdots \leq Adv_{\mathsf{P}_7}^{ake}(\mathcal{A}) + \epsilon_7$$

where $\epsilon_1, \cdots, \epsilon_7$ are all negligible values in k. Adding up the negligible values and counting the success probability of the online attack in P_7 then gives the desired result. The reader can find the proofs of the claims in Appendix C of the full version.

We can assume that n_{ro} and $n_{se} + n_{ex}$ are both ≥ 1. Random oracle queries are answered in the usual way: new queries are answered with uniformly random values, and previously made queries are answered identically to the past response. We further assume that the $H_1(pw)$ query is answered by the simulator by computing the response as $as_h + 2e_h \in R_q$, where (s_h, e_h) is sampled uniformly at random from R_q^2. Finally, if $\mathcal{A}$ makes an $H_l(v)$ query for $l \in \{2, 3, 4\}$ and some v then the corresponding $H_{l'}(v)$ and $H_{l''}(v)$ queries are computed and stored, where $l', l'' \in \{2, 3, 4\} \setminus \{l\}$. $\mathcal{A}$ only sees the output of $H_l(v)$, but the other two queries are still considered to have been made by $\mathcal{A}$.

We now detail our sequence of protocols, and bound $\mathcal{A}$'s advantage difference from each protocol to the next.

Protocol P_0: is just the original protocol P.

Protocol P_1: P_1 is nearly identical to P_0, but is forcefully halted as soon as honest parties randomly choose m or μ values seen previously in the execution.

Specifically, let E_1 be the event that an m value generated in a CA0 or Execute query yields an m value already seen in some previous CA0 or Execute query, an m value already used as input in some previous SA1 query, or an m value from some previous $H_l(.)$ query made by $\mathcal{A}$. Let E_2 be the event that a μ value generated in SA1 or Execute query yields a μ from a previous SA1 or Execute query, a μ value sent as input in some previous CA1 query, or a μ value from a previous $H_l(.)$ query. Setting $E = E_1 \vee E_2$ then P_1 is defined as being identical to P_0 except that the protocol halts and the adversary fails when E occurs.

Claim 1. *For any adversary $\mathcal{A}$,*

$$Adv_{P_0}^{ake}(\mathcal{A}) \leq Adv_{P_1}^{ake}(\mathcal{A}) + \frac{O((n_{se} + n_{ex})(n_{ro} + n_{se} + n_{ex}))}{q^n}$$

Protocol P_2: This protocol is identical to P_1 except that Send and Execute queries are answered without using random oracles. Any random oracle queries $\mathcal{A}$ subsequently makes are answered in such a way as to be consistent with the results of these Send and Execute queries.

In more detail, the queries in P_2 are now answered as follows:

- In an Execute$(\mathcal{C}, i, \mathcal{S}, j)$ query, $m = as_m + 2e_m$ where $s_m, e_m \leftarrow \in R_q$, $\mu = as_S + 2e_S$ where $s_S, e_S \leftarrow \in \chi_\beta$, $w \leftarrow \in \{0,1\}^n$, $k, k' \leftarrow \in \{0,1\}^\kappa$, and $sk_\mathcal{C}^i \leftarrow sk_\mathcal{S}^j \leftarrow \{0,1\}^\kappa$.
- In a CA0 query to instance $\Pi_\mathcal{C}^i$, $m = as_m + 2e_m$ where $s_m, e_m \leftarrow \in R_q$.
- In a SA1 query to instance $\Pi_\mathcal{S}^j$, $\mu = as_S + 2e_S$ where $s_S, e_S \leftarrow \in \chi_\beta$, $w \leftarrow \{0,1\}^n$, and $sk_\mathcal{S}^j, k, k'' \leftarrow \{0,1\}^\kappa$.
- In a CA1 query to instance $\Pi_\mathcal{C}^i$, do the following.
  - If this query causes a $testpw!(\mathcal{C}, i, \mathcal{S}, pw_\mathcal{C})$ event to occur, then set k' to the associated value of the $testpw(\mathcal{C}, i, \mathcal{S}, pw_\mathcal{C}, 3)$ event, and set $sk_\mathcal{C}^i$ to the associated value of the $testpw(\mathcal{C}, i, \mathcal{S}, pw_\mathcal{C}, 4)$ event.
  - Else if $\Pi_\mathcal{C}^i$ is paired with a server instance $\Pi_\mathcal{S}^j$, set $sk_\mathcal{C}^i \leftarrow sk_\mathcal{S}^j$, then $k' \leftarrow \{0,1\}^\kappa$.
  - Otherwise, $\Pi_\mathcal{C}^i$ aborts.
- In a SA2 query to instance $\Pi_\mathcal{S}^j$, if this query causes a $testpw!(\mathcal{S}, j, \mathcal{C}, pw_\mathcal{C})$ event to occur, or if $\Pi_\mathcal{S}^j$ is paired with a client instance $\Pi_\mathcal{C}^i$, terminate. Otherwise, $\Pi_\mathcal{S}^j$ aborts.
- In an $H_l(< \mathcal{C}, \mathcal{S}, m, \mu, \sigma, \gamma' >)$ query, for $l \in \{2,3,4\}$, if this $H_l(.)$ query causes a $testpw(\mathcal{S}, j, \mathcal{C}, pw_\mathcal{C}, l)$ event, or $testexecpw(\mathcal{C}, i, \mathcal{S}, j, pw_\mathcal{C})$ event to occur, then output the associated value of the event. Otherwise, output a random value from $\{0,1\}^\kappa$.

Claim 2. *For any adversary $\mathcal{A}$,*

$$Adv_{P_1}^{ake}(\mathcal{A}) = Adv_{P_2}^{ake}(\mathcal{A}) + \frac{O(n_{ro})}{q^n} + \frac{O(n_{se})}{2^\kappa}$$

Protocol P_3: is identical to P_2 except that in an $H_l(< \mathcal{C}, \mathcal{S}, m, \mu, \sigma, \gamma' >)$ query, for $l \in \{2, 3, 4\}$, it is not checked for consistency against Execute query. So the protocol responds with a random output instead backpatching to preserve consistency with an Execute query. Simply there is no $testexecpw(\mathcal{C}, i, \mathcal{S}, j, pw_\mathcal{C})$ event checking.

Claim 3. *For any adversary $\mathcal{A}$ running in time t, there is a $t' = O(t + (n_{ro} + n_{se} + n_{ex})t_{exp})$ such that,*

$$Adv_{P_2}^{ake}(\mathcal{A}) \leq Adv_{P_3}^{ake}(\mathcal{A}) + Adv_{R_q}^{DRLWE}(t', n_{ro}) + 2Adv_{R_q}^{PWE}(t', n_{ro})$$

Protocol P_4: is identical to P_3 except that if *correctpw* occurs then the protocol halts and the adversary automatically succeeds. This causes theses changes:

1. In a CA1 query to $\varPi_\mathcal{C}^i$, if a $testpw!(\mathcal{C}, i, \mathcal{S}, pw_\mathcal{C})$ event occurs and no Corrupt query has been made, halt and say the adversary automatically succeeds.
2. In an $H_l(< \mathcal{C}, \mathcal{S}, m, \mu, \sigma, \gamma' >)$ query for $l \in \{2, 3, 4\}$, if a $testpw^*(\mathcal{S}, j, \mathcal{C}, pw_\mathcal{C})$ event occurs and no Corrupt query has been made, halt and say the adversary automatically succeeds.

Claim 4. *For any adversary $\mathcal{A}$,*

$$Adv_{P_3}^{ake}(\mathcal{A}) \leq Adv_{P_4}^{ake}(\mathcal{A})$$

Proof. This change can only increase the adversary's chances at winning the game, hence the inequality. □

Protocol P_5: is identical to P_4 except that if the adversary makes a password guess against partnered client and server instances, the protocol halts and the adversary fails. Simply if a *pairedpwguess* event occurs, the protocol halts and the adversary fails. We suppose that when a query is made, the test for *correctpw* occurs after the test for *pairedpwguess*. Note that this causes the following change: if a $testpw(\mathcal{C}, i, \mathcal{S}, pw, l)$ event occurs, this should be checked in a CA1 query, or an $H_l(.)$ query for $l \in \{2, 3, 4\}$ check if a $testpw(\mathcal{C}, i, \mathcal{S}, pw)$ event also occurs.

Claim 5. *For any adversary $\mathcal{A}$ running in time t, there is a $t' = O(t + (n_{ro} + n_{se} + n_{ex})t_{exp})$ such that,*

$$Adv_{P_4}^{ake}(\mathcal{A}) \leq Adv_{P_5}^{ake}(\mathcal{A}) + 2n_{se} Adv_{R_q}^{PWE}(t', n_{ro})$$

Protocol P_6: is identical to P_5 except that if the adversary makes two password guesses against the same server instance, i.e. if a *doublepwserver* event occurs, the protocol halts and the adversary fails. We suppose that when a query is made, the test for *pairedpwguess* or *correctpw* occurs after the test for *doublepwserver*.

Claim 6. *For any adversary $\mathcal{A}$ running in time t, there is a $t' = O(t + (n_{ro} + n_{se} + n_{ex})t_{exp})$ such that,*

$$Adv_{P_5}^{ake}(\mathcal{A}) \leq Adv_{P_6}^{ake}(\mathcal{A}) + 4Adv_{R_q}^{PWE}(t', n_{ro}^2)$$

Protocol P$_7$: is identical to P$_6$ except that this protocol has an internal password oracle that holds all passwords and accepts queries that examine the correctness of a given password. Note that this internal oracle *passwordoracle* is not available to the adversary. So this oracle generates all passwords during initialization. It accepts queries of the form $testpw(\mathcal{C}, pw)$ and returns TRUE if $pw = pw_{\mathcal{C}}$, and FALSE otherwise. It also accepts Corrupt(U) queries whether $U \in \mathfrak{S}$ or $U \in \mathfrak{C}$. When a Corrupt(U) query made in the protocol, it is answered using a Corrupt(U) query to the password oracle. The protocol is also test if *correctpw* occurs, whenever the first $testpw(\mathcal{C}, i, \mathcal{S}, pw)$ event occurs for an instance $\Pi_{\mathcal{C}}^i$ and password pw, or the first $testpw(\mathcal{S}, j, \mathcal{C}, pw)$ event occurs for an instance $\Pi_{\mathcal{S}}^j$ and password pw, a $testpw(\mathcal{C}, pw)$ query is made to the password oracle to see if $pw = pw_{\mathcal{C}}$.

Claim 7. *For any adversary* $\mathcal{A}$,

$$Adv_{\mathsf{P}_6}^{ake}(\mathcal{A}) = Adv_{\mathsf{P}_7}^{ake}(\mathcal{A})$$

Proof. By observation, P$_6$ and P$_7$ are perfectly indistinguishable. $\qquad\qquad\square$

Now we analyze the advantage of an adversary $\mathcal{A}$ against the protocol P$_7$. From the definition of P$_7$, one can easily bounds the probability of adversary $\mathcal{A}$ succeeding in P$_7$ as the following:

$$Pr(Succ_{\mathsf{P}_7}^{ake}(\mathcal{A})) \leq Pr(correctpw) + Pr(Succ_{\mathsf{P}_7}^{ake}(\mathcal{A}) \mid \neg correctpw)Pr(\neg correctpw).$$

Note that $Pr(correctpw) \leq \frac{n_{se}}{L}$ if the passwords are uniformly chosen from a dictionary of size L, because a Corrupt query occurs after at most n_{se} queries were occurred to the password oracle.

Next we compute $Pr(Succ_{\mathsf{P}_7}^{ake}(\mathcal{A}) \mid \neg correctpw)$. Since *correctpw* event does not occur then the only way for $\mathcal{A}$ to succeed is making a Test query to a fresh instance $\Pi_{\mathcal{U}}^i$ and guessing the bit used in the Test query. Note that if we can prove that the view of the adversary is not dependent on sk_U^i then the probability of success is exactly $\frac{1}{2}$ and to do that we have to examine Reveal and $H_4(.)$ queries.

For the first type, we know by definition of Reveal(U, i) query that there could be no one for the fresh instance $\Pi_{\mathcal{U}}^i$. Also there is no Reveal(U', j) query for the instance $\prod_{j}^{U'}$ which is partnered with $\Pi_{\mathcal{U}}^i$. Moreover the adversary fails if more than a single client instance and a single server instance accept with the same *sid* by protocol P$_1$. Thus the output of Reveal queries is independent of sk_U^i.

For the second type, from P$_4$ the unpaired client or server instance will not terminate before a *correctpw* event or a Corrupt query which means an instance may only be fresh and receive a Test query if it is partnered. However if $\Pi_{\mathcal{U}}^i$ is partnered, $H_4(.)$ query will never reveal sk_U^i by P$_5$.

So, the view of the adversary not dependent on sk_U^i then the probability of success is exactly $\frac{1}{2}$. Therefore,

$$Pr(Succ_{\mathsf{P}_7}^{ake}(\mathcal{A})) \leq Pr(correctpw) + Pr(Succ_{\mathsf{P}_7}^{ake}(\mathcal{A}) \mid \neg correctpw)Pr(\neg correctpw)$$
$$\leq Pr(correctpw) + Pr(Succ_{\mathsf{P}_7}^{ake}(\mathcal{A}) \mid \neg correctpw)(1 - Pr(correctpw))$$
$$\leq \frac{n_{se}}{L} + \frac{1}{2}(1 - \frac{n_{se}}{L}))$$
$$\leq \frac{1}{2} + \frac{n_{se}}{2L}.$$

And $Adv_{\mathsf{P}_7}^{ake}(\mathcal{A}) \leq \frac{n_{se}}{L}$. The theorem follows from this and the Claims 1, 2, 3, 4, 5, 6 and 7 above. □

5   Implicit Authentication

In this section, we describe a variant of the protocol that gives implicit authentication, similar to the PPK variant on the PAK protocol. We call it the RLWE-PPK protocol. If either party provides an incorrect password, then the parties' keys will not actually match, and neither party will learn anything about the key held by the other. This effectively prevents communication without explicitly checking for matching passwords.

5.1   RLWE-PPK

The setup is slightly different from that of RLWE-PAK. Here, we need two hash functions H_1 and H_2 from $\{0,1\}^*$ into R_q, and one KDF H_3 from $\{0,1\}^*$ into $\{0,1\}^\kappa$, where κ is again the length of the derived SK. Of course, these are modeled as random oracles. Also, the function f used to compute password verifiers for the server is instantiated as follows: $f(\cdot) = \big(- H_1(\cdot), H_2(\cdot) \big)$.

Initiation. Client $\mathcal{C}$ randomly samples $s_\mathcal{C}, e_\mathcal{C} \leftarrow \chi_\beta$, computes $\alpha = as_\mathcal{C} + 2e_\mathcal{C}$, $\gamma_1 = H_1(pw_\mathcal{C})$, $\gamma_2 = H_2(pw_\mathcal{C})$ and $m = \alpha + \gamma_1$ and sends$< \mathcal{C}, m >$ to party $\mathcal{S}$.

Response. Server $\mathcal{S}$ receives $< \mathcal{C}, m >$ from party $\mathcal{C}$ and checks if $m \in R_q$. If not, abort; otherwise Server $\mathcal{S}$ randomly samples $s_\mathcal{S}, e_\mathcal{S} \leftarrow \chi_\beta$ and computes $\nu = as_\mathcal{S} + 2e_\mathcal{S}$ and recovers $\alpha = m + \gamma_1'$ where $< \gamma_1', \gamma_2 >$. Then compute $\mu = \nu + \gamma_2$ and $k_\mathcal{S} = \alpha \cdot s_\mathcal{S}$.
Next, Server $\mathcal{S}$ computes $w = \mathsf{Cha}(k_\mathcal{S}) \in \{0,1\}^n$ and $\sigma = \mathsf{Mod}_2(k_\mathcal{S}, w)$. Server $\mathcal{S}$ sends μ and w to party $\mathcal{C}$ and computes $sk_\mathcal{S} = H_3(\mathcal{C}, \mathcal{S}, m, \mu, \sigma, \gamma_1')$.

Initiator finish. Client $\mathcal{C}$ receives $< \mu, w >$ from party $\mathcal{S}$ and checks if $\mu \in R_q$. If not, it aborts, and otherwise $\mathcal{C}$ recovers $\nu = \mu - \gamma_2$, computes $k_\mathcal{C} = s_\mathcal{C} \cdot \nu$ and $\sigma = \mathsf{Mod}_2(k_\mathcal{C}, w)$.
Finally, Client $\mathcal{C}$ derives the session key $sk_\mathcal{C} = H_3(\mathcal{C}, \mathcal{S}, m, \mu, \sigma, \gamma_1')$.

5.2   Proof of Security for RLWE-PPK

The proof of security for our implicitly authenticated protocol follows the model of security in the PAK suite paper by Mackenzie [41], and is similar to our proof for the explicitly authenticated protocol above. Therefore we will not go through the proof details. However we give a sketch of the proof in Appendix D of the full version. We first define some similar events to those in Sect. 4, corresponding to the adversary making a password guess against a client instance, against a server instance, and against a client instance and server instance that are partnered. Then we need to show that an adversary attacking the system is unable to determine the session key of a fresh instance with greater advantage than that in an online dictionary attack.

Theorem 5. *Let* P:=RLWE-PPK *as described above, using group* R_q, *and with a password dictionary of size* L. *Fix an adversary* $\mathcal{A}$ *that runs in time* t, *and makes* $n_{se}, n_{ex}, n_{re}, n_{co}$ *queries of type* Send, Execute, Reveal, Corrupt, *respectively, and* n_{ro} *queries to the random oracles. Then for* $t' = O(t + (n_{ro} + n_{se} + n_{ex})t_{exp})$:

$$Adv_P^{ake}(\mathcal{A}) = \frac{n_{se}}{L} + O\left(Adv_{R_q}^{PWE}(t', n_{ro}{}^2) + \frac{(n_{se} + n_{ex})(n_{ro} + n_{se} + n_{ex})}{q^n}\right)$$

For space limitation reasons, the proof of this theorem and more details regarding the security of RLWE-PPK were moved to Appendix D of the full version.

6 Conclusions

We have proposed two new explicitly and implicitly authenticated PAKE protocols. Our protocols are similar to PAK and PPK; however they are based on the Ring Learning with Errors problem. Though our construction is very similar to the classical construction, the security proof is subtle and intricate and it requires novel techniques. We provide a full proof of security of the new protocols in the Random Oracle Model. We also provide a proof of concept implementation and implementation results show our protocols are practical and efficient.

In the proof, we make use of the ROM, which models hash functions as random functions. Our proof is a classical proof of security, and may not hold against a quantum adversary. Against such adversaries, one natural extension of the ROM is to allow the queries to be in *quantum superposition*; this is known as the Quantum Random Oracle Model (QROM) [10]. Unfortunately, many tricks that can be used in the ROM are hard to apply in the QROM. Therefore we leave proving the security of our protocols in the QROM as future work. Although there are some developing proof techniques in the QROM [47–49], more work is needed to adapt classical proofs to this setting.

Acknowledgments. Many thanks to the reviewers for their comments and Peter Ryan for the useful discussions. We would also like to thank the NSF for its partial support. Finally, the third author was supported by the National Research Fund, Luxembourg (CORE project aToMS and INTER project Sequoia).

References

1. Abdalla, M., Benhamouda, F., MacKenzie, P.: Security of the J-PAKE password-authenticated key exchange protocol. In: 2015 IEEE Symposium on Security and Privacy (2015)

2. Abdalla, M., Benhamouda, F., Pointcheval, D.: Public-key encryption indistinguishable under plaintext-checkable attacks. In: Katz, J. (ed.) PKC 2015. LNCS, vol. 9020, pp. 332–352. Springer, Berlin (2015). doi:10.1007/978-3-662-46447-2_15

3. Abdalla, M., Catalano, D., Chevalier, C., Pointcheval, D.: Efficient two-party password-based key exchange protocols in the UC framework. In: Malkin, T. (ed.) CT-RSA 2008. LNCS, vol. 4964, pp. 335–351. Springer, Berlin (2008). doi:10.1007/978-3-540-79263-5_22

4. Abdalla, M., Fouque, P.-A., Pointcheval, D.: Password-based authenticated key exchange in the three-party setting. In: Vaudenay, S. (ed.) PKC 2005. LNCS, vol. 3386, pp. 65–84. Springer, Berlin (2005). doi:10.1007/978-3-540-30580-4_6

5. Alkim, E., Ducas, L., Pöppelmann, T., Schwabe, P.: Post-quantum key exchange-a new hope. In: 25th USENIX Security Symposium, USENIX Security 16, pp. 327–343 (2016)

6. Bellare, M., Pointcheval, D., Rogaway, P.: Authenticated key exchange secure against dictionary attacks. In: Preneel, B. (ed.) EUROCRYPT 2000. LNCS, vol. 1807, pp. 139–155. Springer, Berlin (2000). doi:10.1007/3-540-45539-6_11

7. Bellare, M., Rogaway, P.: Entity authentication and key distribution. In: Stinson, D.R. (ed.) CRYPTO 1993. LNCS, vol. 773, pp. 232–249. Springer, Berlin (1994). doi:10.1007/3-540-48329-2_21

8. Bellare, M., Rogaway, P.: Random oracles are practical: a paradigm for designing efficient protocols. In: Proceedings of the 1st ACM Conference on Computer and Communications Security, CCS 1993, pp. 62–73. ACM, New York (1993)

9. Bellovin, S.M., Merritt, M.: Encrypted key exchange: password-based protocols secure against dictionary attacks. In: 1992 IEEE Computer Society Symposium on Research in Security and Privacy, pp. 72–84, 4–6 May 1992

10. Boneh, D., Dagdelen, Ö., Fischlin, M., Lehmann, A., Schaffner, C., Zhandry, M.: Random oracles in a quantum world. In: Lee, D.H., Wang, X. (eds.) ASIACRYPT 2011. LNCS, vol. 7073, pp. 41–69. Springer, Berlin (2011). doi:10.1007/978-3-642-25385-0_3

11. Bos, J., Costello, C., Ducas, L., Mironov, I., Naehrig, M., Nikolaenko, V., Raghunathan, A., Stebila, D.: Frodo: Take off the ring! practical, quantum-secure key exchange from LWE. In: Proceedings of the 2016 ACM SIGSAC Conference on Computer and Communications Security. ACM (2016)

12. Bos, J.W., Costello, C., Naehrig, M., Stebila, D.: Post-quantum key exchange for the TLS protocol from the ring learning with errors problem. In: 2015 IEEE Symposium on Security and Privacy (SP), pp. 553–570. IEEE (2015)

13. Boyko, V., MacKenzie, P., Patel, S.: Provably secure password-authenticated key exchange using Diffie-Hellman. In: Preneel, B. (ed.) EUROCRYPT 2000. LNCS, vol. 1807, pp. 156–171. Springer, Berlin (2000). doi:10.1007/3-540-45539-6_12

14. Bresson, E., Chevassut, O., Pointcheval, D.: Security proofs for an efficient password-based key exchange. In: ACM Conference on Computer and Communications Security. ACM (2003)

15. Bresson, E., Chevassut, O., Pointcheval, D.: New security results on encrypted key exchange. In: Bao, F., Deng, R., Zhou, J. (eds.) PKC 2004. LNCS, vol. 2947, pp. 145–158. Springer, Berlin (2004). doi:10.1007/978-3-540-24632-9_11

16. Canetti, R., Halevi, S., Katz, J., Lindell, Y., MacKenzie, P.: Universally composable password-based key exchange. In: Cramer, R. (ed.) EUROCRYPT 2005. LNCS, vol. 3494, pp. 404–421. Springer, Berlin (2005). doi:10.1007/11426639_24

17. Cramer, R., Shoup, V.: Universal hash proofs and a paradigm for adaptive chosen ciphertext secure public-key encryption. In: Knudsen, L.R. (ed.) EUROCRYPT 2002. LNCS, vol. 2332, pp. 45–64. Springer, Berlin (2002). doi:10.1007/3-540-46035-7_4

18. Diffie, W., Van Oorschot, P.C., Wiener, M.J.: Authentication and authenticated key exchanges. Des. Codes Crypt. 2, 107–125 (1992)

19. Ding, J., Xie, X., Lin, X.: A simple provably secure key exchange scheme based on the learning with errors problem. Cryptology ePrint Archive, Report 2012/688 (2012)

20. Fujioka, A., Suzuki, K., Xagawa, K., Yoneyama, K.: Strongly secure authenticated key exchange from factoring, codes, and lattices. In: Fischlin, M., Buchmann, J., Manulis, M. (eds.) PKC 2012. LNCS, vol. 7293, pp. 467–484. Springer, Berlin (2012). doi:10.1007/978-3-642-30057-8_28

21. Fujioka, A., Suzuki, K., Xagawa, K., Yoneyama, K.: Practical and post-quantum authenticated key exchange from one-way secure key encapsulation mechanism. In: Proceedings of the 8th ACM SIGSAC Symposium on Information, Computer and Communications Security, ASIA CCS 2013, pp. 83–94. ACM, New York (2013)

22. Gennaro, R.: Faster and shorter password-authenticated key exchange. In: Canetti, R. (ed.) TCC 2008. LNCS, vol. 4948, pp. 589–606. Springer, Berlin (2008). doi:10.1007/978-3-540-78524-8_32

23. Gennaro, R., Lindell, Y.: A framework for password-based authenticated key exchange. In: Biham, E. (ed.) EUROCRYPT 2003. LNCS, vol. 2656, pp. 524–543. Springer, Berlin (2003). doi:10.1007/3-540-39200-9_33

24. Gentry, C., Peikert, C., Vaikuntanathan, V.: Trapdoors for hard lattices and new cryptographic constructions. In: Proceedings of the 40th Annual ACM Symposium on Theory of Computing, STOC 2008, pp. 197–206. ACM, New York (2008)

25. Goldreich, O., Lindell, Y.: Session-key generation using human passwords only. In: Kilian, J. (ed.) CRYPTO 2001. LNCS, vol. 2139, pp. 408–432. Springer, Berlin (2001). doi:10.1007/3-540-44647-8_24

26. Goyal, V., Jain, A., Ostrovsky, R.: Password-authenticated session-key generation on the internet in the plain model. In: Rabin, T. (ed.) CRYPTO 2010. LNCS, vol. 6223, pp. 277–294. Springer, Berlin (2010). doi:10.1007/978-3-642-14623-7_15

27. Groce, A., Katz, J.: A new framework for efficient password-based authenticated key exchange. In: Proceedings of the 17th ACM Conference on Computer and Communications Security, CCS 2010, pp. 516–525. ACM, New York (2010)

28. Halevi, S., Krawczyk, H.: Public-key cryptography and password protocols. ACM Trans. Inf. Syst. Secur. 2, 230–268 (1999)

29. Hao, F., Ryan, P.: J-PAKE: authenticated key exchange without PKI. In: Gavrilova, M.L., Tan, C.J.K., Moreno, E.D. (eds.) Transactions on Computational Science XI. LNCS, vol. 6480, pp. 192–206. Springer, Berlin (2010). doi:10.1007/978-3-642-17697-5_10

30. Jablon, D.P.: Strong password-only authenticated key exchange. ACM SIGCOMM Comput. Commun. Rev. 5, 5–26 (1996)

31. Jiang, S., Gong, G.: Password based key exchange with mutual authentication. In: Handschuh, H., Hasan, M.A. (eds.) SAC 2004. LNCS, vol. 3357, pp. 267–279. Springer, Berlin (2004). doi:10.1007/978-3-540-30564-4_19
32. Katz, J., Ostrovsky, R., Yung, M.: Efficient password-authenticated key exchange using human-memorable passwords. In: Pfitzmann, B. (ed.) EUROCRYPT 2001. LNCS, vol. 2045, pp. 475–494. Springer, Berlin (2001). doi:10.1007/3-540-44987-6_29
33. Katz, J., Vaikuntanathan, V.: Smooth projective hashing and password-based authenticated key exchange from lattices. In: Matsui, M. (ed.) ASIACRYPT 2009. LNCS, vol. 5912, pp. 636–652. Springer, Berlin (2009). doi:10.1007/978-3-642-10366-7_37
34. Katz, J., Vaikuntanathan, V.: Round-optimal password-based authenticated key exchange. In: Ishai, Y. (ed.) TCC 2011. LNCS, vol. 6597, pp. 293–310. Springer, Berlin (2011). doi:10.1007/978-3-642-19571-6_18
35. Krawczyk, H.: HMQV: A high-performance secure Diffie-Hellman protocol. In: Shoup, V. (ed.) CRYPTO 2005. LNCS, vol. 3621, pp. 546–566. Springer, Berlin (2005). doi:10.1007/11535218_33
36. Kwon, T.: Authentication and key agreement via memorable password. In: ISOC Network and Distributed System Security Symposium (2001)
37. Lindner, R., Peikert, C.: Better key sizes (and attacks) for LWE-based encryption. In: Kiayias, A. (ed.) CT-RSA 2011. LNCS, vol. 6558, pp. 319–339. Springer, Berlin (2011). doi:10.1007/978-3-642-19074-2_21
38. Lucks, S.: Open key exchange: how to defeat dictionary attacks without encrypting public keys. In: Christianson, B., Crispo, B., Lomas, M., Roe, M. (eds.) Security Protocols 1997. LNCS, vol. 1361, pp. 79–90. Springer, Berlin (1998). doi:10.1007/BFb0028161
39. Lyubashevsky, V., Peikert, C., Regev, O.: On ideal lattices and learning with errors over rings. In: Gilbert, H. (ed.) EUROCRYPT 2010. LNCS, vol. 6110, pp. 1–23. Springer, Berlin (2010). doi:10.1007/978-3-642-13190-5_1
40. MacKenzie, P.: On the Security of the SPEKE Password-Authenticated Key Exchange Protocol. Cryptology ePrint Archive, Report 2001/057 (2001)
41. MacKenzie, P.: The PAK Suite: Protocols for Password-Authenticated Key Exchange. DIMACS Technical report 2002-46 (2002). p. 7
42. Micciancio, D., Regev, O.: Worst-case to average-case reductions based on Gaussian measures. SIAM J. Comput. 37, 267–302 (2007)
43. Nguyen, M.-H., Vadhan, S.: Simpler session-key generation from short random passwords. In: Naor, M. (ed.) TCC 2004. LNCS, vol. 2951, pp. 428–445. Springer, Berlin (2004). doi:10.1007/978-3-540-24638-1_24
44. NSA: Commercial national security algorithm suite (2015). https://www.iad.gov/iad/programs/iad-initiatives/cnsa-suite.cfm
45. Peikert, C.: Lattice cryptography for the internet. In: Mosca, M. (ed.) PQCrypto 2014. LNCS, vol. 8772, pp. 197–219. Springer, Cham (2014). doi:10.1007/978-3-319-11659-4_12
46. Shoup, V.: On Formal Models for Secure Key Exchange. Cryptology ePrint Archive, Report 1999/012 (1999)
47. Unruh, D.: Quantum position verification in the random oracle model. In: Garay, J.A., Gennaro, R. (eds.) CRYPTO 2014. LNCS, vol. 8617, pp. 1–18. Springer, Berlin (2014). doi:10.1007/978-3-662-44381-1_1
48. Unruh, D.: Revocable quantum timed-release encryption. In: Nguyen, P.Q., Oswald, E. (eds.) EUROCRYPT 2014. LNCS, vol. 8441, pp. 129–146. Springer, Berlin (2014). doi:10.1007/978-3-642-55220-5_8

49. Zhandry, M.: Secure identity-based encryption in the quantum random oracle model. In: Safavi-Naini, R., Canetti, R. (eds.) CRYPTO 2012. LNCS, vol. 7417, pp. 758–775. Springer, Berlin (2012). doi:10.1007/978-3-642-32009-5_44
50. Zhang, J., Zhang, Z., Ding, J., Snook, M., Dagdelen, Ö.: Authenticated key exchange from ideal lattices. In: Oswald, E., Fischlin, M. (eds.) EUROCRYPT 2015. LNCS, vol. 9057, pp. 719–751. Springer, Berlin (2015). doi:10.1007/978-3-662-46803-6_24

Symmetric Key Cryptanalysis

Impossible-Differential and Boomerang Cryptanalysis of Round-Reduced Kiasu-BC

Christoph Dobraunig[1(✉)] and Eik List[2]

[1] Graz University of Technology, Graz, Austria
christoph.dobraunig@iaik.tugraz.at
[2] Bauhaus-Universität Weimar, Weimar, Germany
eik.list@uni-weimar.de

Abstract. KIASU-BC is a tweakable block cipher proposed by Jean et al. at ASIACRYPT 2014 alongside their TWEAKEY framework. The cipher is almost identical to the AES-128 except for the tweak, which renders it an attractive primitive for various modes of operation and applications requiring tweakable block ciphers. Therefore, studying how the additional tweak input affects security compared to that of the AES is highly valuable to gain trust in future instantiations.

This work proposes impossible-differential and boomerang attacks on eight rounds of KIASU-BC in the single-key model, using the core idea that the tweak input allows to construct local collisions. While our results do not threat the security of the full-round version, they help concretize the security of KIASU-BC in the single-key model.

Keywords: Symmetric-key cryptography · Cryptanalysis · Tweakable block cipher

1 Introduction

At ASIACRYPT 2014, Jean et al. [13] proposed the TWEAKEY framework together with three software-efficient tweakable block ciphers based on the AES round function DEOXYS-BC, JOLTIK-BC, and KIASU-BC. Such tweakable block ciphers process, in addition to key and plaintext, an additional public input, called the *tweak*. While the first construction that followed this concept was the AES candidate by Schroeppel and Orman [23], the formal foundations have been laid by Liskov, Rivest, and Wagner [15]. Nowadays, tweakable block ciphers possess various applications in cryptographic schemes, such as compression functions (e.g. [11]), variable-input-length ciphers (e.g. [18]), message-authentication codes (e.g. [20]), or (authenticated) encryption schemes (e.g., [14,22]).

While DEOXYS-BC and JOLTIK-BC use a new linear tweak and key schedule, and in the case of JOLTIK-BC a round function different from AES working on 64 bit blocks, the design of KIASU-BC strictly follows AES-128. KIASU-BC uses

C. Dobraunig—The work has been supported in part by the Austrian Science Fund (project P26494-N15).

H. Handschuh (Ed.): CT-RSA 2017, LNCS 10159, pp. 207–222, 2017.
DOI: 10.1007/978-3-319-52153-4_12

exactly the key schedule, round function, and number of rounds of the AES-128. The only difference is an additional 64-bit tweak that is XORed to the topmost two rows of the state after every round. So, KIASU-BC is identical to the AES-128 if the tweak is set to all-zeroes. Therefore, KIASU-BC may appear attractive as primitive for instantiating ciphers, AE schemes, or MACs based on tweakable block ciphers, for it can reuse existing components of AES implementations. In addition, all the existing and newly done analysis for AES-128 is directly applicable to KIASU-BC. However, the additional tweak input enhances the freedom in attacks. Thus, a comprehensive cryptanalysis of KIASU-BC is necessary to determine possible negative effects.

The designers' analysis in [12] concentrates on differential and meet-in-the-middle attacks. They stress that the size of the tweak and the position where it is XORed to the state has been the result of a careful security analysis and "the current choice in KIASU-BC assures that no such high probability characteristics exist on more than 6 rounds and no boomerang characteristics on more than 7 rounds". Concluding from an automated search, the designers argue that the minimum number of active S-boxes for seven-round KIASU-BC is 22, corresponding to an upper bound of the probability of differential characteristics of $(2^{-6})^{22} = 2^{-132}$. Since the bound is not tight, they conclude in [12, Sect. 4.1] that "in the framework of related-key related-tweak differential attacks [KIASU-BC] has only at most one round security loss compared to AES".

Regarding Meet-in-the-Middle attacks, the designers [12, Sect. 4.2] conclude that "the same [MitM] attacks existing for AES-128 appl[y] to KIASU-BC." Concerning further attacks in the single-key model, [12, Sect. 4.3] states that "the security level of KIASU-BC against the remaining types of attacks stays

Table 1. Selection of existing attacks on the AES-128, KIASU-BC and attacks proposed in this work. Attacks on KIASU-BC are in the chosen-tweak setting. ACC = chosen plaintexts and adaptive chosen ciphertexts; CC = chosen ciphertexts; E = Encryptions; MA = Memory accesses.

Target	Rds.	Attack type	Time	Data (CP)	Memory	Ref.
AES-128	7	Partial sum	2^{120}	$2^{128-\epsilon}$	2^{57}	[10]
	7	MitM	2^{99}	2^{97}	2^{98}	[7]
	7	Imposs. Diff	$2^{107.1}$ E $+2^{117.2}$ MA	$2^{106.2}$	$2^{90.2}$	[17]
	10	Bicliques	$2^{125.98}$	2^{64} (CC)	2^{62}	[5][a]
KIASU-BC	7	Integral	2^{82}	2^{40}	2^{41}	[8]
	7	Integral	$2^{48.5}$	$2^{43.6}$	$2^{41.7}$	[8]
	7	Rectangle	2^{79} E $+2^{80}$ MA	2^{79}	2^{78}	[9]
	7	Boomerang	2^{65} E $+2^{66.6}$ MA	2^{65} (ACC)	2^{60}	[9]
	8	Imposs. Diff	2^{118} E $+2^{125.2}$ MA	$2^{117.6}$	$2^{101.6}$	[1]
	8	Imposs. Diff	$2^{116.1}$ E $+2^{120.2}$ MA	2^{118}	2^{102}	Sect. 3
	8	Boomerang	$2^{103.1}$ E $+2^{103}$ MA	2^{103} (ACC)	2^{60}	Sect. 4

[a] Time complexity corrected in [3]

the same". Recently, Dobraunig, Eichlseder, and Mendel [8] showed that the latter claim does at least not hold in general; the additional degrees of freedom from the choice of the tweak leads to improved attacks on KIASU-BC compared to the AES-128. Dobraunig et al. mounted integral attacks on seven rounds of KIASU-BC and its related nonce-respecting AE scheme KIASU≠.

This work complements the analysis by [8] with differential-based attacks on KIASU-BC on eight rounds of KIASU-BC. Our attacks share the observation that a chosen non-zero tweak difference allows to cancel a difference in the state at the beginning of some round. We propose impossible-differential, and boomerang analysis of KIASU-BC with lower time complexities and/or higher number of covered rounds than comparable attacks on the AES-128. Our detailed results are summarized in Table 1 and compared with existing results on KIASU-BC, and a selection of the best existing attacks on the AES-128. We stress that, while this work was under review, Abdelkhalek, Tolba, and Youssef [1] submitted an impossible-differential attack on KIASU-BC similar to ours (but based on different trails) but independent from us to a journal. In the following, Sect. 2 briefly recalls the basics of KIASU-BC. Section 3 presents our impossible-differential attack, and Sect. 4 a boomerang attack, both on eight-round KIASU-BC. For the interested reader, we provide two further attacks on seven rounds in the full version [9]. Section 5 concludes.

2 Brief Overview of Kiasu-BC

KIASU-BC [13] is a tweakable block cipher that adopts the state size (128 bits), key size (128 bits), round function – consisting of SUBBYTES (SB), SHIFTROWS (SR), MIXCOLUMNS (MC), and ADDKEY (AK) – as well as number of rounds (10), and key schedule from the AES-128. We assume the reader is familiar with the structure of the AES; otherwise, we refer to e.g. [6, 21] for details.

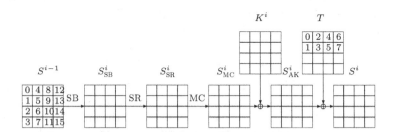

Fig. 1. Round Function of KIASU-BC.

KIASU-BC adds to the AES-128 only an additional public 64-bit tweak $T = (T[0] \| T[1] \| \ldots \| T[7])$. During the en-/decryption, the same tweak is XORed to the topmost two state rows at every occurrence of ADDKEY in every round, as illustrated in Fig. 1. In the following, we denote by

- S^i the state after Round i, for $0 \le i \le 10$.
- K^i the round key of Round i, for $0 \le i \le 10$.
- $S^i[j]$ the j-th byte of the state S^i, for $0 \le j \le 15$, indexed column-wise as usual for the AES, and as illustrated in State S^{i-1} in Fig. 1.
- $K^i[j]$ the j-th byte of the round key K^i, for $0 \le j \le 15$.
- $T[j]$ the j-th byte of the tweak T, for $0 \le j \le 7$. Note that these bytes are enumerated differently than states and keys, as illustrated in Fig. 1.
- S^i_{SB}, S^i_{SR}, S^i_{MC}, and S^i_{AK} the intermediate states in Round i directly after SUBBYTES, SHIFTROWS, MIXCOLUMNS, and ADDKEY, respectively.

For brevity, we also use the notions $\overline{\text{SB}} = \text{SB}^{-1}$, $\overline{\text{SR}} = \text{SR}^{-1}$, and $\overline{\text{MC}} = \text{MC}^{-1}$. Note that the order of some operations can be swapped without affecting the result to simplify the description of attacks. For instance, the order of the MIX-COLUMNS and ADDKEY operations with a round key K^i can be swapped if the key addition with the equivalent round key $\widehat{K}^i = \text{MC}^{-1}(K^i)$ is performed instead. This means, for all $x, K^i \in \{0,1\}^{128}$, it holds that $K^i \oplus \text{MC}(x) = \text{MC}(\widehat{K}^i \oplus x)$. The same argument holds for decryption.

3   Impossible-Differential Attack on 8-Round Kiasu-BC

This section describes an impossible-differential attack on 8-round KIASU-BC. For this attack, we can modify existing differential trails for the AES-128 so that they can be used in attacks, but introduce differences in the tweak. Those introduced differences allow us to extend the key-recovery phase. First, we provide the impossible differential, followed by a detailed description of the attack.

3.1   Impossible Differentials on Kiasu-BC

Influence of the Tweak. The tweak input provides the adversary with additional freedom that can be used for instance to extend the number of covered rounds. In KIASU-BC, the tweak can be used to cancel a difference in the trail, which allows to pass one round for free. Moreover, since the tweak is not modified by a tweak-schedule over the rounds, the subsequent tweak addition will produce exactly the same difference that occurred before the *free* round.

A Concrete Impossible Differential. There exist various impossible differentials – e.g. [2,16,17] on round-reduced AES – which can serve as base of our analysis of KIASU-BC. Though, we have to ensure that the influence of the tweak difference preserves the impossibility of the differential in the middle. Figure 2 shows an impossible differential over 3.5 rounds based on the trail used in [2]. In addition, we use a tweak difference ΔT with a single active byte $T[0]$. In forward direction, the single active byte in the state S^i from the beginning of the trail always activates at least three bytes in the first column of S^{i+1} – depending on whether the tweak difference cancels the difference in $S^{i+1}[0]$ or not. In both cases, the three rightmost columns which correspond to $\overrightarrow{S}^{i+2}[4, \ldots, 15]$ are always active. In backward direction, the fact that only three bytes are active

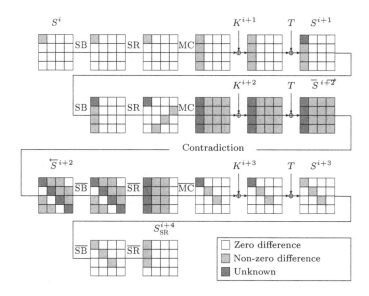

Fig. 2. Impossible differential for 3.5 rounds of KIASU-BC.

in S^{i+3} and the diffusion in the third inverse round will ensure a zero difference in $\overleftarrow{S}^{i+2}[1, 6, 11, 12]$, which contradicts with $\overrightarrow{S}^{i+2}[4, \ldots, 15]$, independent from whether the tweak difference cancels out the diagonal $\overrightarrow{S}^{i+2}[0, 5, 10, 15]$ or not.

3.2   Attack Procedure

We can extend our impossible differential by two rounds at the beginning and the end each to a key-recovery attack over Rounds 3 through 10 of KIASU-BC, using the fact that the final round omits MIXCOLUMNS. Figure 3 shows our differential trail. The following describes the individual steps.

Precomputation. Initially, we precompute a hash table $\mathcal{H}_{\text{precomp}}$ which maps pairs $(S^1_{\text{AK}}, S'^1_{\text{AK}}) \leftarrow (S^1_{\text{MC}}, S'^1_{\text{MC}})$. For all possible pairs $S^1_{\text{MC}}[0, 1, 2, 3]$ and $S'^1_{\text{MC}}[0, 1, 2, 3]$ which differ only in Byte 0, compute

$$S^1_{\text{AK}}[0, 5, 10, 15] = \text{SB}^{-1}(\text{SR}^{-1}(\text{MC}^{-1}(S^1_{\text{MC}}[0, 1, 2, 3])))$$

and $S'^1_{\text{AK}}[0, 5, 10, 15]$ accordingly. Define $\Delta S^1_{\text{AK}}[0, 5, 10, 15] = S^1_{\text{AK}}[0, 5, 10, 15] \oplus S'^1_{\text{AK}}[0, 5, 10, 15]$. Compute $\Delta S^1_{\text{MC}}[0] = S^1_{\text{MC}}[0] \oplus S'^1_{\text{MC}}[0]$ and store the pairs as tuples $(S^1_{\text{AK}}[0, 5, 10, 15], S'^1_{\text{AK}}[0, 5, 10, 15])$ in a hash table $\mathcal{H}_{\text{precomp}}$ indexed by $(\Delta S^1_{\text{AK}} \parallel \Delta S^1_{\text{MC}}[0])$. Since there are 2^{24} values for $S^1_{\text{MC}}[1, 2, 3]$ and $2^8 \cdot (2^8 - 1) \approx 2^{16}$ pairs for $S^1_{\text{MC}}[0]$, there exist about 2^{40} possible pairs. So, $\mathcal{H}_{\text{precomp}}$ has 2^{40} buckets and one element in each on average.

Structures. We will consider sets and structures of plaintexts. A set $\mathcal{S}$ consists of 2^{32} plaintexts P_i which all share equal values in bytes $P_i[1, 2, 3, 4, 6, 7, 8, 9, 11,$

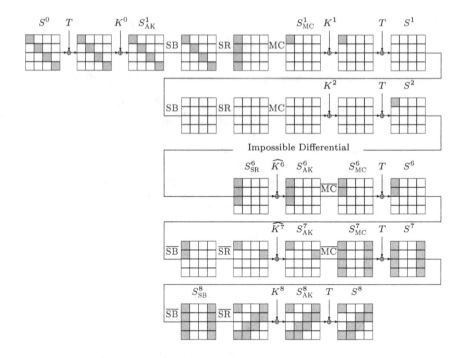

Fig. 3. 8-round impossible-differential attack trail.

$12, 13, 14]$, and are pair-wise distinct in the bytes $P_i[0, 5, 10, 15]$. Assigned to each set is a concrete tweak T. A structure $\mathcal{L}$ consists of 2^8 sets, where each set in $\mathcal{L}$ differs only in the tweak byte $T[0]$. We can build pairs of plaintext-tweak inputs (P_i, T_i) and (P_j, T_j) only inside the same structure. Though, since we want that pairs differ in their tweaks, we have to build pairs across different sets in a structure. Moreover, their bytes 0, 5, 10, and 15 after the initial tweak addition must differ, i.e. $(P_i[0] \oplus T_i[0]) \neq (P_j[0] \oplus T_j[0])$. Given two distinct sets S and S', we obtain $2^{32} \cdot (2^8 - 1)^4 \approx 2^{63.98} \approx 2^{64}$ pairs. Since there are 2^8 sets in a structure, we can build in total $\binom{2^8}{2} \cdot 2^{63.98} \approx 2^{78.97} \approx 2^{79}$ pairs per structure.

Step 1). Choose 2^n structures, which yields about 2^{n+79} possible pairs. For each structure, do the following steps:

1. Ask for the corresponding ciphertexts $C_i \leftarrow E_K^{T_i}(P_i)$ of the structure.
2. Invert the final tweak XOR and insert all states S_{AK}^8 into a hash table $\mathcal{L}$, indexed by bytes $S_{AK}^8[1, 2, 4, 5, 8, 11, 14, 15]$.
3. For each bucket in the hash table that contains more than one entry, consider every combination of pairs therein. We can expect $2^{n+79} \cdot 2^{-8 \cdot 8} = 2^{n+15}$ such pairs that are equal in bytes $S_{AK}^8[1, 2, 4, 5, 8, 11, 14, 15]$.

Step 2). The straight-forward approach would be to guess 32 bits of $K^8[3, 6, 9, 12]$ and partially decrypt these bytes for the remaining pairs to obtain S_{MC}^7 $[12, 13, 14, 15]$. Though, we can use an improvement by Lu et al. [16] to speed up the search. The improvement is based on the following observation: given a random pair of differences $\Delta X, \Delta Y \in \mathbb{F}_2^8$ over the AES S-box, there is on average one pair of $X, X' \in \mathbb{F}_2^8$ with $X \oplus X' = \Delta X$ such that $S(X) \oplus S(X') = \Delta Y$.[1]

For any pair $(S_{AK}^8[3,6,9,12], S'^8_{AK}[3,6,9,12])$, their difference $\Delta S_{SB}^8[12,13,14,15]$ is known. Hence, the knowledge of $\Delta S_{AK}^7[12,13,14,15]$ can be used to derive the values of $S_{SB}^8[12,13,14,15]$ and $S_{SB}'^8[12,13,14,15]$ and thus to derive $K^8[3,6,9,12]$. There exist only $2^8 - 1$ possible values of $\Delta S_{AK}^7[12,13,14,15]$ with exactly one active Byte 13. So, one can perform this step as follows:

1. Initialize 2^{32} empty lists, one for each guess of $K^8[3,6,9,12]$.
2. For each pair $(S_{AK}^8[3,6,9,12], S'^8_{AK}[3,6,9,12])$, and for each of the 255 possible differences $\Delta S_{AK}^7[12,13,14,15] = (0,*,0,0)$, derive $K^8[3,6,9,12]$ that leads this pair to $\Delta S_{AK}^7[12,13,14,15]$ and add this pair to the list corresponding to that key guess.

For each pair and difference guess, we expect one key suggested on average due to the S-box property observed above. These $2^{n+15} \cdot 255 \approx 2^{n+23}$ suggestions are distributed over the 2^{32} possible keys. So, we expect about 2^{n-9} pairs for each guess of $K^8[3,6,9,12]$.

Step 3). In this step, one could guess 32 bits of $K^8[0,7,10,13]$ and partially decrypt these bytes for the remaining pairs to obtain $S_{AK}^7[0,1,2,3]$. Though, this step can be improved in a similar fashion as Step 2):

1. Initialize 2^{32} empty lists, one for each guess of $K^8[0,7,10,13]$.
2. For each pair $(S_{AK}^8[0,7,10,13], S'^8_{AK}[0,7,10,13])$ and for each of the 255 possible differences $\Delta S_{AK}^7[0,1,2,3] = (*,0,0,0)$, derive the key $K^8[0,7,10,13]$ that leads this pair to $\Delta S_{AK}^7[0,1,2,3]$ and add this pair to the list corresponding to that key guess.

Again, we expect one key suggested on average for each pair and each difference guess. These $2^{n-9} \cdot 255 \approx 2^{n-1}$ suggestions are distributed over the 2^{32} possible keys. So, we expect about 2^{n-33} pairs for each guess of $K^8[0,7,10,13]$.

Step 4). The goal of the adversary in this step is to check for all remaining pairs and for the current guess of $K^8[0,3,6,7,9,10,12,13]$ if the difference $\Delta S_{AK}^6[0,1,2,3]$ is zero in exactly one byte, and if the zero byte is Byte 1, 2, or 3. Note that we do not want a zero difference in $S_{AK}^6[0]$ since it could render the impossible differential possible. The straight-forward approach would be to guess the bytes $\widehat{K}^7[0,13]$ and decrypt the states $S^7[0,13]$ of all remaining pairs to obtain $S^6[0,1]$. Again, we use the improvement by Lu et al. [16] instead.

There are $3 \cdot 255^3$ possible differences $\Delta S_{AK}^6[0,1,2,3]$ with exactly three active bytes such that the zero-difference byte is not Byte 0. Among those, $3 \cdot 255$ differences map to a difference $\Delta S_{MC}^6[0,1,2,3]$ where only Bytes 0 and 1 are active. So, the adversary has to check for each pair and each guess of $\widehat{K}^7[0,13]$ whether

---

[1] More precisely, 129 out of 256 trails $\Delta X \rightarrow \Delta Y$ are impossible, about half (126) propose two solutions, and 1 trail proposes four solutions.

$\Delta S^6_{\mathrm{MC}}[0, 1, 2, 3]$ belongs to these $3 \cdot 255$ differences. Again, given the input/output differences of the SubBytes operation, i.e., $\Delta S^6_{\mathrm{MC}}[0, 1]$ and $\Delta S^7_{\mathrm{AK}}[0, 13]$, one can efficiently determine the values $S^6_{\mathrm{MC}}[0, 1]$ and $S'^6_{\mathrm{MC}}[0, 1]$ and therefore determine the value of $\widehat{K}^7[0, 13]$.

The 2^{n-33} pairs and the $3 \cdot 255$ differences yield $3 \cdot 2^{n-25}$ candidates $\widehat{K}^7[0, 13]$ distributed over the 2^{16} possible values. Thus, we expect for a given guess of $\widehat{K}^7[0, 13]$ about $3 \cdot 2^{n-25} / 2^{16} = 3 \cdot 2^{n-41}$ pairs which yield the input difference to the impossible differential for each guess of the considered bytes in K^8 and $\widehat{K}^7$.

Step 5). This step eliminates wrong values of $K^0[0, 5, 10, 15]$ using the precomputed hash table $\mathcal{H}_{\mathrm{precomp}}$. For this purpose, initialize a list $\mathcal{K}$ for the 2^{32} values of $K^0[0, 5, 10, 15]$. For each of the remaining $3 \cdot 2^{n-41}$ pairs (P_i, P_j):

1. Compute $\Delta_{i,j}[0, 5, 10, 15] = (P_i[0, 5, 10, 15] \oplus T_i) \oplus (P_j[0, 5, 10, 15] \oplus T_j)$ and $\Delta T_{i,j}[0] = T_i[0] \oplus T_j[0]$.
2. Access the bucket indexed by $\Delta_{i,j}[0, 5, 10, 15] \| \Delta T_{i,j}[0]$ in $\mathcal{H}_{\mathrm{precomp}}$. For each tuple $(S^1_{\mathrm{AK}}[0, 5, 10, 15], S'^1_{\mathrm{AK}}[0, 5, 10, 15], \Delta S^1_{\mathrm{MC}}[0])$ in that bucket, remove from $\mathcal{K}$ the key entry $K^0[0, 5, 10, 15] = P_i[0, 5, 10, 15] \oplus (T_i[0, 1], 0, 0) \oplus S^1_{\mathrm{AK}}[0, 5, 10, 15]$.

Finally, if $\mathcal{K}$ is not empty, output the remaining value(s) in $\mathcal{K}$ along with the current key guess of $\widehat{K}^7[0, 13]$ and $K^8[0, 3, 6, 7, 9, 10, 12, 13]$.

Wrong-Key Elimination. We can determine the data complexity D such that the following inequality is fulfilled:

$$\left(1 - 2^{-(c_{\mathrm{in}} + c_{\mathrm{out}})}\right)^D < \frac{1}{2^{|k_{\mathrm{in}} \cup k_{\mathrm{out}}|}},$$

where c_{in} and c_{out} denote the number of bit conditions to be fulfilled at the top (in) and bottom (out) parts of the cipher that wrap the impossible differential. $k_{\mathrm{in}} \cup k_{\mathrm{out}}$ denote the number of combined top and bottom key bits that are guessed. Consider that the probability to filter wrong key is 2^{-32} for $K^0[0, 5, 10, 15]$, 2^{-48} for $K^8[0, 3, 6, 7, 9, 10, 12, 13]$, and $3 \cdot 2^{-8}$ for $K^7[0, 13]$. Hence, we have $c_{\mathrm{in}} + c_{\mathrm{out}} = \log_2(2^{-32-48} \cdot 3 \cdot 2^{-8}) \approx 86$ bit conditions. So, the probability that a wrong key passes is about $(1 - 2^{-86})$ per tested pair. The guessed key material sums up to $k_{\mathrm{in}} \cup k_{\mathrm{out}} = 32 + 64 + 16 = 112$ bits. So, we need

$$\left(1 - 2^{-86}\right)^D \leq 2^{-112}$$

which is true for $D \geq 2^{93}$ pairs. Since we can expect about 2^{n+15} pairs from 2^n structures, this method yields also that 2^{78} structures, i.e., 2^{118} chosen plaintexts, are required for the attack.

Complexity Analysis. The time complexity is composed of the following steps:

1. The **precomputation** requires $\approx 2 \cdot 2^{40} \cdot 4/16 = 2^{36}$ single-round decryptions, which is equivalent to $2^{36}/8 = 2^{33}$ eight-round decryptions.
2. **Step 1** requires 2^{n+40} encryptions.

3. **Step 2** can be implemented by a look-up table, as suggested by Lu et al. [16]. By storing the results efficiently, one can fetch a key in one access even if several keys are suggested. Lu et al. state that most queries fail, whereas about $1/16$ (on average) of the queries return 16 options of 32-bit keys each, and a smaller fraction can return more options. In total, this step requires $255 \cdot 2^{n+15} \approx 2^{n+23}$ memory accesses (MA).

4. **Step 3** requires $255 \cdot 255 \cdot 2^{n+15} \approx 2^{n+31}$ memory accesses, since for all 255 differences for the first 32-bit guesses of $K^8[3,6,9,12]$, we consider another 255 differences for the second 32 guessed bits $K^8[0,7,10,13]$.

5. **Step 4** requires $2^{n-33} \cdot 3 \cdot 255 \approx 3 \cdot 2^{n-25}$ memory accesses in a lookup table to determine from the differences $\Delta S^6_{\mathrm{MC}}[0,1,2,3]$ the guess for $\widehat{K}^7[0,13]$. Since we have to perform this step for each of the 2^{64} guesses of $K^8[0,3,6,7,9,10,12,13]$, this step requires in total $2^{64} \cdot 2^{n-33} \cdot 3 \cdot 255 \approx 3 \cdot 2^{n+39}$ MA.

6. **Step 5** analyzes $3 \cdot 2^{n-41}$ remaining pairs, leading in average to one memory access to $\mathcal{H}_{\mathrm{precomp}}$ plus one memory access to $\mathcal{K}$. This step is repeated 2^{80} times (for each guess of $K^8[0,3,6,7,9,10,12,13]$ and $\widehat{K}^7[0,13]$). So, the time complexity of this step is $3 \cdot 2^{n-41} \cdot 2 \cdot 2^{80} = 3 \cdot 2^{n+40}$ memory accesses.

7. In an **exhaustive step**, we can identify the remaining key bytes. This step requires negligible time regarding the total computational complexity.

So, for $n = 78$, the overall time complexity of the attack results from

$$T \approx (2^{n+40} + 2^{33})\, \mathrm{Enc} + (2^{n+23} + 2^{n+31} + 3 \cdot 2^{n+39} + 3 \cdot 2^{n+40})\, \mathrm{MA}$$
$$\approx 2^{118}\, \mathrm{Enc} + 2^{120.2}\, \mathrm{MA}.$$

The precomputation table requires $2 \cdot 2^{40} \cdot (4+4+1) < 2^{45}$ bytes to store the values $S^1_{\mathrm{AK}}[0,5,10,15]$, $S'^1_{\mathrm{AK}}[0,5,10,15]$, and the difference $\Delta S^1_{\mathrm{MC}}[0]$ for each entry. The simple approach would further use $2^{8 \cdot (4+2+8)} = 2^{112}$ bytes of memory for storing the deleted values of the four bytes of K^0, the two bytes of $\widehat{K}^7$, and the eight bytes of K^8. Instead, Lu et al. [16] proposed to perform the attack separately for each key guess, and to immediately append an exhaustive search for the remaining bytes of each guess that is not discarded. So, we have to store instead the about $2^{n+23} = 2^{101}$ suggestions which remain after Step 2, which consist of two plaintexts and two ciphertexts of 16 bytes each. So, the memory complexity of the attack requires $2^{45} + 2^{106} \approx 2^{106}$ bytes of memory, which is equivalent to 2^{102} states.

Several optimizations seem possible to further reduce the attack complexities. For instance, Boura et al. [4] propose to use multiple impossible differentials in order to reduce the data complexity. There are $\binom{4}{2} = 6$ options which two bytes can be chosen to be active in the difference $\Delta S^6[0,1,2,3]$. Each option requires to consider a different set of guessed bytes in $\widehat{K}^7$ and K^8, and a different set of output differences. Moreover, the attack could be executed also with shifted versions of the state differential and tweak difference, or several times in parallel. We omit their description for simplicity.

4   Boomerang Attack on 8-Round Kiasu-BC

In the following, we describe a boomerang attack on the final eight rounds of KIASU-BC, which is an extension of a seven-round rectangle attack and a seven-round boomerang attack. For interested readers, we provide those in the full version of this work [9], together with an short introduction into boomerang attacks. The upper and lower trails are depicted in Fig. 4. We append a key-guessing phase over the final round. Again, we use the fact that the final round omits the MIXCOLUMNS operation so that only four key bytes have to be considered. Figure 5 shows the steps that wrap the upper and lower trails.

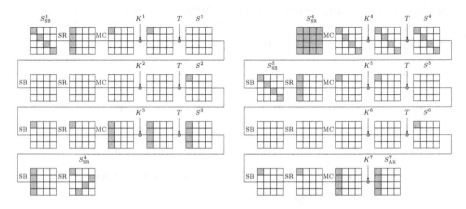

Fig. 4. Upper (**left**) and lower (**right**) trails of our boomerang attack on KIASU-BC.

4.1   Attack Procedure

Differentials. Both trails share the same idea: start from a state directly after the SUBBYTES operation with a difference ΔS^i_{SB} wherein only the bytes on the main diagonal are active, and choose it in a way such that the differences in the state ΔS^i_{MC} and in the tweak ΔT will cancel each other after the first round, i.e., $\Delta S^i = 0$. Then, it follows that $\Delta S^{i+1} = \Delta T$ contains only one and ΔS^{i+2} only four active bytes. In the upper trail, only Bytes $S^{i+2}[0, 1, 2, 3]$ are traced through the final SUBBYTES operation after Round 3. The lower trail then adds the rest of Round 4, i.e. it starts from a fully active state difference ΔS^4_{SR} such that it yields a difference only in the main diagonal after Round 4 with probability one. The remaining Rounds 5–7 follow the same trail as the first three rounds.

Assuming a correct quartet, both its pairs pass the S-box in Round 7 with probability $(2^{-6})^2$. The pairs need not have a specific difference after the final S-box, but only the same difference γ' in the middle. So, the four S-boxes in Round 5 are passed with probability about $(2^{-3.5 \cdot 4})^2 = 2^{-28}$. Concerning the four S-boxes at the bottom of the upper trail, the second pair has a probability of about $(2^{-7})^4 = 2^{-28}$ to pass them with the same trail as the first pair. The final S-box at the beginning of Round 3 is then passed by the second pair

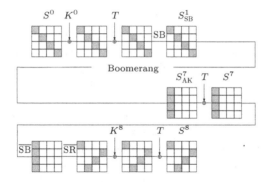

Fig. 5. 8-round boomerang attack trail.

with probability 2^{-7}. So, for the correct guess of $K^8[0,7,10,13]$, the probability of a correct pair to follow our trails is about $2^{-12} \cdot 2^{-28} \cdot 2^{-28} \cdot 2^{-7} = 2^{-75}$. Each structure yields $\binom{2^{40}}{2} \cdot 2^{-32} \approx 2^{47}$ pairs which collide after the first round. Hence, for 2^{31} structures, we can expect $2^{31} \cdot 2^{47} \cdot 2^{-75} = 2^3$ correct quartets, and recover 32 bits of K^8, $K^8[0,7,10,13]$ and/or 32 bits of K^0, $K^0[0,5,10,15]$. The remaining 96 bits of either round key can be found e.g. by exhaustive search.

Structures and Sets. We fix an arbitrary base tweak $\widehat{T}$. We build in total 2^n structures consisting of 2^{40} plaintexts each. For every structure, we choose an arbitrary base plaintext $\widehat{P}$ and derive 2^{32} plaintexts from it by iterating over all values of Bytes 0, 5, 10, and 15, and leaving all other bytes constant. We use the same 2^{32} plaintexts in each of 2^8 sets T^i in the structure, for $0 \le i < 2^8$, where the sets differ only in the first tweak byte. This means, we derive 2^8 tweaks $T^i = (\langle i \rangle, \widehat{T}[1], \dots, \widehat{T}[7])$, for $0 \le i < 2^8$, from $\widehat{T}$ and assign T^i to all texts in T^i.

Steps. Choose $\delta' \in \{0,1\}^8$ so that there exists a differential $\texttt{0x01} \to \delta'$ with probability 2^{-6} through the S-box. Derive $\delta = \mathrm{MC}((\delta', 0, 0, 0))$. Then:

1. Choose 2^n structures of 2^{40} plaintext-tweak tuples (P, T) each. Ask for their ciphertexts C.
2. Initialize a list for all possible values of $K^0[0,5,10,15]$ and $K^8[0,7,10,13]$.
3. For each of 2^n structures and for each key guess $K^8[0,7,10,13]$:
   (a) For each ciphertext and corresponding tweak (C, T_C), derive a tweak T_D with $T_D[1,\dots,7] = T_C[1,\dots,7]$ and $T_D[0] = T_C[0] \oplus \texttt{0x1}$. Then, partially decrypt C under T_D through the inverse final round to obtain $S^7_{\mathrm{AK}}[0,1,2,3]$. Compute $S'^7_{\mathrm{AK}}[0,1,2,3] = S^7_{\mathrm{AK}}[0,1,2,3] \oplus \delta[0,1,2,3]$ and determine $D[0,7,10,13]$ by reencrypting $S'^7_{\mathrm{AK}}[0,1,2,3]$ over the final round, again under T_D. Copy the 12 bytes $D[1,2,3,4,5,6,8,9,11,12,14,15]$ from the corresponding bytes of C.
   (b) Ask for the plaintexts Q of all 2^{n+40} shifted ciphertexts (D, T_D).
   (c) Sort the 2^{40} plaintexts Q together with their tweaks T_Q (we define $T_Q = T_D$), corresponding ciphertext D, and the original plaintext P from which D was derived as tuples (Q, T_Q, D, P) into 2^{96} buckets

indexed by $Q[1, 2, 3, 4, 6, 7, 8, 9, 11, 12, 13, 14]$. Since we search for pairs (Q, Q') with $Q \oplus Q' = \alpha$ and $T_Q \neq T_{Q'}$ from the same structure, we expect in average $2^n \cdot \binom{2^{40}}{2} \cdot 2^{-96} \approx 2^{n-17}$ false-positive colliding pairs (Q, Q') for each candidate K^8. Since $\Pr[T_Q \neq T_{Q'}] = 255/256$, we still have about $2^{n-17} \cdot 255/256 \approx 2^{n-17}$ colliding pairs per key.

(d) For each potential quartet (P, P', Q, Q'), we want to identify the key values $K^0[0, 5, 10, 15]$ which could create the state collisions after the first round. Let $S_P^1, S_{P'}^1, S_Q^1, S_{Q'}^1$ denote their corresponding states after the first round. For a correct quartet, it holds that $S_P^1 = S_{P'}^1$ and analogously, $S_Q^1 = S_{Q'}^1$. There is a unique mapping from the tweak difference $S_{\text{SR},P}^1 \oplus S_{\text{SR},P'}^1$ backwards to $S_{\text{SB},P}^1 \oplus S_{\text{SB},P'}^1$. Since the S-box has two solutions on average, there are on average two key values per byte $K^0[0, 5, 10, 15]$ such that P and P' are mapped to the correct difference ΔS_{SB}^1 which yields $S_P^1 \oplus S_{P'}^1 = 0$. So, we derive those values for $K^0[0, 5, 10, 15]$ for P and P'. We derive those key values analogously for Q and Q'. For each key byte, we have $2 \cdot 2$ combinations on average, i.e., four pairs among which any pair matches with probability about $4 \cdot 2^{-8} = 2^{-6}$. So, an invalid quartet survives this four-byte filter with probability $(2^{-6})^4 = 2^{-24}$. For each surviving pair, we increment the counter for the values of the current $K^0[0, 5, 10, 15]$ and $K^7[0, 7, 10, 13]$. Over all 2^{32} keys K^8, we expect $2^{32} \cdot 2^{n-17} \cdot 2^{-24} = 2^{n-9}$ (false-positive) quartets on average.

(e) For $n = 31$, we expect 2^3 correct quartets, and 2^{22} false positives. While 2^3 can be expected on average, at least three correct quartets occur with significant probability. Correct quartets suggest the same 64 bits $K^8[0, 7, 10, 13]$ and $K^0[0, 5, 10, 15]$. Assuming that the key suggestions from the 2^{22} false-positive quartets are uniformly distributed, we can expect only the correct 64 bits of $K^0[0, 5, 10, 15]$ be suggested at least four times. So, we output the candidate(s) with the highest counters.

4.2   Complexity

The time complexity of the attack consists of the following:

- **Step 1** requires 2^{n+40} full encryptions.
- **Step 3a** requires $2^{32} \cdot 2^{n+40} \cdot 4/16 = 2^{n+70}$ single-round decryptions of four bytes for each ciphertext and key candidate to derive S_{AK}^7 and the same amount of single-round encryptions to derive the shifted ciphertexts D, which is equivalent to $2 \cdot 2^{n+70} \cdot 1/8 = 2^{n+68}$ eight-round encryptions.
- **Step 3b** requires $2^{32} \cdot 2^{n+40} = 2^{n+72}$ full decryptions.
- **Step 3c** requires $2^{32} \cdot 2^{n+40} = 2^{n+72}$ MAs with an efficient data structure.
- **Step 3d** requires $2^{32} \cdot 4 \cdot 2^{n-17} \cdot 4/16 = 2^{n+15}$ single-round decryptions (equivalent to 2^{n+12} eight-round encryptions) of one column backwards through the first round for each text in each potential quartet (P, P', Q, Q') and each key. It requires $2^{n-9} + 2^{n+3}$ memory accesses for the false-positive and the correct quartets, which are negligible in the total computational complexity.
- An **exhaustive** search step requires 2^{96} full encryptions.

The time complexity is given by approximately

$$(2^{n+40} + 2^{n+68} + 2^{n+72} + 2^{n+12} + 2^{96}) \text{ Enc} + 2^{n+72} \text{ MA}$$
$$\approx 2^{103.1} \text{ Enc} + 2^{103} \text{ MA}.$$

The data complexity consists of $2^{n+40} = 2^{71}$ chosen plaintexts and $2^{32} \cdot 2^{71} = 2^{103}$ adaptively chosen ciphertexts. The attack can be run with memory for 2^{40} states plus 2^{64} single-byte key counters, which is equivalent to about 2^{60} states.

4.3 Experimental Verification

Murphy [19] showed that boomerangs and rectangles base on oversimplified conditional systems. He stressed that the techniques need a revised foundation, which is clearly beyond the scope of this work. While the complexity of our attacks prohibits to implement full versions of them at the moment, we implemented variants of our boomerang distinguishers to support our analysis:

1. the 6-rd. distinguisher under our 7-rd. boomerang attack with KIASU-BC,
2. the 6-rd. boomerang distinguisher with a downscaled version of KIASU-BC, called MINI-KIASU-BC hereafter,
3. the 7-rd. distinguisher under our 8-rd. attack, again with MINI-KIASU-BC.
4. the 7-rd. boomerang key-recovery attack with MINI-KIASU-BC.

We defined MINI-KIASU-BC as a nibblewise-operating variant of KIASU-BC that employs the same high-level structure as KIASU-BC in downscaled manner, i.e., the same number and order of operations, equal number of rounds and key schedule, the same SHIFTROWS, ADDKEY, and ADDTWEAK operations, the same MDS matrix, though, with multiplications in $\mathbb{GF}(2^4)$ under the irreducible polynomial $x^4 + x + 1$, operating on nibbles instead of bytes, and with the Piccolo S-box instead of that of the AES. Note that the Piccolo S-box has a maximal probability of 2^{-2} for differential trails. Moreover, for differences $\delta, \gamma' \in \{0, 1\}4$, it holds for differential trails $\delta \rightarrow \gamma'$ through the inverse Piccolo S-box that $\sqrt{\sum_{\gamma' \in \{0,1\}4} \Pr^2[\delta \xrightarrow{S^{-1}} \gamma']}$ is 2^{-1} for two values of δ, $2^{-1.5}$ for five, and about $2^{-1.21}$ for the remaining eight non-zero values of δ.

To verify the intermediate probabilities of our trails, we studied several round-reduced versions of the distinguishers. Since building structures was unnecessary for this purpose, we omitted the first round for those, and started directly from the state S^1 after the first round. Thereupon, we (1) chose a random base plaintext, (2) created plaintexts by iterating over the values of the first column and all possible values of the first tweak byte, (3) encrypted the resulting plaintext-tweak pairs over the remaining rounds, (4) applied the δ-shift, and (5) decrypted 3, 4, 5, or 6 rounds. For each of our experiments, we chose the base plaintexts of our structures and 100 keys randomly from /dev/urandom. The source code of our analysis will be made freely available to the public. The results of our experiments are summarized in Table 2.

Table 2. Probabilities of correct quartets from experiments with our 6- and 7-round boomerang distinguishers on KIASU-BC and MINI-KIASU-BC with 100 random keys per experiment. Probability deviations after subtracting/adding the standard deviation to the average #correct quartets are given in square brackets. Rds. = #rounds; str. = #structure(s).

Rds.	6-round distinguisher		7-round distinguisher
	KIASU-BC	MINI-KIASU-BC	MINI-KIASU-BC
	2^{16} texts, 2^8 sets, 1 str.	2^{16} texts, 2^4 sets, 1 str.	2^{16} texts, 2^4 sets, 2^{10} str.
3	$2^{-12.00}$ $[2^{\pm0.01}]$	$2^{-4.00}$ $[2^{\pm0.01}]$	–
4	$2^{-19.00}$ $[2^{\pm0.03}]$	$2^{-5.88}$ $[2^{\pm0.01}]$	$2^{-14.61}$ $[2^{\pm0.01}]$
5	$2^{-25.42}$ $[2^{-0.22}, 2^{0.19}]$	$2^{-7.14}$ $[2^{\pm0.01}]$	$2^{-26.94}$ $[2^{-0.17}, 2^{0.16}]$
6	–	$2^{-7.14}$ $[2^{\pm0.01}]$	$2^{-30.65}$ $[2^{-0.83}, 2^{0.53}]$
7	–	–	$2^{-30.12}$ $[2^{-0.71}, 2^{0.47}]$

Distinguishers. Concerning the 6-round distinguisher for MINI-KIASU-BC, the active S-box in Round 6 of the lower trail is passed with probability $\approx (2^{-2})^2$ for both pairs. The S-box in Round 4 is passed with probability about 2^{-2}, and that at the beginning of Round 3 is passed with probability about $2^{-1.21}$, both for the second pair only. Concerning the 6-round trail for KIASU-BC, the active S-box in Round 6 of the lower trail is passed with probability $\approx (2^{-6})^2$ for both pairs; the active S-box at the end of the upper trail with probability about 2^{-7}; the S-box at the beginning of Round 3 multiplies a factor of $2^{-6.5}$. From our intuition, this factor results from the matter that some quartets pass the S-box with probability 2^{-6} and some with 2^{-7}. So, the obtained differential probabilities are slightly higher than expected.

Concerning the 7-round trail for MINI-KIASU-BC, the five active S-boxes in the lower trail are passed with a probability about $(2^{-2})^2 \cdot (2^{-1.21\cdot4})^2$. The four active S-boxes at the bottom of the upper trail in Round 4 are passed with slightly lower probability than expected, i.e., $2^{-12.3} \approx (2^{-3})^4$. Again, we antici-pate this to result from quartets that pass the lower trail with lower probability. The final S-box in Round 3 is passed with probability between 2^{-2} and 2^{-3} for the second pair. So, while we do not have figures for KIASU-BC over the 7-round distinguisher yet, the analysis with MINI-KIASU-BC provides us with at least a good indication that we can expect for KIASU-BC that the corresponding prob-abilities are close to $(2^{-6})^2$ for the S-box in Round 7, about $(2^{-3.5\cdot4})^2$ for those in Round 5, $(2^{-7})^4$ for those in Round 4, and about 2^{-7} for that in Round 3, as in our theoretical analysis.

Key-Recovery. In addition, we implemented the 7-round boomerang attack with the key-recovery stage for MINI-KIASU-BC, which yielded practical com-plexity. For this purpose, we created a structure of 2^{16} sets of 2^4 texts each by choosing a random base plaintext P and base tweak T_P, and iterated over all values of $P[0, 5, 10, 15]$ and $T[0]$. We collected the corresponding ciphertexts C.

For all 2^{20} ciphertexts and for all 2^{16} candidates $K^7[0, 7, 10, 13]$, we derived D from the δ-shift and the corresponding shifted tweak $T_D = T_P[0] \oplus \delta$, obtained the corresponding plaintexts Q in a sorted list, and searched for matching quartets. For the correct key, we obtained always more than $55,000 \approx 2^{2 \cdot 16} \cdot \binom{2^4}{2} \cdot 2^{-16} \cdot 2^{-7.14} \approx 2^{15.77}$ quartets. We sorted the list lexicographically and used the first 16 quartets in order for subkey recovery. In total, we tested 100 randomly chosen keys with independently random base plaintext and base tweak. Each run identified the single correct key candidate with more than $55,000$ quartets, whereas the second highest candidate was suggested by only about $1/4$ of that amount. So, we consider our experiments to show that $K^0[0, 5, 10, 15]$ and $K^7[0, 7, 10, 13]$ can reliably be recovered for MINI-KIASU-BC and similar results can be expected for the full KIASU-BC.

5 Conclusion

This work proposed differential-based attacks on eight rounds of KIASU-BC, which share the idea that the tweak input allows to construct a local collision. While the designers already indicated that there exist boomerangs on at most seven rounds, they had to consider attacks in the single- as well as in the related-key setting. Our described boomerang and rectangle attacks do not violate their claims, but concretize the security threats in the single-key model and illustrate that KIASU-BC possesses one round less security than the AES-128. Moreover, the claim that the bounds of existing attacks on the AES for other attacks than boomerangs, conventional differentials, and meet-in-the-middle can be translated without modification to KIASU-BC does not hold in general, which was already observed by [8] and was confirmed by our impossible-differential attack.

Acknowledgments. The authors thank Ralph Ankele, Christof Beierle, and Maria Eichlseder for the fruitful discussions at the DISC workshop in March 2016 at Bochum, and the reviewers for their helpful comments.

References

1. Abdelkhalek, A., Tolba, M., Youssef, A.M.: Impossible Differential Cryptanalysis of 8-round Kiasu-BC (2016, to appear)
2. Bahrak, B., Aref, M.R.: Impossible differential attack on seven-round AES-128. IET Inform. Secur. **2**(2), 28–32 (2008)
3. Bogdanov, A., Chang, D., Ghosh, M., Sanadhya, S.K.: Bicliques with minimal data and time complexity for AES. In: Lee, J., Kim, J. (eds.) ICISC 2014. LNCS, vol. 8949, pp. 160–174. Springer, Cham (2015). doi:10.1007/978-3-319-15943-0_10
4. Boura, C., Naya-Plasencia, M., Suder, V.: Scrutinizing and improving impossible differential attacks: applications to CLEFIA, Camellia, LBlock and SIMON. In: Sarkar, P., Iwata, T. (eds.) ASIACRYPT 2014. LNCS, vol. 8873, pp. 179–199. Springer, Berlin (2014). doi:10.1007/978-3-662-45611-8_10
5. Canteaut, A., Naya-Plasencia, M., Vayssière, B.: Sieve-in-the-middle: improved MITM attacks. In: Canetti, R., Garay, J.A. (eds.) CRYPTO 2013. LNCS, vol. 8042, pp. 222–240. Springer, Berlin (2013). doi:10.1007/978-3-642-40041-4_13

6. Daemen, J., Rijmen, V.: The Design of Rijndael: AES - The Advanced Encryption Standard. Springer, New York (2002)
7. Derbez, P., Fouque, P.-A., Jean, J.: Improved key recovery attacks on reduced-round AES in the single-key setting. In: Johansson, T., Nguyen, P.Q. (eds.) EUROCRYPT 2013. LNCS, vol. 7881, pp. 371–387. Springer, Berlin (2013). doi:10.1007/978-3-642-38348-9_23
8. Dobraunig, C., Eichlseder, M., Mendel, F.: Square attack on 7-round Kiasu-BC. In: Manulis, M., Sadeghi, A.-R., Schneider, S. (eds.) ACNS 2016. LNCS, vol. 9696, pp. 500–517. Springer, Cham (2016). doi:10.1007/978-3-319-39555-5_27
9. Dobraunig, C., List, E.: Impossible-differential and boomerang cryptanalysis of round-reduced KIASU-BC. Cryptology ePrint Archive (2016)
10. Ferguson, N., Kelsey, J., Lucks, S., Schneier, B., Stay, M., Wagner, D., Whiting, D.: Improved cryptanalysis of Rijndael. In: Goos, G., Hartmanis, J., Leeuwen, J., Schneier, B. (eds.) FSE 2000. LNCS, vol. 1978, pp. 213–230. Springer, Berlin (2001). doi:10.1007/3-540-44706-7_15
11. Ferguson, N., Lucks, S., Schneier, B., Whiting, D., Bellare, M., Kohno, T., Callas, J., Walker, J.: The Skein Hash Function Family. Submission to NIST (Round 3) (2010)
12. Jean, J., Nikolić, I., Peyrin, T.: KIASU v1.1. First-round submission to the CAESAR competition (2014). http://competitions.cr.yp.to/caesar-submissions.html
13. Jean, J., Nikolić, I., Peyrin, T.: Tweaks and keys for block ciphers: the TWEAKEY framework. In: Sarkar, P., Iwata, T. (eds.) ASIACRYPT 2014. LNCS, vol. 8874, pp. 274–288. Springer, Berlin (2014). doi:10.1007/978-3-662-45608-8_15
14. Krovetz, T., Rogaway, P.: The software performance of authenticated-encryption modes. In: Joux, A. (ed.) FSE 2011. LNCS, vol. 6733, pp. 306–327. Springer, Heidelberg (2011)
15. Liskov, M., Rivest, R.L., Wagner, D.: Tweakable Block Ciphers. In: Yung, M. (ed.) CRYPTO 2002. LNCS, vol. 2442, pp. 31–46. Springer, Heidelberg (2002)
16. Lu, J., Dunkelman, O., Keller, N., Kim, J.: New impossible differential attacks on AES. In: Chowdhury, D.R., Rijmen, V., Das, A. (eds.) INDOCRYPT 2008. LNCS, vol. 5365, pp. 279–293. Springer, Berlin (2008). doi:10.1007/978-3-540-89754-5_22
17. Mala, H., Dakhilalian, M., Rijmen, V., Modarres-Hashemi, M.: Improved impossible differential cryptanalysis of 7-Round AES-128. In: Gong, G., Gupta, K.C. (eds.) INDOCRYPT 2010. LNCS, vol. 6498, pp. 282–291. Springer, Berlin (2010). doi:10.1007/978-3-642-17401-8_20
18. Minematsu, K.: Building blockcipher from small-block tweakable blockcipher. Des. Code Cryptogr. 74(3), 645–663 (2015)
19. Murphy, S.: The return of the cryptographic boomerang. IEEE Trans. Inform. Theory 57(4), 2517–2521 (2011)
20. Naito, Y.: Full PRF-secure message authentication code based on tweakable block cipher. In: Au, M.-H., Miyaji, A. (eds.) ProvSec 2015. LNCS, vol. 9451, pp. 167–182. Springer, Cham (2015). doi:10.1007/978-3-319-26059-4_9
21. National Institute of Standards and Technology: FIPS 197. National Institute of Standards and Technology, November, pp. 1–51 (2001)
22. Peyrin, T., Seurin, Y.: Counter-in-tweak: authenticated encryption modes for tweakable block ciphers. In: Robshaw, M., Katz, J. (eds.) CRYPTO 2016. LNCS, vol. 9814, pp. 33–63. Springer, Berlin (2016). doi:10.1007/978-3-662-53018-4_2
23. Schroeppel, R., Orman, H.: The hasty pudding cipher. AES candidate submitted to NIST (1998)

Weak Keys for AEZ, and the External Key Padding Attack

Bart Mennink[1,2(✉)]

[1] Department of Electrical Engineering, ESAT/COSIC, KU Leuven, and iMinds,
Leuven, Belgium
bart.mennink@esat.kuleuven.be
[2] Digital Security Group, Radboud University, Nijmegen, The Netherlands
b.mennink@cs.ru.nl

Abstract. AEZ is one of the third round candidates in the CAESAR competition. We observe that the tweakable blockcipher used in AEZ suffers from structural design issues in case one of the three 128-bit sub-keys is zero. Calling these keys "weak," we show that a distinguishing attack on AEZ with weak key can be performed in at most five queries. Although the fraction of weak keys, around 3 out of every 2^{128}, seems to be too small to violate the security claims of AEZ in general, they do reveal unexpected behavior of the scheme in certain use cases. We derive a potential scenario, the "external key padding," where a user of the authenticated encryption scheme pads the key externally before it is fed to the scheme. While for most authenticated encryption schemes this would affect the security only marginally, AEZ turns out to be completely insecure in this scenario due to its weak keys. These observations open a discussion on the significance of the "robustness" stamp, and on what it encompasses.

Keywords: AEZ · Tweakable blockcipher · Weak keys · Attack · External key padding · Robustness

1   Introduction

Authenticated encryption aims to offer both privacy and authenticity of data. The ongoing CAESAR competition [8] targets the development of a portfolio of new, solid, authenticated encryption schemes. It received 57 submissions, 30 candidates advanced to the second round, and recently, 16 of those advanced to the third round.

AEZ is an authenticated encryption scheme by Hoang, Krovetz, and Rogaway [18]. In this work we focus on AEZ v4, the latest version that has been submitted to CAESAR [17]. The addendum "v4" will be omitted for brevity. We remark that our findings can also be generalized to versions v2 and v3, despite the major revisions that have been made in the key scheduling. Our attacks do not apply to v1, because it differs from v4 not only in the key scheduling but also in the encryption mode itself.

© Springer International Publishing AG 2017
H. Handschuh (Ed.): CT-RSA 2017, LNCS 10159, pp. 223–237, 2017.
DOI: 10.1007/978-3-319-52153-4_13

AEZ is designed as a "robust authenticated encryption (RAE) scheme" [18]; this informally means that it achieves privacy and authenticity as good as possible even in the case of nonce-reuse. It moreover implies that it is secure in case of release of unverified plaintext [2]. The designers of AEZ claim that it is a RAE scheme as long as the query complexity does not exceed 2^{55} and the time complexity does not exceed 2^{128} [17].

On the other hand, robustness implies nothing for more "alternative" attacks, such as key recovery attacks, related-key attacks, and others. In [13], Fuhr et al. derived a key recovery attack on AEZ v3 in complexity $2^{n/2}$; not breaking the claimed security, but definitely an unexpected security property. In response to the observation by Fuhr et al., the designers of AEZ performed a major revision from AEZ v3 to AEZ v4 in order to mitigate the attack. Chaigneau and Gilbert [9], however, demonstrated that v4.1 is still vulnerable to a key recovery attack with a similar complexity to that of [13].

Beyond [9,13], no analysis on AEZ has appeared. In this work, we will investigate the underlying tweakable blockcipher of AEZ and notice that it shows remarkable behavior for certain structured sets of keys. We will show how these weak keys can be used to attack the AEZ mode and to distinguish it from a random primitive in constant time. We will additionally discuss a specific use case of AEZ where its weak keys can be exploited.

1.1   Weak Keys

AEZ allows for arbitrarily-sized keys, and transforms them into three subkeys of 128 bits using a key derivation function:

$$I\|J\|L \longleftarrow \begin{cases} K \text{ if } |K| = 384, \\ \text{BLAKE2b}(K) \text{ otherwise.} \end{cases}$$

In other words, if the key is already 384 bits long, it is simply padded into $I\|J\|L$; otherwise, it is first hashed via BLAKE2b [5]. This is done deliberately, as the authors state [17]: "We dispense with calling BLAKE2b if the key K is already $3 \cdot 128$ bits."

We will show that if one of the three subkeys I, J, L equals 0^{128}—call a key K for which this is the case "weak"—AEZ can be distinguished from random in at most two evaluations if it is known which subkey equals 0^{128} and at most five evaluations otherwise. The attack relies on the fact that for weak keys the tweakable blockcipher used in AEZ is completely insecure. In more detail, by explicitly writing out this tweakable blockcipher, as we have done in Sect. 3.1,[1] one finds that if I, J, or L equals 0^{128}, one can identify multiple tweaks for which the tweakable blockcipher collides.

A simple computation shows that, if we consider keys of length 384 bits, $3 \cdot 2^{256} - 3 \cdot 2^{128} + 1 \approx 3 \cdot 2^{256}$ of those are weak. Regarding keys of size different

---

[1] This explicit description may contribute to a better understanding of the primitive used in AEZ, and may be of independent interest.

from 384, assuming that BLAKE2b is a random oracle (see [12,15] for the latest analysis of BLAKE2b) approximately 3 out of 2^{128} keys result in a subkey 0^{128}. Although this in itself does not break the security claims of AEZ, the observation testifies of a more structural weakness in AEZ, namely that *the underlying tweakable blockcipher is not secure (for these weak keys)*.

1.2 External Key Padding

Focusing on keys of length different from 384 bits, a key is weak if $I\|J\|L = $ BLAKE2b(K) satisfies that I, J, or L equals 0^{128}. This set of weak keys is rather unstructured; hitting a weak key is a mere coincidence. As a matter of fact, calling these keys "weak" is debatable in the first place.

For keys of length exactly 384 bits, the situation is completely different. We will illustrate this via a potential use case, which we call the "external key padding." At a high level, this scenario covers the case where the user of AEZ pads the key himself prior to feeding it to the scheme. Partly attributed to the key scheduling of AEZ, this would result in an omission of the evaluation of BLAKE2b. Above-mentioned weak key attacks can then be used to distinguish AEZ from random in case of external key padding. Remarkably, for "ordinary" authenticated encryption schemes (such as [3,7,19,21,22,26]), external key padding would only have a marginal influence, mostly because the scheme already pads the key itself in the first place.

A simple patch for this use case would be to *always* hash the key through BLAKE2b, regardless of the size of K. Unfortunately, this patch does not resolve the structural design issues the tweakable blockcipher of AEZ suffers from, and other problematic use cases may exist.

1.3 Outline

A high-level description of the AEZ mode is given in Sect. 2. We discuss the AEZ tweakable blockcipher primitive, as well as its weak key issues, in Sect. 3. The weak key attacks on AEZ are discussed in Sect. 4. We discuss the external key padding scenario and the corresponding attack in Sect. 5. The work is concluded in Sect. 6.

2 AEZ

We will describe the interface and security model of AEZ in Sect. 2.1, and give a high-level description of AEZ in Sect. 2.2.

2.1 Interface and Security Model

AEZ [17,18] is an authenticated encryption scheme that consists of an encryption function $\mathcal{E}$ and a decryption function $\mathcal{D}$. The encryption $\mathcal{E}$ gets as input a key, nonce, associated data, tag size, and message, and outputs an expanded ciphertext. The decryption $\mathcal{D}$ operates the opposite way; it gets as input a key, nonce, associated data, tag size, and expanded ciphertext, and it outputs either a message or a dedicated $\perp$ symbol. More formally, for some finite key space $\mathcal{K} \subset \{0,1\}^*$,

$$\mathcal{E} : \mathcal{K} \times \{0,1\}^* \times \{0,1\}^* \times \mathbb{N} \times \{0,1\}^* \to \{0,1\}^*,$$
$$(K, N, A, \tau, M) \mapsto C \in \{0,1\}^{|M|+\tau},$$
$$\mathcal{D} : \mathcal{K} \times \{0,1\}^* \times \{0,1\}^* \times \mathbb{N} \times \{0,1\}^* \to \{0,1\}^* \cup \{\perp\},$$
$$(K, N, A, \tau, C) \mapsto M/\perp,$$

where $\mathcal{D}$ is required to satisfy that

$$\mathcal{D}(K, N, A, \tau, \mathcal{E}(K, N, A, \tau, M)) = M$$

for any K, N, A, τ, M.

AEZ is introduced alongside the security model called "robust authenticated encryption (RAE)," and we will describe it in own terminology. Throughout, $x \xleftarrow{\$} \mathcal{X}$ means that x gets sampled uniformly at random from a finite set $\mathcal{X}$. An adversary $\mathcal{A}$ is a probabilistic algorithm that has access to one or more oracles $\mathcal{O}$, denoted $\mathcal{A}^{\mathcal{O}}$. By $\mathcal{A}^{\mathcal{O}} = 1$ we denote the event that $\mathcal{A}$, after interacting with $\mathcal{O}$, outputs 1.

Let $K \xleftarrow{\$} \mathcal{K}$ be a uniformly randomly drawn key. Denote by π a random injection function with the same interface as $\mathcal{E}_K$. More detailed, π is a family of random functions indexed by $(N, A, \tau) \in \{0,1\}^* \times \{0,1\}^* \times \mathbb{N}$, and a query $\pi(N, A, \tau, M)$ is responded with a $C \in \{0,1\}^{|M|+\tau}$. A decryption query $\pi^{-1}(N, A, \tau, C)$ is responded with either the unique M such that $\pi(N, A, \tau, M) = C$, or with $\perp$ if no such M exists. We refer to [18] for the details.

We define the RAE security of AEZ as

$$\mathbf{Adv}_{\mathrm{AEZ}}^{\mathrm{rae}}(\mathcal{A}) = \left| \mathbf{Pr}_K \left(\mathcal{A}^{\mathcal{E}_K, \mathcal{D}_K} = 1 \right) - \mathbf{Pr}_{\pi} \left(\mathcal{A}^{\pi, \pi^{-1}} = 1 \right) \right|,$$

where the probabilities are taken over the randomness of K, π, and the random choices of $\mathcal{A}$. The resources of $\mathcal{A}$ are usually bounded in terms of (q, ℓ, t), where q is the maximum queries to the construction oracle, each query is of length at most ℓ, and $\mathcal{A}$ runs in time t.

2.2   High-Level Description of AEZ

AEZ takes as input an arbitrarily sized key $K \in \{0,1\}^*$, and performs all of its procedures with three keys $I, J, L \in \{0,1\}^{128}$, where

$$I\|J\|L \longleftarrow \begin{cases} K & \text{if } |K| = 384, \\ \mathrm{BLAKE2b}(K) & \text{otherwise.} \end{cases} \tag{1}$$

AEZ then evaluates an algorithm depending on the size of M:[2]

- If $|M| = 0$, it evaluates AEZ-prf$(I\|J\|L, N, A, \tau)$;
- If $|M| > 0$:
  - If $|M| < 256 - \tau$, it evaluates Encipher-AEZ-tiny$(I\|J\|L, N, A, \tau, M)$;
  - If $|M| \geq 256 - \tau$, it evaluates Encipher-AEZ-core$(I\|J\|L, N, A, \tau, M)$.

Each of these algorithms starts with an evaluation of the AEZ-hash algorithm, a multi-layer PMAC-style MAC function that transforms (τ, N, A) into a 128-bit mask Δ. In this work, we are specifically interested in AEZ-hash and Encipher-AEZ-core. In more detail, in Sect. 4, we will describe three weak key attacks on Encipher-AEZ-core: two of which directly concern the Encipher-AEZ-core algorithm, one of which operates via AEZ-hash. The latter attack can be performed equivalently well via AEZ-prf and Encipher-AEZ-tiny, as the three algorithms rely on AEZ-hash in an identical way.

The four sub-algorithms of AEZ internally use a tweakable blockcipher

$$\widetilde{E} : \{0,1\}^{3 \cdot 128} \times \mathcal{T} \times \{0,1\}^{128} \to \{0,1\}^{128}, \tag{2}$$

that gets as input a key $I\|J\|L \in \{0,1\}^{3 \cdot 128}$, a tweak $(j, i) \in \mathcal{T} := (\{-1, 0\} \times [0..7]) \cup (\mathbb{N}^+ \times \mathbb{N})$, and bijectively transforms a plaintext X into a ciphertext $\widetilde{E}_{I\|J\|L}^{j,i}(X)$. We will elaborate on the tweakable blockcipher of AEZ in Sect. 3.

AEZ-Hash. We will use AEZ-hash for the simplified case where $|N| = |A| = 128$; AEZ-hash for this case is given in Algorithm 1. Our attack generalizes to arbitrarily-sized nonces and associated data.

Algorithm 1. AEZ-hash

Input: $(I\|J\|L, \tau, N, A)$ with $|N| = |A| = 128$
Output: $\Delta \in \{0,1\}^{128}$
1: $\Delta_1 \leftarrow \widetilde{E}_{I\|J\|L}^{3,1}(\langle\tau\rangle_{128})$ $\triangleright \langle\tau\rangle_{128}$ is the encoding of τ as an 128-bit string
2: $\Delta_2 \leftarrow \widetilde{E}_{I\|J\|L}^{4,1}(N)$
3: $\Delta_3 \leftarrow \widetilde{E}_{I\|J\|L}^{5,1}(A)$
4: **return** $\Delta = \Delta_1 \oplus \Delta_2 \oplus \Delta_3$

Encipher-AEZ-Core. We will describe our attacks for messages M such that $384 \leq |M| + \tau < 511$, and Encipher-AEZ-core for this case is given in Algorithm 2 and Fig. 1. We remark that the attacks can easily be generalized to any M such that $|M| \geq 256 - \tau$.

---

[2] The interfaces of the underlying algorithms have been slightly modified for the sake of simplicity.

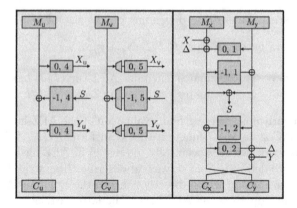

Fig. 1. AEZ for messages such that $384 \leq |M| + \tau < 511$ [17]. Here, the message is padded as $M_u \| M_v \| M_x \| M_y \leftarrow M \| 0^\tau$, where $|M_u| = |M_x| = |M_y| = 128$ and $0 \leq |M_v| < 127$. The mask Δ is computed as $\Delta \leftarrow$ AEZ-hash$(I \| J \| L, \tau, N, A)$. A box with inscription j, i represents an evaluation of $\widetilde{E}_{I\|J\|L}^{j,i}$. A trapezoid represents either chopping or 10^*-padding, depending on the direction.

Algorithm 2. Encipher-AEZ-core

Input: $(I\|J\|L, N, A, \tau, M)$ with $384 \leq |M| + \tau < 511$
Output: $C \in \{0, 1\}^{|M|+\tau}$
1: $\Delta \leftarrow$ AEZ-hash$(I\|J\|L, \tau, N, A)$ ▷ See Algorithm 1
2: $M_u\|M_v\|M_x\|M_y \leftarrow M\|0^\tau$, where $|M_u| = |M_x| = |M_y| = 128$ and $0 \leq |M_v| < 127$
3: $X \leftarrow \widetilde{E}_{I\|J\|L}^{0,4}(M_u) \oplus \widetilde{E}_{I\|J\|L}^{0,5}(M_v 10^*)$
4: $S_x \leftarrow M_x \oplus \Delta \oplus X \oplus \widetilde{E}_{I\|J\|L}^{0,1}(M_y)$; $S_y \leftarrow M_y \oplus \widetilde{E}_{I\|J\|L}^{-1,1}(S_x)$
5: $S \leftarrow S_x \oplus S_y$
6: $C_u \leftarrow M_u \oplus \widetilde{E}_{I\|J\|L}^{-1,4}(S)$; $C_v \leftarrow M_v \oplus \widetilde{E}_{I\|J\|L}^{-1,5}(S)$
7: $Y \leftarrow \widetilde{E}_{I\|J\|L}^{0,4}(C_u) \oplus \widetilde{E}_{I\|J\|L}^{0,5}(C_v 10^*)$
8: $C_y \leftarrow S_x \oplus \widetilde{E}_{I\|J\|L}^{-1,2}(S_y)$; $C_x \leftarrow S_y \oplus \Delta \oplus Y \oplus \widetilde{E}_{I\|J\|L}^{0,2}(C_y)$
9: **return** $C_u\|C_v\|C_x\|C_y$

---

3 AEZ Tweakable Blockcipher

We will elaborate on the tweakable blockcipher used in AEZ in Sect. 3.1, and describe structured sets of weak keys for it in Sect. 3.2.

3.1 Design

The tweakable blockcipher used in AEZ is internally constructed from the AES round function [10]. Define the *keyless* AES round function aesr(X) as

$$\text{aesr}(X) = \text{MixColumns} \circ \text{ShiftRows} \circ \text{SubBytes}(X).$$

AEZ uses the two blockciphers AES4 and AES10, where for $r \in \{4, 10\}$,

$$\text{AESr}_{K_0, K_1, \ldots, K_r}(X) = \text{aesr}(\cdots \text{aesr}(X \oplus K_0) \cdots \oplus K_{r-1}) \oplus K_r.$$

The tweakable blockcipher in AEZ is furthermore built of multiplications. Note that we can represent 128-bit strings as elements of a finite field $\text{GF}(2^{128})$ of order 2^{128}, and vice versa: a 128-bit string $A = a_{127}a_{126}\cdots a_1 a_0 \in \{0,1\}^{128}$ can be seen as a polynomial $A(\mathbf{x}) = a_{127}\mathbf{x}^{127} + \cdots a_1\mathbf{x} + a_0 \in \text{GF}(2^{128})$. We define multiplication of $A, B \in \{0,1\}^{128}$ as multiplication in $\text{GF}(2^{128})$ modulo the irreducible polynomial $f(\mathbf{x})$ used to generate the field:

$$A \cdot B := A(\mathbf{x}) \cdot B(\mathbf{x}) \bmod f(\mathbf{x}).$$

We remark that the multiplications in AEZ usually involve a term A of the form $2^m + n$ for $m \in \mathbb{N}$ and $n \in [0..7]$, which significantly simplifies the computation of $A \cdot B$. We refer to [17] for the details.

The tweakable blockcipher $\widetilde{E}$ of (2) takes as input a key $I\|J\|L \in \{0,1\}^{3 \cdot 128}$, a tweak $(j, i) \in \left(\{-1, 0\} \times [0..7]\right) \cup \left(\mathbb{N}^+ \times \mathbb{N}\right)$, and a plaintext X and computes the ciphertext as

tweak	$\widetilde{E}^{j,i}_{I\|J\|L}(X) =$
$j = -1, i \in [0..7]$	$\text{AES10}_{\mathbf{K}}(X)$ with $\mathbf{K} = (i \cdot J, I, J, L, I, J, L, I, J, L, I)$
$j = 0, \quad i \in [0..7]$	$\text{AES4}_{\mathbf{K}}(X)$ with $\mathbf{K} = (i \cdot I, J, I, L, 0^{128})$
$j = 1, \quad i \in \mathbb{N}$	$\text{AES4}_{\mathbf{K}}(X)$ with $\mathbf{K} = (\Delta_i \cdot I, J, I, L, 0^{128})$
$j = 2, \quad i \in \mathbb{N}$	$\text{AES4}_{\mathbf{K}}(X)$ with $\mathbf{K} = (\Delta_i \cdot I, L, I, J, L)$
$j \geq 3, \quad i = 0$	$\text{AES4}_{\mathbf{K}}(X)$ with $\mathbf{K} = (2^{j-3} \cdot L, J, I, L, 2^{j-3} \cdot L)$
$j \geq 3, \quad i \geq 1$	$\text{AES4}_{\mathbf{K}}(X)$ with $\mathbf{K} = (2^{j-3} \cdot L \oplus \Delta_i \cdot J, J, I, L, 2^{j-3} \cdot L \oplus \Delta_i \cdot J)$

where $\Delta_i = (2^{3 + \lfloor (i-1)/8 \rfloor} + (i{-}1 \bmod 8))$ for brevity. This tweakable blockcipher reminds of the XE(X) tweakable blockcipher used in OCB2 [25], as the "inner keys" are invariant of the tweak, and the "outer keys" depend on the tweak via the powering-up methodology.

Hoang et al. [17] claim that the AEZ construction is secure as long as $\widetilde{E}$ is a secure tweakable blockcipher. The usage of the tweakable blockcipher $\widetilde{E}$ as described above is validated using the so-called proof-then-prune approach: first, it is argued that if the tweakable blockcipher is instantiated with AES10 everywhere, it behaves like XE(X), and then some uses of AES10 are cut down to 4 rounds to speed up AEZ. As it is unreasonable to assume that AES4 behaves like a pseudorandom permutation, the proof-then-prune approach is ultimately a heuristic [17,20]. In this work, we will *not* consider any internal properties of AES4 and AES10, and simply consider both AES4 and AES10 as secure primitives: our attacks are independent of the debated proof-then-prune approach, but rather center around the structural properties of $\widetilde{E}$.

3.2   Weak Keys

The definition of $\widetilde{E}$, and more specifically the generation of the key $\mathbf{K}$ from $I\|J\|L$, reveals peculiar behavior. Particularly, if one of the subkeys I, J, L equals

0^{128}, the tweakable blockcipher allows for trivial collisions among different tweaks and is insecure.

Lemma 1. *The tweakable blockcipher $\widetilde{E}$ satisfies the following properties:*

(i) If $J = 0^{128}$, then $\widetilde{E}^{-1,i}_{I\|0^{128}\|L} = \widetilde{E}^{-1,i'}_{I\|0^{128}\|L}$ for any $i, i' \in [0..7]$;

(ii) If $I = 0^{128}$, then $\widetilde{E}^{0,i}_{0^{128}\|J\|L} = \widetilde{E}^{0,i'}_{0^{128}\|J\|L}$ for any $i, i' \in [0..7]$;

(iii) If $L = 0^{128}$, then $\widetilde{E}^{j,i}_{I\|J\|0^{128}} = \widetilde{E}^{j',i}_{I\|J\|0^{128}}$ for any $j, j' \geq 3$ and $i \in \mathbb{N}$.

Proof. The properties are in fact a direct consequence of the definition of $\widetilde{E}^{j,i}_{I\|J\|L}$ (see Sect. 3.1). Starting with (i): for subkey $J = 0^{128}$ and tweak value $j = -1$, we have

$$\widetilde{E}^{-1,i}_{I\|0^{128}\|L}(X) = \mathrm{AES10}_{\mathbf{K}}(X),$$

with $\mathbf{K} = (i \cdot 0^{128}, I, 0^{128}, L, I, 0^{128}, L, I, 0^{128}, L, I)$. In other words, $\widetilde{E}^{-1,i}_{I\|0^{128}\|L}(X)$ is independent of i, and we obtain that

$$\widetilde{E}^{-1,i}_{I\|0^{128}\|L} = \widetilde{E}^{-1,i'}_{I\|0^{128}\|L}$$

for any $i, i' \in [0..7]$. The proof of (ii) and (iii) is equivalent: for (ii), $\widetilde{E}^{0,i}_{0^{128}\|J\|L}$ is independent of i, and for (iii), $\widetilde{E}^{j,i}_{I\|J\|0^{128}}$ is independent of $j \geq 3$ for all $i \in \mathbb{N}$. □

More properties can be derived in a similar fashion, but these three relations suffice for the discussion of our attacks.

4   Weak Key Attacks on AEZ

We will perform three distinguishing attacks on AEZ, each of which exploits one of the properties of Lemma 1 and distinguishes AEZ from random in at móst two queries. Note that if it is unknown which subkey equals 0^{128}, hence it is unknown which of the properties of Lemma 1 to exploit, all three attacks should be evaluated and the complexity is five queries (at most).

The first two distinguishing attacks rely on weaknesses in Encipher-AEZ-core, while the third one relies on a weakness in AEZ-hash. In these attacks, we will consider an adversary that has access to either $\mathcal{E}_K$ with random key K, or its idealized counterpart π (cf. Sect. 2.1), and denote by $\mathcal{O} \in \{\mathcal{E}_K, \pi\}$ the oracle to which the adversary has access.

4.1   Attack Exploiting Property (i)

Assume that $J = 0^{128}$. Using Lemma 1 property (i), we can perform the following distinguishing attack.

– Let N, A, τ be any nonce, associated data, and tag size;

- Let M be any message such that $384 \le |M| + \tau < 511$. Write $M\|0^\tau = M_u\|M_v\|M_x\|M_y$, where $|M_u| = |M_x| = |M_y| = 128$ and $|M_v| = |M|+\tau-384 =: \ell$;
- Query $C = \mathcal{O}(N, A, \tau, M) \in \{0,1\}^{|M|+\tau}$. Write $C = C_u\|C_v\|C_x\|C_y$, where $|C_u| = |C_x| = |C_y| = 128$, and $|C_v| = \ell$;
- If

$$\mathrm{chop}_\ell\big(M_u \oplus C_u \oplus M_v \oplus C_v\big) = 0^\ell, \tag{3}$$

output 0, otherwise output 1.

If $\mathcal{O} = \mathcal{E}_K$, we have

$$\mathrm{chop}_\ell\big(M_u \oplus C_u\big) = \mathrm{chop}_\ell\big(\widetilde{E}_{I\|0^{128}\|L}^{-1,4}(S)\big)$$

$$\overset{(i)}{=} \mathrm{chop}_\ell\big(\widetilde{E}_{I\|0^{128}\|L}^{-1,5}(S)\big) = \mathrm{chop}_\ell\big(M_v \oplus C_v\big),$$

and (3) is satisfied by construction. Thus, the adversary always outputs 0 in the real world. In the ideal world, if $\mathcal{O} = \pi$, this condition is satisfied with probability $1/2^\ell$. Thus, the success probability of the attack is

$$\mathbf{Adv}_{\mathrm{AEZ}}^{\mathrm{rae}}(\mathcal{A}) = 1 - 1/2^\ell,$$

where $\mathcal{A}$ makes 1 construction query of length $|N|+|A|+|M|$, and has negligible time complexity. Recall that $\ell = |M| + \tau - 384$, where M is a freely chosen message. Hence, by taking $|M| + \tau = 511$ the success probability of the attack is $1 - 1/2^{127}$.

4.2 Attack Exploiting Property (ii)

Assume that $I = 0^{128}$. Using Lemma 1 property (ii), we can perform the following distinguishing attack.

- Let N, A, τ be any nonce, associated data, and tag size;
- Let $0 \le \ell < 128$. Let $M_v, M_v' \in \{0,1\}^\ell$ be any two *distinct* message blocks. Write $M_u = M_v 10^*$ and $M_u' = M_v' 10^*$. Let $M_{xy} \in \{0,1\}^{256-\tau}$ be any message block. Write

$$M = M_u\|M_v\|M_{xy} \text{ and } M = M_u'\|M_v'\|M_{xy};$$

- Query $C = \mathcal{O}(N, A, \tau, M) \in \{0,1\}^{|M|+\tau}$ and $C' = \mathcal{O}(N, A, \tau, M') \in \{0,1\}^{|M'|+\tau}$. Write $C = C_u\|C_v\|C_x\|C_y$ and $C' = C_u'\|C_v'\|C_x'\|C_y'$, where $|C_u| = |C_x| = |C_y| = |C_u'| = |C_x'| = |C_y'| = 128$, and $|C_v| = |C_v'| = \ell$;
- If

$$M_u \oplus C_u \oplus M_u' \oplus C_u' = 0^{128}, \tag{4}$$

output 0, otherwise output 1.

The verification of the attack is a bit more complex than for case (i), and relies on the key observation that in the real world, $S = S'$. In more detail, if $\mathcal{O} = \mathcal{E}_K$, we have

$$X = X_u \oplus X_v = \widetilde{E}^{0,4}_{0^{128}\|J\|L}(M_u) \oplus \widetilde{E}^{0,5}_{0^{128}\|J\|L}(M_v 10^*) \overset{(ii)}{=} 0^{128}, \text{ and}$$

$$X' = X'_u \oplus X'_v = \widetilde{E}^{0,4}_{0^{128}\|J\|L}(M'_u) \oplus \widetilde{E}^{0,5}_{0^{128}\|J\|L}(M'_v 10^*) \overset{(ii)}{=} 0^{128}.$$

In other words, $X = X'$. Furthermore, as (τ, N, A) is the same in both evaluations,

$$\Delta = \text{AEZ-hash}(0^{128}\|J\|L, \tau, N, A) = \Delta'.$$

Finally, the two different queries satisfy $M_{xy} = M'_{xy}$. From Algorithm 2 we obtain that the intermediate value S is a function of $M_{xy}, X,$ and Δ, and thus,

$$S = S'.$$

We consequently obtain

$$M_u \oplus C_u = \widetilde{E}^{-1,4}_{0^{128}\|J\|L}(S) = \widetilde{E}^{-1,4}_{0^{128}\|J\|L}(S') = M'_u \oplus C'_u,$$

and (4) is satisfied by construction. Thus, the adversary always outputs 0 in the real world. In the ideal world, if $\mathcal{O} = \pi$, this condition is satisfied with probability $1/2^{128}$. Thus, the success probability of the attack is

$$\mathbf{Adv}^{\text{rae}}_{\text{AEZ}}(\mathcal{A}) = 1 - 1/2^{128},$$

where $\mathcal{A}$ makes 2 construction queries of length $|N|+|A|+|M|$, and has negligible time complexity.

4.3   Attack Exploiting Property (iii)

Assume that $L = 0^{128}$. Using Lemma 1 property (iii), we can perform the following distinguishing attack.

- Let τ be any tag size and M any message such that $384 \leq |M| + \tau < 511;$[3]
- Let $N, N' \in \{0,1\}^{128}$ be any two *distinct* nonces;
- Query $C = \mathcal{O}(N, N', \tau, M) \in \{0,1\}^{|M|+\tau}$ and $C' = \mathcal{O}(N', N, \tau, M) \in \{0,1\}^{|M|+\tau}$;
- If

$$C \oplus C' = 0^{|M|+\tau}, \tag{5}$$

   output 0, otherwise output 1.

---

[3] The condition on the message length is simply to assure that the attack goes via Encipher-AEZ-core of Algorithm 2. As a matter of fact, AEZ-prf and Encipher-AEZ-tiny use AEZ-hash in an identical way, and the attack applies equally well to messages of a different length.

If $\mathcal{O} = \mathcal{E}_K$, we have

$$
\begin{aligned}
\Delta &= \text{AEZ-hash}(I\|J\|0^{128}, \tau, N, N') \\
&= \widetilde{E}_{I\|J\|0^{128}}^{3,1}(\langle\tau\rangle_{128}) \oplus \widetilde{E}_{I\|J\|0^{128}}^{4,1}(N) \oplus \widetilde{E}_{I\|J\|0^{128}}^{5,1}(N') \\
&\stackrel{\text{(iii)}}{=} \widetilde{E}_{I\|J\|0^{128}}^{3,1}(\langle\tau\rangle_{128}) \oplus \widetilde{E}_{I\|J\|0^{128}}^{4,1}(N') \oplus \widetilde{E}_{I\|J\|0^{128}}^{5,1}(N) \\
&= \text{AEZ-hash}(I\|J\|0^{128}, \tau, N', N) = \Delta'.
\end{aligned}
$$

It follows from Algorithm 2 that $C = C'$, and that (5) is satisfied by construction. Thus, the adversary always outputs 0 in the real world. In the ideal world, if $\mathcal{O} = \pi$, this condition is satisfied with probability $1/2^{|M|+\tau}$. Thus, the success probability of the attack is

$$
\mathbf{Adv}_{\text{AEZ}}^{\text{rae}}(\mathcal{A}) = 1 - 1/2^{|M|+\tau},
$$

where $\mathcal{A}$ makes 2 construction queries of length $256 + |M|$, and has negligible time complexity. Recall that τ can be freely chosen.

5   External Key Padding

We consider a specific scenario, called "external key padding," which shows the potential strength of the attacks of Sect. 4. Consider a user that uses AEZ as a black box. Instead of plugging his key K' into AEZ directly, he naively thinks speed-up could be achieved by artificially extending K' to a 384-bit key in advance:[4]

$$
K \leftarrow K'\|0^{384-|K'|}.
$$

Alternatively, one could consider a scenario where two users communicate, both set a part of the key, K'_a and K'_b, and the final key is established by padding in the middle:

$$
K \leftarrow K'_a\|0^{384-|K'_a|-|K'_b|}\|K'_b.
$$

Although these use cases may sound contrived at first sight, they cover a realistic setting where a user of a scheme "misuses" it to suit the application. More generally, the scenario covers any form of poor key generation where $K \in \{0,1\}^{384}$ is generated according to a very weak key generation function. This could happen due to naive use of the user, key generation regulations enforced by service providers, or whatsoever. See also Fig. 2. Note that in these cases, AEZ works syntactically fine (as would any other authenticated encryption scheme) and will not produce errors due to the abuse.

---

[4] Here, it is implicitly assumed that K' is of size at most 384 bits.

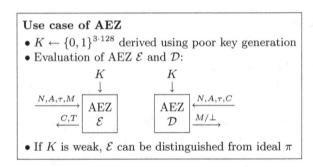

Fig. 2. External key padding scenario. Here, the key K is generated in such a way that it may, inadvertently, contain a 0^{128} subkey.

5.1   How Does AEZ Behave?

It is straightforward to see that in case of external key padding, the attacks of Sect. 4 directly apply. Indeed, if a user of AEZ has a 256-bit key $K' \in \{0,1\}^{256}$, and prematurely pads it to a 384-bit key as $K = K'\|0^{128}$, it obtains a weak key K for which property (iii) of Lemma 1 holds. Thus, the scheme can be broken in at most two queries, making use of the fact that the last subkey equals 0^{128}. A similar reasoning applies to the case $K \leftarrow K'_a\|0^{128}\|K'_b$, where $K'_a, K'_b \in \{0,1\}^{128}$.

5.2   How Do Other Schemes Behave?

Intuitively, one would expect the security of the mode to decrease linearly with the amount of key reduction. In other words, if the security advantage as a function of the key size is $\mathcal{O}(2^{-|K|})$, then in case of the external key padding, the distinguishing advantage would increase to $\mathcal{O}(2^{-|K'|})$.

It turns out that, in fact, the majority of the authenticated encryption schemes show exactly this behavior. For instance, considering Sponge-based authenticated encryption [1,4,7,11,14,19,22,23,27], the key is already padded internally, and the external key padding has no influence. Alternatively, for regular blockcipher-based modes such as OCB [21], SIV [26], and COPA [3], the security of the mode is reduced to the security of the underlying E_K, and the adjustment from K' to K becomes captured in the blockcipher security $\mathbf{Adv}_E^{\mathrm{sprp}}(q,t)$.

6   Conclusion

Given the rarity of weak keys in AEZ (around 3 out of every 2^{128} keys), there is little chance that a randomly selected key is weak, and it seems not possible to break the security claims of AEZ using these weak keys. In addition, a simple mitigation of our attacks consists of imposing that no subkey equals 0^{128}. (But there may be other weak keys as multiplications in the tweakable blockcipher are performed in the finite field $GF(2^{128})$ [16,24,28].)

Nevertheless, the observations *do* show a more peculiar weakness in AEZ, namely that the underlying tweakable blockcipher is not sound. Even if a more complicated key scheduling is used (as was done, for instance, in AEZ v2 and v3), it is still straightforward to see that a certain fraction of the keys allows for collisions in the tweakable blockcipher. In other words, while the issues with the external key padding could be mitigated using a stronger key scheduling, the issues with the tweakable blockcipher in AEZ are more structural.

Regardless of whether or not the external key padding scenario is relevant, it sets the stage for a discussion of what one may expect of a highly secure authenticated encryption scheme. Our observations (as well as the ones by Fuhr et al. [13] and Chaigneau and Gilbert [9]) stand in sharp contrast with the usage of powerful terms like "robustness" and with what high-security authenticated encryption embraces. Barwell et al. [6] already expressed their worries about the usage of the "robustness" term in the context of robust authenticated encryption, and stated: "Robustness characterizes the ability of a construct to be pushed right to the edge of its intended use case (and possibly beyond)." Putting our attacks in this perspective, by padding outside the mode, one incorrectly uses that mode, but on the other hand, "robust authenticated encryption" seems to imply that such modes work properly as long as they are employed in a *syntactically correct manner*. From this point of view, our attacks violate the robustness claims on AEZ.

Acknowledgments. Bart Mennink is supported by a postdoctoral fellowship from the Netherlands Organisation for Scientific Research (NWO) under Veni grant 016.Veni.173.017. The author would like to thank his COSIC colleagues, the attendees of Dagstuhl Symmetric Cryptography 2016, and the reviewers of CT-RSA 2017 for their comments and suggestions.

References

1. Andreeva, E., Bilgin, B., Bogdanov, A., Luykx, A., Mendel, F., Mennink, B., Mouha, N., Wang, Q., Yasuda, K.: PRIMATEs v1.02. (2015, Submission to CAESAR competition)
2. Andreeva, E., Bogdanov, A., Luykx, A., Mennink, B., Mouha, N., Yasuda, K.: How to securely release unverified plaintext in authenticated encryption. In: Sarkar, P., Iwata, T. (eds.) ASIACRYPT 2014. LNCS, vol. 8873, pp. 105–125. Springer, Heidelberg (2014). doi:10.1007/978-3-662-45611-8_6
3. Andreeva, E., Bogdanov, A., Luykx, A., Mennink, B., Tischhauser, E., Yasuda, K.: Parallelizable and authenticated online ciphers. In: Sako, K., Sarkar, P. (eds.) ASIACRYPT 2013. LNCS, vol. 8269, pp. 424–443. Springer, Heidelberg (2013). doi:10.1007/978-3-642-42033-7_22
4. Aumasson, J., Jovanovic, P., Neves, S.: NORX v2.0. (2015, Submission to CAESAR competition)
5. Aumasson, J., Neves, S., Wilcox-O'Hearn, Z., Winnerlein, C.: BLAKE2: simpler, smaller, fast as MD5. In: Jacobson, M., Locasto, M., Mohassel, P., Safavi-Naini, R. (eds.) ACNS 2013. LNCS, vol. 7954, pp. 119–135. Springer, Heidelberg (2013)

6. Barwell, G., Page, D., Stam, M.: Rogue decryption failures: reconciling AE robustness notions. In: Groth, J. (ed.) IMACC 2015. LNCS, vol. 9496, pp. 94–111. Springer, Heidelberg (2015). doi:10.1007/978-3-319-27239-9_6
7. Bertoni, G., Daemen, J., Peeters, M., Assche, G.: Duplexing the sponge: single-pass authenticated encryption and other applications. In: Miri, A., Vaudenay, S. (eds.) SAC 2011. LNCS, vol. 7118, pp. 320–337. Springer, Heidelberg (2012). doi:10.1007/978-3-642-28496-0_19
8. CAESAR: Competition for Authenticated Encryption: Security, Applicability, and Robustness. http://competitions.cr.yp.to/caesar.html
9. Chaigneau, C., Gilbert, H.: Is AEZ v4.1 sufficiently resilient against key-recovery attacks? IACR Trans. Symmetric Cryptology 1(1), 114–133 (2016)
10. Daemen, J., Rijmen, V.: The Design of Rijndael: AES - The Advanced Encryption Standard. Springer, Heidelberg (2002)
11. Dobraunig, C., Eichlseder, M., Mendel, F., Schläffer, M.: Ascon v1.1. (2015, Submission to CAESAR competition)
12. Espitau, T., Fouque, P.-A., Karpman, P.: Higher-order differential meet-in-the-middle preimage attacks on SHA-1 and BLAKE. In: Gennaro, R., Robshaw, M. (eds.) CRYPTO 2015. LNCS, vol. 9215, pp. 683–701. Springer, Heidelberg (2015). doi:10.1007/978-3-662-47989-6_33
13. Fuhr, T., Leurent, G., Suder, V.: Collision attacks against CAESAR candidates. In: Iwata, T., Cheon, J.H. (eds.) ASIACRYPT 2015. LNCS, vol. 9453, pp. 510–532. Springer, Heidelberg (2015). doi:10.1007/978-3-662-48800-3_21
14. Gligoroski, D., Mihajloska, H., Samardjiska, S., Jacobsen, H., El-Hadedy, M., Jensen, R.: π-Cipher v2.0. (2015, Submission to CAESAR competition)
15. Guo, J., Karpman, P., Nikolić, I., Wang, L., Wu, S.: Analysis of BLAKE2. In: Benaloh, J. (ed.) Topics in Cryptology – CT-RSA 2014. LNCS, vol. 8366, pp. 402–423. Springer, Heidelberg (2014)
16. Handschuh, H., Preneel, B.: Key-recovery attacks on universal hash function based MAC algorithms. In: Wagner, D. (ed.) CRYPTO 2008. LNCS, vol. 5157, pp. 144–161. Springer, Heidelberg (2008). doi:10.1007/978-3-540-85174-5_9
17. Hoang, V.T., Krovetz, T., Rogaway, P.: AEZ v4: Authenticated Encryption by Enciphering. (2015, Submission to CAESAR competition)
18. Hoang, V.T., Krovetz, T., Rogaway, P.: Robust authenticated-encryption AEZ and the problem that it solves. In: Oswald, E., Fischlin, M. (eds.) EUROCRYPT 2015. LNCS, vol. 9056, pp. 15–44. Springer, Heidelberg (2015). doi:10.1007/978-3-662-46800-5_2
19. Jovanovic, P., Luykx, A., Mennink, B.: Beyond $2^{c/2}$ security in sponge-based authenticated encryption modes. In: Sarkar, P., Iwata, T. (eds.) ASIACRYPT 2014. LNCS, vol. 8873, pp. 85–104. Springer, Berlin (2014). doi:10.1007/978-3-662-45611-8_5
20. Keliher, L., Sui, J.: Exact maximum expected differential and linear probability for 2-round advanced encryption standard (AES). IET Inf. Secur. 1(2), 53–57 (2007)
21. Krovetz, T., Rogaway, P.: The software performance of authenticated-encryption modes. In: Joux, A. (ed.) FSE 2011. LNCS, vol. 6733, pp. 306–327. Springer, Heidelberg (2011). doi:10.1007/978-3-642-21702-9_18
22. Mennink, B., Reyhanitabar, R., Vizár, D.: Security of full-state keyed sponge and duplex: applications to authenticated encryption. In: Iwata, T., Cheon, J.H. (eds.) ASIACRYPT 2015. LNCS, vol. 9453, pp. 465–489. Springer, Heidelberg (2015). doi:10.1007/978-3-662-48800-3_19

23. Morawiecki, P., Gaj, K., Homsirikamol, E., Matusiewicz, K., Pieprzyk, J., Rogawski, M., Srebrny, M., Wójcik, M.: ICEPOLE v2. (2015, Submission to CAESAR competition)

24. Procter, G., Cid, C.: On weak keys and forgery attacks against polynomial-based MAC schemes. In: Moriai, S. (ed.) FSE 2013. LNCS, vol. 8424, pp. 287–304. Springer, Heidelberg (2014). doi:10.1007/978-3-662-43933-3_15

25. Rogaway, P.: Efficient instantiations of tweakable blockciphers and refinements to modes OCB and PMAC. In: Lee, P.J. (ed.) ASIACRYPT 2004. LNCS, vol. 3329, pp. 16–31. Springer, Heidelberg (2004). doi:10.1007/978-3-540-30539-2_2

26. Rogaway, P., Shrimpton, T.: A provable-security treatment of the key-wrap problem. In: Vaudenay, S. (ed.) EUROCRYPT 2006. LNCS, vol. 4004, pp. 373–390. Springer, Heidelberg (2006). doi:10.1007/11761679_23

27. Saarinen, M.-J.O.: Beyond modes: building a secure record protocol from a cryptographic sponge permutation. In: Benaloh, J. (ed.) CT-RSA 2014. LNCS, vol. 8366, pp. 270–285. Springer, Heidelberg (2014). doi:10.1007/978-3-319-04852-9_14

28. Saarinen, M.-J.O.: Cycling attacks on GCM, GHASH and other polynomial MACs and hashes. In: Canteaut, A. (ed.) FSE 2012. LNCS, vol. 7549, pp. 216–225. Springer, Heidelberg (2012). doi:10.1007/978-3-642-34047-5_13

Symmetric Key Constructions

Full Disk Encryption: Bridging Theory and Practice

Louiza Khati[1,2], Nicky Mouha[3,4,5]([✉]), and Damien Vergnaud[1]

[1] École Normale Supérieure - CNRS - Inria, Paris, France
{Louiza.Khati,Damien.Vergnaud}@ens.fr
[2] Oppida, Montigny-le-Bretonneux, France
[3] Department of Electrical Engineering-ESAT/COSIC,
KU Leuven, Leuven and iMinds, Ghent, Belgium
nicky@mouha.be
[4] Project-team SECRET, Inria, Paris, France
[5] National Institute of Standards and Technology, Gaithersburg, MD, USA

Abstract. We revisit the problem of Full Disk Encryption (FDE), which refers to the encryption of each sector of a disk volume. In the context of FDE, it is assumed that there is no space to store additional data, such as an IV (Initialization Vector) or a MAC (Message Authentication Code) value. We formally define the security notions in this model against chosen-plaintext and chosen-ciphertext attacks. Then, we classify various FDE modes of operation according to their security in this setting, in the presence of various restrictions on the queries of the adversary. We will find that our approach leads to new insights for both theory and practice. Moreover, we introduce the notion of a *diversifier*, which does not require additional storage, but allows the plaintext of a particular sector to be encrypted to different ciphertexts. We show how a 2-bit diversifier can be implemented in the EagleTree simulator for solid state drives (SSDs), while decreasing the total number of Input/Output Operations Per Second (IOPS) by only 4%.

Keywords: Disk encryption theory · Full Disk Encryption · FDE · XTS · IEEE P1619 · Unique first block · Diversifier · Provable security

1 Introduction

The term Full Disk Encryption (FDE) is commonly used when every sector of a disk volume is encrypted. There is typically no space to store any additional data, such as an IV or a MAC. As explained by Ferguson [10], generic solutions to store additional data will at least double the number of read and write operations, and will significantly reduce the available disk space. They also change the disk layout, which makes it extremely complicated to enable FDE on existing disks.

With this restriction, FDE cannot offer authentication, or at best "poorman's authentication" [10], which is to hope that ciphertext changes will result in a plaintext that is random enough to make the application crash. It can also not

H. Handschuh (Ed.): CT-RSA 2017, LNCS 10159, pp. 241–257, 2017.
DOI: 10.1007/978-3-319-52153-4_14

achieve chosen-plaintext indistinguishability: when the same data is encrypted twice at the same sector index, the resulting ciphertexts will be identical.

Additional efficiency constraints may be imposed on FDE as well. For example, it can be desirable to perform encryption and/or decryption in parallel, which is not possible for inherently sequential constructions where the i-th plaintext block of a sector cannot be processed until all previous plaintext blocks are processed. If there is not enough memory available to store an entire sector, it may be required that encryption and decryption are on-line, meaning that the i-th block of a sector may only depend on the preceding blocks.

These implementation constraints impose severe restrictions on the algorithms that can be used for FDE. A general problem in the domain of FDE, however, is that the security properties of the resulting constructions are not always well-understood. On the other hand, cryptographers often complain about the absence of well-defined cryptographic goals for FDE (see e.g. Rogaway [25]), which are prerequisites to find a good-trade-off between security and efficiency.

Our Contributions. Firstly, we want to measure "how much security is left" within the constraints of FDE. In order to do so, we introduce a theoretical framework to capture that a FDE algorithm behaves as "randomly as possible" subject to different practical constraints. We consider settings where the encryption oracle can be random-up-to-repetition, random-up-to-prefix or random-up-to-block.

For each of the attack settings in the framework, we list an efficient construction that achieves security within this setting. We recall existing security results, and provide new proofs, in particular in the *unique-first-block* (ufb) setting where the Operating System (OS) or application ensures that the first n bits of the plaintext will not be repeated for a particular sector number, where n is the block size of the underlying block cipher.

Our model recalls that the modes of operation CBC (Cipher Block Chaining) and IGE (Infinite Garble Extension), even with a secret IV, do not achieve the security properties that developers often wrongly assume for these constructions. As already shown by Bellare et al. [1], CBC and IGE are not IND-CPA secure up-to-prefix. We will prove, however, that both constructions are IND-CPA secure under the ufb constraint.

Regarding chosen-ciphertext attacks, we point out that Added Redundancy Explicit Authentication (AREA) [11] is not secure when used with CBC or IGE, even when the IV is secret. The insecurity of constructions such as AREA was already shown in 2001 by Jutla [20], but has nevertheless not yet been pointed out in the context of FDE. We recall that there exist constructions that are secure in this setting, such as TC2 and TC3 [26].

Secondly, we revisit the FDE constraints from an engineering point of view. We show that it is possible to produce different ciphertexts for the same plaintext at a particular sector index, without storing additional data. Our solution applies to solid state drives (SSDs), where we show how the SSD firmware can be modified to associate a *diversifier* to every sector. This is done without modifying the data structures of the SSD, but by forcing data to be written to a particular Logical Unit Number (LUN).

For any particular sector, the diversifier value must be unique. But as we will explain later, additional requirements are necessary for performance reasons. When looking at all sectors at any particular point in time, each diversifier value should occur roughly the same number of times. Additionally, this diversifier value can typically only be a few bits long. These requirements put the diversifier in a class by itself, and not as a specific case of a random IV or a nonce (i.e. a number that is only used once).

When we benchmarked our solution in a modified EagleTree simulator [8], we found that it increases the average latency by at most 12% for reads and 2% for writes, and that it reduces the SSD throughput (read and write combined) by less than 4%. This paper provides the first efficient FDE solution that can achieve indistinguishability against chosen-plaintext and chosen-ciphertext attacks.

Related Work. The problem of FDE has been researched extensively, see for example Rogaway [25] for a provable security treatment, or Fruhwirth [11] for an implementer's perspective. The formal requirements of disk encryption are often not clearly stated.

FDE is a topic that has gathered significant interest from industry and standardization, and often leads to application-specific solutions due to the special requirements of full disk encryption. Of particular interest are the elephant diffuser used in Microsoft's BitLocker [10], or IEEE P1619's XTS standard [17], which later became a NIST recommendation [9] as well.

Throughout this paper, we assume that adversary has access to the disk volume at any time. The adversary has (partial) knowledge and even control of the plaintext, and can even change the ciphertext as well. We therefore go beyond just "single point-in-time permanent offline compromise" (see e.g. [12,14]). Read and write operations are assumed to be atomic (on a sector level), so we do not consider blockwise adaptive attacks [18].

Sound key management is required to avoid that the plaintext contains the key, or any function of the key [15]. Physical access threats (e.g. cold boot, DMA, evil maid, or hot plug attacks [13,22]) are also outside the scope of this paper.

2   Disk Encryption Methods

Data is read and written in a sector-addressable device by fixed-length units called sectors, usually 512 or 4096 bytes long. The OS can access a specific sector by its *sector number s*. We consider the case of an encrypted disk volume where data is encrypted by the OS before being stored.

We list the modes of operation that frequently appear in the context of FDE, whether it be in academic literature or in practical implementations. We also mention other modes with interesting security or efficiency properties.

ECB (Electronic Codebook). In the simplest encryption mode, the plaintext is divided into blocks of n bits, and each block is encrypted separately using an

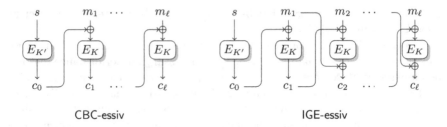

Fig. 1. Description of the CBC-essiv and IGE-essiv modes of operation.

n-bit block cipher. It can readily be used for FDE, even though it is well-known that it does not provide adequate security.

CTR (Counter). This mode uses a counter (incremented for each block) that is encrypted and then XORed with the plaintext block to output the cipher-text block. Typically, the counter is the sector number, bit-shifted to the left over a sufficient number of bits so that the least-significant bits can represent a counter for the number of blocks in one sector. CTR mode is IND-CPA secure [3] under the assumption that the counter is a nonce. In the context of FDE, this assumption does not hold as sectors can be overwritten.

CBC. In this mode, each plaintext block is XORed with the previous ciphertext block (or an IV for the first plaintext block) before being encrypted. To achieve the IND-CPA security notion, it is well-known that the IV has to be a random value [3]. However, for FDE, a first natural idea is to use the sector number as an IV. Fruhwirth [11] proposed to use as an IV the encryption of the sector number by the block cipher keyed with an independent key (see Fig. 1).[1]

IGE. IGE was proposed Campbell [7] as a variant of CBC mode where each block of plaintext is XORed with the *next* ciphertext block (see Fig. 1). For FDE, since the sector number is not secret, we will consider the variant where the IV is the encryption of the sector number s in which s is not XORed to the first ciphertext block. We will refer to this mode as IGE-essiv.

XTS. XTS [17] applies a tweakable block cipher to every n-bit block of a sector, where the tweak depends on the sector number and on the index of the block within the sector. It uses *ciphertext stealing* when the sector size is not a multiple of n bits, however such sectors sizes are not considered in this paper.

TC1, TC2 and TC3. These modes for tweakable block ciphers were defined by Rogaway and Zhang in [26]. The difference between these constructions is the way the tweak is used:

---

[1] Two distinct keys are needed: the message is encrypted with key K, and the IV is encrypted with key $K' \neq K$ (see Fig. 1), in order to avoid an attack by Rogaway [24].

- In TC1, the tweak is the previous ciphertext block as in the HCBC mode [1];
- In TC2, the tweak is the concatenation of the previous ciphertext block and the previous plaintext block as in the HCBC2 mode [1];
- In TC3 mode, the tweak is the XOR of the previous ciphertext block and the previous plaintext block as in the MHCBC2 mode [23].

WTBC (Wide Tweakable Block ciphers). In the context of FDE, the block size of a WTBC is equal to the sector size, and the sector number is used as the tweak input. From a security point of view, any change in the plaintext or ciphertext affects the entire sector. A WTBC is typically realized using smaller (tweakable) block ciphers, as for example in the EME (Encrypt-Mix-Encrypt) [16] mode.

Our goal in this paper is to analyze these constructions and to evaluate their security in different models.

3   Security Notions for FDE

In this section, we formalize several security notions for FDE. We first give a formal syntactic definition of block-cipher-based FDE.

It is assumed that the plaintext of a sector is a multiple of n which is the block cipher size. All plaintexts are ℓ blocks of n bits. m^i denotes the i-th block of the plaintext m such that $m = m^1||m^2||...||m^\ell$ where $||$ denotes concatenation of strings. IND-CPA-xx corresponds to IND-CPA up-to-block, IND-CPA up-to-prefix, IND-CPA up-to-repetition and IND-CPA.

Definition 1. *Let k, n and ℓ be three positive integers. A (k, n, ℓ)-block-cipher-based FDE scheme is a pair of algorithms (Enc, Dec) such that:*

- *Enc is the (deterministic) encryption algorithm which takes as input a key $K \in \{0,1\}^k$, a sector number $s \in \{0,1\}^n$ and a plaintext $m \in \{0,1\}^{\ell \cdot n}$ and outputs a ciphertext $c \in \{0,1\}^{\ell \cdot n}$;*
- *Dec is the (deterministic) decryption algorithm which takes as input a key $K \in \{0,1\}^k$, a sector number $s \in \{0,1\}^n$ and a ciphertext $c \in \{0,1\}^{\ell \cdot n}$ and outputs a plaintext $m \in \{0,1\}^{\ell \cdot n}$,*

such that $\forall (K, s, m) \in \{0,1\}^{k+n+\ell \cdot n} : \mathsf{Dec}(K, s, \mathsf{Enc}(K, s, m)) = m$.

For each security notion, we define two variants: security under Chosen-Plaintext Attack (CPA) where the adversary is given access to the **Encrypt** procedure and security under Chosen-Ciphertext Attack (CCA) where the adversary is also given access to the **Decrypt** procedure. The adversary is not allowed to query the decryption of a ciphertext that was previously returned by **Encrypt** or vice versa.

Indistinguishability up-to-block. Each ciphertext block depends deterministically on the plaintext block, the sector number s and the block position in the plaintext, but behaves as "randomly as possible" subject to this constraint.

The corresponding game described in the full version of this paper [21] uses independent random permutations $\Pi^{(s,i)} : \{0,1\}^n \to \{0,1\}^n$ for each sector number s and each block position $i \in \{1,\ldots,\ell\}$. We specifically introduce this setting to describe the security goal of XTS, see Rogaway [25] for a formal definition (using a filter function) of what is known to leak by the XTS mode.

Indistinguishability up-to-prefix. For $i \in \{1,\ldots,\ell\}$, the i-th ciphertext block depends deterministically on the sector number s and all previous plaintext blocks at position j for $j \in \{1,\ldots,i\}$, but again behave as "randomly as possible" subject to this constraint. The corresponding game described in the full version of this paper [21] uses independent random permutations $\Pi_m^i : \{0,1\}^n \to \{0,1\}^n$ for each sector number s and each plaintext prefix $m \in \{0,1\}^{t \cdot n}$ for $t \in \{1,\ldots,\ell\}$. This notion corresponds to security notion described in [1] for *on-line ciphers*.

Indistinguishability up-to-repetition. Each ciphertext block depends deterministically on the plaintext block and the sector number s, but behaves as "randomly as possible" subject to this constraint. The corresponding game described in the full version of this paper [21] uses independent random permutations $\Pi^s : \{0,1\}^n \to \{0,1\}^n$ for each sector number s. It is the best achievable notion for (length-preserving) deterministic encryption [2].

As different ideal-world encryption oracles are used in the various security notions, it is trivially possible to distinguish between the encryption oracles. For example, for a fixed sector number and position on the plaintext, an IND-CPA up-to-block construction always returns the same ciphertext block. This construction does not reach IND-CPA up-to-prefix security, which requires indistinguishablilty up to the longest common prefix for a fixed sector number. It also does not satisfy IND-CPA up-to-repetition security, which requires indistinguishability up to repetition of the plaintext for a given sector number.

Analysis of Existing Constructions. We now analyze the FDE modes of operation described in Sect. 2 with respect to these security notions. These results are summarized in Table 1. The properties shown in the three last lines of Table 1 are relevant implementation properties, but are not taken into account in the security proofs.

ECB mode. Unsurprisingly, ECB is not IND-CPA for any of our three security notions.

CTR mode. CTR is not IND-CPA for any of our three security notions. An adversary can simply query the encryption of $m_1 = 0^n||m^2||..||m^\ell$ and $m_2 = 1^n||m^2||..||m^\ell$ for the same sector number s (where $m^2,...,m^\ell$ can be any n-bit blocks). The first blocks of the obtained ciphertexts c_1 and c_2 will always satisfy $c_1^1 = c_2^1 \oplus 1^n$, whereas this property holds only with probability 2^{-n} in all three random worlds.

CBC mode. The attack on the CTR mode also applies to the variant of the CBC mode where the sector number is used as an IV. In the context of FDE,

Table 1. The security of FDE modes of operation when no diversifier is used. Here, ✓ means that there is a security proof, and ✗ means that there is an attack. Proofs of the security results can be found in Sect. 3. XTS: see [17]. TC1, TC2 and TC3 [26] are generalizations of the HCBC1 [1], HCBC2 [1] and MHCBC [23] constructions. WTBC: wide tweakable block cipher. The * symbol indicates that the property holds for some constructions, but not for others. Here, $x \geq \log_2(\ell)$.

	ECB	CTR	CBC	CBC	IGE	XTS	TC1	TC2/3	WTBC
	IV → n/a	$s \ll x$	s	$E_{K'}(s)$	$E_{K'}(s)$	s	s	s	s
IND-CPA-block	✗	✗	✗	✗	✗	✓	✗	✗	✗
IND-CPA-prefix	✗	✗	✗	✗	✗	✗	✓	✓	✗
IND-CPA-repetition	✗	✗	✗	✗	✗	✗	✗	✗	✓
IND-CPA	✗	✗	✗	✗	✗	✗	✗	✗	✗
IND-CCA-block	✗	✗	✗	✗	✗	✓	✗	✗	✗
IND-CCA-prefix	✗	✗	✗	✗	✗	✗	✗	✓	✗
IND-CCA-repetition	✗	✗	✗	✗	✗	✗	✗	✗	✓
IND-CCA	✗	✗	✗	✗	✗	✗	✗	✗	✗
on-line enc./dec	✓	✓	✓	✓	✓	✓	✓	✓	✗
parallelizable enc	✓	✓	✗	✗	✗	✓	✗	✗	✓*
parallelizable dec	✓	✓	✓	✓	✗	✓	✓	✗	✓*

this attack is known as a Saarinen's watermarking attack [27]. Bellare et al. [1] describe an attack on the CBC-based online cipher that shows that CBC-essiv is not IND-CPA up-to-prefix. This attack can be adapted to construct an adversary $\mathcal{A}$ for our three security notions:

1. $\mathcal{A}$ arbitrarily chooses $(\ell-1)$ blocks $m^2, .., m^\ell$ and a sector number s. $\mathcal{A}$ builds two plaintexts $m_1 = 0^n||m^2||\ldots||m^\ell$ and $m_2 = 1^n||m^2||\ldots||m^\ell$ that differ only in their first block.
2. $\mathcal{A}$ queries the **Encrypt** procedure to obtain the encryption of m_1 and m_2 on sector number s: $c_b^1||c_b^2||\ldots||c_b^\ell \leftarrow$ **Encrypt**(s, m_b) for $b \in \{1, 2\}$.
3. $\mathcal{A}$ builds $m_3 = 1^n||m_3^2||m^3||\ldots||m^\ell$ where $m_3^2 = m^2 \oplus c_1^1 \oplus c_2^1$ and queries the **Encrypt** procedure to obtain its encryption: $c_3^1||\ldots||c_3^\ell \leftarrow$ **Encrypt**(s, m_3)
4. $\mathcal{A}$ returns 0 to the **Finalize** procedure if $c_3^2 = c_1^2$ and 1 otherwise.

The equality $c_3^2 = c_1^2$ is always satisfied in the real world but holds only with probability 2^{-n} in all three random worlds.

IGE-essiv mode. The previous attack on the CBC-essiv mode can easily be adapted to show that IGE-essiv is not IND-CPA for any of our three security notions. The attack is identical except that the adversary $\mathcal{A}$ checks whether the equality $c_3^2 = c_1^2 \oplus 1^n$ holds or not.

XTS mode. As explained above, in XTS every plaintext block is encrypted separately using a tweakable block cipher, where the tweak is derived from the sector number and index of the block within the sector. As argued by

Rogaway [25], XTS is IND-CPA up-to-block secure. For syntactic reasons, it is not IND-CPA up-to-prefix nor IND-CPA up-to-repetition.

TC1, TC2, TC3 modes. Rogaway and Zhang [26] showed that TC1 is IND-CPA up-to-prefix secure but not IND-CCA up-to-prefix. They also proved that TC2 and TC3 are IND-CCA up-to-prefix (and thus also IND-CPA up-to-prefix) secure. For syntactic reasons, they are not IND-CPA up-to-block nor IND-CPA up-to-repetition.

WTBC modes. Halevi and Rogaway showed that EME is IND-CCA up-to-repetition (and thus IND-CPA up-to-repetition) secure in [16]. For syntactic reasons, these modes are not IND-CPA up-to-block nor IND-CPA up-to-prefix.

4   FDE Security with Unique First Block

Because encryption in the context of FDE is deterministic and length-preserving, encrypting the same plaintext twice will always result in an identical ciphertext. The OS or application may therefore want to use a particular encoding of the plaintext, in order to ensure that the ciphertext will not be repeated. This corresponds to Bellare and Rogaway's Encode-Then-Encipher approach [4] to ensure strong privacy. One thus has to determine which encoding is sufficient to ensure security against CPA. In the context of security "up-to-block," this would require a large overhead, since encoding is then required for every block of a sector (typically 128 bits). However, for schemes that are IND-CPA "up-to-repetition" or "up-to-prefix," it is sufficient to ensure that the beginning of every message is unique. This can be done by prepending either random data or a counter, as suggested by Bellare and Rogaway [4].

In this section, we consider variants of the previous security notions with *unique first block* (ufb) and we prove that IND-CPA security for CBC-essiv and IGE-essiv under this assumption. An application that is aware of this restriction, can therefore format its input such that the first block of every sector is unique. The paper's full version [21] gives a concrete example of such an application.

Relation to Previous Security Notions. The only difference between ufb model and the previous model is that for a given sector number s, $\mathcal{A}$ cannot make two queries to encrypt plaintexts that have same first block. So if a construction is secure under a security notion described in Sect. 3, it is still the case in this model. Furthermore, the IND-CPA up-to-prefix, IND-CPA up-to-repetition and IND-CPA security notions become equivalent. This is easy to see: if the first block of plaintext is not repeated, then this is sufficient to ensure that the plaintext prefix or the entire plaintext is will not be repeated either.

Security Results. In this paragraph, we analyze the FDE modes of operation described in Sect. 2 with respect to these security notions for unique first block. These results are summarized in Table 2. For the security proofs of Theorems 1

Table 2. The security of FDE modes of operation when no diversifier is used, but the first plaintext block unique for any given sector. Here, ✓ means that there is a security proof, and ✗ means that there is an attack.

	ECB	CTR	CBC	CBC	IGE	XTS	TC1	TC2/3	WTBC
	IV → n/a	$s \ll x$	s	$E_{K'}(s)$	$E_{K'}(s)$	s	s	s	s
IND-CPA-block	✗	✗	✗	✗	✗	✓	✗	✗	✗
IND-CPA-prefix	✗	✗	✗	✓	✓	✗	✓	✓	✓
IND-CPA-repetition	✗	✗	✗	✓	✓	✗	✓	✓	✓
IND-CPA	✗	✗	✗	✓	✓	✗	✓	✓	✓
IND-CCA-block	✗	✗	✗	✗	✗	✓	✗	✗	✗
IND-CCA-prefix	✗	✗	✗	✗	✗	✗	✗	✓	✓
IND-CCA-repetition	✗	✗	✗	✗	✗	✗	✗	✗	✓
IND-CCA	✗	✗	✗	✗	✗	✗	✗	✗	✓
on-line enc./dec	✓	✓	✓	✓	✓	✓	✓	✓	✗
parallelizable enc	✓	✓	✗	✗	✗	✓	✗	✗	✓*
parallelizable dec	✓	✓	✓	✓	✗	✓	✓	✗	✓*

and 2, we used the code-based game playing framework [6]. The proofs can be found in the full version of this paper [21].

ECB and CTR. The attacks described in Sect. 3 do not make queries with the same first block and the same sector number, and therefore still apply.

CBC-essiv. For this mode, in the attack of Sect. 3, $\mathcal{A}$ makes forbidden queries with the same first block. The following theorem states that CBC-essiv achieves IND-CPA-ufb security if the underlying block cipher E is a Pseudo-Random Function (PRF) [5].

Theorem 1 [The IND-CPA-ufb Security of CBC-essiv].
*Let $E : \{0,1\}^k \times \{0,1\}^n \to \{0,1\}^n$ be a block cipher. Let $\mathcal{A}$ be an IND-CPA-ufb adversary against the FDE scheme obtained from the CBC-essiv mode on E such that $\mathcal{A}$ runs in time t and makes at most q queries to the **Encrypt** procedure. There exists an adversary $\mathcal{B}$ (attacking the PRF security of E) such that:*

$$\mathbf{Adv}^{ind-cpa-ufb}_{cbc-essiv}(\mathcal{A}) \leq 8 \cdot \mathbf{Adv}^{prf}_E(\mathcal{B}) + \frac{q^2(\ell+1)^2}{2^{n-1}}$$

where $\mathcal{B}$ runs in time at most $t' = t + O(q + nq(\ell+1))$ and makes at most $q' = q(\ell+1)$ queries to its oracle.

The following attack shows that it does not achieve IND-CCA-ufb up-to-prefix security:

1. $\mathcal{A}$ chooses a plaintext $m_1 = m_1^1||m_1^2||..||m_1^\ell \in \{0,1\}^{\ell \cdot n}$ and a sector number s and queries the **Encrypt** procedure with (s, m_1) to obtain $c_1^1||c_1^2||..||c_1^\ell$.

2. $\mathcal{A}$ builds a ciphertext $c_2 = c_2^1||c_2^2||..||c_2^\ell$ with $c_2^1 = c_1^2$, $c_2^2 = c_1^3$ and arbitrary $c_2^i \in \{0,1\}^n$ for $i \in \{3,\ldots,\ell\}$ and query the **Decrypt** procedure with (s,c_2) to obtain a plaintext $m_2 = m_2^1||m_2^2||..||m_2^\ell$.
3. $\mathcal{A}$ outputs 0 if $m_2^2 = m_1^3$ and 1 otherwise.

The equality $m_2^2 = m_1^3$ is always satisfied in the real world but this property holds only with probability 2^{-n} in the random world.

IGE-essiv. The following theorem states that the mode IGE-essiv achieves IND-CPA-ufb security:

Theorem 2 [The IND-CPA-ufb Security of IGE-essiv]. *Let $E : \{0,1\}^k \times \{0,1\}^n \to \{0,1\}^n$ be a block cipher. Let $\mathcal{A}$ be an IND-CPA-ufb adversary against the FDE scheme obtained from the IGE-essiv mode on E such that $\mathcal{A}$ runs in time t and makes at most q queries to the **Encrypt** procedure. There exists an adversary $\mathcal{B}$ (attacking the PRF security of E) such that:*

$$\mathbf{Adv}_{ige-essiv}^{ind-cpa-ufb}(\mathcal{A}) \leq 8 \cdot \mathbf{Adv}_E^{prf}(\mathcal{B}) + \frac{q^2(\ell+1)^2}{2^{n-1}}$$

where $\mathcal{B}$ runs in time at most $t' = t + O(q + nq(\ell+1))$ and makes at most $q' = q(\ell+1)$ queries to its oracle.

The following attack (inspired by Rohatgi [19]) shows that IGE-essiv does not achieve IND-CCA-ufb up-to-prefix security:

1. $\mathcal{A}$ chooses a plaintext $m_1 = m_1^1||m_1^2||..||m_1^\ell \in \{0,1\}^{\ell \cdot n}$ and a sector number s and queries the **Encrypt** procedure with (s,m_1) to obtain $c_1^1||c_1^2||..||c_1^\ell$.
2. $\mathcal{A}$ builds a ciphertext $c_2 = c_2^1||c_2^2||..||c_2^\ell$ with $c_2^1 = c_1^1$, $c_2^2 = m_1^2 \oplus c_1^3 \oplus m_1^1$ and arbitrary $c_2^i \in \{0,1\}^n$ for $i \in \{3,\ldots,\ell\}$ and query the **Decrypt** procedure with (s,c_2) to obtain a plaintext $m_2 = m_2^1||m_2^2||..||m_2^\ell$.
3. $\mathcal{A}$ outputs 0 if $m_2^2 = m_1^3 \oplus c_1^2 \oplus c_1^1$ and 1 otherwise.

The equality $m_2^2 = m_1^3 \oplus c_1^2 \oplus c_1^1$ is always satisfied in the real world but this property holds only with probability 2^{-n} in the random world.

The TC1, TC2, TC3 and WTBC constructions become IND-CPA with the ufb restriction because TC1/2/3 were IND-CPA up-to-prefix and WTBC constructions were IND-CPA up-to-repetition as explained in Sect. 3.

5   FDE Security with a Diversifier

Typically, IND-CPA cannot be reached for FDE, as the deterministic nature of FDE means that identical plaintexts will result in identical ciphertexts. We worked around this problem in the previous section by imposing a restriction on the plaintext: the first plaintext block must be unique.

Now, we introduce another way to achieve IND-CPA, without imposing restrictions on the plaintext, but still without storing additional data. Instead, we will use a diversifier j, which will be associated to every sector.

To stay within the constraints of FDE, it should somehow be possible to assign a diversifier to every sector without using additional storage. Possible candidates in the particular case of SSDs will be considered in Sect. 6. For now, it is enough to consider that for each encryption, a diversifier is picked among $\{0,1\}^d$, in such a way this diversifier is never repeated for a particular sector. Then two identical plaintexts with the same sector number will have different ciphertexts, a property that could previously not be achieved within the context of FDE.

The combination of the sector number s and the diversifier j is used instead of the sector number in FDE constructions. The combination proposed is simply the concatenation between these two values $s\|j$ such as $s \in \{0,1\}^\sigma$, $j \in \{0,1\}^d$ and $n = d + \sigma$.

For the analysis in this section, it suffices that the diversifier is never repeated for a particular sector. As such, the security analysis is the same as if the diversifier were a nonce. However, we will explain in Sect. 6 that efficiency reasons require that at any particular point in time, all diversifier values should occur roughly the same number of times, and that the diversifier must be a rather short value, typically only a few bits.

Security Results. IND-CPA up-to-repetition becomes equivalent to IND-CPA security: the only difference between these notions is that if $\mathcal{A}$ asks to encrypt twice the same query (s, m) the answer will be the same ciphertext but these queries are not allowed any more under the diversifier model. Moreover in IND-CCA game, the adversary is not allowed to query the decryption of a ciphertext what was previously encrypted, or vice versa. It can therefore be seen that IND-CCA up-to-repetition becomes equivalent to IND-CCA. Similarly, since the adversary $\mathcal{A}$ is not allowed to encrypt twice with the same pair $s\|j$, the IND-CPA up-to-block property is also equivalent to the other IND-CPA security notions.

Table 3 summarizes the security properties achieved by the FDE modes of operation when used with a diversifier. The IND-CPA attacks of Sect. 3 still carry over to ECB and CBC with a sector-number IV. However CTR mode becomes secure as the counter value is not repeated [3]. The IND-CPA security of XTS, TC1, TC2, TC3 and WTBC follows from the fact that the tweak is not reused. For CBC-essiv and IGE-essiv, the IND-CPA security follows from the proof of Sect. 4: now the first block may be reused, but the IV is unique.

Let us explain the attacks under IND-CCA in Table 3:

- This following attack shows that CTR is not IND-CCA-xx: $\mathcal{A}$ encrypts $(s\|j, m)$ with m any plaintext and any $s\|j$ and receives c then $\mathcal{A}$ decrypts $(s\|j, c')$ where $c' = c \oplus 0^{n-1}1$. $\mathcal{A}$. Then, $m'^1 = m^1 \oplus 0^{n-1}1$ is always satisfied in the real world, but holds only with probability 2^{-n} in the ideal world.
- CBC-essiv and IGE-essiv are not secure: the attacks of Sect. 4 still apply, as they did not perform two encryptions with the same sector number.
- For syntactical reasons, an encryption scheme can only be IND-CCA up-to-block, IND-CCA up-to-prefix, or IND-CCA up-to-repetition. XTS is only

IND-CCA up-to-block, TC2 and TC3 are only IND-CCA up-to-prefix and WBTC are only IND-CCA up-to-repetition.

As shown in Table 3, the diversifier shows how to reach IND-CPA security for most commonly-used FDE encryption modes. It also succeeds at providing IND-CCA security for WTBC constructions, which not achievable in a "classical" FDE model.

6   Solid State Drive

We now explain the basics of SSD storage, so that we can explain how to modify only the firmware of SSDs to associate a *diversifier* to every sector. This diversifier allows us to encrypt the same plaintext in distinct ways for the same sector number.

Table 3. The security of FDE modes of operation when a diversifier is used. Here, ✓ means that there is a security proof, and ✗ means that there is an attack.

	ECB	CTR	CBC	CBC	IGE	XTS	TC1	TC2/3	WTBC
	IV → n/a	$s\|j \ll x$	$s\|j$	$E_{K'}(s\|j)$	$E_{K'}(s\|j)$	$s\|j$	$s\|j$	$s\|j$	$s\|j$
IND-CPA-block	✗	✓	✗	✓	✓	✓	✓	✓	✓
IND-CPA-prefix	✗	✓	✗	✓	✓	✓	✓	✓	✓
IND-CPA-repetition	✗	✓	✗	✓	✓	✓	✓	✓	✓
IND-CPA	✗	✓	✗	✓	✓	✓	✓	✓	✓
IND-CCA-block	✗	✗	✗	✗	✗	✓	✗	✗	✗
IND-CCA-prefix	✗	✗	✗	✗	✗	✗	✗	✓	✗
IND-CCA-repetition	✗	✗	✗	✗	✗	✗	✗	✗	✓
IND-CCA	✗	✗	✗	✗	✗	✗	✗	✗	✓
on-line enc./dec	✓	✓	✓	✓	✓	✓	✓	✓	✗
parallelizable enc	✓	✗	✓	✗	✗	✓	✗	✗	✓*
parallelizable dec	✓	✓	✓	✓	✗	✓	✓	✗	✓*

SSDs are flash memory devices that are gradually replacing the magnetic Hard Disk Drive (HDD) due to their reliability and performance. Just like HDDs, they are sector-addressable devices. They are indexed by a sector number that is also known as a Logical Block Address (LBA). This ensures that the physical details of the storage device are not exposed to the OS, but are managed by the firmware of the storage device.

For SSDs, the Flash Translation Layer (FTL) stores the mapping between LBAs and Physical Block Addresses (PBAs). This FTL is necessary to ensure an even distribution of writes to every sector number (*wear leveling*) and to invalidate blocks so that they can later be recycled (*garbage collection*). This FTL is necessary due to the physical constraints of flash storage: any physical address can only be written a limited number of times, and rewriting individual sectors is not possible: invalidated sectors can only be recovered in multiples of the *erase block size*.

SSD Components. The flash memory of an SSD is hierarchically organized as a set of flash chips called packages, which are further divided in dies, planes, blocks[2] and pages. Every page consists of one or more sectors, and is the smallest unit that can be written. The smallest unit that can be erased in a block. Invalidated blocks must be erased before writing, and the number of erasures and writes to every block is limited for flash storage.

To abstract away the notion of packages, dies and planes, the Open NAND Flash Interface (ONFI) standard introduces the notion of Logical Unit Number (LUN) as the minimum granularity of parallelism for flash storage. As operations can be issued to several planes in parallel, a LUN corresponds to a plane.

Introducing a Diversifier. In the context of SSD storage, we will show how to associate a *diversifier* to every LBA. This diversifier will not be stored, and will not modify the data structures of the SSD. Instead, the diversifier will impose an additional restriction on the FTL, meaning that the diversifier determines which PBAs can store the data corresponding to a particular LBA.

The intuition is that if the diversifier is selected "randomly" for every write operation, the data will be spread out evenly over the SSD, and the SSD performance should not be affected too much. We will verify this by implementing and benchmarking our proposed solution in Sect. 6.

When a write command is issued, there are various ways to specify a diversifier value for a given LBA. We will prefer to transmit this information in one operation. As such, we do not only avoid the performance drawbacks of issuing several operations to write one sector, but we also do not need to worry about inconsistent states when operations are lost, modified, or reordered.

In particular, we propose to send the diversifier along with the sector data as part of a "fat" sector that is already supported in SATA (Serial ATA) interface for storage devices. This diversifier value will be returned to the OS when a fat sector read command is issued.

In an attempt to make the diversifier as long as possible, we may want to consider the optimal (yet completely unrealistic) scenario: the diversifier uniquely specifies the physical page to which the data must be written. But even then, the size of the diversfier typically be very short: e.g. only 24 bits in case of an 128 GB SSD with 8 kB pages. We will, in fact, choose the diversifier to be much shorter, so that a practical implementation is possible that minimally modifies the SSD firmware. This diversifier cannot be selected at random, as it would be too short to avoid repetitions for a particular sector. But it can also not be a counter, as all diversifier values should be used roughly an equal number of times over all sectors to spread the writes over the entire disk layout.

In our solution, we want to avoid that the SSD needs to send data back to the host for decryption and re-encryption under a different diversifier, as this would affect the SSD performance quite drastically. Therefore, wear leveling and garbage collection operations may not change the diversifier. A straightforward

---

[2] These blocks should not be confused with the blocks of the block cipher, nor with the "block" (actually "sector") in the term Logical Block Address (LBA).

solution is then to make the diversifier correspond to the LUN (or a set of LUNs), which is what we will implement and benchmark in the following section.

The OS can freely select the diversifier values, however they may not be repeated for any particular sector, and all diversifier values should be used roughly the same number of times. In the paper's full version [21], we give a concrete example of an application where the OS implements such diversifier. More specifically, we show how the OS can ensure randomness and uniqueness even for very short diversifier values.

EagleTree Benchmarks. In order to confirm that the concept of a diversifier is not just feasible but also efficient to implement, we implemented this feature in the EagleTree SSD simulator [8] and performed various benchmarks. An overview of the source code modifications can be found in the paper's full version [21].

Table 4. EagleTree Benchmarks for various diversifier sizes.

	diversifier size (bits)			
	0	1	2	3
read latency (μs)	28.292	28.862	31.619	40.640
write latency (μs)	32.009	32.070	32.470	34.560
read throughput (IOPS)	20860	20493	19050	15634
write throughput (IOPS)	31240	31181	30797	28935
garbage collector reads	1284043	1295530	1356043	1489623
garbage collector writes	1284043	1295530	1356041	1489622
erasures	18569	18765	19661	21634

The device that is simulated consists of eight packages, each containing four dies of 256 blocks. Each block consists of 128 pages of 4096 bytes. EagleTree currently does not support multiple planes per die. The page read, page write, bus ctrl, bus data and block erase delays are 5, 20, 1, 10 and 60 μs respectively. We assume that any latencies incurred by the OS (including sector encryption and decryption) are negligible with respect to these numbers. The SSD has an overprovisioning factor of 0.7.

We simulate an SSD configured with DFTL and the greedy garbage collection policy. The base benchmarks use a simple block scheduler that assigns the next write to whichever package is free. We then compare this performance to various choices of the diversifier value, which determines the package for every write.

The workload used in our benchmarks, is the same as the example in Eagle-Tree's `demo.cpp` file: first a large write is made to the entire logical address space. This write is performed in random order, but without writing to the same address twice. Once this large write is finished, two threads are started up: one performs random reads, and the other performs random writes in the address space. After three million I/O operations, the simulation is stopped.

The benchmark results are shown in Table 4. They suggest that not to choose the number of diversifier values to be equal to the number of packages, as the impact on performance is quite significant. Compared to the benchmarks without diversifier, the throughput of the reads and writes drop by 25% and 7% respectively. The average latency increases by 44% for reads, and 8% for writes. In this setup, the reads and writes of each garbage collection operation are restricted to one package, and this affects performance quite significantly. Reads suffer more than writes; this is mainly because there is no significant drop in performance for the initial write operations to fill up the SSD.

To avoid the large impact on performance, we must therefore choose the number of diversifier values to be smaller than the number of packages. When the diversifier is two bits, our simulations show an increase of the average latency of 12% for reads and 1% for writes, and a reduction of throughput of 9% for reads and 1% for writes. The total throughput reduction (reads and writes combined) is at most 4%. For a diversifier of one bit, latency and throughput are affected by less than 2%. We also looked into the number of garbage collection and the number of erasures. They worsen by less than 6% for a diversifier of two bits, but by 16% for a diversifier of three bits.

7   Conclusion

We presented a theoretical framework for disk encryption and we analyzed several existing constructions against chosen-plaintext and chosen-ciphertext attacks, under different notions of the ideal-world encryption oracle: up-to-repetition, up-to-prefix, or up-to-block.

Using this model, we recalled that IGE-essiv does not have chosen-ciphertext-security under any of the notions that we consider, which shows that the AREA construction proposed by Fruhwirth [11] is insecure. Nevertheless, we proved that IGE-essiv and even CBC-essiv can provide security under chosen-plaintext attacks, under the assumption that the first block of a plaintext is never repeated for the same sector number.

We also revisited FDE from an engineering perspective, and showed how to modify the firmware of a solid-state drive to associate a short "diversifier" to every sector-plaintext pair (s, m). This diversifier makes it possible to encrypt the same plaintext into different ciphertexts, something that was previously impossible without additional storage.

Acknowledgments. Nicky Mouha is supported by a Postdoctoral Fellowship from the Flemish Research Foundation (FWO-Vlaanderen), by a JuMo grant from KU Leuven (JuMo/14/48CF), and by FWO travel grant 12F9714N. Certain algorithms and commercial products are identified in this paper to foster understanding. Such identification does not imply recommendation or endorsement by NIST, nor does it imply that the algorithms or products identified are necessarily the best available for the purpose. Damien Vergnaud is supported in part by the French ANR JCJC ROMAnTIC project (ANR-12-JS02-0004).

We thank Matias Bjørling, Luc Bouganim, Niv Dayan and Javier Gonzalez for their useful comments and suggestions on SSD technology.

References

1. Bellare, M., Boldyreva, A., Knudsen, L.R., Namprempre, C.: On-line ciphers and the hash-CBC constructions. J. Cryptology **25**(4), 640–679 (2012)
2. Bellare, M., Boldyreva, A., O'Neill, A.: Deterministic and efficiently searchable encryption. In: Menezes, A. (ed.) Advances in Cryptology - CRYPTO 2007. LNCS, vol. 4622, pp. 535–552. Springer, Heidelberg (2007). doi:10.1007/978-3-540-74143-5_30
3. Bellare, M., Desai, A., Jokipii, E., Rogaway, P.: A concrete security treatment of symmetric encryption. In: 38th FOCS, pp. 394–403. IEEE Computer Society Press (1997)
4. Bellare, M., Rogaway, P.: Encode-then-encipher encryption: how to exploit nonces or redundancy in plaintexts for efficient cryptography. In: Okamoto, T. (ed.) ASIACRYPT 2000. LNCS, vol. 1976, pp. 317–330. Springer, Heidelberg (2000). doi:10.1007/3-540-44448-3_24
5. Bellare, M., Rogaway, P.: Introduction to Modern Cryptography. In: UCSD CSE 207 Course Notes, 207 pages (2005). http://cseweb.ucsd.edu/~mihir/cse207/
6. Bellare, M., Rogaway, P.: The security of triple encryption and a framework for code-based game-playing proofs. In: Vaudenay, S. (ed.) EUROCRYPT 2006. LNCS, vol. 4004, pp. 409–426. Springer, Berlin (2006). doi:10.1007/11761679_25
7. Campbell, C.M.: Design and specification of cryptographic capabilities. IEEE Commun. Soc. Mag. **16**(6), 15–19 (1978)
8. Dayan, N., Svendsen, M.K., Bjørling, M., Bonnet, P., Bouganim, L.: EagleTree: exploring the design space of SSD-based algorithms. PVLDB **6**(12), 1290–1293 (2013)
9. Dworkin, M.: Recommendation for Block Cipher Modes of Operation: The XTS-AES Mode for Confidentiality on Storage Devices. NIST SP 800–38E (2010)
10. Ferguson, N.: AES-CBC + Elephant diffuser: A Disk Encryption Algorithm for Windows Vista (2006). http://www.microsoft.com/en-us/download/details.aspx?id=13866
11. Fruhwirth, C.: New methods in hard disk encryption. Master's thesis, Vienna University of Technology (2005)
12. Gjøsteen, K.: Security notions for disk encryption. In: Vimercati, S.C., Syverson, P., Gollmann, D. (eds.) ESORICS 2005. LNCS, vol. 3679, pp. 455–474. Springer, Heidelberg (2005). doi:10.1007/11555827_26
13. Götzfried, J., Müller, T.: Analysing android's full disk encryption feature. JoWUA **5**(1), 84–100 (2014)
14. Halcrow, M., Savagaonkar, U., Ts'o, T., Muslukhov, I.: EXT4 Encryption Design Document (public version). Google Technical report (2015)
15. Halevi, S.: Re: Lrw key derivation (formerly pink-herring). IEEE P1619 Mailing List, May 2006
16. Halevi, S., Rogaway, P.: A parallelizable enciphering mode. In: Okamoto, T. (ed.) CT-RSA 2004. LNCS, vol. 2964, pp. 292–304. Springer, Heidelberg (2004). doi:10.1007/978-3-540-24660-2_23
17. IEEE: IEEE Standard for Cryptographic Protection of Data on Block-Oriented Storage Devices. IEEE Std 1619–2007, pp. 1–32 (2008)
18. Joux, A., Martinet, G., Valette, F.: Blockwise-adaptive attackers revisiting the (in)security of some provably secure encryption modes: CBC, GEM, IACBC. In: Yung, M. (ed.) CRYPTO 2002. LNCS, vol. 2442, pp. 17–30. Springer, Berlin (2002). doi:10.1007/3-540-45708-9_2

19. Jutla, C.: Attack on Free-MAC (2000). https://groups.google.com/d/msg/sci.crypt/4bkzm_n7UGA/5cDwfju6evUJ
20. Jutla, C.S.: Encryption modes with almost free message integrity. In: Pfitzmann, B. (ed.) EUROCRYPT 2001. LNCS, vol. 2045, pp. 529–544. Springer, Heidelberg (2001). doi:10.1007/3-540-44987-6_32
21. Khati, L., Mouha, N., Vergnaud, D.: Full Disk Encryption: Bridging Theory and Practice. Cryptology ePrint Archive, Report 2016/1114, full version of this paper (2016)
22. Müller, T., Freiling, F.C.: A systematic assessment of the security of full disk encryption. IEEE Trans. Dependable Sec. Comput. **12**(5), 491–503 (2015)
23. Nandi, M.: Two new efficient CCA-secure online ciphers: MHCBC and MCBC. In: Chowdhury, D.R., Rijmen, V., Das, A. (eds.) INDOCRYPT 2008. LNCS, vol. 5365, pp. 350–362. Springer, Heidelberg (2008). doi:10.1007/978-3-540-89754-5_27
24. Rogaway, P.: Nonce-based symmetric encryption. In: Roy, B., Meier, W. (eds.) FSE 2004. LNCS, vol. 3017, pp. 348–358. Springer, Heidelberg (2004). doi:10.1007/978-3-540-25937-4_22
25. Rogaway, P.: Evaluation of Some Blockcipher Modes of Operation. Technical report, CRYPTREC Investigation Report (2011)
26. Rogaway, P., Zhang, H.: Online ciphers from tweakable blockciphers. In: Kiayias, A. (ed.) CT-RSA 2011. LNCS, vol. 6558, pp. 237–249. Springer, Heidelberg (2011). doi:10.1007/978-3-642-19074-2_16
27. Saarinen, M.-J.O.: Encrypted watermarks and linux laptop security. In: Lim, C.H., Yung, M. (eds.) WISA 2004. LNCS, vol. 3325, pp. 27–38. Springer, Heidelberg (2005). doi:10.1007/978-3-540-31815-6_3

Revisiting Full-PRF-Secure PMAC and Using It for Beyond-Birthday Authenticated Encryption

Eik List[1($\boxtimes$)] and Mridul Nandi[2]

[1] Bauhaus-Universität Weimar, Weimar, Germany
eik.list@uni-weimar.de
[2] Applied Statistics Unit, Indian Statistical Institute, Kolkata, India
mridul.nandi@gmail.com

Abstract. This paper proposes an authenticated encryption scheme, called SIVx, that preserves BBB security also in the case of unlimited nonce reuses. For this purpose, we propose a single-key BBB-secure message authentication code with $2n$-bit outputs, called PMAC2x, based on a tweakable block cipher. PMAC2x is motivated by PMAC_TBC1k by Naito; we revisit its security proof and point out an invalid assumption. As a remedy, we provide an alternative proof for our construction, and derive a corrected bound for PMAC_TBC1k.

Keywords: Symmetric cryptography · Message authentication codes · Authenticated encryption · Provable security

1 Introduction

Nonce-Based Authenticated Encryption. Authenticated encryption (AE) schemes aim at protecting both the privacy and the integrity of submitted messages. Authenticated encryption schemes that allow to authenticate not only the encrypted message, but also associated data, are commonly known as AEAD schemes [22]. The common security notions for AE schemes concern the prevention of any leakage about the encrypted messages except for their lengths. Since stateless schemes would enable adversaries to detect a duplicate encryption of the same associated data and message under the current key, Rogaway proposed nonce-based encryption [24], where the user provides an additional nonce for every message she wants to process. In theory, the concept of nonces is simple. However, the practice has shown numerous examples of implementation failures, and settings that render it difficult to almost impossible to prevent nonce reuse (cf. [8]). Before the CAESAR competition, the majority of widely used AE schemes protected neither the confidentiality nor the integrity of messages in the case of nonce repetitions. As a consequence, a considerable number of CAESAR candidates aimed a certain level of security if nonces repeat (e.g., [1,9,10,15]).

Parallelizable MACs in Authenticated Encryption. Block-cipher-based message authentication codes (MACs) are important components not only for

© Springer International Publishing AG 2017
H. Handschuh (Ed.): CT-RSA 2017, LNCS 10159, pp. 258–274, 2017.
DOI: 10.1007/978-3-319-52153-4_15

authentication, but also as part of AE schemes, where they are used to derive an initialization vector (IV) that is then used for encryption. In particular, parallelizable MACs like PMAC [6] allow to process multiple blocks in parallel, which is beneficial for software performance on current x64 processors. Since PMAC has several further desirable properties, e. g. being online and incremental, it is not a surprise that all the CAESAR candidates cited above essentially combine a variant of PMAC (or its underlying hash function) with a block-cipher-based mode of operation for efficiently processing associated data and message.

Beyond-Birthday-Bound AE. Besides performance, the quantitative security guarantees are important aspects for AE schemes. The privacy and authenticity guarantees of the AE schemes cited above are limited by the birthday bound of $O(\ell^2/2^n)$, where n denotes the state size of the underlying primitive, and ℓ the number of blocks processed over all queries. Since the schemes above possess an n-bit state, a state collision that leads to attacks has significant probability after approximately $2^{n/2}$ blocks have been processed under the same key.

To address this issue, Peyrin and Seurin presented Synthetic Counter in Tweak (SCT) [20], a beyond-birthday-bound (BBB) AE scheme based on a tweakable block cipher under a single key. Internally, SCT is a MAC-then-Encrypt composition: the MAC part is a PMAC-like construction, called EPWC. The encryption part is Counter-in-Tweak (CTRT), an efficient mode of operation that takes an n-bit nonce and an n-bit IV. Both EPWC and CTRT guarantee BBB security as long as nonces never repeat. However, the security of the nonce-IV-based CTRT degrades to the birthday bound with an increasing number of nonce reuses; even worse, the security of EPWC (and consequently that of SCT) immediately reduces to the birthday bound if a single nonce repeats once. In [21, p. 7], the extended version of [20], the authors remarked therefore (among others) the following open problem: "[...] *to construct an AE scheme which remains BBB-secure even when nonces are arbitrarily repeated. The main difficulty is to build a deterministic, stateless, BBB-secure MAC, which is known to be notably hard*". Intuitively, an efficient block-cipher-based BBB-secure MAC with $2n$ bit output length could allow to construct such a deterministic AE scheme.[1] Thus, this work will put a large focus on the construction of BBB-secure MACs.

Previous Work. Naito [17] proposed two MACs with full PRF security based on a tweakable block cipher: PMAC_TBC3K and PMAC_TBC1K. While the former requires three keys, the latter uses tweak-based domain separation to require only a single key. Extending the latter seemed a well-suited starting point for our work since such a MAC could be combined in straight-forward manner with a BBB-secure mode of encryption. Though, during our studies, we found that the analysis in [17] assumed internal values to be independent, which—as we will show—cannot always be guaranteed. Since the proof depended largely

---

[1] We stress that BBB-secure AE is not new if one considers schemes with multiple primitives and keys. For the sake of space limitations, a discussion can be found in the full version of this work [11].

Table 1. Previous parallel BBB-secure MACs. (T)BC = (tweakable) block cipher, q = max. #queries, m = max. #blocks per query, ℓ = max. total #blocks.

Primitive	Construction	Keys	Output size	Advantage	Reference
BC	PMAC$^+$	3	n	$O(q^3 m^3 / 2^{2n} + qm/2^n)$	[25]
	1K_PMAC$^+$	1	n	$O(qm^2/2^n + q^3 m^4 / 2^{2n})$	[7]
TBC	PMAC_TBC3K	3	n	$O(q^2/2^{2n})$	[17]
	PMAC_TBC1K	1	n	$O(q/2^n + q^2/2^{2n})$	[17]
	PMACx	1	n	$O(q^2/2^{2n} + q^3/2^{3n})$	Section 5
	PMAC2x	1	$2n$	$O(q^2/2^{2n} + q^3/2^{3n})$	Section 4

on this aspect, we developed an alternative analysis for our construction and derived a corrected bound for a PMAC_TBC1K-like variant with n-bit output. So, despite the assumption in the original proof, we confirm that Naito's MAC is secure for close to 2^{n-2} blocks processed under the same key.

Contribution. Our contributions are threefold: first, we propose a BBB-secure parallelizable MAC, called PMAC2x, which produces $2n$-bit outputs and bases on a tweakable block cipher. Figure 1 provides a schematic illustration. Our MAC differs from PMAC_TBC1K mainly in the fact of the extended output, and minorly in the point that we add support for inputs whose length is not a multiple of n. As our second contribution, we briefly revisit the analysis by Naito on PMAC_TBC1K and show that we can easily adapt our proof for PMAC2x and derive a secure variant that we call PMACx which XORs both its outputs and produces only n-bit tags. Table 1 compares our constructions to earlier parallelizable BBB-secure MACs. As our third contribution, we combine PMAC2x with the purely IV-based variant of Counter-in-Tweak to a single-primitive, single-key deterministic authenticated encryption scheme, which we call SIVx, and which provides BBB-security without assumptions about nonces.

Earlier Parallelizable MACs. A considerable amount of works considered parallel MACs; parallel XOR-MACs have already been introduced in 1995 by Bellare et al. [2]; their constructions fed the message blocks together with a counter into a primitive to obtain stateful and randomized MACs. Bernstein [5] published the Protected Counter Sum (PCS), which transformed an XOR-MAC with an independent PRF into a stateless deterministic MAC. PMAC was described by Black and Rogaway first in [6], and was slightly modified to PMAC1 in [23]. Since then, the security of PMAC has been rigorously studied in various works [12,14,16,18,19]. The first BBB-secure parallelizable MAC was proposed by Yasuda [25]; His PMAC$^+$ construction is a three-key version of PMAC which possesses two n-bit state values, which are processed by two independently keyed PRPs, and are XORed to produce the tag. Datta et al. [7] derived a single-key version thereof, called 1K_PMAC$^+$. While those are rate-1 designs with larger internal state, there also exist slightly less efficient proposals with smaller state. Yasuda [26] introduced PMAC with parity (PMAC/P), which processes each sequence of r consecutive message blocks in PMAC-like

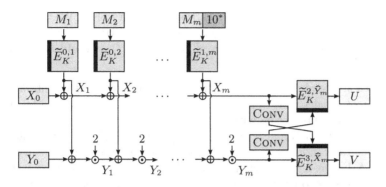

Fig. 1. Processing an m-block message with a partial final block in PMAC2x. $\widetilde{E}$: $\{0,1\}^k \times \{0,1\}^d \times \{0,1\}^t \times \{0,1\}^n \to \{0,1\}^n$ denotes a tweakable block cipher, Conv : $\{0,1\}^n \to \{0,1\}^t$ a regular function, and $\odot$ multiplication in Galois-Field $\mathbb{GF}(2^n)$.

manner, but inserts the XOR sum of those r blocks as an additional block. Zhang's PMACX construction [27] generalized PMAC/P by multiplying the input with an MDS matrix before authentication. In a similar direction goes LightMAC [13], a lightweight variant similar to Bernstein's PCS. However, the security guarantees of all earlier parallelizable MACs in this paragraph are far from the optimal PRF bound.

2   Preliminaries

General Notation. We use lowercase letters x, y for indices and integers, upper-case letters X, Y for binary strings and functions, calligraphic uppercase letters $\mathcal{X}, \mathcal{Y}$ for sets; $X \parallel Y$ for the concatenation of binary strings X and Y, and $X \oplus Y$ for their bitwise XOR. We indicate the length of X in bits by $|X|$, and write X_i for the i-th block, $X[i]$ for the i-th most-significant bit of X, and $X[i..j]$ for the bit sequence $X[i], \ldots, X[j]$. We denote by $X \twoheadleftarrow \mathcal{X}$ that X is chosen uniformly at random from the set $\mathcal{X}$. We define $\mathsf{Func}(\mathcal{X}, \mathcal{Y})$ for the set of all functions $F : \mathcal{X} \to \mathcal{Y}$, $\mathsf{Perm}(\mathcal{X})$ for the set of all permutations $\pi : \mathcal{X} \to \mathcal{X}$, and $\widetilde{\mathsf{Perm}}(\mathcal{T}, \mathcal{X})$ for the set of tweaked permutations over $\mathcal{X}$ with tweak space $\mathcal{T}$. We define by $X_1, \ldots, X_j \xleftarrow{x} X$ an injective splitting of a string X into blocks of x-bit such that $X = X_1 \parallel \cdots \parallel X_j$, $|X_i| = x$ for $1 \le i \le j-1$, and $|X_j| \le x$. For two sets $\mathcal{X}$ and $\mathcal{Y}$, let $\mathcal{X} \xleftarrow{\cup} \mathcal{Y}$ denote $\mathcal{X} \leftarrow \mathcal{X} \cup \mathcal{Y}$. A uniform random function $\rho : \mathcal{X} \to \mathcal{Y}$ is a random variable uniformly distributed over $\mathsf{Func}(\mathcal{X}, \mathcal{Y})$. Given a function $F : \mathcal{X} \to \mathcal{Y}$, we write $\mathsf{domain}(F)$ for the set of all inputs $X \in \mathcal{X}$ to F that occurred *before* (i.e., excluding) the current query; similarly, we write $\mathsf{range}(F)$ for the set of all outputs $Y \in \mathcal{Y}$ that occurred before the current query. We borrow the notation for a restriction on a set from [8]: let $\mathcal{Q} \subseteq (\mathcal{X} \times \mathcal{Y} \times \mathcal{Z})^*$, then we denote by $\mathcal{Q}_{|Y,Z} = \{(Y, Z) \mid \exists X : (X, Y, Z) \in \mathcal{Q}\}$ the restriction of $\mathcal{Q}$ to values $Y \in \mathcal{Y}$ and $Z \in \mathcal{Z}$. This generalizes in the obvious way.

For an event E, we denote by $\Pr[E]$ the probability of E. We write $\langle x \rangle_n$ for the binary representation of an integer x as an n-bit string, or short $\langle x \rangle$ if n is clear from the context, in big-endian manner, e. g., $\langle 1 \rangle_4$ would be encoded to $\langle 0001 \rangle$.

Adversaries. An adversary $\mathbf{A}$ is an efficient Turing machine that interacts with a given set of oracles that appear as black boxes to $\mathbf{A}$. We denote by $\mathbf{A}^{\mathcal{O}}$ the output of $\mathbf{A}$ after interacting with some oracle $\mathcal{O}$. We write $\Delta_{\mathbf{A}}(\mathcal{O}^1; \mathcal{O}^2)$ for the advantage of $\mathbf{A}$ to distinguish between oracles $\mathcal{O}^1$ and $\mathcal{O}^2$. All probabilities are defined over the random coins of the oracles and those of the adversary, if any. We write $\mathbf{Adv}_F^X(q, \ell, \theta) := \max_{\mathbf{A}} \{ \mathbf{Adv}_F^X(\mathbf{A}) \}$ for the maximal advantage over all X-adversaries $\mathbf{A}$ on F that run in time at most θ and pose at most q queries of at most ℓ blocks in total to its oracles. Wlog., we assume that $\mathbf{A}$ never asks queries to which it already knows the answer.

We will provide pseudocode descriptions of the oracles, which will be referred to as games, according to the game-playing framework by Bellare and Rogaway [3]. Each game consists of a set of procedures. We define $\Pr[G(\mathbf{A}) \Rightarrow x]$ as the probability that the game G outputs x when given $\mathbf{A}$ as input.

Definition 1 (TPRP Advantage). *Let $\widetilde{E} : \mathcal{K} \times \mathcal{T} \times \mathcal{X} \to \mathcal{X}$ be a tweakable block cipher with non-empty key space $\mathcal{K}$ and tweak space $\mathcal{T}$. Let $\mathbf{A}$ a computationally bounded adversary with access to an oracle, where $K \twoheadleftarrow \mathcal{K}$ and $\widetilde{\pi} \twoheadleftarrow \widetilde{\mathsf{Perm}}(\mathcal{T}, \mathcal{X})$. Then, the TPRP advantage of $\mathbf{A}$ on $\widetilde{E}$ is defined as $\mathbf{Adv}_{\widetilde{E}}^{\mathrm{TPRP}}(\mathbf{A}) := \Delta_{\mathbf{A}}(\widetilde{E}_K; \widetilde{\pi})$.*

A MAC is a tuple of functions $F : \mathcal{K} \times \mathcal{X} \to \mathcal{Y}$ with non-empty key space $\mathcal{K}$, and a generic verification function $\mathrm{VERIFY} : \mathcal{K} \times \mathcal{X} \times \mathcal{Y} \to \{1, \bot\}$, where for all $K \in \mathcal{K}$ and $X \in \mathcal{X}$, $\mathrm{VERIFY}_K(X, Y)$ returns 1 iff $F_K(X) = Y$ and $\bot$ otherwise. We use $\bot$, when in place of an oracle, to always return the invalid symbol $\bot$. It is well-known that if F is a secure PRF, it is also a secure MAC; however, the converse statement is not necessarily true.

Definition 2 (PRF Advantage). *Let $F : \mathcal{K} \times \mathcal{X} \to \mathcal{Y}$ be a function with non-empty key space $\mathcal{K}$, and $\mathbf{A}$ a computationally bounded adversary with access to an oracle, where $K \twoheadleftarrow \mathcal{K}$ and $\rho \twoheadleftarrow \mathsf{Func}(\mathcal{X}, \mathcal{Y})$. Then, the PRF advantage of $\mathbf{A}$ on F is defined as $\mathbf{Adv}_F^{\mathrm{PRF}}(\mathbf{A}) := \Delta_{\mathbf{A}}(F_K; \rho)$.*

3   Definition of PMAC2x and PHASHx

This section defines the generic PMAC2x construction and its underlying hash function PHASHx. Fix integers k, n, t, d, with $d \geq 2$. Let $\mathcal{K} = \{0,1\}^k$ and $\mathcal{T} = \{0,1\}^t$ be non-empty sets of keys and tweaks, respectively. Moreover, derive a set of domains $\mathcal{D} := \{0, 1, 2, 3\} = \{0,1\}^d$ which are encoded as their respective d-bit values, and a domain-tweak set $\mathcal{T}' := \mathcal{D} \times \mathcal{T}$. Let $\mathcal{M} \subseteq (\{0,1\}^n)^*$ denote an input space. Further, let $\widetilde{E} : \mathcal{K} \times \mathcal{T}' \times \{0,1\}^n \to \{0,1\}^n$ denote a tweakable block cipher. We will often write $\widetilde{E}_K^{D,T}(\cdot)$ as short form of $\widetilde{E}(K, D, T, \cdot)$. $K \in \mathcal{K}$, $D \in \mathcal{D}$, and

Algorithm 1. Definition of PMAC2x$[\widetilde{E}]$ and its internal hash function PHASHx$[\widetilde{E}]$ with a tweakable block cipher $\widetilde{E} : \mathcal{K} \times \{0,1\}^d \times \{0,1\}^t \times \{0,1\}^n \to \{0,1\}^n$. n, t, and d denote state, tweak and domain sizes, respectively.

11: **function** PMAC2x$[\widetilde{E}_K](M)$	41: **function** PHASHx$[\widetilde{E}_K](M)$
12: $\quad (X_m, Y_m) \leftarrow$ PHASHx$[\widetilde{E}_K](M)$	42: $\quad X_0 \leftarrow 0^n$; $Y_0 \leftarrow 0^n$
13: $\quad \widehat{X}_m \leftarrow$ Conv(X_m)	43: $\quad (M_1, \ldots, M_m) \xleftarrow{n} M$
14: $\quad \widehat{Y}_m \leftarrow$ Conv(Y_m)	44: $\quad$ **for** $i \leftarrow 1$ **to** $m-1$ **do**
15: $\quad U \leftarrow \widetilde{E}_K^{2,\widehat{Y}_m}(X_m)$	45: $\quad\quad Z_i \leftarrow \widetilde{E}_K^{0,\langle i \rangle}(M_i)$
16: $\quad V \leftarrow \widetilde{E}_K^{3,\widehat{X}_m}(Y_m)$	46: $\quad\quad X_i \leftarrow X_{i-1} \oplus Z_i$
17: $\quad$ **return** $(U \,\|\, V)$	47: $\quad\quad Y_i \leftarrow 2 \cdot (Y_{i-1} \oplus Z_i)$
	48: $\quad$ **if** $|M_m| = n$ **then**
21: **function** Conv(X)	49: $\quad\quad Z_m \leftarrow \widetilde{E}_K^{0,\langle m \rangle}(M_m)$
22: $\quad$ **if** $t \geq n$ **then**	50: $\quad$ **else**
23: $\quad\quad$ **return** X	51: $\quad\quad M_m^* \leftarrow M_m \,\|\, 10^{n-|M_m|-1}$
24: $\quad$ **return** $X[1..t]$	52: $\quad\quad Z_m \leftarrow \widetilde{E}_K^{1,\langle m \rangle}(M_m^*)$
	53: $\quad X_m \leftarrow X_{m-1} \oplus Z_m$
31: **function** $\widetilde{E}_K^{D,T}(X)$	54: $\quad Y_m \leftarrow 2 \cdot (Y_{m-1} \oplus Z_m)$
32: $\quad \widetilde{T} \leftarrow \langle D \rangle_d \,\|\, T[1..t]$	55: $\quad$ **return** (X_m, Y_m)
33: $\quad$ **return** $\widetilde{E}_K^{\widetilde{T}}(X)$	

$T \in \mathcal{T}$ denote key, domain, and tweak, respectively. Conv $: \{0,1\}^n \to \{0,1\}^t$ be a regular function[2] which is used to convert the outputs of PHASHx, X_m and Y_m, so they can be used as tweaks of $\widetilde{E}$ in the finalization step. We denote by PMAC2x$[\widetilde{E}]$ and PHASHx$[\widetilde{E}]$ the instantiation of PMAC2x and PHASHx with $\widetilde{E}$. Both are defined, with a default instantiation of Conv, in Algorithm 1.

4   Security Analysis of PMAC2x

Theorem 1. *Let $\widetilde{E}$ and* PMAC2x$[\widetilde{E}]$ *be defined as in Sect. 3. Let $d + t = n$, and let $m < 2^t$ denote the maximum number of n-bit blocks of any query. Then*

$$\mathbf{Adv}_{\text{PMAC2x}[\widetilde{E}]}^{\text{PRF}}(q, \ell, \theta) \leq \frac{2^{2d}q^2}{2 \cdot (2^n - q)^2} + \frac{2^d q^3}{3 \cdot 2^{2n}(2^n - q)} + \frac{2^d q^2}{2^n(2^n - q)}$$
$$+ \mathbf{Adv}_{\widetilde{E}}^{\text{TPRP}}(\ell + 2q, O(\theta + \ell + 2q)).$$

The final term results from a standard argument after replacing the tweakable block cipher $\widetilde{E}$ by a random tweakable permutation $\widetilde{\pi} \twoheadleftarrow \widetilde{\text{Perm}}(\mathcal{T}', \{0,1\}^n)$. Let **A** be an adversary that makes at most q queries of at most m blocks each and of at most ℓ blocks in total. We assume, **A** does not ask duplicate queries and has the goal to distinguish between a PMAC2x$[\widetilde{\pi}]$ oracle with an internally sampled tweaked permutations $\widetilde{\pi}$ and a random function $\rho : \{0,1\}^* \to \{0,1\}^{2n}$.

We consider the game described in Algorithm 2. The game *without* the boxed statements coincides with a random function ρ, whereas the game *with* them

---

[2] A function is called regular iff all outputs are produced by an equal number of inputs.

Algorithm 2. Main Game, initialization, finalization, and subroutines. Boxed statements belong exclusively to the real world.

1: **procedure** INITIALIZE	11: **function** FINALIZE(b')
2: $\text{bad}_U \leftarrow$ false; $\text{bad}_V \leftarrow$ false; $\mathcal{Q} \leftarrow \emptyset$	12: $\text{bad} \leftarrow \text{bad}_U \vee \text{bad}_V$
3: $X_0 \leftarrow 0^n$; $Y_0 \leftarrow 0^n$; $b \twoheadleftarrow \{0,1\}$	13: **return** $b' = b \vee \text{bad}$

21: **function** ORACLE(M)	51: **function** CASE1($X_m, Y_m, \widehat{X}_m, \widehat{Y}_m$)
22: $(X_m, Y_m) \leftarrow \text{PHASHx}[\widetilde{\pi}](M)$	52: **if** $X_m \in \text{domain}(\widetilde{\pi}^{2,\widehat{Y}_m})$ **then**
23: $\widehat{X}_m \leftarrow \text{CONV}(X_m)$	53: $\boxed{U \leftarrow \widetilde{\pi}^{2,\widehat{Y}_m}[X_m]}$
24: $\widehat{Y}_m \leftarrow \text{CONV}(Y_m)$	54: **else**
25: $U \twoheadleftarrow \{0,1\}^n$	55: $\boxed{U \twoheadleftarrow \{0,1\}^n \setminus \text{range}(\widetilde{\pi}^{2,\widehat{Y}_m})}$
26: $V \twoheadleftarrow \{0,1\}^n$	56: **if** $Y_m \in \text{domain}(\widetilde{\pi}^{3,\widehat{X}_m})$ **then**
27: **if** $(\widehat{X}_m, \widehat{Y}_m) \in \mathcal{Q}_{|\widehat{X}_m, \widehat{Y}_m}$ **then**	57: $\boxed{V \leftarrow \widetilde{\pi}^{3,\widehat{X}_m}[Y_m]}$
28: $(U, V) \leftarrow \text{Case1}(X_m, Y_m, \widehat{X}_m, \widehat{Y}_m)$	58: **else**
29: **else if** $U \in \text{range}(\widetilde{\pi}^{2,\widehat{Y}_m}) \wedge$	59: $\boxed{V \twoheadleftarrow \{0,1\}^n \setminus \text{range}(\widetilde{\pi}^{3,\widehat{X}_m})}$
30: $V \in \text{range}(\widetilde{\pi}^{3,\widehat{X}_m})$ **then**	60: $\text{bad}_U \leftarrow \text{bad}_V \leftarrow$ true
31: $(U, V) \leftarrow \text{Case2}(X_m, Y_m, \widehat{X}_m, \widehat{Y}_m)$	61: **return** (U, V)
32: **else if** $U \in \text{range}(\widetilde{\pi}^{2,\widehat{Y}_m}) \wedge$	71: **function** CASE2($X_m, Y_m, \widehat{X}_m, \widehat{Y}_m$)
33: $V \notin \text{range}(\widetilde{\pi}^{3,\widehat{X}_m})$ **then**	72: $\boxed{U \twoheadleftarrow \{0,1\}^n \setminus \text{range}(\widetilde{\pi}^{2,\widehat{Y}_m})}$
34: $(U, V) \leftarrow \text{Case3}(\widehat{X}_m, \widehat{Y}_m, U, V)$	73: $\boxed{V \twoheadleftarrow \{0,1\}^n \setminus \text{range}(\widetilde{\pi}^{3,\widehat{X}_m})}$
35: **else if** $U \notin \text{range}(\widetilde{\pi}^{2,\widehat{Y}_m}) \wedge$	74: $\text{bad}_U \leftarrow \text{bad}_V \leftarrow$ true
36: $V \in \text{range}(\widetilde{\pi}^{3,\widehat{X}_m})$ **then**	75: **return** (U, V)
37: $(U, V) \leftarrow \text{Case4}(\widehat{X}_m, \widehat{Y}_m, U, V)$	81: **function** CASE3($\widehat{X}_m, \widehat{Y}_m, U, V$)
38: **else if** $U \notin \text{range}(\widetilde{\pi}^{2,\widehat{Y}_m}) \wedge$	82: $\boxed{U \twoheadleftarrow \{0,1\}^n \setminus \text{range}(\widetilde{\pi}^{2,\widehat{Y}_m})}$
39: $V \notin \text{range}(\widetilde{\pi}^{3,\widehat{X}_m})$ **then**	83: $\text{bad}_U \leftarrow$ true
40: $(U, V) \leftarrow \text{Case5}(\widehat{X}_m, \widehat{Y}_m, U, V)$	84: **return** (U, V)
41: $\mathcal{Q} \overset{\cup}{\leftarrow} \{(\widehat{X}_m, \widehat{Y}_m, U, V)\}$	91: **function** CASE4($\widehat{X}_m, \widehat{Y}_m, U, V$)
42: $\widetilde{\pi}^{2,\widehat{Y}_m}[X_m] \leftarrow U$	92: $\boxed{V \twoheadleftarrow \{0,1\}^n \setminus \text{range}(\widetilde{\pi}^{3,\widehat{X}_m})}$
43: $\widetilde{\pi}^{3,\widehat{X}_m}[Y_m] \leftarrow V$	93: $\text{bad}_V \leftarrow$ true
44: **return** $(U \parallel V)$	94: **return** (U, V)
95: **function** CASE5($\widehat{X}_m, \widehat{Y}_m, U, V$)	
96: **return** (U, V)	

exactly represents PMAC2x$[\widetilde{\pi}]$, performing lazy sampling for the permutations $\widetilde{\pi}^{2,\widehat{Y}_m}(\cdot)$ and $\widetilde{\pi}^{3,\widehat{X}_m}(\cdot)$, for all $\widehat{X}_m, \widehat{Y}_m \in \{0,1\}^t$. Both algorithms differ only when bad events occur. Hence, by the fundamental lemma of game playing [4], it holds

$$\Pr[\mathbf{A}^{\text{PMAC2x}[\widetilde{\pi}](\cdot)} \Rightarrow 1] - \Pr[\mathbf{A}^{\rho(\cdot)} \Rightarrow 1] \le \Pr[\mathbf{A} \text{ sets bad}].$$

In the remainder, we consider five cases which cover all possibilities:

- Case1: $(\widehat{X}_m, \widehat{Y}_m) \in \mathcal{Q}_{|\widehat{X}_m, \widehat{Y}_m}$.

- **Case2**: $(\widehat{X}_m, \widehat{Y}_m) \notin \mathcal{Q}_{|\widehat{X}_m, \widehat{Y}_m} \wedge U \in \text{range}(\widetilde{\pi}^{2,\widehat{Y}_m}) \wedge V \in \text{range}(\widetilde{\pi}^{3,\widehat{X}_m})$.
- **Case3**: $(\widehat{X}_m, \widehat{Y}_m) \notin \mathcal{Q}_{|\widehat{X}_m, \widehat{Y}_m} \wedge U \in \text{range}(\widetilde{\pi}^{2,\widehat{Y}_m}) \wedge V \notin \text{range}(\widetilde{\pi}^{3,\widehat{X}_m})$.
- **Case4**: $(\widehat{X}_m, \widehat{Y}_m) \notin \mathcal{Q}_{|\widehat{X}_m, \widehat{Y}_m} \wedge U \notin \text{range}(\widetilde{\pi}^{2,\widehat{Y}_m}) \wedge V \in \text{range}(\widetilde{\pi}^{3,\widehat{X}_m})$.
- **Case5**: $(\widehat{X}_m, \widehat{Y}_m) \notin \mathcal{Q}_{|\widehat{X}_m, \widehat{Y}_m} \wedge U \notin \text{range}(\widetilde{\pi}^{2,\widehat{Y}_m}) \wedge V \notin \text{range}(\widetilde{\pi}^{3,\widehat{X}_m})$.

We list Case 5 only for the sake of completeness. It is easy to see that in Case 5, the output of the game is indistinguishable between the worlds. We use M^i, $\widehat{X}_m^i$, $\widehat{Y}_m^i$, U^i, V^i to refer to the respective values of the i-th query, where $i \in [1, q]$, and assume it is the current query of the adversary. Additionally, we will use indices j and k, where $j, k \in [1, i-1]$, to refer to previous queries.

Case1: For this case, we revisit the fact that for two fixed disjoint queries M^i and M^j, the values X_m and Y_m are results of two random variables. A variant of the proof is given e.g. in [17, Sect. 3.3], and revisited in the following only for the sake of completeness. Fix query indices $i \in [1, q]$ and $j \in [1, i-1]$. Let m and m' denote the number of blocks of the i-th and j-th query, respectively; moreover, let X_m^i, Y_m^i denote the values X_m and Y_m of the i-th query and $X_{m'}^j$, $Y_{m'}^j$ those of the j-th query, respectively. X_m^i, Y_m^i, X_m^j, and Y_m^j result from:

$$X_m^i = C_1^i \oplus C_2^i \oplus \cdots \oplus C_m^i \qquad Y_m^i = 2^m C_1^i \oplus 2^{m-1} C_2^i \oplus \cdots \oplus 2 \cdot C_m^i,$$
$$X_{m'}^i = C_1^j \oplus C_2^j \oplus \cdots \oplus C_{m'}^j \qquad Y_{m'}^j = 2^{m'} C_1^j \oplus 2^{m'-1} C_2^j \oplus \cdots \oplus 2 \cdot C_{m'}^j.$$

So, we want to bound the probability for the following equational system:

$$C_1^i \oplus C_2^i \oplus \cdots \oplus C_m^i = C_1^j \oplus C_2^j \oplus \cdots \oplus C_{m'}^j$$
$$2^m C_1^i \oplus 2^{m-1} C_2^i \oplus \cdots \oplus 2 \cdot C_m^i = 2^{m'} C_1^j \oplus 2^{m'-1} C_2^j \oplus \cdots \oplus 2 \cdot C_{m'}^j.$$

There exist three distinct subcases which cover all possibilities:

- **Subcase 1: $m \neq m'$.** In this case, the equations above provide a unique solution set for two random variables; thus, the probability that the equations hold for two fixed queries is upper bounded by $1/(2^n - (i-1))^2$.
- **Subcase 2: $m = m'$ and there exists $\alpha \in [1, m]$ s.t. $C_\alpha^i \neq C_\alpha^j$ and for all $\beta \in [1, m]$ with $\beta \neq \alpha$: $C_\beta^i = C_\beta^j$.** In this case, both messages share only a single different output block. Thus, the equations above never hold.
- **Subcase 3: $m = m'$ and there exist distinct $\beta \in [1, m]$ with $\beta \neq \alpha$: $C_\beta^i = C_\beta^j$.** In this case, both messages share only a single different output block. Thus, the equations above can never hold.

So, the probability for two fixed disjoint queries M^i and M^j that $(X_m^i, Y_m^i) = (X_{m'}^j, Y_{m'}^j)$ holds, is bounded by $1/(2^n - q)^2$. Since $\widehat{X}_m^i$ and $\widehat{Y}_m^i$ are derived from X_m^i and Y_m^i, respectively, by a regular function (and so are $\widehat{X}_m^j$ and $\widehat{Y}_m^j$ derived from $X_{m'}^j$ and $Y_{m'}^j$, respectively), it follows that the probability of $(\widehat{X}_m^i, \widehat{Y}_m^i) = (\widehat{X}_{m'}^j, \widehat{Y}_{m'}^j)$ to hold for the i-th and j-th query, with $j \in [1, i-1]$, is at most

$$(i-1) \cdot \frac{2^d}{(2^n - q)} \cdot \frac{2^d}{(2^n - q)} = \frac{2^{2d}(i-1)}{(2^n - q)^2}.$$

Case2: In this case, there exists some previous query $(\widehat{X}_{m'}^j, \widehat{Y}_{m'}^j, U^j, V^j)$ s.t. $U = U^j \wedge \widehat{Y}_m = \widehat{Y}_{m'}^j$, and a distinct previous query $(\widehat{X}_{m''}^k, \widehat{Y}_{m''}^k, U^k, V^k)$ s.t. $V = V^k \wedge \widehat{X}_m = \widehat{X}_{m''}^k$. From our assumption $(\widehat{X}_m, \widehat{Y}_m) \notin \mathcal{Q}_{|\widehat{X}_m, \widehat{Y}_m}$, it follows that $j \neq k$; otherwise, the current query would have stepped into Case1 instead. We can bound the probability by

$$\Pr\left[(U = U^j \wedge \widehat{Y}_m = \widehat{Y}_{m'}^j) \wedge (V = V^k \wedge \widehat{X}_m = \widehat{X}_{m''}^k)\right]$$

$$\leq \Pr\left[U = U^j \wedge \widehat{Y}_m = \widehat{Y}_{m'}^j \wedge V = V^k \mid \widehat{X}_m = \widehat{X}_{m''}^k\right].$$

U and V are chosen independently and uniformly at random from $\{0,1\}^n$ each, and can collide with at most $i-1$ previous values U^j and at most $i-1$ previous values V^k, respectively. For fixed j and k, the probability for U to collide with U^j is upper bounded by $1/2^n$, and independently, the probability for V to collide with V^k is also $1/2^n$. Since the game chooses U and V independently from Y_m, the probability that $\widehat{Y}_m$ collides with $\widehat{Y}_{m'}^j$ is at most $2^d/(2^n - q)$ since we assumed that the adversary poses no duplicate queries, and therefore, Y_m and $Y_{m'}^j$ are results of two random variables. Since the collision of $U = U^j$ already fixes the colliding query pair, there is no additional factor $(i-1)$ for the choice of which pairs of $\widehat{Y}_m$ and $\widehat{Y}_{m'}^j$ to collide. It follows that the probability for this case to occur at the i-th query is upper bounded by

$$\frac{i-1}{2^n} \cdot \frac{i-2}{2^n} \cdot \frac{2^d}{2^n - q} \leq \frac{2^d(i-1)^2}{2^{2n}(2^n - q)}.$$

Case3: In this case, there exists some previous query $(\widehat{X}_{m'}^j, \widehat{Y}_{m'}^j, U^j, V^j)$ s.t. $U = U^j \wedge \widehat{Y}_m = \widehat{Y}_{m'}^j$. From our assumption $(\widehat{X}_m, \widehat{Y}_m) \notin \mathcal{Q}_{|\widehat{X}_m, \widehat{Y}_m}$ and $\widehat{Y}_m = \widehat{Y}_{m'}^j$ follows that $\widehat{X}_m \neq \widehat{X}_{m'}^j$ holds, like in Case2. We can bound the probability by

$$\Pr\left[U = U^j \wedge \widehat{Y}_m = \widehat{Y}_{m'}^j \wedge V \notin \mathrm{range}(\widetilde{\pi}^{3,\widehat{X}_m})\right]$$

$$\leq \Pr\left[U = U^j \wedge \widehat{Y}_m = \widehat{Y}_{m'}^j \mid V \notin \mathrm{range}(\widetilde{\pi}^{3,\widehat{X}_m})\right].$$

For a fixed j-th query, the probability that $\widehat{Y}_m$ collides with $\widehat{Y}_{m'}^j$ is at most $2^d/(2^n - q)$. Since U is chosen uniformly at random from $\{0,1\}^n$ and independently from Y_m, U can collide with U^j with probability $1/2^n$. So, the probability of this case to occur for the i-th query can be upper bounded by

$$\frac{2^d}{2^n - q} \cdot \frac{i-1}{2^n} = \frac{2^d(i-1)}{2^n(2^n - q)}.$$

Case4: In this case, it holds that $V = V^j \wedge \widehat{X}_m = \widehat{X}_{m'}^j$. From $(\widehat{X}_m, \widehat{Y}_m) \notin \mathcal{Q}_{|\widehat{X}_m, \widehat{Y}_m}$ and $\widehat{X}_m = \widehat{X}_{m'}^j$ follows here that $\widehat{Y}_m \neq \widehat{Y}_{m'}^j$ holds, analogously to

Case2 and Case3. We can bound the probability by

$$\Pr\left[V = V^j \wedge \widehat{X}_m = \widehat{X}^j_{m'} \wedge U \notin \mathsf{range}(\widetilde{\pi}^{2,\widehat{Y}_m})\right]$$
$$\leq \Pr\left[V = V^j \wedge \widehat{X}_m = \widehat{X}^j_{m'} \mid U \notin \mathsf{range}(\widetilde{\pi}^{2,\widehat{Y}_m})\right].$$

Obviously, this case can be handled similarly as Case3. For a fixed j-th query, the probability that $\widehat{X}_m$ collides with $\widehat{X}^j_{m'}$ is at most $2^d/(2^n - q)$. Since V is chosen uniformly at random from $\{0,1\}^n$ and independently from X_m, V can collide with V^j with probability $1/2^n$. So, the probability of this case to occur for the i-th query can also be upper bounded by

$$\frac{2^d(i-1)}{2^n(2^n - q)}.$$

Taking the terms from all cases and the union bound over at most q queries gives

$$\Pr\left[\mathbf{A} \text{ sets bad}\right] \leq \sum_{i=1}^{q} \left(\frac{2^{2d}(i-1)}{(2^n - q)^2} + \frac{2^d(i-1)^2}{2^{2n}(2^n - q)} + 2 \cdot \frac{2^d(i-1)}{2^n(2^n - q)}\right)$$
$$\leq \frac{2^{2d}q^2}{2 \cdot (2^n - q)^2} + \frac{2^d q^3}{3 \cdot 2^{2n}(2^n - q)} + \frac{2^d q^2}{2^n(2^n - q)}.$$

5   Security Analysis of PMACx

This section considers a variant of PMAC2x, PMACx, that adds a final XOR to produce only an n-bit tag, following the design of PMAC_TBC1K. A schematic illustration is given in Fig. 2. We revisit the assumption by Naito, and show that our proof of PMAC2x needs only a slight adaption for PMACx.

Previous Analysis. Theorem 2 in [17] proves the security of PMAC_TBC1K with the help of an analysis of multi-collisions of the final chaining values (X_m and Y_m in our notation). Note that our notation differs from [17] to be consistent to our previous section. Define two monotone events mcoll$_1$ and mcoll$_2$. Let ρ and ξ denote positive integers and define three sets $\mathcal{X}$, $\mathcal{Y}$, and $\mathcal{Q}$ which store the values $\widehat{X}^i_m$, $\widehat{Y}^i_m$, and the tuples $(X^i_m, \widehat{Y}^i_m)$, respectively, of the queries $1 \leq i \leq q$.

$$\mathsf{mcoll}_1 := (\exists\, \widehat{X}^1_m, \ldots, \widehat{X}^\rho_m \in \mathcal{X} \text{ s.t. } \widehat{X}^1_m = \ldots = \widehat{X}^\rho_m)$$
$$\vee\ (\exists\, \widehat{Y}^1_m, \ldots, \widehat{Y}^\rho_m \in \mathcal{Y} \text{ s.t. } \widehat{Y}^1_m = \ldots = \widehat{Y}^\rho_m),$$
$$\mathsf{mcoll}_2 := \exists\, (X^1_m, \widehat{Y}^1_m), \ldots, (X^\xi_m, \widehat{Y}^\xi_m) \in \mathcal{Q} \text{ s.t. } (X^1_m, \widehat{Y}^1_m) = \ldots = (X^\xi_m, \widehat{Y}^\xi_m).$$

The original proof further defined a monotone compound event mcoll := mcoll$_1 \vee$ mcoll$_2$ and used the fact that

$$\Pr\left[\mathbf{A} \text{ sets bad}\right] = \Pr\left[\mathbf{A} \text{ sets bad} \wedge \mathsf{mcoll}\right] + \Pr\left[\mathbf{A} \text{ sets bad} \wedge \neg\mathsf{mcoll}\right]$$
$$\leq \Pr\left[\mathsf{mcoll}_1\right] + \Pr\left[\mathsf{mcoll}_2\right] + \Pr\left[\mathbf{A} \text{ sets bad}\mid\neg\mathsf{mcoll}\right].$$

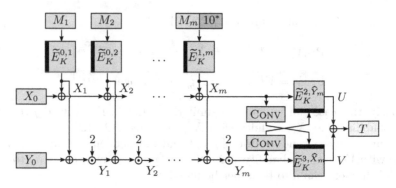

Fig. 2. PMACx, the variant of PMAC2x with n-bit output, following the design of PMAC_TBC1k.

The analysis in [17] bounds $\Pr[\mathsf{mcoll}_1]$ as

$$\Pr[\mathsf{mcoll}_1] \leq 2 \cdot 2^t \cdot \binom{q}{\rho} \cdot \left(\frac{2^{n-t}}{2^n - q} \right)^{\rho} \leq 2^{t+1} \cdot \left(\frac{2^{n-t} \cdot eq}{\rho(2^n - q)} \right)^{\rho},$$

using Stirling's approximation $x! \geq (x/e)^x$ for any x. Note, in PMAC_TBC1k, the domain size in PMAC2x is fixed to $d = 2$ bits. The bound above consists of the probability that ρ values are all equal, $(2^{n-t}/(2^n - q))^{\rho}$; the fact that there are 2^t tweak values; and the $\binom{q}{\rho}$ possible ways to choose ρ out of q values. However, the bound holds *only if* the ρ colliding tweaks stem from ρ linearly independent random variables, which is *not* necessarily the case. Imagine a sequence of 2^m queries which combine pair-wise distinct blocks $\{M_i, M_i'\}$ with $M_i \neq M_i'$, for $1 \leq i \leq m$ position-wise, i. e., we have 2^m queries of m blocks each: $Q^0 = (M_1, M_2, M_3, \ldots, M_m)$, $Q^1 = (M_1', M_2, M_3, \ldots, M_m)$, $Q^2 = (M_1, M_2', M_3, \ldots, M_m)$, $\ldots$, $Q^{2^m-1} = (M_1', M_2', M_3', \ldots, M_m')$. When used as queries to PMAC_TBC1k, the 2^m resulting values X_m^i, for $0 \leq i \leq 2^m - 1$, depend on the linear combination of only $2m$ random variables. A similar argument holds for the values Y_m^i, as well as for the similarly treated bound of mcoll_2. Thus, the multi-collision bound demands a significantly more detailed analysis.

Fixing the Analysis. From our proof for PMAC2x, we can now derive a corollary for a similar security bound for PMACx, which again can be easily transformed into a bound for PMAC_TBC1k.

Corollary 1. *Let $\widetilde{E}$ and* PMAC2x$[\widetilde{E}]$ *be defined as in Sect. 3. Let $d + t = n$, and let $m < 2^t$ denote the maximum number of n-bit blocks of any query. Then, it holds that* $\mathbf{Adv}_{\mathrm{PMACx}[\widetilde{E}]}^{\mathrm{PRF}}(q, \ell, \theta) \leq \mathbf{Adv}_{\mathrm{PMAC2x}[\widetilde{E}]}^{\mathrm{PRF}}(q, \ell, \theta)$.

The proof can use a game almost identical to that in Algorithm 2, where we only modify Line 44 to return the XOR of U and V. This is shown in Algorithm 3. All further procedures and functions remain identical to those in Algorithm 2.

If $(U \| V)$ is indistinguishable from outputs of a $2n$-bit random function ρ, then each of the n-bit outputs U and V can be considered random. It follows,

Algorithm 3. The updated game for the security proof of PMACx. Only the double-boxed statement changes compared with the game in Algorithm 2.

21: **function** ORACLE(M)
22: ...
42:   $\widetilde{\pi}^{2,\widehat{Y}_m}[X_m] \leftarrow U$
43:   $\widetilde{\pi}^{3,\widehat{X}_m}[Y_m] \leftarrow V$
44:   **return** $\boxed{\boxed{T \leftarrow U \oplus V}}$

if U is indistinguishable from n-bit values, then the XOR sum of $U \oplus V$ is also indistinguishable from a random n-bit value. Hence, the PRF advantage of **A** on PMACx is upper bounded by that of an adversary **A**$'$ on PMAC2x with an equal amount of resources as **A**; hence, the corollary follows.

PMACx and PMAC_TBC1k differ in three aspects: (1) PMACx allows messages whose length is not a multiple of n bits by padding the final block with 10^* and using a distinct tweak for it; (2) PMACx defines a generic d-bit domain encoding and defines a conversion function CONV for deriving the inputs for the finalization; and (3) PMACx adds a final doubling for a simpler and consistent description. Clearly, none of the differences affects the distribution of final chaining values $\widehat{X}_m$ and $\widehat{Y}_m$. Hence, when fixing $d = 2$, a security result for PMACx can be easily carried over to a bound for PMAC_TBC1k.

6   Definition and Security Analysis of SIVx

Next, we define the deterministic AE scheme SIVx, which combines PMAC2x and the IV-based Counter-in-Tweak mode IVCTRT. We recall the definitions of IV-based encryption and Deterministic AE security in the full version of this work [11]. Note, that it is straight-forward to derive a nonce-based AE scheme by fixing the nonce length and appending the nonce to the associated data.

IVCTRT. IVCTRT denotes the version of Counter in Tweak [21, Appendix C], which takes a $2n$-bit random IV plus the message for each encryption. Let $\mathcal{T} = \{0,1\}^t$, and $\mathcal{T}' = \{0,1\} \times \mathcal{T}$. The mode first splits $(U, V) \xleftarrow{n} IV$, and uses a given tweakable block cipher $\widetilde{E} : \mathcal{K} \times \mathcal{T}' \times \{0,1\}^n \to \{0,1\}^n$ in counter mode, with V as cipher input. Next, it derives $T \leftarrow \text{CONV}'(U)$ from U with a regular function $\text{CONV}' : \{0,1\}^n \to \mathcal{T}$ and increments T for every call to $\widetilde{E}$ using addition modulo 2^t. We denote by IVCTRT[$\widetilde{E}$] the instantiation of IVCTRT with $\widetilde{E}$; from Theorem 1 and Appendix C in [21], we recall the following theorem:

Theorem 2 (ivE Security of IVCTRT). *Let* $\widetilde{\pi} \xleftarrow{} \widetilde{\text{Perm}}(\mathcal{T}', \{0,1\}^n)$ *be an ideal tweakable block cipher. Let* **A** *be an adversary which asks at most q queries of at most $8 \leq \ell \leq |\mathcal{T}|$ blocks in total. Then*

$$\mathbf{Adv}^{\text{IVE}}_{\text{IVCTRT}[\widetilde{\pi}]}(\mathbf{A}) \leq \frac{1}{2^n} + \frac{1}{|\mathcal{T}|} + \frac{4\ell \log q}{|\mathcal{T}|} + \frac{\ell \log^2(\ell)}{2^n}.$$

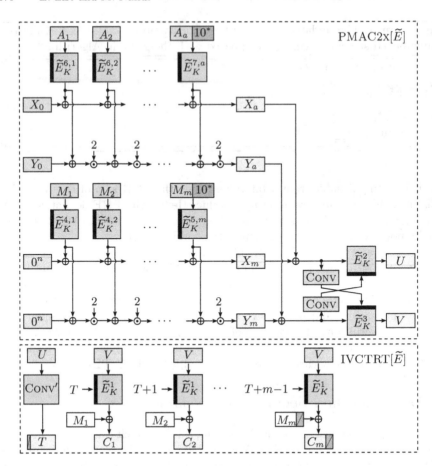

Fig. 3. The deterministic AE scheme SIVx from the composition of $\text{PMAC2x}[\widetilde{E}]$ (**top**), and the $\text{IVCTRT}[\widetilde{E}]$ mode of encryption (**bottom**) [20]. The figure starts the message processing in PMAC2x from 0^n only to prevent that X_0 and Y_0 cancel out.

***Definition of* SIVx.** We define the deterministic AE scheme $\text{SIVx}[\widetilde{E}]$ as the composition of $\text{PMAC2x}[\widetilde{E}]$ and $\text{IVCTRT}[\widetilde{E}]$, as given in Algorithm 4. A schematic illustration of the encryption process is depicted in Fig. 3. In general, we denote by $\text{SIVx}[F, \Pi]$ the instantiation of SIVx with a function F and an IV-based encryption scheme Π in SIVx. To use the same key in all calls to $\widetilde{E}$, we parametrize PHASHx to separate the domains. We use the domains $2 = (0010)_2$ and $3 = (0011)_2$ for the finalization steps, $4 = (0100)_2$ and $5 = (0101)_2$ for processing the associated data, as well as $6 = (0110)_2$ and $7 = (0111)_2$ for processing the message in PMAC2x. We encode them into the $d = 4$ most significant bits of the tweak. Inside IVCTRT, however, we use the single-bit domain 1 in all calls to $\widetilde{E}$ for we lose only a single bit from the IV. For concreteness, we define the initial values in PMAC2x as $X_0 = Y_0 = 0^n$.

Algorithm 4. Definition of SIVx$[\widetilde{E}]$. Note that IVCTRT$[\widetilde{E}]$ and its inverse IVCTRT$^{-1}[\widetilde{E}]$ are identical operations. Moreover, we define PHASHx$[\widetilde{E}]^{D_1,D_2}$ to denote PHASHx$[\widetilde{E}]$ using D_1 as domain for processing full blocks, and using D_2 for a potential partial final block of the input to PHASHx.

11: **function** SIVx$[\widetilde{E}_K](A,M)$	41: **function** SIVx$^{-1}[\widetilde{E}_K](A,C,\text{TAG})$
12: TAG $\leftarrow$ PMAC2x$[\widetilde{E}_K](A,M)$	42: $(U,V) \xleftarrow{n}$ TAG
13: $(U,V) \xleftarrow{n}$ TAG	43: $IV \leftarrow$ CONV$'(U) \,\|\, V$
14: $IV \leftarrow$ CONV$'(U) \,\|\, V$	44: $M \leftarrow$ IVCTRT$^{-1}[\widetilde{E}_K](IV,C)$
15: $C \leftarrow$ IVCTRT$[\widetilde{E}_K](IV,M)$	45: TAG$'$ $\leftarrow$ PMAC2x$[\widetilde{E}_K](A,M)$
16: **return** (C,TAG)	46: **if** TAG $=$ TAG$'$ **then**
	47: **return** M
21: **function** CONV$'(U)$	48: **return** $\perp$
22: **return** $T \leftarrow U[1..t]$	
	51: **function** PMAC2x$[\widetilde{E}_K](A,M)$
31: **function** IVCTRT$[\widetilde{E}_K](IV,M)$	52: $(X_a,Y_a) \leftarrow$ PHASHx$[\widetilde{E}_K]^{4,5}(A)$
32: $(T,V) \leftarrow IV$	53: $(X_m,Y_m) \leftarrow$ PHASHx$[\widetilde{E}_K]^{6,7}(M)$
33: $(M_1,\ldots,M_m) \xleftarrow{n} M$	54: $\widehat{X}_m \leftarrow$ CONV$(X_m \oplus X_a \oplus X_0)$
34: **for** $i \leftarrow 1$ to $m-1$ **do**	55: $\widehat{Y}_m \leftarrow$ CONV$(Y_m \oplus Y_a \oplus Y_0)$
35: $C_i \leftarrow \widetilde{E}_K^{1,T+(i-1)}(V) \oplus M_i$	56: $U \leftarrow \widetilde{E}_K^{2,\widehat{Y}_m}(X_m)$
36: $S_m \leftarrow \widetilde{E}_K^{1,T+(m-1)}(V)[1..|M_m|]$	57: $V \leftarrow \widetilde{E}_K^{3,\widehat{X}_m}(Y_m)$
37: $C_m \leftarrow S_m \oplus M_m$	58: **return** $IV \leftarrow (U \,\|\, V)$
38: **return** $C \leftarrow (C_1 \,\|\, \ldots \,\|\, C_m)$	

Theorem 3 (DAE Security of SIVx). *Let $F : \mathcal{K}_1 \times \mathcal{A} \times \mathcal{M} \to \{0,1\}^{2n}$, and let $\Pi = (\widetilde{\mathcal{E}}, \widetilde{\mathcal{D}})$ be an IV-based encryption scheme with key space $\mathcal{K}_2$ and IV space $\mathcal{IV}$. Let $K_1 \twoheadleftarrow \mathcal{K}_1$ and $K_2 \twoheadleftarrow \mathcal{K}_2$ be independent. Let* CONV$' : \{0,1\}^n \to \mathcal{IV}$ *be a regular function. Let **A** be a DAE adversary running in time at most θ, asking at most q queries of at most $8 \le \ell < 2^t$ blocks in total. Then, it holds that*

$$\mathbf{Adv}^{\mathrm{DAE}}_{\mathrm{SIVx}[F,\Pi]}(\mathbf{A}) \le \mathbf{Adv}^{\mathrm{IVE}}_{\Pi}(\theta + O(\ell), q, \ell) + \mathbf{Adv}^{\mathrm{PRF}}_{F}(\theta + O(\ell), q, \ell) + \frac{q}{2^n}.$$

We defer the proof of Theorem 3 to the full version of this work [11]. Inserting the bounds from Theorems 1 and 2, we obtain the corollary below, where F denotes PMAC2x$[\widetilde{E}]$ and Π represents IVCTRT$[\widetilde{E}]$.

Corollary 2. *Fix positive integers k, n, t and $d = 4$. Define $d + t = n$ and let $\mathcal{T} = \{0,1\}^t$ and $\mathcal{T}' = \{0,1\}^d \times \{0,1\}^t$, and $\mathcal{IV} = \{0,1\}^{n-1}$. Let $\widetilde{E} : \mathcal{K} \times \mathcal{T}' \times \{0,1\}^n \to \{0,1\}^n$, and* CONV$' : \{0,1\}^n \to \mathcal{IV}$ *be a regular function. Let $K \twoheadleftarrow \mathcal{K}$ and **A** be a DAE adversary that runs in time at most θ, and asks at most q queries of at most $8 \le \ell < 2^t$ blocks in total. Then*

$$\mathbf{Adv}^{\mathrm{DAE}}_{\mathrm{SIVx}[\widetilde{E}]}(\mathbf{A}) \le \frac{2^{2d}q^2}{2 \cdot (2^n - q)^2} + \frac{2^d q^3}{3 \cdot 2^{2n}(2^n - q)} + \frac{2^d q^2}{2^n(2^n - q)} + \frac{4\ell \log q + 1}{2^{n-1}} + \frac{q + 1 + \ell \log^2(\ell)}{2^n} + \mathbf{Adv}^{\mathrm{TPRP}}_{\widetilde{E}}(\theta + O(2\ell + 2q), 2\ell + 2q).$$

7   Conclusion

This work revisited the PMAC_TBC1K construction by Naito for constructing a MAC with beyond-birthday-bound (BBB) security and $2n$-bit outputs,
called PMAC2x. We identified a critical assumption in the previous analysis of
PMAC_TBC1K and circumvented it by a new proof for PMAC2x; moreover,
we could easily derive a proof for PMACx, a variant of our PMAC2x construction with n-bit outputs. So, we also provided a corrected bound for Naito's
construction. We obtained the positive result that all three constructions provide
PRF security for up to $O(q^2/2^{2n} + q^3/2^{3n})$ queries. With the help of PMAC2x,
we constructed a BBB-secure AE scheme from a tweakable block cipher whose
security is independent of nonces and which depends on a single primitive under
a single key. We are aware that the $2n$-bit tag of SIVx requires still as many
bits to be transmitted as for the $2n$-bit nonce-IV in SCT; future work could
study how an appropriate truncation could reduce the transmission overhead
while retaining BBB security.

Acknowledgments. The authors would like to thank Yusuke Naito and the anonymous reviewers for fruitful comments that helped improve our work.

References

1. Andreeva, E., Bogdanov, A., Datta, N., Luykx, A., Mennink, B., Nandi, M.,
   Tischhauser, E., Yasuda, K.: COLM v1 (2016). Submission to the CAESAR competition. http://competitions.cr.yp.to/caesar-submissions.html
2. Bellare, M., Guérin, R., Rogaway, P.: XOR MACs: new methods for message
   authentication using finite pseudorandom functions. In: Coppersmith, D. (ed.)
   CRYPTO 1995. LNCS, vol. 963, pp. 15–28. Springer, Heidelberg (1995). doi:10.
   1007/3-540-44750-4_2
3. Bellare, M., Rogaway, P.: Code-based game-playing proofs and the security of triple
   encryption. IACR Cryptology ePrint Archive, 2004:331 (2004)
4. Bellare, M., Rogaway, P.: The security of triple encryption and a framework for code-based game-playing proofs. In: Vaudenay, S. (ed.) EUROCRYPT
   2006. LNCS, vol. 4004, pp. 409–426. Springer, Heidelberg (2006). doi:10.1007/
   11761679_25
5. Bernstein, D.J.: How to stretch random functions: the security of protected counter
   sums. J. Crypt. **12**(3), 185–192 (1999)
6. Black, J., Rogaway, P.: A block-cipher mode of operation for parallelizable message
   authentication. In: Knudsen, L.R. (ed.) EUROCRYPT 2002. LNCS, vol. 2332, pp.
   384–397. Springer, Heidelberg (2002). doi:10.1007/3-540-46035-7_25
7. Datta, N., Dutta, A., Nandi, M., Paul, G., Zhang, L.: Building single-key beyond
   birthday bound message authentication code. IACR Cryptology ePrint Archive,
   2015/958 (2015)
8. Fleischmann, E., Forler, C., Lucks, S.: McOE: a family of almost foolproof on-line
   authenticated encryption schemes. In: Canteaut, A. (ed.) FSE 2012. LNCS, vol.
   7549, pp. 196–215. Springer, Heidelberg (2012). doi:10.1007/978-3-642-34047-5_12

9. Hoang, V.T., Krovetz, T., Rogaway, P.: Robust authenticated-encryption AEZ and the problem that it solves. In: Oswald, E., Fischlin, M. (eds.) EUROCRYPT 2015. LNCS, vol. 9056, pp. 15–44. Springer, Heidelberg (2015). doi:10.1007/978-3-662-46800-5_2

10. Jean, J., Nikolić, I., Peyrin, T.: Deoxys v1.4 (2016). Third-round submission to the CAESAR competition. http://competitions.cr.yp.to/caesar-submissions.html

11. List, E., Nandi, M.: Revisiting Full-PRF-Secure PMAC and using it for beyond-birthday authenticated encryption. Cryptology ePrint Archive (2016, to appear)

12. Luykx, A., Preneel, B., Szepieniec, A., Yasuda, K.: On the influence of message length in PMAC's security bounds. In: Fischlin, M., Coron, J.-S. (eds.) EURO-CRYPT 2016. LNCS, vol. 9665, pp. 596–621. Springer, Heidelberg (2016). doi:10.1007/978-3-662-49890-3_23

13. Luykx, A., Preneel, B., Tischhauser, E., Yasuda, K.: A MAC mode for lightweight block ciphers. In: Peyrin, T. (ed.) FSE 2016. LNCS, vol. 9783, pp. 43–59. Springer, Heidelberg (2016). doi:10.1007/978-3-662-52993-5_3

14. Mandal, A., Nandi, M.: An improved collision probability for CBC-MAC and PMAC. IACR Cryptology ePrint Archive, 2007:32 (2007)

15. Minematsu, K.: Parallelizable rate-1 authenticated encryption from pseudorandom functions. In: Nguyen, P.Q., Oswald, E. (eds.) EUROCRYPT 2014. LNCS, vol. 8441, pp. 275–292. Springer, Berlin (2014). doi:10.1007/978-3-642-55220-5_16

16. Minematsu, K., Matsushima, T.: New bounds for PMAC, TMAC, and XCBC. In: Biryukov, A. (ed.) FSE 2007. LNCS, vol. 4593, pp. 434–451. Springer, Berlin (2007). doi:10.1007/978-3-540-74619-5_27

17. Naito, Y.: Full PRF-secure message authentication code based on tweakable block cipher. In: Au, M.-H., Miyaji, A. (eds.) ProvSec 2015. LNCS, vol. 9451, pp. 167–182. Springer, Cham (2015). doi:10.1007/978-3-319-26059-4_9

18. Nandi, M.: A unified method for improving PRF bounds for a class of blockcipher based MACs. In: Hong, S., Iwata, T. (eds.) FSE 2010. LNCS, vol. 6147, pp. 212–229. Springer, Berlin (2010). doi:10.1007/978-3-642-13858-4_12

19. Nandi, M., Mandal, A.: Improved security analysis of PMAC. J. Math. Crypt. 2(2), 149–162 (2008)

20. Peyrin, T., Seurin, Y.: Counter-in-tweak: authenticated encryption modes for tweakable block ciphers. In: Robshaw, M., Katz, J. (eds.) CRYPTO 2016. LNCS, vol. 9814, pp. 33–63. Springer, Berlin (2016). doi:10.1007/978-3-662-53018-4_2

21. Peyrin, T., Seurin, Y.: Counter-in-tweak: authenticated encryption modes for tweakable block ciphers. IACR Cryptology ePrint Archive, 2015:1049, Version, 27 May 2016

22. Rogaway, P.: Authenticated-encryption with associated-data. In: ACM Conference on Computer and Communications Security, pp. 98–107 (2002)

23. Rogaway, P.: Efficient instantiations of tweakable blockciphers and refinements to modes OCB and PMAC. In: Lee, P.J. (ed.) ASIACRYPT 2004. LNCS, vol. 3329, pp. 16–31. Springer, Berlin (2004). doi:10.1007/978-3-540-30539-2_2

24. Rogaway, P.: Nonce-based symmetric encryption. In: Roy, B., Meier, W. (eds.) FSE 2004. LNCS, vol. 3017, pp. 348–358. Springer, Berlin (2004). doi:10.1007/978-3-540-25937-4_22

25. Yasuda, K.: A new variant of PMAC: beyond the birthday bound. In: Rogaway, P. (ed.) CRYPTO 2011. LNCS, vol. 6841, pp. 596–609. Springer, Berlin (2011). doi:10.1007/978-3-642-22792-9_34

26. Yasuda, K.: PMAC with parity: minimizing the query-length influence. In: Dunkelman, O. (ed.) CT-RSA 2012. LNCS, vol. 7178, pp. 203–214. Springer, Berlin (2012). doi:10.1007/978-3-642-27954-6_13
27. Zhang, Y.: Using an error-correction code for fast, beyond-birthday-bound authentication. In: Nyberg, K. (ed.) CT-RSA 2015. LNCS, vol. 9048, pp. 291–307. Springer, Cham (2015). doi:10.1007/978-3-319-16715-2_16

2017 Selected Topics

Publish or Perish: A Backward-Compatible Defense Against Selfish Mining in Bitcoin

Ren Zhang$^{(\boxtimes)}$ and Bart Preneel

KU Leuven, ESAT/COSIC and IMEC, Leuven, Belgium
{ren.zhang,bart.preneel}@esat.kuleuven.be

Abstract. The Bitcoin mining protocol has been intensively studied and widely adopted by many other cryptocurrencies. However, it has been shown that this protocol is not incentive compatible, because the selfish mining strategy enables a miner to gain unfair rewards. Existing defenses either demand fundamental changes to block validity rules or have little effect against a resourceful attacker. This paper proposes a backward-compatible defense mechanism which outperforms the previous best defense. Our fork-resolving policy neglects blocks that are not published in time and appreciates blocks that incorporate links to competing blocks of their predecessors. Consequently, a block that is kept secret until a competing block is published contributes to neither or both branches, hence it confers no advantage in winning the block race. Additionally, we discuss the dilemma between partition recovery time and selfish mining resistance, and how to balance them in our defense.

Keywords: Bitcoin · Selfish mining · Incentive compatibility

1 Introduction

Bitcoin [15], a decentralized cryptocurrency system, attracts not only many users but also significant attention from academia. Bitcoin critically relies on an incentive mechanism named *mining*. Participants of the mining process, called *miners*, compete in producing and broadcasting *blocks*, hoping to get their blocks into the *main chain* and receive block rewards. When more than one block extends the same preceding block, the main chain is decided by a *fork-resolving policy*: a miner adopts and mines on the chain with the most work, which is typically the longest chain, or the first received block when several chains are of the same length. We refer to this *forked* situation as a *block race*, and to an equal-length block race as a *tie*. Bitcoin's designer implicitly assumed the *fairness* of the mining protocol: as long as more than half of the mining power follows the protocol, the chance that a miner can earn the next block reward is proportional to the miner's computational power [15].

Unfortunately, this assumption has been disproven by a *selfish mining attack* highlighted by Eyal and Sirer [10]. In this attack, the *selfish miner* keeps discovered blocks secret and continues to mine on top of them, hoping to gain a

© Springer International Publishing AG 2017
H. Handschuh (Ed.): CT-RSA 2017, LNCS 10159, pp. 277–292, 2017.
DOI: 10.1007/978-3-319-52153-4_16

larger lead on the public chain, and only publishes the selfish chain to claim the rewards when the public chain approaches the length of the selfish chain. Though risking the rewards of some secret blocks, once the selfish chain is longer than its competitor, the selfish miner can securely invalidate honest miners' competing blocks. Accordingly, the overall expectation of the selfish miner's relative revenue increases. An attacker with faster block propagation speed than honest miners can profit from this attack no matter how small the mining power share is. Furthermore, since the revenue of a malicious miner rises superlinearly with the computational power, rational miners would prefer to act collectively for a higher input-output ratio, damaging the decentralized structure of Bitcoin.

Existing defenses can be categorized into two approaches: making fundamental changes to the block validity rules, as suggested by Bahack [6], Shultz [21], Solat and Potop-Butucaru [22]; or lowering the chance of honest miners working on the selfish miner's chain during a tie, as suggested by Eyal, Sirer [10] and Heilman [12]. The former approach demands a backward-incompatible upgrade on both miners and non-miner users; while the latter approach, which we refer to as *tie breaking defenses*, has no effectiveness when the selfish chain is longer than the public chain, therefore cannot defend against resourceful attackers.

This paper proposes another defense against selfish mining. We observe that two policies are involved in deciding which blocks receive mining rewards: the fork-resolving policy demands that the main chain is the longest chain, and the reward distribution policy demands that all blocks in the main chain and no other block receive rewards. Because changing the latter leaves the protocol backward-incompatible, our defense aims to change only the fork-resolving policy.

Our defense replaces the original Bitcoin fork-resolving policy, denoted by *length FRP*, with a *weighted FRP*. By asking miners to compare the *weight* of the chains instead of their length, weighted FRP puts the selfish miner in a dilemma: if the selfish miner keeps a block secret after a competing block is published, the secret block does not contribute to the weight of its chain; if the secret block is published with the competing block, the next honest block gains a higher weight by embedding a proof of having seen this block. In both scenarios, the secret block does not help the selfish miner win the block race. Consequently, our scheme is the first backward-compatible defense that is able to disincentivize block withholding behavior when the selfish chain is longer.

Comparing with existing defenses, our defense has the three-fold advantage of backward-compatible, decentralized and effective. For evaluation, we extend the method developed by Sapirshtein et al. [18] to compute the optimal selfish mining strategy and its relative revenue within our defense. For a selfish miner with more than 40% of mining power, our defense outperforms *the optimal tie breaking defense*: an imaginary defense in which the selfish miner loses every tie.

As an additional contribution, we point out that the core reason of selfish miner's profitability is the high tolerance of the Bitcoin protocol towards network partition. To resolve this dilemma, our scheme introduces a fail-safe parameter k and demands miners to adopt the longest chain if it is at least k blocks ahead of its competitors. Adjusting this parameter allows us to balance between partition recovery time and selfish mining resistance. In the extreme forms, when $k = \infty$, although sacrificing the ability to recover from temporary network split, our defense is effective against a 51% attacker; and when $k = 1$, the partition recovery time remains unaffected and the performance of our defense, which is reduced to a tie breaking defense, is close to the optimal tie breaking defense against a resourceful attacker. Through evaluating both the effectiveness of our scheme and its partition recovery performance in the presence of an attacker, we recommend the value $k = 3$, which can effectively defend against selfish mining attack and keep the partition recovery time within acceptable range.

We also reflect upon our modeling of the Bitcoin network and discuss the limitations and possible future work.

2 Preliminaries

2.1 Bitcoin Blockchain and Mining

We summarize here essential characteristics of Bitcoin for our discussion and refer to Nakamoto's original paper [15] and the textbook by Narayanan et al. [16] for a complete view of the system. To ensure participants have a consensus on valid transactions, all nodes follow the same *block and transaction validity rules*. A typical transaction in Bitcoin consists of at least one input and one output. The difference between the total amount of inputs and outputs in a transaction, the *transaction fee*, goes to the miner who includes the transaction in the blockchain.

The *blockchain* is a ledger containing all transactions organized as a chain of blocks. Each block contains its distance from the first block, called *height*, the hash value of the preceding *parent block*, a set of transactions, and a nonce. Information about the parent block guarantees that a miner must choose which chain to mine on before starting. A special empty-input *coinbase* transaction in the block allocates a fixed amount of new bitcoins to the miner, thus incentivizing miners to contribute their resources to the system. The miner can embed an arbitrary string in this transaction. To construct a valid block, miners work on finding the right nonce so that the hash of the block is smaller than the *block difficulty target*. This target is adjusted every 2016 blocks so that on average a block is generated every ten minutes. The protocol demands miners to publish valid blocks to the overlay network the moment they are found. If a block ends up being in the longest chain, the coinbase output and all transaction fees in the block belong to the miner. The discarded blocks do not receive any block rewards. To decrease the variance of mining revenues, miners often form *mining pools* to work on the same puzzle and split the rewards according to their contributions.

2.2 A History of Selfish Mining Strategies

The idea of selectively delaying publication of blocks to gain an unfair advantage of block rewards appears as early as 2010 [7]. In 2013, Eyal and Sirer [10] defined and analyzed a specific strategy which they called "selfish mining".

Bahack presented a family of selfish mining strategies and evaluated their relative revenue [6]. Sapirshtein et al. [18] and Nayak et al. [17] showed that under certain conditions, the selfish miner can obtain higher expected relative revenue by working on the selfish chain even when the public chain is longer.

2.3 Existing Defenses Against Selfish Mining

Backward-Incompatible Defenses. Bahack proposed a *fork-punishment rule*: competing blocks receive no block reward. The first miner who incorporates a proof of the block fork in the blockchain gets half of the forfeited rewards [6]. However, honest miners suffer collateral damage of this defense, which constitutes another kind of attack. Shultz [21], Solat and Potop-Butucaru [22] recommended that each solved block be accompanied by a certain number of signatures or dummy blocks, proving that the block is witnessed by the network and a competing block is absent, before miners are able to work on it. However they did not provide a mechanism to evaluate whether the number of proofs is adequate to continue working. Neither did they mention how to prevent the selfish miner from generating a dominant number of proofs and releasing them when necessary. In addition, these three defenses require fundamental changes on the block validity and reward distribution rules, consequently network participants who do not upgrade their clients will not understand the new protocol.

Tie Breaking Defenses. Eyal and Sirer proposed that a miner chooses which chain to mine on uniformly at random in a tie [10]. The defense is referred to as *uniform tie breaking* in [18]. The authors showed that this defense raises the *profit threshold*, namely the minimum mining power share to earn unfair block rewards, to 25% within their selfish mining strategy. The threshold was later shown to be 23.21% under the optimal selfish mining strategy [18]. Heilman suggested each miner incorporate the latest unforgeable timestamp issued by a trusted party into the working block [12]. The publicly accessible and unpredictable timestamp is issued with a suggested interval of 60 s. When two competing blocks are received within 120 s, a miner prefers the block whose timestamp is fresher. The author claimed that this *freshness preferred* mechanism can raise the profit threshold to 32%. However, introducing an extra trusted party is inconsistent with Bitcoin's decentralized philosophy. At last, we note that the rules of tie breaking defenses do not apply when the selfish miner's chain is longer than the public chain, rendering the defenses ineffective against resourceful attackers. A selfish miner with 48% of total mining power earns 89% and 83% of mining rewards within uniform tie breaking and freshness preferred, respectively, given that all other miners follow honest mining strategy.

2.4 Properties of an Ideal Defense

Acknowledging the weaknesses of existing defenses, we enumerate here the desirable properties of an ideal defense.

Decentralization. Introducing a trusted server would open a new single point of failure. Moreover, it violates Bitcoin's fundamental philosophy.

Incentive compatibility. The expected relative revenue of a miner should be proportional to the mining power.

Backward compatibility. Non-miners who cannot upgrade their clients can still participate in the network. This is important for hardware products such as Bitcoin ATMs. Specifically, the following rules should not be changed:

Block validity rules. A valid block in the current Bitcoin protocol should also be valid within the defense, and vice versa.

Reward distribution policy. All blocks in the main chain and no other block receive block rewards.

Eventual consensus. Even when an attack happens, old and new clients should reach a consensus on the main chain eventually. We will further discuss this notion in Sects. 5 and 6.

3   Our Defense Mechanism

3.1   Threat Model

We follow the threat model of most selfish mining studies [6,10,12,18,21,22]. In this model, there is only one colluding pool of selfish miners. This is considered to be the strongest form of attack because malicious miners can achieve higher input-output ratio by acting together. For brevity we use "the selfish miner" or "the attacker" instead of "the colluding pool of miners". The other miners follow the honest mining strategy. The goal of the attacker is to maximize the expected relative revenue. The selfish miner controls less than half of the total mining power so that it is infeasible for the attacker to simply generate a longer chain to invalidate the work of all the other miners.

In terms of network connectivity, we assume the selfish miner receives and broadcasts blocks with no propagation delay. However, the attacker does not have the power to downgrade the propagation speed of blocks found by other miners. Moreover, we assume there is an upper bound on honest miners' block propagation time among each other. We believe this assumption is realistic for the following reasons. First, the Bitcoin developers and the community have investigated a substantial effort to defend against network attackers and to decrease the block propagation time. This can be seen from the number of dedicated Bitcoin Improvement Proposals [5] and the fast response time to new attacks [1]. Second, miners endeavor to receive new blocks as quickly as possible, because a few more seconds of delay may render the mining effort unprofitable. The majority of large miners utilize Corallo's relay network to send and receive blocks faster [8], which consists of eight well-connected Internet traffic hubs around the globe. Other methods are observed by researchers, such as using multiple gateways for more reliable connection to the public Bitcoin network [14]. Third, Miller et al. discovered some very well-connected nodes that can increase network propagation speed [14]. In early September 2016, 50% of publicly reachable Bitcoin nodes receive a new block within 10 s of its creation [2].

At last, although transaction fees are supposed to substitute mining rewards in the long run, we do not consider them in our model because the amount is still quite small comparing with block rewards at this moment.

3.2   Mining Algorithm and Fork-Resolving Policy

We use τ to denote the upper bound on the block propagation time. In reality we expect miners to have a rough consensus on its value. The value can be computed either by a deterministic mechanism like block difficulty adjustment or from a miner's local information similar to network adjusted time [3].

In line with [6,10,12,21,22], we believe nodes should broadcast all competing valid blocks, instead of just the first one they receive, as in the current implementation.

Definition 1. *From a miner's perspective, a valid block is considered **in time** if (1) its height value is bigger than the miner's local chain head or (2) its height is the same as the local chain head and it is received no later than τ after receiving the first block of this height. Conversely, a valid block is **late** if the receiving miner has received a block of the same height τ before receiving this block.*

Definition 2. *A block B_1 is considered to be the **uncle** of another block B_2 if B_1 is a competing in time block of B_2's parent block.*

Notably, our definition of an uncle has two differences with the better-known uncle definition of Ethereum [4], the cryptocurrency with the second largest market capitalization: (1) the uncle has to be in time; (2) the height of a block B's uncle must be one less than the height of B.

Mining Algorithm Modification. A miner should embed in the working block the hashes of all its uncles.

We now introduce our fork-resolving policy *weighted FRP*. Since two competing chains always have a common prefix, our weight calculation only considers the last part of the chains, i.e., excluding the common prefix.

Definition 3. *From a miner's perspective, the **weight** of a chain is the number of its in time blocks plus the number of in time uncle hashes embedded in these in time blocks. Whether a block is in time is evaluated from the miner's local perspective.*

Figure 1 illustrates two different choices of the selfish miner in the same mining sequence. In the left graph, both chains have the same weight three. Although the honest miners have only two blocks, the second block contains the hash of its uncle S because S is published in time. While in the right graph, both chains are weighted two, because the selfish miner does not publish S in time.

Weighted FRP. In a block race,

1. if one chain is longer than the others by no less than k blocks, a miner mines on the longest chain;

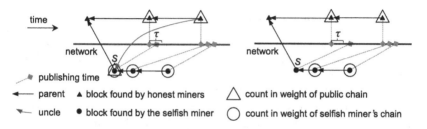

Fig. 1. Two choices of the selfish miner in the same mining sequence. In each graph, blocks are mined from left to right. A blue square indicates the time its connecting block is published. Uncle relation is represented by a red curve with an arrow. No matter whether the selfish miner publishes the first secret block in time or not, the two chains have the same weight. (Color figure online)

2. otherwise the miner chooses the chain with the largest weight;
3. if the largest weight is achieved by multiple chains simultaneously, the miner chooses one among them randomly.

A fail-safe parameter k is introduced here. When $k = 1$, our defense is reduced to a tie breaking defense: honest miners mine on the heavier chain during a tie. When $k = \infty$, the first rule in weighted FRP never applies. The implication of this parameter will be discussed further in Sect. 5.2.

It can be seen from Fig. 1 that weighted FRP puts the selfish miner in a dilemma. When a competitor of the first secret block S is released, the selfish miner has two options: if publishing S, it will be an uncle of the next honest block; if not, it will be considered a late block by honest miners. In neither way could S contribute only to the weight of the selfish chain. Furthermore, because the second selfish block is mined before the first honest block, it is impossible for the former to embed the hash of this uncle. Hence the latter block is guaranteed to contribute only to the honest chain. As a result, our defense lowers the selfish miner's incentive to withhold a discovered block. This defense is fully decentralized. As for backward compatibility, keeping the current block validity rules and reward distribution policy allows a smooth transition; non-miners who cannot upgrade their clients can also be tolerated. Miners and most publicly reachable network participants need to upgrade in order to implement our defense. Next we will show that we are the closest defense to achieving incentive compatibility to date.

4   Evaluation

Sapirshtein et al. developed an algorithm to convert a mining model into an undiscounted average reward Markov decision process (MDP), which makes it possible to compute the optimal selfish mining policy and its relative revenue [18]. In this part we first formally model the mining process of a selfish miner within our defense, then present the results output by the MDP solver, in the end compare our results with uniform tie breaking and a variant of freshness preferred.

4.1 Modeling a Block Race

We consider only the simple case of one honest miner and one selfish miner. This is based on the assumption that τ is carefully calibrated so that all honest miners have the same view on whether a block is in time. We do not consider an attacker who publishes blocks right before they are late to create inconsistent views here and discuss this attack later in Sect. 6. Mining proceeds in steps. In each step, the selfish miner first makes a decision on how many secret blocks to publish. Then the publicly visible weight of the selfish chain is updated and both miners start mining. The honest miner follows weighted FRP and compares the length and weight of both chains before starting, whereas the selfish miner always works on the selfish chain before giving up. A new block is then mined with the probability α by the selfish miner, and with $1 - \alpha$, the honest miner. The honest miner publishes the new block immediately and updates the weight, whereas the selfish miner always decides whether to publish the new block at the beginning of the next step. At the end of each step, if the block race is concluded or partially concluded, the block rewards are allocated. The rationale behind this publish-mine-found-reward sequence is that rational decisions may only change when a new block is available [6,18].

Actions. The selfish miner has five possible rational actions in our model.

Adopt. Give up the selfish chain and mine on the honest chain. This action is always available.

Override. Publish enough blocks so that the published selfish chain is k blocks longer than the honest chain or the selfish chain's public weight is heavier than the honest chain's. The honest miner would start mining on the published selfish chain. Feasible when the selfish miner has enough blocks.

Match. If the weight difference is 0, the selfish miner keeps mining without publishing anything; otherwise publishing enough blocks so that two published chains are of the same weight. If both chains are non-empty, half of the honest mining power, namely $(1 - \alpha)/2$, would mine on the published selfish chain, while the other half works on the honest chain. Feasible when the selfish miner has enough blocks except the following scenario: there is a secret block that contains a hash of an honest uncle; if published, the selfish chain's public weight would exceed the honest chain's weight by one, otherwise the former would be smaller than the latter by one.

Even. Publish enough blocks so that the published selfish chain is no shorter than the honest chain but the selfish chain's public weight is smaller than that of the honest chain. Feasible when the selfish chain is no shorter than the honest chain, but publishing until the chains are the same length does not result in a *match*.

Hide. Do not publish anything new so that the next honest block will not embed an uncle hash. Feasible when the published selfish chain is strictly shorter than the honest chain.

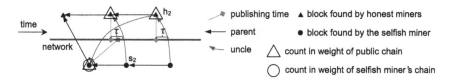

Fig. 2. A block race. Selfish block s_2 is not published in time after its competitor h_2 is mined, thus becomes a late block.

These five actions are adequate because they cover all combinations of selfish miner's chain selection and reasonable public weight that might affect the honest miner's chain selection. All other actions either result in the same weight calculation and chain selection or are obviously irrational. Notably, our definitions of the actions are different from those of [18]. A *wait* action in their model can either be *match, even* or *hide* in our model. Moreover, sometimes a *match* or *even* operation do not lead to the publication of any block.

State Space. A state is represented as a 6-tuple (B_h, B_s, $Diff_w$, *luck, last, published*). B_h and B_s denote the total length of the honest and selfish chain, respectively. $Diff_w = W_h - W_s$ is the weight difference between two chains. The Boolean value *luck* indicates whether there is a secret non-late block with an honest uncle. We refer to this block as *the lucky block*. There can be at most one lucky block because if there are two, the uncle with larger height would force the first lucky block to be published or convert it into a late block. There are two possible values of *last*: h or s, indicating the miner of the block mined in the last step. Finally, *published* denotes the number of published selfish blocks.

In our notation, the state in Fig. 2 is described as (2, 3, 2, 1, s, 1). The lucky block is the last selfish block because s_2 is already late. In this state, a *hide* action publishes no more block; choosing *even* publishes s_2, the resulting temporary state before mining is (2, 3, 2, 1, s, 2); publishing the entire chain would be *match*, the resulting state before mining is (2, 3, 0, 0, s, 3). The *luck* value is updated to 0 because the lucky block is no longer secret.

Reward Allocation and State Transition. The reward of each step is a 2-tuple (R_h, R_s), indicating the number of block rewards for the honest and the selfish miner, respectively. The honest miner gets $R_h = B_h$ only in *adopt*. In two situations the selfish miner gets rewards: *override* and at the end of *match* if the honest miner finds a block on the published selfish chain. In both situations, all selfish blocks that are published before and in this step goes to R_s.

Values in the transition matrix are assigned straight-forwardly according to weighted FRP and the action definitions. We only highlight two tricky details here. First, the selfish miner compares which rule in weighted FRP requires less blocks published before *override*. Second, the state tuple indicates which selfish block is the lucky block: if the latest block is honest, the lucky block is the competitor of this block; otherwise the last honest block is an uncle of the lucky block.

Solving for the Optimal Policy. We compute the optimal policy and its relative revenue with the goal of maximizing the objective function

$$\frac{\sum R_s}{\sum R_s + \sum R_h} .$$

Due to the limitation of computational resources, we set the truncating threshold of B_h and B_s to 13. Once this number is reached, the selfish miner can only choose between *adopt* and *override*. We believe this only affects the results marginally, because most block races end long before reaching this threshold. For $\alpha \leq 0.4$, lowering the threshold to 12 alters the result less than 2×10^{-4}; for $\alpha = 0.48$, lowering the threshold to 12 changes the result no more than 10^{-3}.

4.2 The Optimal Selfish Mining Strategy and Its Relative Revenue

The left part of Fig. 3 displays the optimal strategy within our defense when $k = 3$, $\alpha = 0.48$. All the states of the strategy and the transitions among them are shown when the first four blocks are mined in a block race. Each circle represents a state. The string above the horizontal line in a circle describes the mining history resulted in this state. For example, "SH" means that so far two blocks are mined, the first by the selfish miner, the second by the honest miner. An apostrophe after H means the published selfish and honest chains are of the same weight, and the honest miner mined a new block on the selfish chain. The optimal action is written below the horizontal line in each state. The transition probabilities are omitted. The resulted reward is listed next to the transition arrow if it is not zero. A black dot indicates a temporary state which deteriorates into one or several other states.

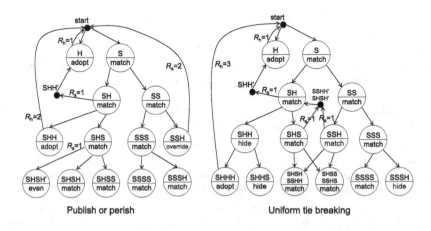

Fig. 3. The optimal selfish mining strategy when $\alpha = 0.48$.

For comparison, we listed the optimal selfish mining strategy within uniform tie breaking for $\alpha = 0.48$ next to our result. The action names are changed to match our terminology. In three colored mining sequences, uniform tie breaking encourages more risky behavior for the selfish miner. Specifically, if the honest miner mines two blocks on the honest chain after a selfish block is found, a rational selfish miner would be forced to *adopt* in our scheme whereas in uniform tie breaking, the miner tries to catch up from behind. This is because in our scheme, if the selfish miner keeps working on the selfish chain, the next selfish block would be late and an extra selfish block is required to catch up the honest chain's weight. However in uniform tie breaking, only one selfish block is enough. Similarly, if the honest miner finds one block after the selfish miner finds two blocks, a rational selfish miner chooses *override* in our scheme and claim the block rewards instead of risking them.

For a complete picture on the performance, we computed the relative revenue for α between 0.20 and 0.45 with interval 0.05 within our defense, plus a powerful selfish miner with $\alpha = 0.48$. Four different k values are chosen: $1, 2, 3$ and ∞.

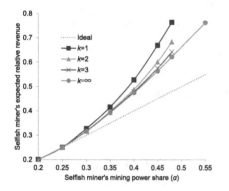

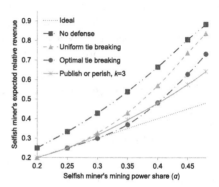

Fig. 4. Relative revenue of the selfish miner within our defense

Fig. 5. Comparison with other defenses

The results can be found in Fig. 4. In all four settings, the profit threshold, minimum α to gain unfair rewards, is larger than 0.25. The relative revenue for $\alpha = 0.48$ is 0.764, 0.684, 0.642, 0.622 when $k = 1, 2, 3$ and ∞, respectively. The effectiveness of our defense increases as k grows.

An interesting result is that when $k = \infty$, our defense can prevent a malicious miner with more than 50% of mining power from taking over the network. In Fig. 4, the selfish miner with 55% of mining power only earns 76.3% of block rewards. When the blockchain integrity is more important than partition recovery time, this variant of our protocol can be useful.

4.3 Comparison with Other Defenses

The optimal selfish mining strategy with no defense is used as the base line for the comparison. Two other defenses are implemented besides our own. The first is uniform tie breaking. The second is an imaginary defense called *optimal tie breaking*, in which the selfish miner loses every tie. This defense can be considered as the strongest form of freshness preferred, in which timestamps are issued with unlimited granularity. For our own defense we choose $k = 3$. In all defenses the selfish miner follows the optimal strategy with truncating threshold 13 to facilitate the comparison. We do not consider the defenses due to Bahack [6], Schultz [21], Solat and Potop-Butucaru [22] because the authors provided no guideline on choosing the parameters or evaluating the performance.

The results are displayed in Fig. 5. The relative revenue for uniform tie breaking and optimal tie breaking when $\alpha = 0.48$ is 0.837 and 0.731, respectively. The numbers become 0.891 and 0.831 if we set the truncating threshold to 160. The difference is because block races with a resourceful attacker usually last for dozens of blocks in these defenses. Neither defense has any effect for $\alpha \geq 0.5$. Our defense has the best performance for all α values except when $\alpha = 0.3$ and 0.35. The performance of our defense can be boosted by including a trusted timestamp server or using local time to identify potential selfish miner's blocks, however we gave up these ideas to maintain the decentralized nature of Bitcoin and avoid opening new attack vectors such as the time jacking attack [9]. Moreover, optimal tie breaking is just imaginary.

5   Balancing Partition Recovery Time and Selfish Mining Resistance

5.1   The Dilemma

If the Bitcoin network is partitioned for at least a moderate amount of time, say several hours, every part of the network would continue to mine new blocks and maintain its own version of the blockchain during the partitioned period. Upon reunion, a malicious miner may strategically mine and publish blocks to keep the network partitioned, in order to lower the quality of service, or to perform double-spending attacks. Deploying such an attack is relatively difficult in Bitcoin, since honest miners would converge to the same history as soon as one chain becomes longer than the others.

Unfortunately, Bitcoin's high tolerance of network partition and fast recovery time is one of the main causes of the selfish miner's profitability. Bitcoin allows the longer chain to claim all the block rewards after the reunion. Withholding blocks receives no punishment because of the indistinguishability between selfish mining and network partition. As a result, in the default Bitcoin protocol, there is no mechanism incentivizing the selfish miner to publish blocks as long as the selfish chain is longer.

Confronted with the dilemma, weighted FRP addresses the selfish mining problem at the price of sacrificing partition recovery time. In our scheme, previously separated groups of miners would consider blocks mined by the other groups during the partitioned period late and only recognize the weight of their own blocks. Consequently, every group works on its own chain until the first rule in weighted FRP applies.

This problem is not fundamental for the following reasons. First, large scale network partition is relatively easy to detect and hard to forge. Therefore it is possible for the protocol to switch to a different fork-resolving policy when such an incident happens. Second, the partition attack is not deemed a substantial threat in the cryptocurrency community. The GHOST protocol [23], whose variant is adopted by Ethereum, is also vulnerable to this attack. In both length FRP and weighted FRP, a newly released malicious block must be on top of its chain in order to affect honest miners' decisions, whereas in GHOST, honest miners may switch to a different chain when a secret uncle is released. Many popular cryptocurrencies, including Stellar [13] and Ripple [20], do not tolerate network partition at all. Third, our defense can achieve a good balance between effectiveness and partition recovery time. Next we explain how our protocol allows a designer to fine-tune between these two goals.

5.2   A Tradeoff

We demonstrate here the resistance of our scheme towards a network partition attack in a setting highly favorable to the attacker. In this setting, the malicious miner controls α fraction of mining power, while the honest network was partitioned into two parts with the same mining power, $(1 - \alpha)/2$ each. The two versions of main chains have the same length upon reunion. The partition lasts long enough so that the second rule of weighted FRP persistently results in different choices on the main chain. We use the Monte Carlo method to compute how long it takes on average for the two chains to converge by the first rule of weighted FRP.

At the beginning of each simulation, the lengths of the two chains are both 0. A block is generated in each round according to the mining power on both sides. The malicious miner always works on the shorter chain. The goal of the attacker is to keep the honest mining power segregated as long as possible. The simulation terminates when one chain is longer than the other by k blocks. For every k, the simulation is repeated 10^5 times and the average numbers of blocks generated on both sides before convergence are recorded.

The results of our simulation are shown in Table 1. The number of blocks can be roughly converted to time, if we assume on average a block is generated every 10 min. For example, when $k = 2$, a malicious miner with 30% of mining power can keep the network segregated for less than one hour after reunion. It can be seen that although a larger k value means better effectiveness of our defense, the

Table 1. Average number of blocks mined before two parts of the network converge

$k \setminus \alpha$	0	0.1	0.2	0.3	0.4	0.5
2	4.00	4.45	4.99	5.69	6.67	7.98
3	12.97	15.35	18.48	23.02	29.90	40.96
4	28.96	36.27	47.08	65.03	95.12	152.14

partition recovery time also grows exponentially. However, this simulation only reflects the worst case. If the partition period is short, honest miners are more likely to converge according to the second rule of weighted FRP; if the partition period is long, there is a large probability that one chain will be longer than the other by at least k blocks shortly after reunion.

We believe when $k = 3$, a good balance is achieved: as can be seen from Fig. 4, the effectiveness of our defense is close to optimal; the partition recovery time is no more than a few hours. For a higher-level protocol that has a contingency plan against network partition, a few hours of recovery time should not cause extra problems.

6   Limitations and Future Work

We acknowledge our scheme's limitation as it is designed and evaluated in a specific threat model, while the reality could be far more complicated.

First, Bitcoin is designed with the goal of functioning in an asynchronous network, yet we designed our defense within the synchronous model by assuming an upper bound of block propagation time. Designing secure protocols in an asynchronous network is a known hard problem. The only research that proves the security properties of Bitcoin [11] is conducted within the synchronous model. Several other selfish mining defenses also require a fixed upper bound on the block propagation time in order to be effective [12,21,22]. Recognizing this possible divergence between this wide-used assumption and the reality, we are unable to solve this conundrum and remove it from our model.

Second, when the fail-safe parameter $k > 1$, an attacker may broadcast blocks right before they are late to cause inconsistent views among honest miners. This is a common problem for all protocols in the synchronous model [13]. One solution is to ask miners of, e.g., 100 predecessor blocks to broadcast signatures that a block is in time, and the successor block's miner to incorporate them in the block to justify the choice of parent block. Another solution is to establish an open trust network similar to Stellar [13] to reach consensus among miners on controversial blocks. The network can be silent when no attack happens.

Third, although eventual consensus is achieved, an attacker may still utilize the temporary inconsistency between weighted FRP and old clients who follow length FRP to launch double-spending attacks. Therefore non-miners who are susceptible to this attack should also upgrade their clients. For example, a merchant running SPV client should receive and verify information about uncle blocks in order to calculate the weight value. However, to deploy this attack, the attacker needs to create a competing chain longer than the public chain and discard it, which costs at least two block rewards, around \$14,000 in early September 2016.

Fourth, our model of the mining process neglects some real-world factors. Our model does not permit the occurrence of natural forks, neither did we consider the influence of transaction fees on the selfish miner's strategy or how multiple selfish miners colluding and competing with each other.

At last we note that our protocol does not achieve incentive compatibility, though it is the closest scheme to date. Achieving incentive compatibility is not an impossible task [19], we hope to reach this point in the near future.

7 Conclusion

The selfish mining attack is a fundamental challenge faced by Bitcoin, for it not only breaks the fairness assumption in the original analysis by the Bitcoin designer, but also posts potential threats to the decentralized structure of Bitcoin. Existing backward-compatible defenses can only deal with equal-length block race, but are powerless when facing a selfish chain longer than the public one. In this study, we proposed a decentralized backward-compatible defense by replacing the current length with our weighted FRP. It can defend against selfish mining when the selfish miner's private chain is longer than the honest miner's chain. Under our weighted FRP, miners evaluate which chain to mine on based on the number of blocks that are published in time and the knowledge of competing blocks that are published in time. As a result, the selfish miner's chain would have disadvantages either because the blocks are not published in time or the lack of knowledge of competing blocks. Our defense outperforms existing defenses under the optimal selfish mining strategy. Additionally, we observed that the selfish mining attack is made possible in Bitcoin as a result of Bitcoin's high tolerance towards network partition. Reflecting upon this dilemma, our scheme introduced a fail-safe parameter k. Through adjusting k, we achieved a balance between effectiveness and partition recovery time.

In contrast to existing defenses, our work attempts to defend against selfish mining through revising the fork-resolving policy rather than the reward distribution policy. This direction promises the advantage of backward-compatibility. We believe it contributes to the discussion on defending against selfish mining as it generates an alternative approach and therefore more possibilities. Our study also contributes to an in-depth understanding on the origins of selfish mining attack, namely Bitcoin's high partition tolerance. By highlighting this dilemma, we hope to raise the awareness on the trade-off between service availability and security, and therefore to open discussions on a series of choices in front of us in designing and improving Bitcoin and other blockchain technologies.

We also acknowledge the limitations of our scheme. Particularly, our defense is still only a mitigation solution towards selfish mining. A truly incentive compatible proof-of-work mining protocol is yet to be discovered.

Acknowledgements. This work was supported in part by the Research Council KU Leuven: C16/15/058. In addition, this work was supported by the imec High Impact initiative Distributed Trust project on Blockchain and Smart contracts. The authors would like to thank Yonatan Sompolinsky for pointing out several potential attacks on an earlier version of the protocol. We would also like to thank Kaiyu Shao, Güneş Acar, Alan Szepieniec, Danny De Cock, Michael Herrmann and the anonymous reviewers for their valuable comments and suggestions.

References

1. Bitcoin core version 0.10.1 released. https://bitcoin.org/en/release/v0.10.1
2. Bitcoin stats - data propagation. http://bitcoinstats.com/network/propagation/
3. Block timestamp. https://en.bitcoin.it/wiki/Block_timestamp
4. Ethereum white paper: Modified Ghost implementation. https://github.com/ethereum/wiki/wiki/White-Paper#modified-ghost-implementation
5. Bitcoin improvement proposals (2016). https://github.com/bitcoin/bips/blob/master/README.mediawiki
6. Bahack, L.: Theoretical Bitcoin attacks with less than half of the computational power (draft). arXiv preprint arxiv:1312.7013 (2013)
7. Btchris, Bytecoin: Mtgox, RHorning: Mining cartel attack (2010). https://bitcointalk.org/index.php?topic=2227
8. Corallo, M.: Bitcoin relay network. http://bitcoinrelaynetwork.org/
9. culubas: Timejacking & bitcoin (2011). http://culubas.blogspot.be/2011/05/timejacking-bitcoin_802.html
10. Eyal, I., Sirer, E.G.: Majority is not enough: bitcoin mining is vulnerable. In: Christin, N., Safavi-Naini, R. (eds.) FC 2014. LNCS, vol. 8437, pp. 436–454. Springer, Berlin (2014). doi:10.1007/978-3-662-45472-5_28
11. Garay, J., Kiayias, A., Leonardos, N.: The bitcoin backbone protocol: analysis and applications. In: Oswald, E., Fischlin, M. (eds.) EUROCRYPT 2015. LNCS, vol. 9057, pp. 281–310. Springer, Heidelberg (2015). doi:10.1007/978-3-662-46803-6_10
12. Heilman, E.: One weird trick to stop selfish miners: Fresh Bitcoins, a solution for the honest miner. Cryptology ePrint Archive, Report 2014/007 (2014). https://eprint.iacr.org/2014/007
13. Mazieres, D.: The stellar consensus protocol: A federated model for internet-level consensus. Stellar Development Foundation (2015)
14. Miller, A., Litton, J., Pachulski, A., Gupta, N., Levin, D., Spring, N., Bhattacharjee, B.: Discovering bitcoins public topology and influential nodes (2015)
15. Nakamoto, S.: Bitcoin: a peer-to-peer electronic cash system (2008). http://www.bitcoin.org/bitcoin.pdf
16. Narayanan, A., Bonneau, J., Felten, E., Miller, A., Goldfeder, S.: Bitcoin and cryptocurrency technologies. Princeton University Press (2016)
17. Nayak, K., Kumar, S., Miller, A., Shi, E.: Stubborn mining: generalizing selfish mining and combining with an eclipse attack. In: IEEE European Symposium on Security and Privacy (EuroS&P), pp. 305–320. IEEE (2016)
18. Sapirshtein, A., Sompolinsky, Y., Zohar, A.: Optimal selfish mining strategies in Bitcoin. Financial Cryptography and Data Security (2016)
19. Schrijvers, O., Bonneau, J., Boneh, D., Roughgarden, T.: Incentive Compatibility of Bitcoin Mining Pool Reward Functions. In: Financial Cryptography and Data Security (2016)
20. Schwartz, D., Youngs, N., Britto, A.: The Ripple protocol consensus algorithm. Ripple Labs White Paper (2014)
21. Shultz, B.L.: Certification of witness: Mitigating blockchain fork attacks (2015). http://bshultz.com/paper/Shultz_Thesis.pdf
22. Solat, S., Potop-Butucaru, M.: Zeroblock: Preventing selfish mining in bitcoin. arXiv preprint arXiv:1605.02435 (2016)
23. Sompolinsky, Y., Zohar, A.: Secure high-rate transaction processing in Bitcoin. Financial Cryptography and Data Security (2015)

WEM: A New Family of White-Box Block Ciphers Based on the Even-Mansour Construction

Jihoon Cho[1], Kyu Young Choi[1], Itai Dinur[2]([✉]), Orr Dunkelman[3],
Nathan Keller[4], Dukjae Moon[1], and Aviya Veidberg[4]

[1] Security Research Group, Samsung SDS, Inc., Seoul, Republic of Korea
[2] Computer Science Department, Ben-Gurion University, Beersheba, Israel
dinuri@cs.bgu.ac.il
[3] Computer Science Department, University of Haifa, Haifa, Israel
[4] Department of Mathematics, Bar-Ilan University, Ramat Gan, Israel

Abstract. White-box cryptosystems aim at providing security against an adversary that has access to the encryption process. As a countermeasure against code lifting (in which the adversary simply distributes the code of the cipher), recent white-box schemes aim for 'incompressibility', meaning that any useful representation of the secret key material is memory-consuming.

In this paper we introduce a new family of white-box block ciphers relying on incompressible permutations and the classical Even-Mansour construction. Our ciphers allow achieving tradeoffs between encryption speed and white-box security that were not obtained by previous designs. In particular, we present a cipher with reasonably strong space hardness of 2^{15} bytes, that runs at less than 100 cycles per byte.

1 Introduction

The white-box threat model in cryptography, introduced by Chow et al. [6] in 2002, assumes that the adversary is accessible to the entire information on the encryption process, and can even change parts of it at will. The initial scenario-in-mind behind the model was the Digital Rights Management (DRM) realm, where an authorized user, who of course has full access to the encryption process, may be adversarial. The model has gained more relevance in recent years due to additional applications, such as mitigation of mass surveillance.

Numerous primitives claiming for security at the white-box model (in short: white-box primitives) were proposed in the last few years. These primitives can be roughly divided into two classes.

O. Dunkelman—The fourth author was supported in part by the Israeli Science Foundation through grant No. 827/12 and by the Commission of the European Communities through the Horizon 2020 program under project number 645622 PQCRYPTO.
N. Keller—The fifth author was supported by the Alon Fellowship.

© Springer International Publishing AG 2017
H. Handschuh (Ed.): CT-RSA 2017, LNCS 10159, pp. 293–308, 2017.
DOI: 10.1007/978-3-319-52153-4_17

The first class includes algorithms which take an existing block cipher (usually AES or DES), and use various methods to 'obfuscate' the encryption process, so that a white-box adversary will not be able to extract the secret key. Pioneered by Chow et al. [6], this approach was followed by quite a few designers. The more common way to fortify the encryption process is using large tables and *random encodings*, as proposed in [6]. Unfortunately, most of these designs were broken by practical attacks a short time after their presentation (see [1,16,19]), and the remaining ones are very recent and have not been subjected to extensive cryptanalytic efforts yet. Another disadvantage of the designs in this class is their performance – all of them are orders of magnitude slower than the 'black-box' primitives they are based upon.

The second class includes new block ciphers designed with white-box protection in mind. Usually such designs are based on *key-dependent components* (e.g., S-boxes), designed in such a way that even if a white-box adversary can recover the full dictionary of such a component, he still cannot use this knowledge to recover the secret key. Recent designs of this class include the ASASA family [3], the SPACE family [5], and the WhiteKey and WhiteBlock ciphers [13]. An important advantage of these designs is their better performance and higher security (though, some instantiations of ASASA were broken, see [14,17]).

A common property of the new white-box designs is *incompressibility* [8] (also called *weak white-box security* [3] and *space hardness* [5]), meaning that an adversary with access to the white-box implementation cannot produce a functionally equivalent program of significantly smaller size. This property is needed, as a white-box adversary can perform *code lifting*, i.e., extract the entire code and use it as an equivalent secret key. While incompressibility does not make code lifting impossible, it does make it harder to implement in practice, especially when the adversary wants to attack multiple targets, e.g., for mass surveillance purposes. The previous designs SPACE and WhiteBlock achieved incompressibility by using *key-dependent pseudo-random functions*.

In this paper we propose a new family of white-box block ciphers in which the basic S-box component is a *pseudo-random permutation*, rather than a pseudo-random function. The new ciphers are based on iterates of the classical Even-Mansour construction [11], in which instead of each key addition one applies an S-box layer, where the S-boxes are key-dependent incompressible permutations. The size of the incompressible S-box is flexible, and can be adjusted to the desired level of incompressibility, without slowing up the encryption process significantly. While the new family proposes similar security level as the SPACE and WhiteBlock ciphers, we show that it allows for additional tradeoffs between performance and white-box security level that were not achievable in previous designs. In particular, we achieve encryption speed of less than 100 cycles per byte with a reasonably strong space hardness of 2^{15} bytes.

This paper is organized as follows. In Sect. 2 we present the WEM family of white-box block ciphers and explain the rationale behind its design. In Sect. 3 we analyze the security of the new ciphers in the black-box model. In Sect. 4 we analyze the security of the new ciphers in the white-box model and compare them with the SPACE and WhiteBlock ciphers. We conclude the paper in Sect. 5.

2   A New Family of White-Box Block Ciphers Based on Incompressible Permutations

In this section we present WEM – a new family of white-box block ciphers based on iterates of the classical Even-Mansour construction [11] and on a key-less variant of a given block cipher. In order to be specific, we present the scheme with AES as the basic block cipher, but any other iterated block cipher can be used instead.

We begin this section with a brief recap of the Even-Mansour construction. Then we present the new family of block ciphers, and finally we explain the rationale behind its design.

2.1   The Even-Mansour Construction

The Even-Mansour (EM) construction was designed by Even and Mansour [11] in 1991, as an attempt to design the 'simplest possible' block cipher based upon a single public permutation. It uses a publicly-known permutation $P : \{0,1\}^n \rightarrow \{0,1\}^n$, and two independent n-bit keys K_0, K_1. The encryption function is defined simply as $EM_{K_0,K_1}(X) = K_1 \oplus P(K_0 \oplus X)$, for $X \in \{0,1\}^n$. Even and Mansour [11] showed that any attack on EM that requires D queries to the entire scheme and T queries to the permutation P must satisfy $DT = \Omega(2^n)$. On the other hand, attacks on the scheme were presented by Daemen [7], Biryukov and Wagner [4], and Dunkelman et al. [10] who showed that the lower bound of [11] is tight by devising a known-plaintext attack that requires D queries to EM and T queries to P, for any (D, T) such that $DT = \Omega(2^n)$.

As a security level of $2^{n/2}$ is considered insufficient for an n-bit block cipher, several authors proposed to enhance the security level by considering *iterates* of the EM construction. For $r \geq 1$, the r-round EM scheme is defined as

$$rEM_{K_0,K_1,\ldots,K_r}(X) = K_r \oplus P_r(K_{r-1} \oplus P_{r-1}(\cdots (P_1(K_0 \oplus X)))),$$

where $P_1, P_2, \ldots, P_r : \{0,1\}^n \rightarrow \{0,1\}^n$ are public permutations, and $K_0, K_1, \ldots, K_r$ are independent n-bit keys. The iterated EM scheme was studied in numerous papers, and multiple upper and lower bounds on its security level were obtained (see, e.g., [9]). The analysis conducted so far indicates that even for small values of r, the security level of the scheme is high. In particular, for the single-key variant in which $K_0 = K_1 = \ldots = K_r$, no attack faster than $2^n/n$ is known even for 2EM (i.e., iterated EM with 2 rounds).

2.2   The New Family of Block Ciphers

The new family of block ciphers, WEM (standing for white-box Even-Mansour), is based on an iterated EM construction, in which the key additions are replaced by layers of incompressible key-dependent S-boxes. In order to allow flexibility, the scheme uses several parameters: n denotes the block size of the cipher, m denotes the size of the incompressible S-box, where $m|n$ is required. r denotes the number of rounds in the underlying iterated EM construction, E denotes the 'name' of the underlying block cipher (e.g., AES), and d denotes the number of rounds we take in its key-less version.

The Overall Structure of the Cipher. The $\text{WEM}(n, m, r, E, d)$ encryption scheme is a modification of the r-round EM scheme, in which:

- A d-round reduced variant of E with the all-zero key is used as the 'public permutation' P. (The same permutation can be used in all rounds of WEM.)
- Each key addition is replaced by an S-box layer, which consists of parallel application of n/m incompressible m-to-m bit S-boxes. For this, we generate $(r+1)n/m$ independent incompressible S-boxes[1] $S_1, S_2, \ldots, S_{(r+1)n/m}$ and use S-boxes $S_{(i-1)n/m+1}, \ldots, S_{in/m}$ in the i'th S-box layer. (The generation of the S-boxes is presented below.)

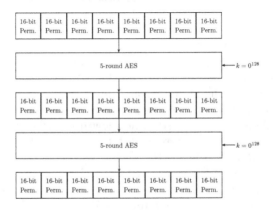

Fig. 1. The cipher WEM(128,16,2,AES-128,5)

A specific instantiation of the scheme, with $n = 128$, $m = 16$, $r = 2$, $E =$ AES-128, and $d = 5$, is presented in Fig. 1. As can be seen in the figure, the 128-bit plaintext is divided into eight 16-bit values. These values enter the first S-box layer. The outputs of the S-box layer are treated as a 128-bit state, to which 5-round AES with the zero key is applied.[2] Then, the value is split again and enters another S-box layer. The outputs of the S-box layer are again unified and processed with 5-round AES with the zero key, and the resulting values are passed through a final S-box layer.

Due to the choice of parameters, the cipher presented in the figure has 24 S-boxes and each encryption has time complexity roughly equivalent to a single AES encryption plus 3 sequences of 8 parallel table lookups.

---

[1] We may also reuse S-boxes to obtain greater flexibility, as noted below.

[2] We note that instead, a per-domain fixed key can be used, e.g., each country gets its own key, or even a per-user key. However, we assume this key to be publicly known.

The Structure of the S-Box. Since the S-box can be isolated by a white-box adversary, it must be a stand-alone primitive that ensures n-bit security to the key even against an attacker that has the full S-box code-book in his disposal.

To obtain this goal, we instantiate the S-box using a two-step procedure. First, we use the secret key to generate a long sequence of pseudo-random bits, and then we use the Fisher-Yates shuffle algorithm [12] to instantiate an S-box from m bits to m bits from the pseudo-random sequence. We note that similar methods were used to generate a pseudo-random function in the SPACE family [5] and in WhiteBlock [13].

The Fisher-Yates algorithm gets an array a of 2^m entries of m bits each, and outputs the designed S-box (where the S-box value on input i is simply $a[i]$). It has the following simple structure:

for $i = 0 \ldots 2^m - 1$ **do**
   $a[i] \leftarrow i$
end for
for $i = 2^m - 1 \ldots 0$ **do**
   $j \leftarrow$ random integer modulo i
   exchange $a[j]$ and $a[i]$
end for

The Fisher-Yates shuffle was shown in [12] to provide perfect randomness: when instantiated with a truly random sequence of bits, it generates a truly random permutation over the range $0, 1, \ldots, 2^m - 1$, meaning that each permutation $\sigma \in S_{2^n}$ is obtained with probability $1/(2^n)!$.

The pseudo-random sequence is generated using the block cipher E (keyed with the master key) in counter mode. For example, in the case $E =$ AES-128, we set the key of AES-CTR as our 128-bit secret master key and generate pseudo-random numbers by encrypting 128-bit plaintexts $0, 1, \ldots$ (as many numbers as required). Thus, the value of the encrypted plaintext functions as the state of the pseudo-random generator, and is incremented as more pseudo-random numbers are required.

Our construction requires generating several such S-boxes (e.g., $8 \cdot 3 = 24$ S-boxes in the above example), and the only difference between the generation of these S-boxes is in the state of the generator (the value of the plaintext encrypted). This value is initialized to 0 for the first S-box and incremented as long as pseudo-random numbers are required (namely, the state is preserved across the initializations of the different S-boxes).

2.3 Design Rationale

The design aims at achieving good performance, while at the same time providing strong security both in the black-box and white-box models (with an appropriate choice for the number of rounds).

1. Good performance and strong security in the black-box model are achieved by using the iterated EM construction as the basis of the cipher. The numerous

works published so far on iterated EM suggest that even with only two rounds, the security level of the scheme (in the black-box model) is close to 2^n, and of course, the scheme becomes even stronger when simple key addition is replaced by a secret S-box layer. Furthermore, by taking a round-reduced variant of E as the public permutation of 2EM we obtain good performance without sacrificing security, as a round-reduced variant of the cipher with sufficiently many rounds already provides sufficient randomness (even if it is not secure as a stand-alone cipher).

2. The white-box security is obtained by making it very hard for an adversary to extract the master key, even if the full code-book of all S-boxes is known. (Note that the user gets the S-boxes in the form of look-up tables, so that even a white-box adversary does not have access to the generation process of the S-box.) Hence, the generation of the S-box must be 'very secure'. On the other hand, as the S-boxes are generated only rarely, we can opt for security in their generation, allowing some performance overhead. Our S-box generation satisfies the required security criterion: while an adversary that knows the full code-book can reverse the Fisher-Yates process and find the pseudo-random string that was used in the S-box generation, this only gives him knowledge of a few plaintext/ciphertext pairs of AES-CTR. Those cannot be used to recover the secret key, unless AES-CTR is insecure.

3. A main idea behind the design is to base it upon thoroughly-analyzed components, in order to gain confidence in its security. This can be seen in the previous two points, where the security of WEM is 'reduced' to the security of iterated EM (though, not in a provable manner) and AES-CTR.

4. The S-box generation process also ensures incompressibility. Indeed, recall that the Fisher-Yates shuffle provably generates a random permutation if the initial sequence is random. Hence, if an S-box (given in a form of a lookup table) has a compressed representation, this representation can be used to distinguish the pseudo-random initial sequence from a truly random sequence, or in other words, to provide a distinguisher for AES-CTR.

5. Given a desired level of incompressibility, one can choose the parameter m appropriately to obtain it. More flexibility can be obtained by allowing re-use of S-boxes. For example, one may use a single S-box for the full scheme, but then the public permutations must be made slightly different (e.g., by using another fixed key instead of the zero key) in order to avoid slide attacks.

6. As noted above, one of the main differences between WEM and the SPACE [5] and WhiteBlock [13] families is that we use secret permutation S-boxes (rather than secret pseudo-random function S-boxes) in our iterated Even-Mansour scheme.

2.4   Performance

We implemented the cipher WEM(128,16,12,AES-128,5) (which is our main instance for white-box security) using AES rounds which are based on tables (i.e., without using the AES-NI instruction set), thus offering a relatively portable code which offers decent performance figures on 32-bit platforms.

The code was compiled under g++ 4.8.4 and was run on an Intel(R) Core(TM) i7-5500U CPU @ 2.40 GHz (after being compiled using the -O2 flag). The running speed we obtained for this basic code was 96.8 cycles per byte.

Compared to WEM(128,16,12,AES-128,5), the related WhiteBlock instance HOUND with 16-bit S-boxes requires a bit more than 140 cycles per byte. However, we note that the comparison is not completely fair. First, the authors of WhiteBlock use a different platform, and in particular, they use AES-NI. On the other hand, the memory consumption of HOUND with 16-bit S-boxes is about 4 times larger than WEM(128,16,12,AES-128,5), as it uses an expanding S-box.

3   Security in the Black-Box Model

In this section we present security analysis of the WEM family in the black-box model. Our conclusion is that *two* rounds of the scheme are sufficient for providing strong security, and in particular, WEM(128,16,2,AES-128,5) is expected to provide 128-bit security, basing on previous extensive analysis of its components. On the other hand, we show that *one* round of the scheme is not sufficient, by devising an attack with complexity slightly higher than $2^{n/2}$ for an n-bit block size. Due to space constraints, the more technical parts of the analysis are presented in the full version of this paper.

In order to be specific, we assume throughout the analysis that the underlying block cipher is AES-128, and focus on the variants WEM(128,8,2,AES-128,5) and WEM(128,16,2,AES-128,5) described above. When WEM is used with another underlying block cipher instead of AES-128, a separate security analysis should be conducted. For brevity, we abbreviate WEM(128,8,2,AES-128,5) and WEM(128,16,2,AES-128,5) to WEM-8 and WEM-16, respectively.

As justified in Sect. 2.2, for the sake of black-box analysis we may view the secret S-boxes of our construction as random permutations. The security of WEM-8 and WEM-16 is related to the security of several previously studied constructions:

1. 2-round Iterated Even-Mansour construction [9],
2. Standard AES with 128-bit key,
3. AES with secret S-boxes [18],
4. 10-round AES with random S-boxes,
5. Known-key round-reduced AES,

and results on these five constructions can be used to obtain evaluation of the security of WEM-8 and WEM-16, as described below.

When considering round-reduced variants of WEM-8 and WEM-16, we note that any such round-reduced variant employs a final secret S-box layer. For most of the attacks described below, we claim that our round-reduced construction is at least as strong as a previously studied round-reduced construction, thus establishing confidence in the security of our design. For sake of convenience, in this section we count the rounds in units of AES rounds, so WEM-8 and

WEM-16 have 10 rounds each. The following is a brief assessment of the security with respect to various attack techniques.

Key Recovery Attacks in General. The secret S-box layers at the beginning and the end of the encryption make round-reduced variants of WEM-8 and WEM-16 significantly stronger than corresponding variants of AES with respect to key-recovery attacks. This is due to the fact that an adversary cannot 'peel off' the first/last rounds without guessing a very significant amount of key material. In this respect, the security of WEM-8 and WEM-16 can be derived from the security of AES with secret S-boxes, studied in [18]. The best currently known attack on this version of AES is on 6 rounds [18], with time complexity of 2^{96}, and it translates to an attack on 5-round WEM-8/WEM-16 with the same complexity (due to the additional MixColumns and secret S-box layers at the end of the cipher). This is clearly far from endangering our 10-round construction. (It should be noted however that there is no direct reduction from WEM-8 or WEM-16 to AES with secret S-boxes, since in WEM, only three layers of S-boxes are secret and not all of them).

Differential and Linear Characteristics. The analysis here is somewhat technical, and thus, is presented in the full version of this paper. The conclusion is that for WEM-8, it is expected that any 4-round differential has probability of less than 2^{-90}, and any 4-round linear hull has a bias of less than 2^{-45}. For WEM-16, we prove that the number of active S-boxes in any 4-round characteristic is at least 15 (and this is tight), and expect that any 4-round differential has probability of less than 2^{-75} and any 4-round linear hull has a bias of less than $2^{-37.5}$. As in addition, for 4 rounds of our construction that contain a secret S-box layer, the actual best differential characteristics and linear approximations are unknown to the adversary, we conclude that the full 10-round WEM-8 and WEM-16 are expected to be immune to both differential and linear cryptanalysis.

Boomerang Attacks. The boomerang attack of Biryukov [2] on round-reduced AES (with at most 6 rounds) can be adapted to WEM-8 and WEM-16, with the same probability as in AES. However, the key recovery part of the attack becomes significantly more expensive, and thus, even on 6-round WEM-8/WEM-16 its complexity is expected to be extremely high.

Square Attacks. The classical Square attacks on round-reduced AES are applicable to WEM-8/WEM-16 with at most 5 rounds, but their complexity becomes much higher. Actually, this is the class of attacks considered in [18] (on AES with secret S-boxes), and the best current attack of this class requires 2^{96} time for 5-round WEM-8, and a similar amount for 5-round WEM-16.

Impossible Differentials. Similarly to the previous cases, the classical impossible differentials apply to our construction but with a significantly more expensive key recovery phase for the full attacks. Therefore, we do not expect these attacks to break more rounds of our construction compared to AES (where the best attack requires more than 2^{100} data and time for 7 rounds).

Collision Attacks (Demirci-Selçuk attacks). It is expected that reduced WEM-8 and WEM-16 are much stronger than reduced AES with respect to these attacks, due to the secret S-box layer in the middle (which increases significantly the number of possible multisets) and the outer secret S-box layers that make key recovery more expensive. As the best known collision attack on round-reduced AES requires 2^{98} data and time for 7 rounds, we expect that WEM-8 and WEM-16 with at least 7 rounds are secure with respect to collision attacks.

Attacks on the EM Construction. The added S-box layer in the middle makes the cipher a 2-round EM construction (rather than the relatively weak 1-round EM construction). The best currently known attacks on 2-round EM [9] are only slightly faster than exhaustive key search. Furthermore, WEM-8 and WEM-16 are stronger than 2-round EM, since the key-additions of EM are replaced in WEM-8/WEM-16 with secret S-box layers, which make all current attacks on 2-round EM inapplicable to WEM-8/WEM-16.

Related-Key Attacks. WEM-8 and WEM-16 are expected to be immune to related-key attacks, due to the key generation procedure. Indeed, as no related-key properties are known for *full AES-128*, it is expected that two related keys (even with relation chosen by the adversary) lead to two unrelated output streams of AES-CTR, and thus, the sets of secret S-boxes generated for the two keys do not have any easy-to-exploit relation. Therefore, it is expected that no related-key attacks on WEM-8/WEM-16 can target more rounds that the single-key attacks (i.e., not more than 7 of the 10 rounds).

WEM(128,8,1,AES-128,10) Does Not Supply 128-Bit Security

The schemes WEM-8 and WEM-16 considered above are 'minimal', in the sense that if the underlying iterated EM construction of WEM has only one round, then the security level of the scheme is much weaker than 2^{128}. This is similar to the situation with iterated EM schemes, where the security level of 1-round EM is only $2^{n/2}$ while with $r \geq 2$ rounds the security increases significantly (so that no attack faster than $2^{128}/128$ is known).

To show this, we present a structural attack on 1-round WEM, which is a variant of the chosen plaintext attack on the Even-Mansour scheme by Daemen [7]. In the attack, we consider pools of 2^m chosen plaintexts that assume all possible values in the input of one S-box, and the same value in the input to all other S-boxes. This property is, of course, preserved by the first secret S-box layer. Then, we look at the corresponding ciphertexts, and in each S-box output of the final S-box layer, we count the number of values that occur 0 times, the number of values that occur 1 time, etc. As this property is also preserved by the S-box layer, it allows us to use comparison between 1-round WEM and key-less AES to recover the secret S-boxes. The full attack is presented in the full version of this paper. The complexity of the attack is only slightly higher than $2^{n/2}$, thus showing that 1-round WEM is not secure and should not be used.

4 Space-Hardness of the WEM Ciphers

The notion of *space-hardness* was introduced in [5] as a generalization of the notion of *weak white-box security* introduced in [3].

Definition 1. *A cipher is said to be (M, Z)-space hard if it infeasible for an adversary to encrypt (decrypt) a randomly chosen plaintext with probability more that 2^{-Z} given code (table) size less than M.*

There are several motivations behind this definition. One of them is that a space-hard cipher makes it more difficult for a DRM attacker in the white-box setting to distribute meaningful attack code (whose size is large). Additionally, a space-hard cipher may make it more challenging for malware (limited by communication) to leak meaningful secrets from an infected network.

In this section, we compute the number of rounds required for the WEM ciphers to achieve space-hardness, and compare the space-hardness security and performance of WEM to that of the schemes proposed in [5,13]. For sake of simplicity and for comparison with previous work [5,13], we only consider in this section instantiations of our schemes with a single secret S-box. We note that a more formal treatment of space-hardness was published in [13] by Fouque et al., using the notion of *weak incompressibility* (formulated as a cryptographic game with the aim of obtaining provable security). Since our motivation is more practical, we will refer to the less formal space-hardness definition of [5]. Hence, our security analysis will be cryptanalytic in nature (focusing on or the best algorithm for breaking our scheme rather than on provable security).[3]

Nevertheless, we point out two issues mentioned in [13] that are relevant for our space-hardness security analysis. First, for an n-bit block cipher, given T words of memory, one cannot hope for space-hardness security better than $Z = n - \log(T)$. The reason is that the memory can simply be utilized to store plaintext/ciphertext pairs, allowing the adversary to correctly encrypt (or decrypt) a fraction at least $2^{\log(T)-n}$ of the code-book. Even when we restrict the adversary's memory to contain entries of the secret S-box in our scheme, it is still possible to store the particular entries that are accessed in the encryption procedure of about T plaintexts (up to some multiplicative factor which depends on the number of times the S-box is accessed in an encryption). A second issue is that our analysis will indeed assume that the adversary's memory contains only secret S-box entries. While we are not aware of significantly better attacks that store other types of information, these attacks are generally much harder to analyze.[4]

---

[3] Interestingly, it is shown in [13] that for certain types of schemes, the gap between the number of rounds required to resist the best known attack and the the number of rounds required to obtain provable security is not large.

[4] Resistance to such attacks is addressed by the *strong incompressibility* definition of [13], which also gives a scheme (called WHITEKEY) that provably achieves this security notion. However, WHITEKEY is a key generator rather than a block cipher, and hence is incomparable to our scheme.

4.1   Previous Space-Hard Block Ciphers

There are two previous space-hard block cipher designs. The first one is SPACE, introduced in [5]. SPACE is a 128-bit generalized Feistel structure with a secret expanding S-box of input size m (where m is a parameter of the block cipher instance) and output size $128 - m$.[5]

The second space-hard block cipher design is WhiteBlock, introduced in [13]. The general structure of WhiteBlock is more similar to our scheme. It is a 128-bit block cipher family, where each round contains a secret S-box layer followed by several rounds of standard AES (an AES layer). There are several differences between our scheme and WhiteBlock. The most relevant one in terms of space-hardness is the structure of the secret S-box layer. In WhiteBlock, the secret S-box layer is a single-round Feistel-like structure. For a secret S-box of input size m bits (where m is a parameter of the block cipher instance) and $k = \lfloor 64/m \rfloor$, the 128-bit state is partitioned into two parts, where the 'right part' contains km bits and the 'left part' contains the remaining $128 - km$ bits. The output size of each S-box is $128 - km$ bits (equal to the size of the left part), hence it is an expanding S-box. The secret S-box layer applies k parallel S-boxes to the km bits of the right part of the state and XORs their outputs (in some arbitrary order) to the right part (hence the km bits of the right part are left unchanged).

WhiteBlock has two variants: PUPPYCIPHER, which was designed with provable security in mind, and HOUND which is optimized for performance. The difference between the two schemes is in the AES layer (but not in the secret S-box layer). As our main goal is to resist cryptanalysis while optimizing performance, our scheme is more comparable to HOUND. We note that it should be possible to tweak our AES layer and apply similar provable security arguments to our scheme as in PUPPYCIPHER, but this is out of the scope of this paper.

4.2   Space-Hardness of Our Scheme

We evaluate the space-hardness of our proposal and show that it can be achieved using less rounds than the previous schemes of [5,13] (thus, resulting in a faster cipher with the same level of white-box security). We start by analyzing our scheme WEM(128,16,r,AES-128,5) and assume that the adversary obtained a fraction of $1/4$ of the S-box entries (the value $1/4$ is chosen to be comparable to the analysis of [5,13]). We then generalize the analysis.

We consider r rounds of our scheme and determine the minimal value of r such that it achieves $(T/4, 112)$-space hardness, where T is the size of the 16-bit S-box in 16-bit words (and we aim for the maximal achievable security of 112 bits for a 128-bit cipher with a 16-bit S-box). Our analysis is related to the one of [13] for WhiteBlock, although less formal. The goal is to show that given an arbitrary set of (only) $1/4$ of the S-box entries, the adversary cannot guess the encryption of any plaintext with probability which is significantly higher than 2^{-128}.

---

[5] For the sake of convenience, we rename the block cipher instance parameters for both previous space-hard designs [5,13].

Note that the set of S-box entries is chosen by the adversary and is not arbitrary (in particular, it can correlate with the encryption/decryption procedure of several plaintexts/ciphertexts). However, roughly speaking, a set of $2^{16}/4$ of the S-box entries can be chosen to reveal information about the encryption of (no more than) 2^{16} plaintexts, whereas for the rest of the plaintexts the analysis below will apply.[6]

The encryption procedure of WEM(128,16,r,AES-128,5) contains $8r$ S-boxes of 16 bits. Therefore, an adversary can encrypt a random plaintext if he is given the corresponding $8r$ S-box entries, which occurs with probability $2^{-2 \cdot 8r}$ (assuming that the known S-box entries are arbitrary). Hence, taking $r = 9$ such that $2^{-2 \cdot 8r} < 2^{-128}$ should prevent the adversary from correctly encrypting any of the 2^{128} plaintexts. However, the adversary can still miss the entries of several S-boxes and succeed in encrypting the plaintext with probability better than 2^{-128} simply by guessing the S-box outputs. A guess for an S-box output is correct with probability $1/(2^{16} - 2^{14}) < 2^{-15}$ (the adversary has 2^{14} entries of the permutation). Hence, we require that the adversary misses only 8 S-box entries[7] with very low probability (but we do not mind if the adversary misses 9 S-box entries, as $2^{15 \cdot 9} < 2^{-128}$).

Overall, to predict the encryption of a plaintext with probability better than 2^{-128}, the adversary should have $8r - 8$ S-box entries which can occur at $\binom{8r}{8}$ places. Therefore, we require that $2^{-2(8r-8)} \cdot \binom{8r}{8} < 2^{-128}$, which is satisfied for $r \geq 12$.

More generally, for a block cipher with an m-bit S-box and $k = n/m$ S-boxes in a round, we apply a similar line of arguments to analyze the number of rounds required to obtain $(2^{-\alpha} \cdot T, n - \log(T))$-space hardness (where T is the S-box size). If the adversary has a $2^{-\alpha}$ fraction of the 2^m possible S-box entries, then we require $2^{-\alpha \cdot k(r-1)} \cdot \binom{k \cdot r}{k} < 2^{-k \cdot m}$ (namely, the adversary misses only k S-box inputs with very low probability). Since $\binom{k \cdot r}{k} < (k \cdot r)^k$, it is sufficient to require

$$-\alpha \cdot k(r - 1) + k \log(k) + k \log(r) < -k \cdot m.$$

Dividing by αk, we get $-r + 1 + \log(r)/\alpha + \log(k)/\alpha < -m/\alpha$ or

$$r - \log(r)/\alpha > m/\alpha + \log(k)/\alpha + 1.$$

In other words, the required number of rounds r is larger than m/α by an additive logarithmic factor.

---

[6] We note that in terms of provable security, it was shown in [13] for WhiteBlock (and similar arguments can be applied to our scheme) that the analysis for an arbitrary set of S-box entries should give a close estimation to the number of rounds required to achieve the desired security level of 112 bits.

[7] We point out that the adversary can reduce the number of guesses in case of common missed S-boxes entries. We do not expect this to give the adversary a significant advantage, as the adversary can only miss a small number of S-box entries in the encryption which are likely to be distinct. Nevertheless, this is a shortcoming of our analysis (which is also present in the analysis of [13]).

Next, we compare our scheme to the previous proposals of [5,13]. For the sake of simplicity, we focus on S-box sizes m which divide the block size n.[8]

Comparison to WhiteBlock [13]. According to the provable security analysis of [13], for an S-box size of m bits and $\alpha = 2$ (namely, assuming that the adversary has $1/4$ of the code) WhiteBlock should have $r = m + 2$ rounds (assuming that m divides the block size n). Therefore, for $\alpha = 2$, our scheme requires half the number of rounds, up to an additive logarithmic factor. However, this comparison is not completely fair since it was obtained using different analysis methods (even though they are related). Hence, we redo our analysis for WhiteBlock, and show that it gives similar results as the related analysis [13].

As in Sect. 4.1, we denote the number of S-boxes in a round of WhiteBlock by k, giving $km = 64$, namely each S-box maps m bits to 64 bits. Similarly to the previous section, we require that the adversary cannot guess the encryption of any plaintext with probability which is significantly higher than 2^{-128} given an arbitrary set of the S-box entries of size $2^{m-\alpha}$. The encryption procedure of r rounds contains kr S-boxes of m bits, and since the output of each S-box is 64 bits, we require that for all plaintexts, the adversary misses at least one S-box entry in at least 2 different rounds. As $2^{-64 \cdot 2} \leq 2^{-128}$, this should suffice to prevent the adversary from predicting the output of an encryption. Note that we require that the missed entries occur in distinct rounds, since even if the adversary misses several S-box entries in a single round, he can directly guess the 64-bit output of the left part of the secret S-box layer (rather than guessing the output of each S-box separately).

To predict the encryption of a plaintext with probability better than 2^{-128}, the adversary should have all the $k(r-1)$ S-box entries which can occur in $r-1$ rounds, and there are $\binom{r}{r-1} = r$ options to choose this round. Overall, we require that $2^{-\alpha k(r-1)} \cdot r < 2^{-2km}$ or $-\alpha k(r-1) + \log(r) < -2ms$. Rearranging, we obtain

$$r - \log(r)/(\alpha k) > 2m/\alpha + 1.$$

In other words, the required number of rounds r is about $2m/\alpha$ and is twice the number of rounds required by our scheme up to additive logarithmic factors. This may seem obvious since the S-boxes of WhiteBlock encrypt only half of the state in each S-box layer, whereas they cover the full state in our scheme. However, this simplistic argument does not take into account the fact that each S-box of WhiteBlock is expanding and hence in order to predict the encryption of a plaintext with good probability, the adversary is allowed to miss less S-box entries (while guessing their values) in WhiteBlock compared to our scheme. Our analysis shows that the use of half as many expanding S-boxes in WhiteBlock compared to our scheme increases the number of rounds required to achieve the same space-hardness security, and thus, generally leads to slower encryption speed. Nevertheless, if one seeks to minimize the number of secret S-box look-ups

---

[8] It is also possible to instantiate our scheme for values of m that do not divide n, as briefly discussed in Sect. 4.4.

in the encryption process, then WhiteBlock is superior to our scheme (which has more table look-ups, but evaluated in parallel).

Comparison to SPACE [5]. Unlike the case of WhiteBlock whose structure is similar to WEM (both using interleaved applications of a secret S-box layer and an AES layer), the SPACE family differs from WEM significantly, having a generalized Feistel structure. Of course, we can directly compare performance figures, but SPACE was designed with a large security margin and hence, is expectedly much slower. Thus, comparing design strategies will be more interesting.

If we ignore the fact that there are no AES layers in SPACE and redo the security analysis presented above, we get that the SPACE design strategy requires the smallest number of table lookups to achieve space-hardness, but the largest number of secret S-box layers. This is a direct continuation of the trend we previously observed: as we reduce the number of S-boxes applied in a single round, we can use S-boxes with larger output sizes, and thus we need fewer secret table look-ups in the cipher to achieve space-hardness. On the other hand, we still need more secret S-box layers since the reduction in the number of S-boxes is not sufficient to reduce the number of rounds.

4.3   Space-Hardness Using Permutation S-Boxes

While previous space-hard designs were built using randomly chosen S-boxes, our scheme was built using permutation S-boxes. This has some impact on the space hardness of our scheme, as a permutation on m-bit words can be represented using less memory compared to a random function mapping m-bit words to m-bit words. However, the difference is only by a small multiplicative factor of about $1 - 1.44/m$, since by Stirling's approximation, $\log((2^m)!) > m \cdot 2^m - \log(e)2^m \approx (1 - 1.44/m)(m \cdot 2^m)$. For example, representing a 16-bit random function requires $16 \cdot 2^{16}$ bits, while a 16-bit random permutation requires about $16 \cdot 2^{16} - (1.44/16)(16 \cdot 2^{16}) = 14.56 \cdot 2^{16}$ bits.[9]

4.4   Concrete Instances

Our main instance uses 16-bit S-boxes. It has 12 rounds and is claimed to have $(1/4 \cdot 14.56 \cdot 2^{16}, 112)$-space hardness or $(2^{14.86}, 112)$-space hardness in bytes.

We note that additional instances can be picked by choosing additional S-boxes sizes (e.g., we can define an instance with a 21-bit S-box, where the S-box layer contains 6 S-boxes and 2 bits are left unchanged), although that requires a slightly more technical security analysis.

---

[9] This factor is even smaller when considering representation of a fraction of the S-box entries.

5   Conclusions

In this paper we presented a new family of white-box block ciphers, called WEM, which combines the iterated Even-Mansour construction with incompressible S-boxes and a round-reduced key-less variant of a 'standard' block cipher (e.g., the AES). The structure of WEM allows obtaining good performance, while basing the security confidence in the black-box model on the extensive analysis of the cipher's components, and the security in the white-box model on the provable randomness of the Fisher-Yates shuffle algorithm.

Our cipher is an SP network, in which the incompressible S-boxes are random permutations. This is in contrast with the previous SPACE and WhiteBlock designs, in which the secret S-boxes are expanding, and the cipher is either a generalized Feistel construction (SPACE) or interleaving of Feistel layers with SPN layers (WhiteBlock). We showed that using an SP network allows reducing the number of rounds in the scheme (for the same space-hardness level), and thus, making the scheme faster if application of S-boxes in parallel is possible. In particular, we present a specific scheme called WEM(128,16,12,AES-128,5) with space hardness of $(2^{14.86}, 112)$ bytes and encryption speed of less than 100 cycles per byte.

References

1. Billet, O., Gilbert, H., Ech-Chatbi, C.: Cryptanalysis of a white box AES implementation. In: Handschuh, H., Hasan, M.A. (eds.) SAC 2004. LNCS, vol. 3357, pp. 227–240. Springer, Berlin (2004). doi:10.1007/978-3-540-30564-4_16

2. Biryukov, A.: The boomerang attack on 5 and 6-round reduced AES. In: Dobbertin, H., Rijmen, V., Sowa, A. (eds.) AES 2004. LNCS, vol. 3373, pp. 11–15. Springer, Berlin (2005). doi:10.1007/11506447_2

3. Biryukov, A., Bouillaguet, C., Khovratovich, D.: Cryptographic schemes based on the ASASA structure: black-box, white-box, and public-key (extended abstract). In: Sarkar, P., Iwata, T. (eds.) ASIACRYPT 2014. LNCS, vol. 8873, pp. 63–84. Springer, Berlin (2014). doi:10.1007/978-3-662-45611-8_4

4. Biryukov, A., Wagner, D.: Advanced slide attacks. In: Preneel, B. (ed.) EUROCRYPT 2000. LNCS, vol. 1807, pp. 589–606. Springer, Berlin (2000). doi:10.1007/3-540-45539-6_41

5. Bogdanov, A., Isobe, T.: White-box cryptography revisited: space-hard ciphers. In: Ray, I., Li, N., Kruegel, C. (eds.) Proceedings of the 22nd ACM SIGSAC Conference on Computer and Communications Security, Denver, CO, USA, 12–6 October 2015, pp. 1058–1069. ACM (2015). http://doi.acm.org/10.1145/2810103.2813699

6. Chow, S., Eisen, P., Johnson, H., Oorschot, P.C.: White-box cryptography and an AES implementation. In: Nyberg, K., Heys, H. (eds.) SAC 2002. LNCS, vol. 2595, pp. 250–270. Springer, Berlin (2003). doi:10.1007/3-540-36492-7_17

7. Daemen, J.: Limitations of the Even-Mansour construction. In: Imai, H., Rivest, R.L., Matsumoto, T. (eds.) ASIACRYPT 1991. LNCS, vol. 739, pp. 495–498. Springer, Berlin (1993). doi:10.1007/3-540-57332-1_46

8. Delerablée, C., Lepoint, T., Paillier, P., Rivain, M.: White-box security notions for symmetric encryption schemes. In: Lange, T., Lauter, K., Lisoněk, P. (eds.) SAC 2013. LNCS, vol. 8282, pp. 247–264. Springer, Berlin (2014). doi:10.1007/978-3-662-43414-7_13

9. Dinur, I., Dunkelman, O., Keller, N., Shamir, A.: Key recovery attacks on iterated Even-Mansour encryption schemes. J. Cryptology **29**(4), 697–728 (2016)

10. Dunkelman, O., Keller, N., Shamir, A.: Slidex attacks on the Even-Mansour encryption scheme. J. Cryptology **28**(1), 1–28 (2015)

11. Even, S., Mansour, Y.: A construction of a cipher from a single pseudorandom permutation. J. Cryptology **10**(3), 151–162 (1997)

12. Fisher, R.A., Yates, F.: Statistical Tables for Biological, Agricultural and Medical Research. Oliver and Boyd, London (1938)

13. Fouque, P., Karpman, P., Kirchner, P., Minaud, B.: Efficient and Provable White-Box Primitives. IACR Cryptology ePrint Archive 2016, 642 (2016). http://eprint.iacr.org/2016/642

14. Gilbert, H., Plût, J., Treger, J.: Key-recovery attack on the ASASA cryptosystem with expanding S-boxes. In: Gennaro, R., Robshaw, M. (eds.) CRYPTO 2015. LNCS, vol. 9215, pp. 475–490. Springer, Berlin (2015). doi:10.1007/978-3-662-47989-6_23

15. Lange, T., Lauter, K.E., Lisonek, P.: Selected Areas in Cryptography – SAC 2013. LNCS, vol. 8282. Springer, Berlin (2014). doi:10.1007/978-3-662-43414-7

16. Lepoint, T., Rivain, M., Mulder, Y., Roelse, P., Preneel, B.: Two attacks on a white-box AES implementation. In: Lange, T., Lauter, K., Lisoněk, P. (eds.) SAC 2013. LNCS, vol. 8282, pp. 265–285. Springer, Berlin (2014). doi:10.1007/978-3-662-43414-7_14

17. Minaud, B., Derbez, P., Fouque, P.-A., Karpman, P.: Key-recovery attacks on ASASA. In: Iwata, T., Cheon, J.H. (eds.) ASIACRYPT 2015. LNCS, vol. 9453, pp. 3–27. Springer, Berlin (2015). doi:10.1007/978-3-662-48800-3_1

18. Tiessen, T., Knudsen, L.R., Kölbl, S., Lauridsen, M.M.: Security of the AES with a secret S-box. In: Leander, G. (ed.) FSE 2015. LNCS, vol. 9054, pp. 175–189. Springer, Berlin (2015). doi:10.1007/978-3-662-48116-5_9

19. Wyseur, B., Michiels, W., Gorissen, P., Preneel, B.: Cryptanalysis of white-box DES implementations with arbitrary external encodings. In: Adams, C., Miri, A., Wiener, M. (eds.) SAC 2007. LNCS, vol. 4876, pp. 264–277. Springer, Heidelberg (2007). doi:10.1007/978-3-540-77360-3_17

Improved Key Recovery Algorithms

A Bounded-Space Near-Optimal Key Enumeration Algorithm for Multi-subkey Side-Channel Attacks

Liron David$^{(\boxtimes)}$ and Avishai Wool$^{(\boxtimes)}$

School of Electrical Engineering, Tel Aviv University, 69978 Ramat Aviv, Israel
lirondavid@gmail.com, yash@eng.tau.ac.il

Abstract. Enumeration of cryptographic keys in order of likelihood based on side-channel leakages has a significant importance in cryptanalysis. The best optimal-order key enumeration algorithms have a huge space complexity of $\Omega(n^{d/2})$ when there are d subkeys and n candidate values per subkey. In this paper, we present a parallelizable algorithm that enumerates the keys in near-optimal order but enjoys a much better space complexity of $O(d^2w + dn)$ for a design parameter w which can be tuned to available RAM.

Before presenting our algorithm, we provide lower and upper bounds on the guessing entropy of the full key in terms of the easy-to-compute guessing entropies of the individual subkeys. We use these results to quantify the near-optimality of our algorithm's ranking, and to bound its guessing entropy. Finally, we evaluate our algorithm through extensive simulations, to show the advantages of our new algorithm in practice, on realistic SCA scenarios. We show that our algorithm continues its near-optimal-order enumeration far beyond the rank at which the optimal algorithm fails due to insufficient memory.

1 Introduction

1.1 Background

Side-channel attacks (SCA) represent a serious threat to the security of cryptographic hardware products. As such, they reveal the secret key of a cryptosystem based on leakage information gained from physical implementation of the cryptosystem on different devices. Information provided by sources such as timing [13], power consumption [12], electromagnetic emulation [20], electromagnetic radiation [1,9] and other sources, can be exploited by SCA to break cryptosystems.

Most of the attacks that have been published in the literature are based on a "divide-and-conquer" strategy. In the first "divide" part, the cryptanalyst recovers multi-dimensional information about different parts of the key, usually called subkeys (e.g., each of the $d = 16$ AES key bytes can be a subkey). In the "conquer" part the cryptanalyst combines the information all together in an efficient way. In the attacks we consider in this paper, the information that the SCA provides for each subkey is a probability distribution over the n candidate values for that subkey.

© Springer International Publishing AG 2017
H. Handschuh (Ed.): CT-RSA 2017, LNCS 10159, pp. 311–327, 2017.
DOI: 10.1007/978-3-319-52153-4_18

Much attention has been paid to the "divide" part of side channel analysis, aiming to optimize its performance: Kocher et al.'s Differential Power Analysis (DPA) [12], Brier et al.'s Correlation Power Analysis (CPA) [5] and Chari et al.'s Template Attacks [6] are some examples. In contrast, less attention has been paid to the "conquer" part.

1.2 Related Work

The problem of merging two lists of subkey candidates was encountered by Junod and Vaudenay [11]. The simple approach of merging and sorting the subkeys lists was tractable thanks to the small size of the lists (up to 2^{13}). By decreasing the order of the probabilities, given partial information obtained for each key bit individually, Dichtl [8] considered a faster enumeration of key candidates. A more general and challenging problem is enumerating keys from lists that cannot be merged, exploiting any partial information on subkeys. For this, a probabilistic algorithm was proposed in [15]. In this work the attacker has no access to the subkey distributions but is able to generate subkeys according to them. The proposed solution is to enumerate keys by randomly choosing subkeys according to these distributions. This implementation requires $O(1)$ memory but keys may be chosen many times, leading to useless repetitions.

A deterministic enumeration algorithm was described by Pan et al. [17]. It enumerates key candidates in the optimal order, but large memory requirements prevent the application of this, when the number of keys to enumerate increases.

The currently best optimal algorithm was proposed by Veyrat-Charvillon, Gérard, Renauld and Standaert, [22], which we denote by OKEA. This algorithm significantly improves the time and memory complexity thanks to clever data structures and a recursive decomposition of the problem. However, its worst case space complexity is $\Omega(n^{d/2})$ when d is the number of subkey dimensions and n is the number of candidates per subkey - and the space complexity is $\Omega(r)$ when enumerating up to a key at rank $r \leq n^{d/2}$. Thus its space complexity becomes a bottleneck on real computers with bounded RAM in realistic SCA attacks.

To tackle this problem, two improved key enumeration algorithms were proposed by Bogdanov et al. [4] and Martin et al. [14]. Similar to us, both papers improve upon OKEA [22] by suggesting bounded-memory algorithms.

Bogdanov et al. [4] uses a score-based enumeration, rather than the probability-based enumeration that OKEA and our algorithm use, producing an enumeration that is suboptimal in terms of output order, and can be parallelized. The algorithm of Martin et al. [14] also uses a score-based enumeration, focuses on rank estimation via a reduction to counting knapsack and utilizes it to enumerate the B keys with the highest scores in a parallel manner, for any B. Like [4] they also manipulate the side-channel leakages, but into different weights. Both use additive scoring (the scores of different subkeys are added to score a full key): [4] suggests scores that are scaled-and-truncated probabilities, whereas [14] skirts this issue. This makes it difficult to compare apples with apples: the quality of their order would have been comparable to the optimal (OKEA) order and to our order only if they had used log-probabilities (whose addition is semantically equivalent to multiplication of probabilities). Moreover, with scores, standard metrics such as the Guessing Entropy, which we analyze, cannot be computed, since they require probabilities. Finally, giving our algorithm more memory greatly improves both its order quality and its runtime, whereas their algorithms do not enjoy this benefit.

The most similar work to ours was developed in parallel to our technical report [7] by Poussier et al. [19]. The authors use a very different, histogram-based method, to enumerate the keys in parallelizable sub-optimal order. Like us they also use probabilities (technically, log-probabilities). Using our notation, their algorithm has a $\Omega(d^2 N_b + nd)$ space complexity—when N_b (number of bins) is a design parameter, i.e., the same asymptotic space complexity as our method. However, like both [4,14] Poussier et al. [19] did not provide any analytical bounds on the distance between their rank and the optimum, nor did they analyze the guessing entropy of their algorithm—they only provide empirical evidence based on one dataset.

Ye et al. [24] take a different approach: they limit the key enumeration to a hypercube of the top e candidates for every subkey. Their KSF fails if the true key is outside this hypercube. This is unlike all previously mentioned papers, which always find the correct key if given enough time. In some sense KSF is analogous to the first step of our algorithm: instead of giving up, our algorithm continues to adjacent volumes wrapping the hypercube, and uses the OKEA inside the hypercube and in the adjacent volumes, while maintaining a bound on the memory complexity.

The paper of Poussier et al. [18] is primarily a taxonomy and comparison of rank estimation algorithms, suggesting new algorithmic combinations. It continues the work of Veyrat [23], Bernstein [3], Glowacz [10] and also of Martin et al. [14]. Rank estimation is a closely related, yet different, question, to the key enumeration we address: It doesn't necessarily require to enumerate all the key candidates ranked before the correct key, as it is only necessary to estimate how many there are.

1.3 Contributions

In this paper, we propose a parallelizable key enumeration algorithm, with bounded memory requirement of $O(d^2 w + dn)$ for a design parameter w which can be tuned to available RAM and allows the enumeration of a large number of keys without exceeding the available memory. Our algorithm enumerates in near-optimal order with a bounded ratio between optimal and near-optimal ranks.

Before presenting our algorithm, we utilize the evaluation framework of [21], providing lower and upper bounds on the guessing entropy of the full key in terms of the easy-to-compute guessing entropies of the individual subkeys. We use these results to quantify the near-optimality of our algorithm's ranking, and to bound its guessing entropy.

Finally, we evaluate our algorithm through extensive simulations, to show the advantages of our new algorithm in practice, on realistic SCA scenarios. On our lab environment we found that the optimal algorithm fails due to insufficient memory when attempting to enumerate beyond rank 2^{33}, while our bounded-space algorithm continued its near-optimal-order enumeration unhindered.

Organization: In Sect. 2 we describe the optimal-order key enumeration algorithm of [22]. In Sect. 3 we introduce some bounds on the guessing entropy of the full key based on the guessing entropies of the individual subkeys. In Sect. 4 we introduce our w-layer key enumeration algorithm and analyze its properties. In Sect. 5 we present our performance analysis, and we conclude in Sect. 6.

2   Preliminaries

The key enumeration problem: The cryptanalyst obtains d independent subkey spaces $k_1, ..., k_d$, each of size n, and their corresponding probability distributions $P_{k_1}, ..., P_{k_d}$. The problem is to enumerate the full-key space in decreasing probability order, from the most likely key to the least, when the probability of a full key is defined as the product of its subkey's probabilities.

The best key enumeration algorithm so far, in terms of optimal-order, was presented by Veyrat-Charvillon, Gérard, Renauld and Standaert in [22], which we denote by OKEA. To explain the algorithm, we will use a graphical representation of the key space—the case of $d = 2$ is depicted in Fig. 1. In this figure, we see two subkeys k_1 and k_2 along the axes of the graph, both sorted by decreasing order of probability. The width and the height of the rows and columns correspond to the probability of the corresponding subkey. Let $k_i^{(j)}$ denote the j'th likeliest value for the i'th subkey. Then, the intersection of row j_1 and column j_2 is a rectangle corresponding to the key $(k_1^{(j_1)}, k_2^{(j_2)})$ whose probability is equal to the area of the rectangle.

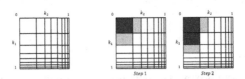

Fig. 1. Left: geometric representation of the key space. Right: geometric representation of the first two steps of key enumeration.

The algorithm outputs the keys in decreasing order of probability. The algorithm maintains a data structure F of candidates to be the next key in the sorted order. In each step the algorithm extracts the most likely candidate from F, $(k_1^{(j_1)}, k_2^{(j_2)})$, and outputs it. F is then updated by inserting the potential successors of this candidate: $(k_1^{(j_1+1)}, k_2^{(j_2)})$ and $(k_1^{(j_1)}, k_2^{(j_2+1)})$. An important observation made by [22] is that F should never include 2 candidates in the same column, or in the same row: one candidate will clearly dominate the other. Thus the algorithm maintains auxiliary data structures ("bit vectors") to indicate which rows and columns currently have a member in F. This observation has a crucial effect on the size of the data structure, $|F|$.

We can see in Fig. 1 the first steps of the algorithm: the most likely key is $(k_1^{(1)}, k_2^{(1)})$, therefore this is the key that is output first (represented in dark gray in step 1). Now, the only possible next key candidates are the successors (represented in light gray in step 1) $(k_1^{(2)}, k_2^{(1)})$ and $(k_1^{(1)}, k_2^{(2)})$, which are inserted into F. Then in step 2, the most likely key is extracted, but this time only one successor is inserted because there is already a key in column 2.

In general, we need to enumerate over more than two lists of subkeys ($d > 2$). For AES, typically $d = 16$ for byte-level side channels or $d = 4$ for 32-bit subkeys as in [16]. To do this, [22] suggested a recursive decomposition of the problem. The algorithm described above is only used for merging two lists, and its outputs are used to form larger subkey lists which are in turn merged together. In order to minimize the storage and the enumeration effort, these lists are generated only as far as required by the key enumeration. Therefore, whenever a new subkey is inserted into the candidate set, its value is obtained by applying the enumeration algorithm to the lower level, (for example 64-bit subkeys obtained by merging two 32-bit subkeys), and so on.

3   Bounding the Guessing Entropy

An important security metric for the evaluation of a side channel attack [21] is the Guessing Entropy, which intuitively corresponds to the average number of keys to test before reaching the correct one, based on the probabilities assigned to key candidates by the side channel attack.

Definition 1 (Guessing Entropy). *For a random variable X with n values, denote the elements of its probability distribution P_X by $P_X(x_i)$ for $x_i \in X$ such that $P_X(x_1) \geq P_X(x_2) \geq ... \geq P_X(x_n)$. The guessing entropy of X is:*

$$G(X) = \sum_{i=1}^{n} i \cdot P_X(x_i).$$

The case $d = 2$: Let the key be split into 2 independent subkey spaces X and Y, each of size n, thus a key is a vector xy s.t. $x \in X$ and $y \in Y$. A side channel attack produces 2 separate distributions $P_X(x_i)$ for $x_i \in X$ and $P_Y(y_j)$ for $y_j \in Y$. Assume that the subkey distributions are sorted: $P_X(x_1) \geq P_X(x_2) \geq ... \geq P_X(x_n)$ and similarly for P_Y, then $G(X)$ and $G(Y)$ are well defined.

Let XY denote the list of (full) keys sorted in decreasing order of probability, where $P_{XY}(x_i, y_j) = P_X(x_i)P_Y(y_j)$ since the subkeys are independent. Thus $G(XY)$ is well defined. However, calculating $G(XY)$ requires a time and space complexity of $\Omega(n^2)$. Therefore bounding $G(XY)$ in terms of the easy-to-compute $G(X)$ and $G(Y)$ is a useful goal. To this end, let $rank(x_i, y_j)$ be the position of key (x_i, y_j) in XY. Clearly, $rank(x_1, y_1) = 1$ and $rank(x_n, y_n) = n^2$. By definition we get:

$$G(XY) = \sum_{i=1}^{n} \sum_{j=1}^{n} rank(x_i, y_j) \cdot P_X(x_i) P_Y(y_j). \tag{1}$$

Theorem 1. *The guessing entropy of XY, $G(XY)$, is bounded by:*

$$G(X)G(Y) \le G(XY) \le n(G(X) + G(Y)) - G(X)G(Y). \tag{2}$$

Proof. Appears in the extended version of this paper [7].

We can see that in general $G(XY)$ is not multiplicative:

Corollary 1. $G(X)G(Y) \le G(XY) \le 2n \cdot \max\big(G(X), G(Y)\big).$

Proof. Appears in the extended version of this paper [7].

These bounds can be expanded for $d > 2$. In this case it holds:

$$\prod_{m=1}^{d} i_m \le rank(x_{i_1}^{(1)}, x_{i_2}^{(2)}, ..., x_{i_d}^{(d)}) \le n^d - \prod_{m=1}^{d} (n - i_m).$$

Therefore we obtain

Theorem 2. *The guessing entropy $G(X^{(1)} X^{(2)} ... X^{(d)})$, is bounded by:*

$$\prod_{m=1}^{d} G(X^{(m)}) \le G(X^{(1)} X^{(2)} ... X^{(d)}) \le n^d - \prod_{m=1}^{d} (n - G(X^{(m)})).$$

As an example of using these bounds, with byte-level SCA on AES we have $d = 16$. If the SCA discards 128 values per byte and returns a probability distribution over the remaining 128 candidates we have $n = 128$. Assuming that $G(X^{(m)}) = 8$ for all 16 subkeys we get that

$$2^{48} = 8^{16} \le G(X^{(1)} X^{(2)} ... X^{(d)}) \le 128^{16} - (128 - 8)^{16} = 2^{111.36}.$$

Reducing the gap between the lower and the upper bounds is left as an open question.

4   The Key Enumeration Algorithm

The key enumeration in [22] enumerates the key candidates in optimal order, but has a significant drawback, its memory requirements may exceed the available memory. Its worst-case space complexity is $\Omega(n^{d/2})$ since it needs to store the full sorted distribution of the 2 top-level dimensions (in addition to the data structure F), for each dimension. Moreover, in order to enumerate until a key of rank $r \leq n^{d/2}$ it has a space complexity of $\Omega(r)$. In this section, we present a new key enumeration algorithm with bounded memory requirements, which therefore allows to enumerate a large number of key candidates.

To achieve the desired memory bound, we relax the "optimal order" requirement: our algorithm enumerates the keys in near-optimal order, and we are able to bound the ratio between the optimal rank of a key and our algorithm's rank of that key.

4.1   The Layering Approach

In order to explain our algorithm, we start with the case of two dimensions, $d = 2$. We divide the key-space $(n \times n)$ into layers of width w, as depicted in Fig. 2. The first layer contains the keys $(k_1^{(i)}, k_2^{(j)})$ such that $(i, j) \in \{1, ..., w\} \times \{1, ..., w\}$. The second layer contains the keys $(k_1^{(i)}, k_2^{(j)})$ such that $(i, j) \in \{1, ..., 2w\} \times \{1, ..., 2w\} \setminus \{1, ..., w\} \times \{1, ..., w\}$ and so on. More formally:

Definition 2. *Given $w > 0$ and $l > 0$, let*

$$layer_l^w = \{(k_1^{(i)}, k_2^{(j)}) | (i, j) \in \{1, ..., l \cdot w\} \times \{1, ..., l \cdot w\} \setminus \{1, ..., (l-1) \cdot w\} \times \{1, ..., (l-1) \cdot w\}\}.$$

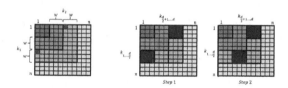

Fig. 2. Left: geometric representation of the key space divided into layers of width $w = 3$. The keys in cells $(1, 7)$ and $(7, 1)$ are the algorithm's seeds for $layer_3^{(3)}$. Right: geometric representation of the key enumeration at $layer_3^3$.

A key observation is that we can run the optimal enumeration algorithm of [22] *within* a layer: we seed the algorithm data structure F by inserting the two "corners" (see Fig. 2), and then extract candidates and insert their successors as usual - limiting ourselves not to exceed the boundaries of the layer. Moreover, within a layer of width w, we can bound the space used by F:

Proposition 1. *For every $l > 0$ and $w > 0$, applying the optimal key enumeration of [22] on $layer_l^w$, the number of next potential key candidates is bounded by $2w$, i.e., $|F| \leq 2w$.*

Proof. Appears in the extended version of this paper [7].

Importantly, the bound on $|F|$ is independent of n, and depends only on the design parameter w which we can tune.

4.2   The Two-Dimensional Algorithm

Proposition 1 leads us to our *w-layer key enumeration* algorithm: Divide the key-space into layers of width w. Then, go over the $layer^w$s, one by one, in increasing order. For each $layer_l^w$, enumerate its key candidates, by applying the optimal key enumeration [22]. Following the proposition, the number of potential next candidates, F, that our algorithm should store is bounded by $2w$.

4.3   Generalization to a Multi-dimensional Algorithm

For $d > 2$, similarly to [22] we apply a recursive decomposition of the problem. Whenever a new subkey is inserted into the candidate set, its value is obtained by applying the enumeration algorithm to the lower level. For example, let's look at $d = 4$. In order to generate the ordered full-key, we need to generate the 2 ordered lists of the lower level $L_{1,2}$ and $L_{3,4}$ on the fly as far as required. For this, we maintain a set of next potential candidates, for each dimension - $F_{1,2}$ and $F_{3,4}$, so that each next subkey candidate we get from $F_{1,2}$ (or $F_{3,4}$) we store at $L_{1,2}$ (or $L_{3,4}$). The length of these generated subkey lists, $L_{1,2}$ and $L_{3,4}$ is $\Omega(n^2)$. For general d, the sizes of the data structures $F_{1,...,d/2}$ and $F_{d/2+1,...,d}$ are bounded by $2w$, however, we still have a bottleneck of $\Omega(n^{d/2})$ because of $L_{1,...,d/2}$ and $L_{d/2+1,...,d}$. Therefore, instead of naively storing the full subkey order of $L_{1,...,d/2}$ and $L_{d/2+1,...,d}$, we only store the $O(w)$ candidates which were computed "recently".

To do this, we divide each $layer^w$ in the geometrical representation, into squares of size $w \times w$, as depicted in Fig. 2 (right side). Our algorithm still enumerates the key candidates in $layer_1^w$ first, then in $layer_2^w$ and so on, but in each $layer_l^w$ the enumeration will be square-by-square.

More specifically, let $S_{x,y}^w$ be a set of the key candidates in the square $S_{x,y}^w = \{(k_{1,...,d/2}^{(i)}, k_{d/2+1,...,d}^{(j)})|(x-1) \cdot w < i \leq x \cdot w$ and $(y-1) \cdot w < j \leq y \cdot w\}$. We say that two squares, $S_{x,y}$ and $S_{z,w}$ are *in the same row* if $y = w$, and are *in the same column* if $x = z$.

This in-layer split into squares reduces the space complexity, since instead of storing the full ordered lists of the lower levels, we store only the relevant subkeys candidates for enumerating the current two squares, i.e., $2w$ subkey candidates for each dimension. However, these subkey candidates which are redundant for enumerating the current squares, might be useful later in the enumeration of the next layer. In that case we will need to recompute them.

Now let's describe the enumeration at each $layer_l^w$. We know that the most likely candidate in $layer_l^w$ is either at $S_{1,l}$ or $S_{l,1}$. Therefore, we enumerate first the key candidates in $S_{1,l} \cup S_{l,1}$ by applying the key enumeration in [22] on them (represented in dark gray in step 1 in Fig. 2). Let S denote the set of squares that contain potential next candidates in this layer. At some point, one of the two squares is completely enumerated. Without loss of generality, we assume this is $S_{1,l}$. At this point, the only square that contains the next key candidates after $S_{1,l}$ is the successor $S_{2,l}$ (represented in dark gray in step 2 of Fig. 2).

Algorithm 1. w-Layer Key Enumeration Algorithm.

Input: Subkey distributions $\{k_i\}_{1\leq i\leq d}$.
Output: The correct key, if exists, NOT-FOUND otherwise.

1   $found = false$; initialize($F_{1,...,d}$);
2   **while** $(F_{1,...,d} \neq \emptyset)$ **do**
3      $candidate$ = nextCandidate($F_{1,...,d}$, $\{k_i\}_{1\leq i\leq d}$);
4      $found$ = isCorrectKey($candidate$);
5   **if** *(found)* **then**
6   return $candidate$;
7   return NOT-FOUND;

In the general case, the successor of $S_{x,y}$ is either $S_{x+1,y}$ or $S_{x,y+1}$, only one of which is in $layer_l^w$. Therefore, when one of the squares is completely enumerated, it is extracted from S, and its successor is inserted, as long as S doesn't contain a square in the same row or column.

Notice that only after a square is completed we continue to its successor. Without loss of generality, we assume that the successor is in the same row as the current one. Therefore, for all candidates $(k_{1,...,d/2}^{(i)}, k_{d/2+1,...,d}^{(j)})$ we intend to check next, the j index is higher than the j index of any candidate in the current square, therefore these j indexes of the current square are redundant and we do not need to store them.

It is simple to see that we maintain at most 2 squares of size $w \times w$ each time, therefore we need to maintain sets of next potential candidates and ordered lists for each square, i.e., $F_{1,...,d/2}^1$, $L_{1,...,d/2}^1$, $F_{d/2+1,...,d}^1$, $L_{d/2+1,...,d}^1$ and $F_{1,...,d/2}^2$, $L_{1,...,d/2}^2$, $F_{d/2+1,...,d}^2$, $L_{d/2+1,...,d}^2$.

4.4   Bounding the Rank and the Guessing Entropy

Let v^w denote the vector resulting from enumerating all key candidates, applying our w-layer key enumeration, for fixed w, and let v denote the vector resulting from applying the optimal order enumeration. Additionally, let $rank^w(i_1, i_2, .., i_d)$ denote the order statistic of key $(k_1^{(i_1)}, k_2^{(i_2)}, ..., k_d^{(i_d)})$ in v^w, and $rank(i_1, i_2, .., i_d)$ be the order statistic of key $(k_1^{(i_1)}, k_2^{(i_2)}, ..., k_d^{(i_d)})$ in v. Now, we want to bound the rank of the w-layer algorithm, and the guessing entropy of v^w, $G(v^w)$, related to $G(v)$.

Theorem 3. *Consider a key* $(k_1^{(i_1)}, ..., k_d^{(i_d)})$. *Let* $i^* = \max\{i_1, ..., i_d\}$, *and let* $\alpha_m = i_m/i^*$ *for* $m = 1, ..., d$ $(\alpha_m \leq 1)$. *Then,*

$$rank^w(i_1, ..., i_d) \leq \prod_{m=1}^{d} \left(\frac{2}{\alpha_m}\right) \cdot rank(i_1, ..., 1_d).$$

Proof. Appears in the extended version of this paper [7].

Algorithm 2. nextCandidate.

Input: $F_{p,...,r}$ and subkey distributions $\{k_i\}_{p \leq i \leq r}$.
Output: The next key candidate in $F_{p,...,r}$.

1 $q \triangleq \lfloor p + r \rfloor / 2;\ x \triangleq \{p, ..., q\};\ y \triangleq \{q + 1, ..., r\}$;
2 $(k_x^{(i)}, k_y^{(j)}) \leftarrow$ most likely candidate in $F_{p,...,r}$;
3 $F_{p,...,r} \leftarrow F_{p,...,r} \setminus \{(k_x^{(i)}, k_y^{(j)})\}$;
4 $I \triangleq \lceil i \rceil / w;\ J \triangleq \lceil j \rceil / w;\ t \triangleq (I \geq J)\ ?\ 1 : 2;\ //\ (k_x^{(i)}, k_y^{(j)})$ is in $S_{I,J}$;
5 **if** $S_{I,J}$ *is completely enumerated* **then**
6   **if** $I == J$ **then**
7   **if** $r - p > 1$ **then**
8   nextCandidate$(F_x^1);\ k_x^{(i+1)} \leftarrow L_x^1[(i+1)\%w]$;
9   nextCandidate$(F_y^2);\ k_y^{(j+1)} \leftarrow L_y^2[(j+1)\%w]$;
10           $F_x^2 \leftarrow k_x^{(1)};\ F_y^1 \leftarrow k_y^{(1)}$;
11           $F_{p,...,r} \leftarrow \{(k_x^{(1)}, k_y^{(j+1)})\} \cup \{(k_x^{(i+1)}, k_y^{(1)})\}$;
12   **else**
13   **if** *no candidates are in same row/column as* $Successor(S_{I,J})$ **then**
14           $F_{p,...,r} \leftarrow F_{p,...,r} \cup \{$most likely candidate in $Successor(S_{I,J})\}$;
15 **else**
16   **if** $(k_x^{(i+1)}, k_y^{(j)}) \in S_{I,J}$ *and no candidate in row* $i{+}1$ **then**
17   **if** $r - p > 1$ **then**
18   **if** $k_x^{(i+1)}$ *is not stored at* L_x^t **then**
19   nextCandidate(F_x^t);
20           $k_x^{(i+1)} \leftarrow L_x^t[(i+1)\%w]$;
21        $F_{p,...,r} \leftarrow F_{p,...,r} \cup \{(k_x^{(i+1)}, k_y^{(j)})\}$;
22   **if** $(k_x^{(i)}, k_y^{(j+1)}) \in S_{I,J}$ *and no candidate in column* $j{+}1$ **then**
23   **if** $r - p > 1$ **then**
24   **if** $I == J$ **then**
25               $k_y^{(j+1)} \leftarrow L_y^2[(j+1)\%w]$
26   **else**
27   **if** $k_y^{(j+1)}$ *is not stored at* L_y^t **then**
28   nextCandidate(F_y^t);
29               $k_y^{(j+1)} \leftarrow L_y^t[(j+1)\%w]$;
30        $F_{p,...,r} \leftarrow F_{p,...,r} \cup \{(k_x^{(i)}, k_y^{(j+1)})\}$;
31 $L[(L.size + 1)\%w] \leftarrow$ most likely candidate in $F_{p,...,r}$;
32 **return** $(k_x^{(i)}, k_y^{(j)})$;

Theorem 4. *The bound of the guessing entropy of* v^w, $G(v^w)$, *related to* $G(v)$ *is:*

$$G(v^w) \leq 2^d n^{d-1} \cdot G(v).$$

Proof. Appears in the extended version of this paper [7].

It is somewhat counter-intuitive that the bound on the approximation factors does not depend on the size of the layer w, while, as we will see in Sect. 5, the experimental analysis suggests a much better (yet w-dependent) behavior. We leave for further work to find w-dependent theoretical bounds.

4.5 Parallelization of w-Layer Algorithm

We parallelize our algorithm by parallelizing the OKEA [22] inside each square. OKEA is an inherently serial algorithm, so by parallelizing it we lose the enumeration order's optimality inside the square. However, our bounds on the rank (Theorem 3) and the guessing entropy (Theorem 4) are independent of the internal enumerating order in each layer. Therefore our parallel algorithm retains the same guaranteed performance.

According to Proposition 1, when enumerating a whole layer, the number of next potential candidates, $|F|$, is bounded by $2w$, and within a single $w \times w$ square we have $|F| \leq w$. Hence, enumerating each square can be parallelized between at most w cores, protecting the access to the structure F with concurrency controls. Each core extracts the most likely candidate from F, and let s_1 be one of its two successors. The algorithm inserts s_1 back into F only if there is no candidates in the same row/column as s_1, and if all the candidates in the same row/column, before s_1, were already enumerated. For this, we need to replace the simple "bit vector" implementation of [22] by a "greatest index vector". This vector stores for any row/column the greatest enumerated index in that row/column.

4.6 Space Complexity Analysis

The algorithm needs to store the candidates of the 2 top-level dimensions, for each dimension. However, it doesn't need to store the whole candidate list, but only two lists (L) of size w for each dimension. For this, it needs to store 2 sets of potential candidates (F) for each dimension, each one of these sets is bounded by $2w$. Moreover, it needs to store 2 data structures ("bit vectors") for each F to indicate which rows and columns currently have a member in it. All together, we get the following recurrence relation for the space complexity:

$$S(d) = 4S(d/2) + cw,$$

for some constant c, which sums to $O(d^2 w)$. Taking into account the input, whose space is $O(dn)$, we get a total space complexity of $O(d^2 w + dn)$.

5 Performance Analysis

We evaluated the performance of our w-layer key enumeration algorithm through an extensive simulation study. We implemented the optimal algorithm [22] and our algorithm in Java, and ran both algorithms on a 3.07 GHz PC with 24 GB RAM running Microsoft windows 7, 64 bit. Note that the code of the optimal algorithm is used as a subroutine in the w-layer algorithm, thus any potential improvement in the former's implementation would automatically translate into an analogous improvement in the latter.

We used synthetic SCA distributions with $d = 8$ subkeys and $n = 2^{12}$ candidates per subkey for a total enumeration space of $O(n^d) = 2^{96}$. We chose $d = 8$ and $n = 2^{12}$ since for a key whose rank is 'deep', the optimal-order algorithm takes space of $\Omega(n^{d/2}) = \Omega(2^{48})$ which exceeds the available memory. We generated the synthtic SCA distributions according to Pareto distributions, with $\alpha = 0.575$ and $\beta = 0.738$. The choice of the Pareto distribution and these specific parameters is based on empirical evidence we discovered, see the extended version of this paper [7]. For the simulations, we chose two different values of w to limit our space complexity $O(d^2w + dn)$. The first one is $w = n$ which gives a linear space complexity of $O(d^2n + dn)$ and the second one is $w = 2^{25}$ which gives an $O(2^{31})$ space complexity which is about 1 Gb.

We also evaluated the algorithm's performance for $d = 16$ subkeys and $n = 2^6$, again for a total enumeration space of 2^{96}. The probability distributions were Pareto distributions with $\alpha = 0.3$ and $\beta = 1.1197$, see the extended version of this paper [7]. We analyzed our w-layer algorithm for two different values of w: $w = n = 2^6$ and $w = 2^{25}$. The obtained results are similar to those with $d = 8$. Graphs are omitted.

We conducted the experiments as follows. We ran the optimal algorithm on different (optimal) ranks starting from 2^{12}, and measured its time and space consumption. For each optimal rank, 2^x, we extracted the key corresponding to this rank, and ran each of our w-layer key enumeration algorithm variants until it reached the same key, and measured its rank, time and space. We repeated this simulation for 64 different ranks near 2^x — the graphs below display the median of the measured values.

Because of time consumption, we decided to stop each w-layer run after 2 h - if it didn't find the given key by then. We marked the timed-out runs.

5.1   Runtime Analysis

Figure 3 illustrates the time (in minutes) of the 3 algorithms: OKEA (optimal-order) (green triangles), w-layer with $w = n = 2^{12}$ (red squares) and w-layer with $w = 2^{25}$ (blue diamonds) for different ranks. The figure shows that, crucially, the optimal-order key enumeration stops at 2^{33}. This is because of high memory consumption which exceeds the available memory. The w-layer key enumeration, in contrast, keeps running.

For ranks beyond 2^{22} we noticed that the w-layer enumeration with $w = n = 2^{12}$ became significantly slower than the others. The red squares ($w = n$) in Fig. 3 are misleadingly low, since as Fig. 4 shows, a large fraction of runs timed-out at the 2 h mark, and we stopped experimenting with this setting beyond rank 2^{32}. It is important to remark that we chose to stop because of the time consumption - the algorithm doesn't stop till it finds the correct key.

For the w-layer algorithm with $w = 2^{25}$, however, we see excellent results. For small ranks it takes exactly the same time consumption as OKEA, (hidden by the green triangles in Fig. 3), and for high ranks, its bounded space complexity enables it to enumerate in reasonable time.

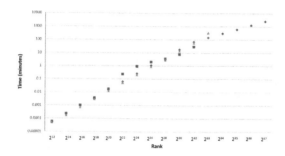

Fig. 3. Median run time, in minutes, of OKEA (green triangles), w-layer key enumeration with $w = 2^{25}$ (blue diamonds) and w-layer key enumeration with $w = n$ (red squares) on different ranks. (Color figure online)

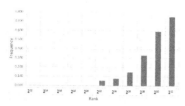

Fig. 4. Frequency of the keys whose time consumption applying the w-layer key enumeration with $w = n$ is higher than 2 h.

Note that for ranks beyond 2^{33}, the optimal algorithm failed to run, so we could not identify the keys with those ranks. In order to demonstrate the w-layer algorithm's ability to continue its enumeration we let it run until it reached a rank r in its own near-optimal order (for $r = 2^{34}, .., 2^{37}$) - and for those experiments we removed the 2 h time out.

We can see that bigger values of w lead to more candidates in each w-layer which leads to less recomputing and therefore a lower running time.

5.2 Space Utilization

Figure 5 illustrates the space (in bytes) used by the 3 algorithms' data structures for different ranks. As we can see again, OKEA stops at 2^{33} because of memory shortage, while the w-layer algorithm keeps running. For the w-layer algorithm with $w = n$ we can clearly see the bounded space consumption leveling at around 1 MB. For the w-layer algorithm with $w = 2^{25}$ we see that its space consumption levels around 4 GB and remains steady, allowing the algorithm to enumerate further into the key space, limited only by the time the cryptanalyst is willing to spend.

5.3 The Difference in Ranks

Figure 6 illustrates the ranks detected by the 3 algorithms as a function of the optimal rank. By definition the optimal algorithm finds the correct ranks. Despite

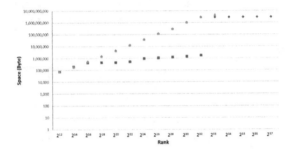

Fig. 5. Median space, counting the data structure elements, of OKEA (green triangles), w-layer key enumeration with $w = 2^{25}$ (blue diamonds) and w-layer key enumeration with $w = n$ (red squares) on different ranks. (Color figure online)

the somewhat pessimistic bounds of Theorem 4, the figure shows that with $w = n$ the ratio between the optimal rank and $rank^w$ is approximately 2.32 (again, for those runs that complete faster than 2 h running time). Beyond 2^{28} too many runs timed out for meaningful data. For $w = 2^{25}$ the discovered ranks are almost identical to the optimal ranks (the symbols in the figure overlap) - and beyond 2^{33} the optimal algorithm failed so comparison is not possible.

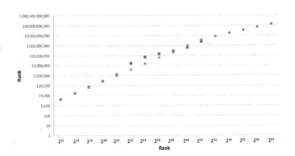

Fig. 6. Median rank of OKEA (green triangles), w-layer key enumeration with $w = 2^{25}$ (blue diamonds) and w-layer key enumeration with $w = n$ (red squares) on different ranks. (Color figure online)

5.4   Influence of w on Space Complexity and Enumeration Accuracy

The trade-off between the space complexity and the accuracy of the enumeration order is summed up in Table 1. As we can see, for $w = 2^{12}$ our enumeration uses space of 1 MB. We see the maximum rank for which 80% of the simulations take less than 2 h is 2^{26}, and up to this rank the rank accuracy is at most 2.32 times the optimal rank. For $w = 2^{25}$ our enumeration uses more space (4 GB), but the maximum rank for which 80% of the simulations take less than 2 h is 2^{33}, and accuracy is at most 1.007 times the optimal rank. As a consequence, we recommend to increase w as much as possible without exceeding the available

Table 1. Influence of w on space complexity and enumeration accuracy

w	Space	Max rank of 80% in 2 h	Accuracy
$w = 2^{12}$	1 MB	2^{26}	$\leq 2.32 \cdot$ OPT
$w = 2^{25}$	4 GB	2^{33}	$\leq 1.007 \cdot$ OPT

memory. This bounds the space complexity, and therefore enables to enumerate more keys, with better accuracy.

6 Conclusion

In this paper, we investigated the side channel attack improvement obtained by adversaries with non-negligible computation power to exploit physical leakage. For this purpose, we presented a new parallelizable w-layer key enumeration algorithm, that trades-off the optimal enumeration order in favor of a bounded memory consumption. We analyzed the algorithm's space complexity, guessing entropy, and rank distribution. We also evaluated its performance by extensive simulations. As our simulations show, our w-layer key enumeration allows stronger attacks than the order-optimal key enumeration [22], whose space complexity grows quickly with the rank of the searched key—and exceeds the available RAM in realistic scenarios. Since our algorithm can be configured to use as much RAM as available (but no more) it can continue its near optimal enumeration unhindered.

Along the way, we also provided bounds on the full key guessing entropy in terms of the guessing entropies of the individual subkeys.

Finally, an open-source Java implementation for both our w-layer key enumeration and the order-optimal enumeration [22] are available via the authors' home pages.

References

1. Agrawal, D., Archambeault, B., Rao, J.R., Rohatgi, P.: The EM side—channel(s). In: Kaliski, B.S., Koç, K., Paar, C. (eds.) CHES 2002. LNCS, vol. 2523, pp. 29–45. Springer, Berlin (2003). doi:10.1007/3-540-36400-5_4
2. Anonymous: Anonymous (2015)
3. Bernstein, D.J., Lange, T., van Vredendaal, C.: Tighter, faster, simpler side-channel security evaluations beyond computing power. Cryptology ePrint Archive, Report 2015/221 (2015). http://eprint.iacr.org/
4. Bogdanov, A., Kizhvatov, I., Manzoor, K., Tischhauser, E., Witteman, M.: Fast and memory-efficient key recovery in side-channel attacks. In: Dunkelman, O., Keliher, L. (eds.) SAC 2015. LNCS, vol. 9566, pp. 310–327. Springer, Cham (2016). doi:10.1007/978-3-319-31301-6_19
5. Brier, E., Clavier, C., Olivier, F.: Correlation power analysis with a leakage model. In: Joye, M., Quisquater, J.-J. (eds.) CHES 2004. LNCS, vol. 3156, pp. 16–29. Springer, Berlin (2004). doi:10.1007/978-3-540-28632-5_2

6. Chari, S., Rao, J.R., Rohatgi, P.: Template attacks. In: Kaliski, B.S., Koç, K., Paar, C. (eds.) CHES 2002. LNCS, vol. 2523, pp. 13–28. Springer, Berlin (2003). doi:10.1007/3-540-36400-5_3

7. David, L., Wool, A.: A bounded-space near-optimal key enumeration algorithm for multi-dimensional side-channel attacks. Cryptology ePrint Archive, Report 2015/1236 (2015). http://eprint.iacr.org/2015/1236

8. Dichtl, M.: A new method of black box power analysis and a fast algorithm for optimal key search. J. Crypt. Eng. 1(4), 255–264 (2011)

9. Gandolfi, K., Mourtel, C., Olivier, F.: Electromagnetic analysis: concrete results. In: Koç, Ç.K., Naccache, D., Paar, C. (eds.) CHES 2001. LNCS, vol. 2162, pp. 251–261. Springer, Berlin (2001). doi:10.1007/3-540-44709-1_21

10. Glowacz, C., Grosso, V., Poussier, R., Schüth, J., Standaert, F.-X.: Simpler and more efficient rank estimation for side-channel security assessment. In: Leander, G. (ed.) FSE 2015. LNCS, vol. 9054, pp. 117–129. Springer, Berlin (2015). doi:10.1007/978-3-662-48116-5_6

11. Junod, P., Vaudenay, S.: Optimal key ranking procedures in a statistical cryptanalysis. In: Johansson, T. (ed.) FSE 2003. LNCS, vol. 2887, pp. 235–246. Springer, Berlin (2003). doi:10.1007/978-3-540-39887-5_18

12. Kocher, P., Jaffe, J., Jun, B.: Differential power analysis. In: Wiener, M. (ed.) CRYPTO 1999. LNCS, vol. 1666, pp. 388–397. Springer, Berlin (1999). doi:10.1007/3-540-48405-1_25

13. Kocher, P.C.: Timing attacks on implementations of Diffie-Hellman, RSA, DSS, and other systems. In: Koblitz, N. (ed.) CRYPTO 1996. LNCS, vol. 1109, pp. 104–113. Springer, Berlin (1996). doi:10.1007/3-540-68697-5_9

14. Martin, D.P., O'Connell, J.F., Oswald, E., Stam, M.: Counting keys in parallel after a side channel attack. In: Iwata, T., Cheon, J.H. (eds.) ASIACRYPT 2015. LNCS, vol. 9453, pp. 313–337. Springer, Berlin (2015). doi:10.1007/978-3-662-48800-3_13

15. Meier, W., Staffelbach, O.: Analysis of pseudo random sequences generated by cellular automata. In: Davies, D.W. (ed.) EUROCRYPT 1991. LNCS, vol. 547, pp. 186–199. Springer, Berlin (1991). doi:10.1007/3-540-46416-6_17

16. Oren, Y., Weisse, O., Wool, A.: A new framework for constraint-based probabilistic template side channel attacks. In: Batina, L., Robshaw, M. (eds.) CHES 2014. LNCS, vol. 8731, pp. 17–34. Springer, Berlin (2014). doi:10.1007/978-3-662-44709-3_2

17. Pan, J., Woudenberg, J.G.J., Hartog, J.I., Witteman, M.F.: Improving DPA by peak distribution analysis. In: Biryukov, A., Gong, G., Stinson, D.R. (eds.) SAC 2010. LNCS, vol. 6544, pp. 241–261. Springer, Berlin (2011). doi:10.1007/978-3-642-19574-7_17

18. Poussier, R., Grosso, V., Standaert, F.-X.: Comparing approaches to rank estimation for side-channel security evaluations. In: Homma, N., Medwed, M. (eds.) CARDIS 2015. LNCS, vol. 9514, pp. 125–142. Springer, Cham (2016). doi:10.1007/978-3-319-31271-2_8

19. Poussier, R., Standaert, F.-X., Grosso, V.: Simple key enumeration (and rank estimation) using histograms: an integrated approach. In: Gierlichs, B., Poschmann, A.Y. (eds.) CHES 2016. LNCS, vol. 9813, pp. 61–81. Springer, Berlin (2016). doi:10.1007/978-3-662-53140-2_4

20. Quisquater, J.-J., Samyde, D.: ElectroMagnetic analysis (EMA): measures and counter-measures for smart cards. In: Attali, I., Jensen, T. (eds.) E-smart 2001. LNCS, vol. 2140, pp. 200–210. Springer, Berlin (2001). doi:10.1007/3-540-45418-7_17

21. Standaert, F.-X., Malkin, T.G., Yung, M.: A unified framework for the analysis of side-channel key recovery attacks. In: Joux, A. (ed.) EUROCRYPT 2009. LNCS, vol. 5479, pp. 443–461. Springer, Berlin (2009). doi:10.1007/978-3-642-01001-9_26
22. Veyrat-Charvillon, N., Gérard, B., Renauld, M., Standaert, F.-X.: An optimal key enumeration algorithm and its application to side-channel attacks. In: Knudsen, L.R., Wu, H. (eds.) SAC 2012. LNCS, vol. 7707, pp. 390–406. Springer, Berlin (2013). doi:10.1007/978-3-642-35999-6_25
23. Veyrat-Charvillon, N., Gérard, B., Standaert, F.-X.: Security evaluations beyond computing power. In: Johansson, T., Nguyen, P.Q. (eds.) EUROCRYPT 2013. LNCS, vol. 7881, pp. 126–141. Springer, Berlin (2013). doi:10.1007/978-3-642-38348-9_8
24. Ye, X., Eisenbarth, T., Martin, W.: Bounded, yet sufficient? how to determine whether limited side channel information enables key recovery. In: Joye, M., Moradi, A. (eds.) CARDIS 2014. LNCS, vol. 8968, pp. 215–232. Springer, Cham (2015). doi:10.1007/978-3-319-16763-3_13

Improved Key Recovery Algorithms from Noisy RSA Secret Keys with Analog Noise

Noboru Kunihiro[✉] and Yuki Takahashi

The University of Tokyo, Kashiwa, Japan
kunihiro@k.u-tokyo.ac.jp

Abstract. From the proposal of key-recovery algorithms for RSA secret key from its noisy version at Crypto2009, there have been considerable researches on RSA key recovery from discrete noise. At CHES2014, two efficient algorithms for recovering secret keys are proposed from noisy analog data obtained through physical attacks such as side channel attacks. One of the algorithms works even if the noise distributions are unknown. However, the algorithm is not optimal especially if the noise distribution is imbalanced. To overcome this problem, we propose new algorithms to recover from such an imbalanced analog noise. We first present a generalized algorithm and show its success condition. We then construct the algorithm suitable for imbalanced noise under the condition that the variances of noise distributions are a priori known. Our algorithm succeeds in recovering the secret key from much more noise. We present the success condition in the explicit form and verify that our algorithm is superior to the previous results. We then show its optimality. Note that the proposed algorithm has the same performance as the previous one in the balanced noise. We next propose a key recovery algorithm that does not use the values of the variances. The algorithm first estimates the variance of noise distributions from the observed data with help of the EM algorithm and then recover the secret key by the first algorithm with their estimated variances. The whole algorithm works well even if the values of the variance is unknown in advance. We examine that our proposed algorithm succeeds in recovering the secret key from much more noise than the previous algorithm.

Keywords: RSA · Key-recovery · Side channel attack · EM algorithm

1 Introduction

1.1 Background and Motivation

RSA [14] is the most widely used cryptosystem and its security is based on the difficulty of factoring a large composite. Furthermore, the side-channel attacks are a real threat to RSA scheme. This kind of attack can be executed by physically observing cryptographic devices and recovering internal information. Side channel attacks are important concerns for security analysis in the both of public key cryptography and symmetric cryptography. In the typical scenario of the

© Springer International Publishing AG 2017
H. Handschuh (Ed.): CT-RSA 2017, LNCS 10159, pp. 328–343, 2017.
DOI: 10.1007/978-3-319-52153-4_19

side channel attacks, an attacker tries to recover the full secret key when he can measure some leaked information from cryptographic devices.

In this paper, we focus on the side channel attacks on RSA cryptosystem. In the RSA cryptosystem [14], a public modulus N is the product of two distinct primes p and q. The public and secret exponents are (e, d), which satisfy $ed \equiv 1 \pmod{(p-1)(q-1)}$. In the textbook RSA, the secret key is only d. However, the PKCS#1 standard [13] specifies that the RSA secret key includes $(p, q, d, d_p, d_q, q^{-1} \bmod p)$ in addition to d, which allows for fast decryption using the Chinese Remainder Theorem (CRT). It is important to analyze the security of CRT-RSA in addition to that of the original RSA.

Halderman et al. [4] presented the cold boot attack at USENIX Security 2008, which is classified as a practical side channel attack. They demonstrated that DRAM remanence effects make possible practical, nondestructive attacks that recover a noisy version of secret keys stored in a computer's memory. They showed how to reconstruct the full of the secret key from the noisy variants for some encryption schemes including RSA scheme. How to recover the correct secret key from a noisy version of the secret key is an important question concerning the cold boot attack situation.

Inspired by cold boot attacks [4], there have been considerable researches on RSA secret key recovery from discrete noise [5,6,8,12]. In contrast, Kunihiro and Honda introduced an analog leakage model and proposed two efficient key recovery algorithm (ML-based algorithm and DPA-like algorithm) from the observed analog data [9].

Observing Analog Data and Motivation. Consider the simple power analysis [7] for CRT-RSA, which is conducted by observing power consumption while executing decryption process. Power consumption trace depends on the bit value of d_p (and d_q). We can obtain analog data from the observed trace through some adequate functions. Further, the distributions of such analog data for the bit 0 and 1 differ from each other due to the difference of power consumption trace. In the same manner, we can obtain the analog data for the bit of d. Note that we cannot obtain those of p and q in the scenario.

In another attack scenario, analog data may be obtained from a (discrete) measurement value of bit value with an analog value of *confidence*. This will be done by some side-channel attacks such as cold boot attack where different pieces of RAM have different preference to flip towards 0 or to flip towards 1.

Our main research aim is to propose efficient algorithms when such analog data, especially imbalanced analog data, are obtained.

1.2 Our Contributions

This paper discusses secret key recovery algorithms from noisy analog data. In our noise model, the observed value is output according to some fixed probability distribution depending on the corresponding correct secret key bit. Unlike [9], we do not assume that the probability density functions are known. Our strategy for constructing the algorithms is summarized as follows: (i) estimate the probability density functions and (ii) run the key-recovery algorithm with the score function designed by the estimated one. We present the success condition (Theorem 3) in adapting the strategy, which shows that we can recover the secret key from more noisy keys if we could succeed to obtain a closer estimation of the probability density functions.

Next, we propose an efficient algorithm (V-based algorithm) to improve the success condition from that of [9]. We propose a new score function (Variance-based Score) by modifying the DPA-like score function introduced in [9] to suit for imbalanced noise. Concretely, we incorporate the variances of the probability distributions into the DPA-like score. By this modification, we succeed in improving the success condition compared to the DPA-like algorithm in [9]. We then present the success condition in the explicit form (Theorem 4), which significantly improves the previously shown bounds. We then prove that Variance-based score is optimal in the weighted variant of DPA-like score. Moreover, we then verify that our algorithm is superior to the previous results by both of theoretical analysis and numerical experiments for various noise distributions. Note it has the same performance as the DPA-like algorithm in the balanced noise.

Although our first algorithm improves the bound, it requires the values of the variances as additional inputs, which is a significant disadvantage to the DPA-like algorithm. To overcome this problem, we use the help of the Expectation-Maximization (EM) algorithm [1,3], which is a well-known algorithm in the area of machine learning, to estimate the variances from the observed data. The second algorithm (KRP algorithm) is constructed by combining V-based algorithm and the EM algorithm. In our combined algorithm, we first run the EM algorithm for the estimation of the variances and run the V-based algorithm with the estimated variances as additional inputs. The KRP algorithm works under the same condition as the DPA-like algorithm, that is, that we can use only the observed data. The numerical results show that our KRP algorithm is superior to the DPA-like algorithm. For example, when the standard deviation of noises (a precise noise model is discussed in Sect. 2.2) is given 0.4 and 2.2, DPA-like algorithm succeeds with probability 0.16, but KRP algorithm succeeds with probability 0.65 (see Table 2). We also verify the effectiveness of our algorithms by numerical experiments on several noise distributions: Gaussian, Laplace, and Uniform distributions, which are shown in the full version [11].

2   Preliminaries

This section presents an overview of the methods [5,6,9,12] using binary trees to recover the secret key of the RSA cryptosystem. We use similar notations to those in [5]. For an n-bit sequence $\mathbf{x} = (x_{n-1}, \ldots, x_0) \in \{0,1\}^n$, we denote the i-th bit of $\mathbf{x}$ by $x[i] = x_i$, where $x[0]$ is the least significant bit of $\mathbf{x}$. Let $\tau(M)$ denote the largest exponent such that $2^{\tau(M)} | M$. We denote by $\ln n$ the natural logarithm of n to the base e and by $\log n$ the logarithm of n to the base 2. We denote the expectation of random variable X by $\mathrm{E}[X]$. We remind readers the Gaussian distribution $\mathcal{N}(\mu, \sigma^2)$. The probability density function of this distribution is $f_N(x; \mu, \sigma^2) = \frac{1}{\sqrt{2\pi\sigma^2}} \exp\left(-\frac{(x-\mu)^2}{2\sigma^2}\right)$, where μ and σ^2 are the mean and variance of the distribution, respectively.

2.1 Recovering the RSA Secret Key Using a Binary Tree

An explanation of this subsection is almost the same as previous works [5,6,9,12]. We first explain how to set the keys of the RSA cryptosystem [14], especially of the PKCS #1 standard [13]. Let (N, e) be the RSA public key and $\mathbf{sk} = (p, q, d, d_p, d_q, q^{-1} \bmod p)$ be the RSA secret key. We denote the bit length of N by n. As in the previous works, we ignore the last component $q^{-1} \bmod p$ in the secret key. The public and secret keys follow four equations: $N = pq$, $ed \equiv 1 \pmod{(p-1)(q-1)}$, $ed_p \equiv 1 \pmod{p-1}$, $ed_q \equiv 1 \pmod{q-1}$. Then, there exist integers k, k_p and k_q such that

$$N = pq, \ ed = 1 + k(p-1)(q-1), \ ed_p = 1 + k_p(p-1), \ ed_q = 1 + k_q(q-1). \quad (1)$$

A small public exponent e is usually used in practical applications [15], so we suppose that e is small enough such that $e = 2^{16} + 1$ as is the case in [5,6,8,9,12]. See [5] for how to compute k, k_p and k_q. Then there are five unknowns (p, q, d, d_p, d_q) in the four equations in Eq. (1).

In the same manner as previous methods, our new methods recover secret key $\mathbf{sk}$ by using a binary tree based technique. We explain how to recover secret keys, considering $\mathbf{sk} = (p, q, d, d_p, d_q)$ as an example.

First we discuss the generation of the tree. Since p and q are $n/2$ bit prime numbers, there exist at most $2^{n/2}$ candidates for each secret key in (p, q, d, d_p, d_q). Heninger and Shacham [6] introduced the concept of **slice**. We define the i-th bit slice for each bit index i as $\mathbf{slice}(i) := (p[i], q[i], d[i + \tau(k)], d_p[i + \tau(k_p)], d_q[i + \tau(k_q)])$. Assume that we have computed a partial solution $\mathbf{sk}'$ up to $\mathbf{slice}(i-1)$. Heninger and Shacham [6] applied Hensel's lemma to Eq. (1) and obtained the following identities

$$p[i] + q[i] = (N - p'q')[i] \bmod 2,$$
$$d[i + \tau(k)] + p[i] + q[i] = (k(N+1) + 1 - k(p'+q') - ed')[i + \tau(k)] \bmod 2,$$
$$d_p[i + \tau(k_p)] + p[i] = (k_p(p'-1) + 1 - ed_p')[i + \tau(k_p)] \bmod 2,$$
$$d_q[i + \tau(k_q)] + q[i] = (k_q(q'-1) + 1 - ed_q')[i + \tau(k_q)] \bmod 2.$$

This means that we have four linearly independent equations in the five unknowns $p[i], q[i], d[i + \tau(k)], d_p[i + \tau(k_p)]$, and $d_q[i + \tau(k_q)]$ of $\mathbf{slice}(i)$. Each Hensel lift, therefore, yields exactly two candidate solutions. Then, the total number of candidates is given by $2^{n/2}$.

Henecka et al.'s algorithm [5] and Paterson et al.'s algorithm (in short, the PPS algorithm) [12] perform t Hensel lifts for some fixed parameter t. For each surviving candidate solution on $\mathbf{slice}(0)$ to $\mathbf{slice}(it-1)$, a tree with depth t and whose 2^t leaf nodes represent candidate solutions on $\mathbf{slice}(it)$ to $\mathbf{slice}((i+1)t-1)$, is generated. This causes $5t$ new bits. For each new node generated, a pruning phase is carried out. A solution is kept for the next iteration if the likelihood of the corresponding noisy variants of the secret key for the $5t$ new bits is in the highest L nodes of the $L2^t$ nodes as for the PPS algorithm [12]. Kunihiro and Honda [9] adopted a similar approach to the PPS algorithm [12]. They introduced a concept of score function, and their algorithms keep the top L nodes with the highest score.

2.2   Our Noise Model

Let F_0 and F_1 be probability distributions of an observed value when the correct secret key bits are 0 and 1, respectively. That means we assume that each the observed value x follows the *fixed* probability distribution F_b. Though this assumption comes from simplification, it is frequently considered and verified in the practice of side channel attacks. In this paper, we assume that F_0 and F_1 have probability densities f_0 and f_1, respectively. Without loss of generality, we assume that the means of F_b are $(-1)^b$. Throughout this paper, we assume that f_0 and f_1 are unknown to the attackers. That implies that we do not use any knowledge about explicit forms of probability density functions in designing algorithms.

   We say that the probability density functions f_0 and f_1 are *imbalanced* when $f_0(x)$ and $f_1(-x)$ are (very) different. Suppose that $f_0 = \mathcal{N}(+1, \sigma_0^2)$ and $f_1 = \mathcal{N}(-1, \sigma_1^2)$. We say that f_0 and f_1 are imbalanced when $\sigma_0 \ll \sigma_1$. (Note that $f_1(-x) = \mathcal{N}(x; +1, \sigma_1^2)$). In this paper, we mainly focus on the case that f_0 and f_1 are imbalanced.

2.3   Previous Works on Key-Recovery for Analog Observed Data

A score function is introduced in [9], that is calculated with observed data and a candidate sequence (if necessary, additional information such as the probability density functions of noise). A framework of key-recovery algorithm that uses the score function in Pruning phase is then proposed. Definition 1 gives the syntax of the score function.

Definition 1 (Syntax of Score Function). The score function receives a candidate sequence $\mathbf{b} = (b_1, \ldots, b_n) \in \{0, 1\}^n$ and the corresponding observed sequence $\mathbf{x} = (x_1, \ldots, x_n) \in \mathbb{R}^n$ and outputs a real number. We use the notation: $\mathbf{Score}_n(\mathbf{b}, \mathbf{x})$.

The score function $\mathbf{Score}_n(\mathbf{b}, \mathbf{x})$ is designed so that the following properties hold for any fixed $\mathbf{x}$: the score will be large if $\mathbf{b}$ is a correct candidate; the score will be small if $\mathbf{b}$ is incorrect.

   We review a framework shown in [9,12] for the RSA key-recovery algorithm. Our proposed algorithms are based on the same framework. It is pointed out in [9] that the setting $t = 1$ is enough for gaining high success rates. We use slightly different notations of generalized PPS algorithm from [9]. Revising the algorithm framework itself is not our target. This paper mainly focuses on designing the score function.

The following two score functions have been proposed in [9]. Denote a candidate sequence $\mathbf{b} = (b_1, \ldots, b_n)$ and an observed sequence $\mathbf{x} = (x_1, \ldots, x_n)$. The first one is defined by

$$\mathrm{ML}(\mathbf{b}, \mathbf{x}) := \sum_{i=1}^{n} \log \frac{f_{b_i}(x_i)}{g(x_i)}, \tag{2}$$

where $g(x) = (f_0(x) + f_1(x))/2$. The second one is defined by

$$\mathrm{DPA}(\mathbf{b}, \mathbf{x}) := \sum_{i=1}^{n} (-1)^{b_i} x_i. \tag{3}$$

Equations (2) and (3) are called as ML-based score and DPA-like score, and the algorithms employing Eqs. (2) and (3) are called as ML-based algorithm and DPA-like algorithm, respectively. Note that the ML-based algorithm requires the complete information about probability density functions as inputs. In contrast, the DPA-like algorithm does not require them as shown in Eq. (3).

We summarize the success condition for the ML-based algorithm, and the DPA-like algorithm [9]. First, we introduce a differential entropy [2].

Definition 2. The differential entropy $h(f)$ of a probability density function f is defined as

$$h(f) = -\int_{-\infty}^{\infty} f(y) \log f(y) dy.$$

Theorem 1 (Corollary 1, [9]). Assume that the probability density functions for $b = 0, 1$ are given by f_b. The error probability of the ML-based Algorithm converges to zero as $L \to \infty$ if

$$h\left(\frac{f_0 + f_1}{2}\right) - \frac{h(f_0) + h(f_1)}{2} > \frac{1}{5}. \tag{4}$$

Theorem 2 (Theorem 2, [9]). Assume that the probability density functions for $b = 0, 1$ are given by f_b. Denote the variance of F_b by σ_b^2. The error probability of the DPA-like Algorithm converges to zero as $L \to \infty$ if

$$h\left(\frac{f_0 + f_1}{2}\right) - \log \sqrt{\pi e(\sigma_0^2 + \sigma_1^2)} > \frac{1}{5}. \tag{5}$$

Consider the case that f_0 and f_1 are imbalanced. Without loss of generality, we assume that $\sigma_0 \ll \sigma_1$. In this case, the left-hand side of Eq. (5) heavily depends on only the variance σ_1, which is unnatural. We will give improvement of the success condition by incorporating the values σ_b^2 to the score function in Sect. 4.

Remark 1. Throughout the paper, we only consider the case that we employ (p, q, d, d_p, d_q) as secret key tuple. However, we can easily extend to more general case. For the (p, q), (d_p, d_q), (p, q, d), and (d, d_p, d_q) cases, we just replace $1/5$ with $1/2, 1/2, 1/3$, and $1/3$, respectively in Theorems 1–4 and Eq. (11).

3   Generalized Algorithm via Estimation of Distributions

In actual attack situations, the attacker does not know the exact form of f_b. Then, we cannot apply the ML-based score directly. On the other hands, if one could obtain a closer estimation of probability density functions, one can hope to attain the key-recovery from larger noise. The second best strategy is then (i) to estimate f_b in some way (discussed in Sect. 5) and (ii) to run the key recovery algorithm with the score function designed by estimated probability density functions. In this section, we will derive the success condition under the condition that we have learned the estimation of the distributions. We denote the estimated distributions of f_0 and f_1 by $f_0^{(E)}$ and $f_1^{(E)}$, respectively.

Before giving the detailed analysis, we introduce the Kullback-Leibler divergence [2].

Definition 3. For the probability density functions p and q, the Kullback–Leibler divergence $D(p||q)$ of p and q is defined as

$$D(p||q) = \int_{-\infty}^{\infty} p(y) \log \frac{p(y)}{q(y)} dy.$$

It is well-known that the Kullback–Leibler divergence $D(p||q)$ is non-negative and it is zero if and only if $p = q$. It is considered as some kind of the distance between p and q.

We introduce a new notion of the score function based on the estimated probability density functions, which would be a natural modification of ML-based score. We define the new score as

$$R^{(E)}(\mathbf{b}, \mathbf{x}) = \sum_i \log f_{b_i}^{(E)}(x_i). \tag{6}$$

In the modification, we replace the true densities f_b with their estimations $f_b^{(E)}$ (and ignore the denominator). Using $R^{(E)}(\mathbf{b}, \mathbf{x})$ as a score function, we have the following theorem.

Theorem 3. Assume that the probability density functions for $b = 0, 1$ are given by f_b. The error probability of Algorithm with the score $R^{(E)}(\mathbf{b}, \mathbf{x})$ converges to zero as $L \to \infty$ if

$$\left(h\left(\frac{f_0 + f_1}{2} \right) - \frac{h(f_0) + h(f_1)}{2} \right) - \frac{D(f_0||f_0^{(E)}) + D(f_1||f_1^{(E)})}{2} > \frac{1}{5}. \tag{7}$$

Proof. A proof strategy is almost the same as that of Theorem 2 in [10]. The score $R^{(E)}(\mathbf{b}, \mathbf{x})$ is essentially equivalent to the score

$$R'^{(E)}(\mathbf{b}, \mathbf{x}) = \sum_i \log \frac{f_{b_i}^{(E)}(x_i)}{g(x_i)}$$

since $g(x_i)$ does not depend on $\mathbf{b}$. It is enough for proving the theorem to calculate

$$I^{(\mathrm{E})} := \sum_{b \in \{0,1\}} \frac{1}{2} \int_x \left(\log \frac{f_b^{(\mathrm{E})}(x)}{g(x)} \right) f_b(x) \mathrm{d}x.$$

The exact form of $I^{(\mathrm{E})}$ is calculated as follows.

$$I^{(\mathrm{E})} = h \left(\frac{f_0 + f_1}{2} \right) - \frac{h(f_0) + h(f_1)}{2} - \frac{D(f_0 \| f_0^{(\mathrm{E})}) + D(f_1 \| f_1^{(\mathrm{E})})}{2}$$

The full calculation of $I^{(\mathrm{E})}$ is shown in the full version [11]. The rest of the proof is the same as that of Theorem 2 in [10]. Then, we have the theorem. □

The former half of the left hand side in Eq. (7), $h((f_0 + f_1)/2) - (h(f_0) + h(f_1))/2$, is equivalent to the condition when the true distributions are known (see Theorem 1). Its latter half $(D(f_0 \| f_0^{(\mathrm{E})}) + D(f_1 \| f_1^{(\mathrm{E})}))/2$ corresponds to the *information loss* or *penalty* caused by mis-estimations. From the definition, it is always non-negative. If the probability density function is correctly estimated (which means that the both of $f_0^{(\mathrm{E})} = f_0$ and $f_1^{(\mathrm{E})} = f_1$ hold), the information loss vanishes since $D(f_0 \| f_0^{(\mathrm{E})}) = D(f_1 \| f_1^{(\mathrm{E})}) = 0$. Conversely, if the accurate estimation fails, the success condition is much worse than expected due to the information loss caused by mis-estimation of f_0 and f_1.

4   New Score Function with a Priori Known Variances

In this section, we propose an effective score function when the noise distributions are unknown but their average and variances are a priori known. Note that we remove this requirement in Sect. 5. Our score function explicitly uses the values of the variances of the noise distributions. Specifically, the proposed score is much more effective than previous one when the variance of F_0 and F_1 are different.

First, we point out drawbacks of DPA-like algorithm introduced in [9]. The DPA-like algorithm works with only observed data even if the probability density functions are not known. From the nature of the DPA-like score, it can not use any other side information of probability density function such as variances even if they are available.

We try to incorporate the side information into the DPA-like function. It is natural to consider the weighted variant of DPA-like score, which is defined by

$$\text{w-DPA}(\mathbf{b}, \mathbf{x}) := \sum_{i=1}^{n} w_{b_i} (-1)^{b_i} x_i \tag{8}$$

for some kind of weights w_0 and w_1. The performance on weighted variant of DPA-like score heavily relies on how to set w_0 and w_1. If the observed value is *reliable*, the corresponding weight should be large. We propose a new score by following this idea.

4.1   New Score Function: Variance-Based Score

We consider the case where F_0 and F_1 (and hence also f_0 and f_1) are unknown, but, their variances are known a priori. We denote by σ_0^2 and σ_1^2 the variances of F_0 and F_1. Under the situation, we have a chance to choose an adequate score function including the explicit values of the variances.

We introduce a new score function (Variance-based Score):

$$V(\mathbf{b}, \mathbf{x}) := \sum_i \frac{(-1)^{b_i} x_i}{\sigma_{b_i}^2}. \tag{9}$$

It can be considered that $w_0 = 1/\sigma_0^2$ and $w_1 = 1/\sigma_1^2$ in the context of weighted variant of DPA. We denote Key Recover Algorithm employing Variance-based Score $V(\mathbf{b}, \mathbf{x})$ as a score function by *V-based algorithm*. Note that in evaluating the score function by Eq. (9), we explicitly use the variances σ_0^2 and σ_1^2. Consider the case when $\sigma_0^2 = \sigma_1^2 = \sigma^2$. Then, the score function can be transformed into

$$V(\mathbf{b}, \mathbf{x}) = \frac{\sum_i (-1)^{b_i} x_i}{\sigma^2} = \frac{1}{\sigma^2} \sum_i (-1)^{b_i} x_i = \frac{1}{\sigma^2} \mathrm{DPA}(\mathbf{b}, \mathbf{x}).$$

Since the part $1/\sigma^2$ does not affect the order of score value, we can ignore it and recover the DPA-like score. The Variance-based score then includes the DPA-like score in the special case.

Our strategy for designing a score function can be interpreted as follows: The observed data from the distribution with larger variance will not be reliable. Then, its contribution is set to be small if the variance is large, and vice versa.

4.2   Theoretical Analysis for V-Based Algorithm

In this section, we discuss the success condition of the V-based algorithm for recovering the secret key. The following theorem shows the success condition on f_0 and f_1 when we use V-based algorithm for recovering the RSA secret key.

Theorem 4. Assume that the probability density function for $b = 0, 1$ are given by f_b. The error probability of the V-based algorithm converges to zero as $L \to \infty$ if

$$h\left(\frac{f_0 + f_1}{2}\right) - \log \sqrt{2\pi e \sigma_0 \sigma_1} > \frac{1}{5}. \tag{10}$$

Proof. A proof is almost the same as the proof [10] of Theorem 2 in [9]. We denote by $f_b^{(\mathrm{G})}$ the probability density function of Gaussian distributions with average $(-1)^b$ and σ_b^2, respectively. The Variance-based score is essentially equivalent to the score

$$R^{(\mathrm{G})}(\mathbf{b}, \mathbf{x}) = \sum_i \log \frac{f_{b_i}^{(\mathrm{G})}(x_i)}{g(x_i)}.$$

As the same discussion in [10], it is enough to calculate

$$I^{(G)} := \sum_{b \in \{0,1\}} \frac{1}{2} \int_x \left(\log \frac{f_b^{(G)}(x)}{g(x)} \right) f_b(x) dx.$$

According to Theorem 1 in [9], the condition is given by $I^{(G)} > 1/5$. The exact form of $I^{(G)}$ is calculated as follows.

$$I^{(G)} = \sum_{b \in \{0,1\}} \frac{1}{2} \int_x \left(\log \frac{f_b^{(G)}(x)}{g(x)} \right) f_b(x) dx$$

$$= -\int_x (\log g(x)) g(x) dx + \frac{1}{2} \sum_{b \in \{0,1\}} \int_x \left(\log f_b^{(G)}(x) \right) f_b(x) dx$$

$$= h(g) - \frac{1}{2} \sum_{b \in \{0,1\}} \left\{ \frac{\log(2\pi\sigma_b^2)}{2} + \frac{1}{2(\ln 2)\sigma_b^2} \int_x (x - (-1)^b)^2 f_b(x) dx \right\}$$

$$= h(g) - \log(\sqrt{2\pi e \sigma_0 \sigma_1}).$$

Then, we have the theorem. $\qquad\square$

We give a comparison between the DPA-like algorithm and the V-based algorithm. The difference between the left hand side of two inequalities: Eqs. (5) and (10) is given by

$$\log \sqrt{\pi e (\sigma_0^2 + \sigma_1^2)} - \log \sqrt{2\pi e \sigma_0 \sigma_1} = \frac{1}{2} \log \frac{\sigma_0^2 + \sigma_1^2}{2\sigma_0\sigma_1}.$$

Since the arithmetic mean is always larger than or equal to the geometric mean, it holds that $\frac{\sigma_0^2 + \sigma_1^2}{2} \geq \sqrt{\sigma_0^2 \sigma_1^2} = \sigma_0\sigma_1$. Then, the difference is always non-negative. Furthermore, the difference is 0 if and only if $\sigma_0 = \sigma_1$. It shows that V-based algorithm is superior to the DPA-like algorithm except the case that $\sigma_0 = \sigma_1$. As the ratio between σ_0 and σ_1 becomes larger, our improvement is more significant.

4.3 Optimality of Variance-Based Score

We show that our proposed variance-based score is optimal in the framework of weighted variant of DPA-score. If we adopt w_0 and w_1 as weights, the success condition is given by

$$h\left(\frac{f_0 + f_1}{2}\right) - \log \sqrt{2\pi e} - \frac{\log e}{4}(-\ln w_0 - \ln w_1 + \sigma_0^2 w_0 + \sigma_1^2 w_1 - 2) > \frac{1}{5}. \quad (11)$$

Denote the the left hand side of Eq. (11) by $H(w_0, w_1)$. By solving a simultaneous equation $\frac{\partial H}{\partial w_0} = \frac{\partial H}{\partial w_1} = 0$, we obtain $w_0 = 1/\sigma_0^2$ and $w_1 = 1/\sigma_1^2$, which maximizes $H(w_0, w_1)$. We recover the Variance-based score introduced in Sect. 4.1. This shows its optimality.

4.4   Experimental Results for V-Based Algorithm

We give experiment results on DPA-like algorithm [9] and our proposed V-based algorithm, which uses Eq. (9) as a score function. We implemented our algorithm in gcc with NTL 6.0, GMP 5.1.3 library and tested it on Intel Xeon 6-Core processor at 2.66 GHz with 32 GB memory. We set the public exponent to $e = 2^{16} + 1$. In all experiments shown in this section, we generated the output $\overline{\mathbf{sk}}$ for each $\mathbf{sk}$ from the Gaussian distribution. Denoting the correct secret bit in $\mathbf{sk}$ by b, we concretely generated $\overline{\mathbf{sk}}$ as follows: the observed value follows $\mathcal{N}(-1, \sigma_1^2)$ if $b = 1$; and the observed value follows $\mathcal{N}(+1, \sigma_0^2)$ if $b = 0$. In our experiments on 1024 bit RSA, we prepared 200 different tuples of secret keys $\mathbf{sk}$, e.g., $\mathbf{sk} = (p, q, d, d_p, d_q)$. We set a parameters L as $L = 2^{12}$.

We especially focus on the case where the $\sigma_1^2 \neq \sigma_0^2$. Figure 1 shows the success rates of DPA-like algorithm and V-based algorithm for $\sigma_0 = 0.4$ and $\sigma_0 = 1.0$. The vertical axis represents the success rates, and the horizontal axis shows the value of σ_1.

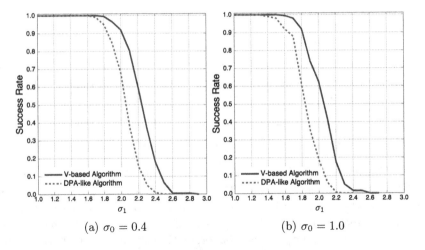

(a) $\sigma_0 = 0.4$ (b) $\sigma_0 = 1.0$

Fig. 1. Comparison between DPA-like and V-based algorithms

We give some discussion for the case of $\sigma_0 = 0.4$ from Fig. 1(a). When $\sigma_1 \leq 1.6$, the both algorithms succeed in recovering the secret key with success rate 1. Further, when $\sigma_1 \geq 2.6$, the both algorithms fail to do that for all trials. Meanwhile, when $1.7 \leq \sigma_1 \leq 2.5$, the two algorithms show the different behavior. For example, when $\sigma_1 = 2.1$, our algorithm recovers the secret key with success rate 0.8; while DPA-like algorithm recovers one with success rate 0.35. For another example, when $\sigma_1 = 2.4$, our algorithm recovers one with success rate 0.20; while DPA-like algorithm fails to recover the keys for all trials. These observations show that our V-based algorithm has superior performance to the DPA-like algorithm. The running time for the V-based algorithm to find the secret key is at most 36.9 s under our computer circumstance for any cases. More experimental results are shown in the full version [11].

5   Estimation of Variances by the EM Algorithm

Our new score function $V(\mathbf{b}, \mathbf{x})$ requires the additional inputs: variances σ_0^2 and σ_1^2 of F_0 and F_1. It is a significant disadvantage against the DPA-like algorithm. To solve this problem, we will use the help of the EM algorithm [1,3] in estimating the variances from the observed data. The EM algorithm is a popular algorithm in the area of machine learning and is used to estimate hidden parameters of mixture distributions.

We will use the EM algorithm to estimate the variances σ_0^2 and σ_1^2 as a preprocessing of the V-based algorithm. That means, we first run the EM algorithm to estimate the variances and then run the V-based algorithm with the estimated variances to recover the secret key. It enables us to recover the secret key by using only the observed data, as well as the DPA-like algorithm. Unlike DPA-like algorithm, we succeed in taking account of the values of the variances in the combined algorithm. It can lead to a significant improvement of the bound for key-recovery against DPA-like algorithm, which will be examined in Sect. 5.2.

5.1   Variance Estimation by the EM Algorithms

Before giving the detailed explanation of the EM algorithm, we present another view of our noise model. It can be regarded as follows:

- The probability density functions $f_b(x; \boldsymbol{\theta_b})$ are defined by hidden parameters $\boldsymbol{\theta_b}$ for $b = 0, 1$.
- The observed value follows the mixture distribution $p(x)$ of f_0 and f_1, where $p(x) = \alpha f_0(x; \boldsymbol{\theta_0}) + (1 - \alpha) f_1(x; \boldsymbol{\theta_1})$ for $0 \leq \alpha \leq 1$

In the usual setting in the EM algorithm, the form of f_0 and f_1 are known (say, f_0 is the Gaussian distribution, etc.), but, the set of parameters $\Theta = \{\alpha, \boldsymbol{\theta_0}, \boldsymbol{\theta_1}\}$ are a priori unknown (or hidden). The EM algorithm is usually used to estimate these parameters from the observed data.

We show the EM algorithm in more details. We denote by D a set of the observed values. Assume that all the observed value $x_i \in D$ follows the mixture density: $p(x) = \alpha_0 f_0(x; \boldsymbol{\theta_0}) + \alpha_1 f_1(x; \boldsymbol{\theta_1})$. We introduce Membership Weight γ_{ik} for $x_i \in D$ given parameters Θ as follows:

$$\gamma_{ik} = \frac{\alpha_k \bar{f}_k(x; \boldsymbol{\theta_k})}{\alpha_0 \bar{f}_0(x; \boldsymbol{\theta_0}) + \alpha_1 \bar{f}_1(x; \boldsymbol{\theta_1})} \tag{12}$$

for $1 \leq i \leq |D|$ and $k = 0, 1$. Note that $\alpha_0 + \alpha_1 = 1$ and $\alpha_0, \alpha_1 \geq 0$. Intuitively, the γ_{ik} corresponds to a probability that x_i comes from the bit k. If we know the exact form of f_k, we use f_k itself for $\bar{f}_k$ for $k = 0$ and 1. However, in our attack scenario, we have no knowledge about f_k as described before. Then, we cannot use the EM algorithm as-is. We use the Gaussian distribution in place of true unknown distribution, which enables the EM algorithm to work. We adopt the probability density function of Gaussian distribution for $\bar{f}_k = \mathcal{N}(\mu_k, \sigma_k^2)$. In this setting, the purpose of the EM algorithm will estimate means and variances for mixture distributions.

Next, we focus on our attack scenario. The attacker now wants to know the means μ_0 and μ_1, and variances σ_0^2 and σ_1^2 by using the EM algorithm. In this scenario, it is implicitly assumed that the noise distributions f_0 and f_1 are the Gaussian. Then, we can explicitly write Θ as $\Theta = \{\alpha_0, \alpha_1, \mu_0, \mu_1, \sigma_0, \sigma_1\}$.

It is proved that the log-likelihood of the mixture distribution monotonically decreases by using the EM algorithm. On the other hand, it is hard to estimate precisely in advance the number of iteration required until the log-likelihood converges. We will verify that the computational time for the estimation phase for variances is negligible to the total time for the whole key-recovery by measuring an actual running time of the EM algorithm.

Algorithm 1 shows the whole proposed algorithm. This algorithm is composed of two phase: Parameter Estimation Phase and Key-Recovery Phase. That means we use the EM algorithm as a pre-processing of the key-recovery algorithm. We call the whole algorithm KRP algorithm.

Algorithm 1. KRP algorithm (Key Recovery with Pre-processing Algorithm)

Input: Public Key (N, e) observed noisy sequences $\overline{\mathbf{sk}}$
Output: Correct Secret Key $\mathbf{sk}$
Parameter: $L \in \mathbb{N}$
Parameter Estimation Phase Run the EM algorithm to estimate the variances σ_0^2 and σ_1^2 from the observed sequence.
Key-Recovery Phase: Run V-based algorithm with inputs (estimated) σ_0^2 and σ_1^2, the observed sequence, and L.

5.2 Experimental Results for KRP Algorithm

We first examine the running time of the EM algorithm for various input length of the observed sequence. We repeat the EM algorithm 100 times given an initial parameter for Θ and calculate the average of the running time. The environment for computation is the same as that in Sect. 4.4. In the experiments, we iterate E-step and M-step until convergence.

Table 1 shows the average time for the RM algorithm.

In general, we can estimate parameters with higher accuracy if we use more data for estimation. On the other hand, it causes more running time. We can see that the running time is at most 21.8 ms even if we use full sequences for estimating the variances. Since the running time for the V-based algorithm is in average 36.9 s as shown in Sect. 4.4, the running time of the EM-algorithm is negligible to the whole running time. From now on, we ignore the running time of the EM Algorithm.

Table 1. Average of running time for the EM algorithm

Input Data Length	500	1000	1500	2000	2500	3000	3500	4000	4500	5000
Computational Time (ms)	1.90	3.94	6.22	8.38	10.5	12.8	15.1	17.5	19.9	21.8

Comparison Between KRP Algorithm and V-Based Algorithm. Next, we compare KRP algorithm and V-based algorithm. Remember that V-based algorithm requires additional input: σ_0 and σ_1 but KRP algorithm works without them. In the experiments, we consider the case that both of f_0 and f_1 are the Gaussian distributions: $f_k = \mathcal{N}((-1)^k, \sigma_k^2)$ for $k = 0$ and 1. Here, we run the experiments under the same environments as in Sect. 4.4.

Figure 2 shows the success rates of the KRP algorithm and the V-based algorithm for $\sigma_0 = 0.4$ and 1.0. We give some discussion for the case of $\sigma_0 = 0.4$. Figure 2(a) shows that the success rates of the both algorithms are almost 1 if σ_1 is less than or equal to 1.8. We can see that when $1.8 \leq \sigma_1 \leq 2.6$, their success rates decrease gradually, but, they are almost the same. Hence, we can say that there is no difference between their performance. The success rates for $\sigma_0 = 1.0$ denote the same tendency as for $\sigma_0 = 0.4$, that is, there is no difference in performance between the two algorithms. The above discussion shows that the EM algorithm succeeds in estimating the variances with enough accuracy and KRP algorithm. Consequently, the KRP algorithm, which does not receive σ_0 and σ_1 as inputs, has almost the same performance as the V-based algorithm. More experimental results are shown in the full version [11].

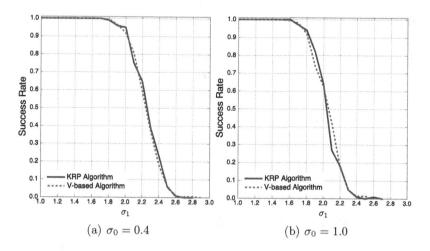

(a) $\sigma_0 = 0.4$ (b) $\sigma_0 = 1.0$

Fig. 2. Success Rates of KRP algorithm and V-based algorithm

Comparison Between KRP and DPA-like Algorithm. Finally, we compare KRP and DPA-like algorithms [9]. Note that the both algorithms work given only the observed data, which means that they do not require additional information about the probability density functions.

We consider the case that the both of f_0 and f_1 are the Gaussian distributions: $f_k = \mathcal{N}((-1)^k, \sigma_k^2)$ for $k = 0$ and 1. Here, we execute the experiments under the same environments as in Sect. 4.4.

Figure 3 shows the success rates of KRP algorithm and DPA-like algorithm for $\sigma_0 = 0.4$ and $\sigma_0 = 1.0$. We can see that KRP algorithm attains higher success rates than the DPA-like algorithm from Figs. 3(a) and (b). Further, their computational time for recovering the keys are almost the same because the running time of the EM algorithm is negligible as described before. Summing up, we can conclude that our proposed KRP algorithm is superior to the DPA-like algorithm.

Table 2 summarizes the success rates of the DPA-like algorithm, the V-based algorithm, and the KRP algorithm (more results are given in full version [11]). We can see that the proposed algorithms in this paper are superior to DPA-like algorithm [9]. Moreover, the proposed two algorithms have almost the same performance; while V-based algorithm requires the variances of the noise distributions and KRP algorithm does not.

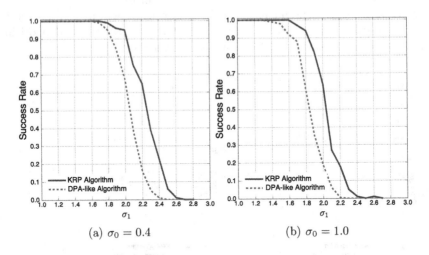

(a) $\sigma_0 = 0.4$ (b) $\sigma_0 = 1.0$

Fig. 3. Success Rates of KRP algorithm and DPA-like algorithm

Table 2. Success rates of three algorithms for $\sigma_0 = 0.4$

σ_1	$0\cdots1.6$	1.7	1.8	1.9	2.0	2.1	2.2	2.3	2.4	2.5	2.6	2.7	2.8
DPA-like [9]	1	0.99	0.95	0.85	0.68	0.38	0.16	0.05	0.01	0	0	0	0
V-based (this paper)	1	1	1	0.97	0.92	0.81	0.60	0.37	0.18	0.07	0.01	0.01	0.01
KRP (this paper)	1	1	0.99	0.96	0.95	0.75	0.65	0.39	0.23	0.06	0.01	0	0

We present the performance of KRP algorithm for non-Gaussian distributions is the full version [11].

Acknowledgement. This research was supported by CREST, JST and supported by JSPS KAKENHI Grant Number 25280001 and 16H02780.

References

1. Bishop, C.M.: Pattern Recognition and Machine Learning. Springer, New York (2006)
2. Cover, C.M., Thomas, J.A.: Elements of Information Theory, 2nd edn. Wiley, Hoboken (2006)
3. Dempster, A.P., Laird, N.M., Rubin, D.B.: Maximum likelihood from incomplete data via the EM algorithm. J. Roy. Stat. Soc. Ser. B **39**(1), 1–38 (1977)
4. Halderman, J.A., Schoen, S.D., Heninger, N., Clarkson, W., Paul, W., Calandrino, J.A., Feldman, A.J., Appelbaum, J., Felten, E.W.: Lest we remember: cold boot attacks on encryption keys. In: Proceedings of USENIX Security Symposium 2008, pp. 45–60 (2008)
5. Henecka, W., May, A., Meurer, A.: Correcting errors in RSA private keys. In: Rabin, T. (ed.) CRYPTO 2010. LNCS, vol. 6223, pp. 351–369. Springer, Berlin (2010). doi:10.1007/978-3-642-14623-7_19
6. Heninger, N., Shacham, H.: Reconstructing RSA private keys from random key bits. In: Halevi, S. (ed.) CRYPTO 2009. LNCS, vol. 5677, pp. 1–17. Springer, Berlin (2009). doi:10.1007/978-3-642-03356-8_1
7. Kocher, P., Jaffe, J., Jun, B.: Differential power analysis. In: Wiener, M. (ed.) CRYPTO 1999. LNCS, vol. 1666, pp. 388–397. Springer, Heidelberg (1999). doi:10.1007/3-540-48405-1_25
8. Kunihiro, N., Shinohara, N., Izu, T.: Recovering RSA secret keys from noisy key bits with erasures and errors. In: Kurosawa, K., Hanaoka, G. (eds.) PKC 2013. LNCS, vol. 7778, pp. 180–197. Springer, Heidelberg (2013). doi:10.1007/978-3-642-36362-7_12
9. Kunihiro, N., Honda, J.: RSA meets DPA: recovering RSA secret keys from noisy analog data. In: Batina, L., Robshaw, M. (eds.) CHES 2014. LNCS, vol. 8731, pp. 261–278. Springer, Heidelberg (2014). doi:10.1007/978-3-662-44709-3_15
10. Kunihiro, N., Honda, J.: RSA meets DPA: recovering RSA secret keys from noisy analog data. In: IACR eprint: 2014/513 (2014)
11. Kunihiro, N., Takahashi, Y.: Improved key recovery algorithms from noisy RSA secret keys with analog noise. In: IACR eprint: 2016/1095 (2016)
12. Paterson, K.G., Polychroniadou, A., Sibborn, D.L.: A coding-theoretic approach to recovering noisy RSA keys. In: Wang, X., Sako, K. (eds.) ASIACRYPT 2012. LNCS, vol. 7658, pp. 386–403. Springer, Heidelberg (2012). doi:10.1007/978-3-642-34961-4_24
13. PKCS #1: RSA Cryptography Specifications Version 2.0. http://www.ietf.org/rfc/rfc2437.txt
14. Rivest, R., Shamir, A., Adleman, L.: A method for obtaining digital signatures and public-key cryptosystems. Commun. ACM **21**(2), 120–126 (1978)
15. Yilek, S., Rescorla, E., Shacham, H., Enright, B., Savage, S.: When private keys are public: results from the 2008 Debian OpenSSL vulnerability. In: IMC 2009, pp. 15–27. ACM Press (2009)

Side-channel Analysis

Ridge-Based Profiled Differential Power Analysis

Weijia Wang[1], Yu Yu[1,3,4]($\boxtimes$), François-Xavier Standaert[2], Dawu Gu[1], Xu Sen[1], and Chi Zhang[1]

[1] School of Electronic Information and Electrical Engineering,
Shanghai Jiao Tong University, Shanghai, China
{aawwjaa,yyuu,dwgu,push.beni,liujr,guozheng}@sjtu.edu.cn
[2] ICTEAM/ELEN/Crypto Group,
Université catholique de Louvain, Louvain-la-Neuve, Belgium
fstandae@uclouvain.be
[3] State Key Laboratory of Cryptology, P.O. Box 5159, Beijing 100878, China
[4] Westone Cryptologic Research Center, Beijing, China

Abstract. Profiled DPA is an important and powerful type of side-channel attacks (SCAs). Thanks to its profiling phase that learns the leakage features from a controlled device, profiled DPA outperforms many other types of SCA and are widely used in the security evaluation of cryptographic devices. Typical profiling methods (such as linear regression based ones) suffer from the overfitting issue which is often neglected in previous works, i.e., the model characterizes details that are specific to the dataset used to build it (and not the distribution we want to capture). In this paper, we propose a novel profiling method based on ridge regression and investigate its generalization ability (to mitigate the overfitting issue) theoretically and by experiments. Further, based on cross-validation, we present a parameter optimization method that finds out the most suitable parameter for our ridge-based profiling. Finally, the simulation-based and practical experiments show that ridge-based profiling not only outperforms 'classical' and linear regression-based ones (especially for nonlinear leakage functions), but also is a good candidate for the robust profiling.

Keywords: Side-channel attack · Profiled DPA · Linear regression · Ridge regression · Cross-validation

1 Introduction

Side-channel attacks (SCAs) exploit the physical information leaked from the implementation of a cryptographic algorithm, and they are usually more powerful than brute-force attacks or classical cryptanalytic techniques that target at the mathematical weakness of the underlying algorithm. Differential power analysis (DPA), proposed by Kocher et al. [15], is a form of side-channel attack that efficiently recovers the secret key from multiple (typically noisy) power consumption measurements (on different plaintexts). Profiled DPA (e.g., [3,20,24]) adds a profiling phase (prior to the online exploitation phase) to the original

© Springer International Publishing AG 2017
H. Handschuh (Ed.): CT-RSA 2017, LNCS 10159, pp. 347–362, 2017.
DOI: 10.1007/978-3-319-52153-4_20

DPA and can be considered as a powerful class of power analysis. The profiling phase learns the leakage function from the power consumption of a training device, and it can significantly enhance the performance of the subsequent online exploitation phase, namely, the key recovery attack mounted against a similar target device. We will focus on the profiling phase in this paper.

Chari et al. [3] proposed the first profiled DPA called template attacks, whose profiling phase is based on multivariate Gaussian templates. We refer to the profiling phase of templates attacks as classical profiling (following the terminology in [24]). Later Schindler et al. [20] proposed a very promising profiled DPA that uses linear regression (LR) as its profiling method (referred to as LR-based profiling hereafter). Compared with classical profiling, LR-based profiling builds up a model more efficiently with less number of measurements and it allows a trade-off between the profiling and online exploitation phases: more measurements used in the profiling phase, less measurements needed in the exploitation phase [8,22,24]. However, the LR-based profiling suffers from the overfitting issue in practice. That is, noisy measurements in the profiling phase can result in a model that describes mostly the noise instead of the actual leakage function. Thus, the LR-based profiling may need more measurements than necessary. We mention other profiling methods those based on agglomerative hierarchical clustering [25], K-means [25] and different machine learning methods such as SVM [12,14,16], random forests [16,17], neural networks [18,19], which enjoy additional features or are more useful for specific data structures or have an overhead for the time complexity. We are not extending this line of research any further.

In this paper, we propose a new profiling method (named ridge-based profiling) based on ridge regression. By imposing a constraint on the coefficients of linear regression, ridge regression is a good alternative to linear regression with better performance on noisy data [11]. As the constraint (described by a parameter) affects the performance of ridge-based profiling, we apply the K-fold cross-validation to find out the most suitable constraint (i.e., the optimized parameter) for ridge-based profiling. We also conduct experiments of the above parameter optimization in settings of various noise levels. Our results suggest that the optimized parameter is related to the noise level of measurements (i.e., the optimized parameter increases with respect to the noise level).

We analyze the ridge-based profiling both in theory and by experiments. Our theoretical investigation aims to answer the question:

Why, how and when is ridge-based profiling better?

where 'why' aims to justify the improvement of ridge-based profiling over LR-based one, 'how' and 'when' analyze to which extent and under what condition an improvement can be achieved. Then for a comprehensive comparison, we evaluate the performances of classical, LR-based and ridge-based profiling in simulation-based experiments on various settings, which shows the improvement of ridge-based profiling and confirm the theoretical analysis, At last, we conduct the practical experiments on the FPGA implementation. The results are consistent to the ones of simulation-based experiments, and furthermore, they show that the ridge-based profiling can tolerate (some) differences between profiling and exploitation traces, resulting in a type of robust profiling [25]. Therefore, on one hand, our results can be considered as an improvement of [3,20,24]. And on the other hand, we extend the related works which applied the stepwise and ridge regressions to the non-profiled setting [23,26].

2   Background

Following the 'divide-and-conquer' strategy, a profiled DPA attack breaks down a secret key into a number of subkeys of small length and recovers them independently. Let X be a vector of some (partial) plaintext in consideration, i.e., $X = (X_i)_{i \in \{1,...,n\}}$, where n is the number of measurements and X_i corresponds to the (partial) plaintext of i-th measurement. Let k be a hypothesis subkey, let $F_k : \mathbb{F}_2^m \to \mathbb{F}_2^m$ be a target function, where m is the bit length of X_i, and thus the intermediate value $Z_{i,k} = F_k(X_i)$ is called a target and $Z_k = F_k(X) = (Z_{i,k})_{i \in \{1,...,N\}}$ is the target vector obtained by applying F_k to X component-wise.

The leakage of a target can be scattered over several points in a measurement's power consumption. Let $L^j : \mathbb{F}_2^m \to \mathbb{R}$ be the leakage function at jth point and let T_i be a vector of power consumption points whose target is Z_{i,k^*}. We have $T_i^j = L^j \circ Z_{i,k^*} + \varepsilon_j$ and $T^j = L \circ Z_{k^*} + \varepsilon_j$, where $\circ$ denotes function composition, k^* is the correct subkey key and ε_j denotes probabilistic noise. A trace t_i is the combination of power consumption T_i and plaintext X_i, i.e., $t_i = (T_i, X_i)$. Let the function $M^j : \mathbb{F}_2^m \to \mathbb{R}$ be the model that approximates the determinate part of leakage function L^j, namely, $T_i^j \approx M^j \circ F_{k^*}(X_i) + \varepsilon_j$.[1] The model is obtained by learning from the profiled device in the profiling phase.

Profiled DPA can be divided into two phases: profiling phase and online exploitation phase. In the rest of this section, we recall these two phases. Our presentation is largely based on the (excellent) introduction provided in [24].

2.1   Profiling Phase

The aim of the profiling phase is to 'learn' the leakage functions L^j and the noises ε_j for all the points. We briefly introduce classical and LR-based profilings below.

Classical profiling. Classical profiling is the profiling phase of template attacks [3] and it views the leakage of each intermediate value as a vector of random values following the multivariate Gaussian distribution, i.e., $T_z \sim N(\mu_z, \Sigma_z)$, where T_z is the power consumption (points) given the associated intermediate target being z. The adversary 'learns' the physical leakages by finding the $p \times 1$ sample mean $\hat{\mu}_z$ and the $p \times p$ sample covariance $\hat{\Sigma}_z$ for all the target z on the profiling device. Finally, the intermediate value-conditioned leakages is $N(\hat{\mu}_z, \hat{\Sigma}_z)$ for the intermediate value z. As suggested in [4], we assume the noise distribution of

---

[1] We often omit the superscript 'j' in L^j, M^j and ε^j for succinctness.

different intermediate targets to be equal and use the same covariance estimates (across all intermediate targets).

Linear regression-based profiling. LR-based profiling [20] uses the stochastic model of the following form: $M(Z_i) = \alpha_0 + \sum_{u \in \mathbb{F}_2^m} \alpha_u Z_i^u + \varepsilon$, where coefficients $\alpha_u \in \mathbb{R}$, $Z_i = Z_{i,k^*}$, z^u denotes monomial $\prod_{j=1}^m z_j^{u_j}$, and z_j (resp., u_j) refers to the jth bit of z (resp., u). The degree of the model is the highest degree of the non-zero terms in polynomial $M(Z_i)$. Define the set $\mathbb{U}_d = \{u | u \in \mathbb{F}_2^m, \mathrm{HW}(u) \leq d\}$ (where $\mathrm{HW} : \mathbb{F}_2^m \to \mathbb{Z}$ is the Hamming weight function), then we denote $\boldsymbol{\alpha}_d = (\alpha_u)_{u \in \mathbb{U}_d}$ as the vector of coefficients with degree d, which is estimated from $\boldsymbol{U}_d = (Z_i^u)_{i \in \{1,2,...,N\}, u \in \mathbb{U}_d}$ and T using ordinary least squares, i.e., $\boldsymbol{\alpha}_d = (\boldsymbol{U}_d^\mathrm{T} \boldsymbol{U}_d)^{-1} \boldsymbol{U}_d^\mathrm{T} T$, where $(Z_i^u)_{i \in \{1,2,...,N\}, u \in \mathbb{U}}$ is a matrix with (i,u) being row and column indices respectively, and $\boldsymbol{U}_d^\mathrm{T}$ is the transposition of $\boldsymbol{U}_d$.

In the LR-based profiling phase, the adversary chooses the degree of model and calculates the coefficients $\boldsymbol{\alpha}$ of the profiling device. Then, the $p \times p$ sample covariance $\hat{\Sigma}$ is computed assuming the noise distributions are identical for various values of intermediate. Finally, the intermediate value-conditioned leakages is $\mathrm{N}(\hat{\alpha}_0 + \sum_{u \in \mathbb{U}_d} \hat{\alpha}_u z_i^u, \hat{\Sigma})$ for the intermediate value z.

2.2 Online Exploitation Phase

Bayesian key recovery. If the covariance matrix is symmetric and positive definite, a p-dimensional multivariate Gaussian distribution $\mathrm{N}(\mu, \Sigma)$ has the following density function:

$$f(x) = \frac{1}{(2\pi)^{d/2} |\Sigma|^{1/2}} \exp\left(-\frac{1}{2}(x - \mu)^\mathrm{T} \Sigma^{-1}(x - \mu)\right) . \tag{1}$$

Therefore, we can describe Bayesian key recovery as follows:

1. Acquire n traces (T_i, X_i), each of p points, for $1 \leq i \leq n$ from the target device.
2. Make a subkey guess k and compute the corresponding intermediate target $Z_{i,k} = F_k(X_i)$ for $1 \leq i \leq n$.
3. Calculate the log likelihood: $\prod_{i=1}^n \log(f_{i,k}(T_i))$, where $f_{i,k}(\cdot)$ is the density function associated with the intermediate target $Z_{i,k}$.
4. The log likelihood should be maximum upon correct key guess (which can be decided after repeating the above for all possible subkey guesses).

Correlation DPA. Correlation DPA employs a simple (univariate) online exploitation strategy, and it finds the subkey guess under which the correlation between the determinate part of the template (e.g., $\mathrm{M}_{classical}(z) = \hat{\mu}_z$ in 'classical' profiling and $\mathrm{M}_{LR}(z) = \hat{\alpha}_0 + \sum_{u \in \mathbb{F}_2^m} \hat{\alpha}_u z_i^u$ in LR-based profiling) and the (univariate) leakage is maximized, namely,

$$k_{guess} = \underset{k}{\mathrm{argmax}}\, \rho(\mathrm{M}(Z_{i,k}), T_i) \tag{2}$$

where ρ is the Pearson's coefficient.

3 Ridge-Based Profiling

In this section, we introduce our ridge-based profiling and give a formal analysis. We consider only the deterministic part of the model, and meanwhile the sample variance $\hat{\Sigma}$ is considered the same way as LR-based profiling.

3.1 Construction

Our new profiling (for each power consumption point) can be see as a generalization of LR-based profiling by explicitly imposing penalty on the coefficients' size, formally,

$$\hat{\alpha}_d^{ridge} \overset{\text{def}}{=} \underset{\alpha}{\text{argmin}} \sum_{i=1}^{N} \left(T_i - M_d^{ridge}(Z_i) \right)^2 ,$$

$$\text{subject to} \sum_{u \in \mathbb{U}_d} \alpha_u^2 \le s. \tag{3}$$

An equivalent formulation to above is (see [11] for detailed derivation):

$$\hat{\alpha}_d^{ridge} = \underset{\alpha}{\text{argmin}} \left(\sum_{i=1}^{N} (T_i - M_d^{ridge}(Z_i))^2 + \lambda \sum_{u \in \mathbb{U}_d} \alpha_u^2 \right), \tag{4}$$

whose optimal solution is given by:

$$\hat{\alpha}_d^{ridge} = (U_d^{\mathrm{T}} U_d + \lambda I_d)^{-1} U_d^{\mathrm{T}} T, \tag{5}$$

where $\mathbb{U}_d$, U_d and Z_i are defined in Sect. 2.1, matrix I_d is the $|\mathbb{U}_d| \times |\mathbb{U}_d|$ identity matrix and $|\mathbb{U}_d|$ denotes the cardinality of $\mathbb{U}_d$.

Parameter optimization. As illustrated above, there is an undetermined parameter (i.e., λ), the choice of which affects the performance of the profiling. For each power consumption point, we propose a method to choose the optimized parameter based on the K-fold[2] cross-validation technique from statistical learning. We mention that cross-validation was already used in the field of side-channel attack (for different purposes), such as evaluation of side-channel security [6] and unprofiled DPA [23]. Algorithm 1 finds the optimized parameter using cross-validation, where we omit the subscript d (the degree) for succinctness.

We sketch the algorithm below. We first choose a set of candidate parameters (up to some accuracy), and then split profiling traces into K parts $\mathcal{C}_{\{1...K\}}$ of roughly equal size. For each part $\mathcal{C}_i$, we compute the coefficients $\alpha_{\lambda,i}$ using the remaining $K-1$ parts from the trace set, and calculate the goodness-of-fit $R_{\lambda,i}$ using the traces in $\mathcal{C}_i$, which is a measurement of similarity between estimated power consumption and the actual power consumption T.[3] We then get the

---

[2] We shall not confuse K with k in online exploitation phase, where K is a parameter as in the "K-fold cross-validation" and k is a subkey hypothesis.

[3] We use the coefficient of determination to measure the goodness-of-fit in this paper, i.e., $R = \sum_{i=1}^{N_t} (\hat{T}_i - T_i)^2 / \sum_{i=1}^{N_t} (T_i - \sum_{i=1}^{N_t} T_i)^2$, where $\hat{T}$ is the estimated power consumption and N_t is the trace number in $\mathcal{C}_i$.

Algorithm 1. finding the optimized parameter

Require: profiling traces $t_i = \{T_i, x_i\}$ where $i \in \{1, ..., N\}$; the number of parts K;
    the true key k^*; the set of candidate parameters Λ;
Ensure: $\hat{\lambda}$ as the optimized parameter for the subkey;
  1: **for** $i = 1$; $i <= K$; $i{+}{+}$ **do**
  2:    $\mathcal{C}_i = \{t_{K*(i-1)+1}, ..., t_{K*i}\}$
  3: **end for**
  4: **for all** λ such that $\lambda \in \Lambda$ **do**
  5:    **for** $i = 1$; $i <= K$; $i{+}{+}$ **do**
  6:        Compute the $\alpha_{\lambda,i}$ using the traces in $\mathcal{C}_j$, where $j \in \{1...K\} \setminus \{i\}$
  7:        Calculate the goodness-of-fit $R_{\lambda,i}$ from $\mathcal{C}_i$
  8:    **end for**
  9:    $R_\lambda = (\sum_{i=1}^{K} R_{\lambda,i})/K$
10: **end for**
11: $\hat{\lambda} = \underset{\lambda}{\mathrm{argmax}}\, R_\lambda$

---

average goodness-of-fit $R_\lambda = (\sum_{i=1}^{K} R_{\lambda,i})/K$ for the each candidate parameter λ in consideration. Finally, we return the parameter with the highest averaged goodness-of-fit.

3.2   Theoretical Analysis

In this sub-section, we investigate the improvement of ridge-based profiling (over LR-based one) theoretically. We first answer the 'why' and 'how' questions by analyze the sampling variance of model's coefficients. Then we answer the 'when' question by studying the way that the coefficients shrink in the ridge-based profiling.

Why and How is Ridge-Based Profiling Better? For simplicity we consider the univariate leakage, where the leakage of the i-th trace is $T_i = \mathrm{L} \circ Z_{i,k^*} + \varepsilon$. Since the coefficients learned from the LR-based (resp., ridge-based) profiling determine the model (by definition), varying the coefficients will affect stability of the performance. The variance-covariance matrix of the coefficients learned from the LR-based (resp., ridge-based) profiling are given by [13, Eq. 4.8]:

$$\mathrm{Var}(\alpha_d^{lr}) = (U_d^{\mathrm{T}} U_d)^{-1} \sigma^2 \tag{6}$$

$$\mathrm{Var}(\alpha_d^{ridge}) = W U_d^{\mathrm{T}} U_d W \sigma^2 \tag{7}$$

where $W = (U_d^{\mathrm{T}} U_d + \lambda I_d)^{-1}$ and σ^2 is the variance of noise ε, which is identical for both LR-based and ridge-based profilings.

Without loss of generality, we fix $\sigma^2 = 1$ and the target values to be bytes, then compare $\mathrm{Var}(\alpha_d^{lr})$ to $\mathrm{Var}(\alpha_d^{ridge})$. Figure 1 shows that the variances goes up with the increase of d and the decrease of λ. For the same degree and parameter, the variance learned from ridge-based profiling are much lower than the ones from

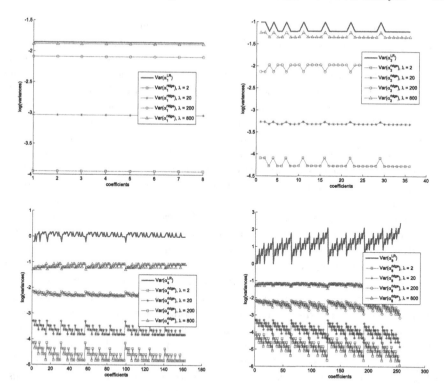

Fig. 1. The variances of the coefficients for different degrees (of the model) and λ. The upper-left, upper-right, lower-left, and lower-right figures correspond to the cases for $d = 1$, $d = 2$, $d = 4$, and $d = 8$ respectively.

LR-based profiling, thus the former has a more stable performance and is less prone to noise. Thus, to avoid overfitting one may use a large λ, but then it may result in a biased model, i.e., the difference between the leakage function and the model becomes more significant, which also decreases performance. Therefore, for best performance we need to choose a judicious value for λ by reaching a tradeoff between bias and coefficients' variance. To this end, we propose to use the cross-validation method in parameter optimization (see Sect. 3.1).

How the Coefficients Shrink in the Ridge-Based Profiling? As described before, the ridge-based profiling enforces a general constraint $\sum_{u \in \mathbb{U}} \alpha_u^2 < s$ on the coefficients of M_k, but it is not clear how each individual coefficient α_u shrinks (e.g., which coefficient shrinks more than the others). In [23], an interesting connection between the degree of a term $Z_{i,k}^u$ in M_k (i.e., the Hamming Weight of u) and the amount of shrinkage of its coefficient α_u is shown. See the following for a brief introduction and a conclusion of the analysis, and we refer to [23] for more details.

The principal components of U_d are a set of linearly independent vectors obtained by applying an orthogonal transformation to U_d, i.e., $P_d = U_d V_d$, where the columns of matrix V_d are called directions of the (respective) principal components. An interesting property is that among the columns of V_d, the first one, denoted V_d^1 (the direction of P_d^1), has the maximal correlation to coefficient vector α_d. Figure 2(a) and (b) depict the direction of the first principal component V_8^1 and the degrees of terms in U_8 respectively, and they represent a high similarity (albeit in a converse manner). Quantitatively, the Pearson's coefficient between V_8^1 and the corresponding vector of degrees is -0.9704, which is a nearly perfect negative correlation. Therefore, we establish the connection that α_u is conversely proportional to the Hamming weight of u. Above analysis is based on the $d = 8$ setting, for the other degrees (1 to 7), similar results can be obtained. To summarize, the more Hamming weight that u has, the less α_u contributes to the model. Therefore, ridge-based distinguisher is consistent with the leakage functions that consist of more low degree terms.

Another observation is that the improvement of ridge-based profiling (over LR-based one) is significant only for non-linear models (used for profiling). We can see that for the model of degree 1 the $u(s)$ of all coefficients have same Hamming weight, and thus every coefficient contributes equally to the model. That is, the coefficients shrink equally in this setting, which leads to comparable performance for both ridge-based and LR-based profilings. However, we stress that the degree of the model (for profiling) is not the same as (and typically no less than) that of the leakage function, and ridge-based profiling can just still enjoy performance improvement for linear leakage functions by setting the degree of model to be greater than 1. We refer to Sect. 4.1, where we will show that the ridge-based profiling outperforms the LR-based one for leakage function of degree 1 and model of degree 4.

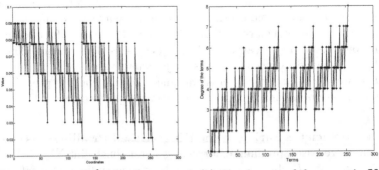

(a) The value of V_8^1 (the direction of the first principal component of U_8).

(b) The degrees of the terms in U_8.

Fig. 2. The similarity between the direction of the first principle component V_8^1 and the degrees of terms in U_8

4 Experimental Results

4.1 Simulation-Based Experiments

In this section, we evaluate the ridge-based, LR-based and classical profiling for univariate leakage functions with different degrees and randomized coefficients in the setting of simulated traces. We target at AES-128's first S-box of the first round with an 8-bit subkey (recall that AES-128's first round key is the same as its encryption key). We do the following trace pre-processing to facilitate the profiling: we average the traces based on their the input (an 8-bit plaintext) and use the resulting 256 mean power traces to mount the profiling. This reduces noise and the number of traces needed for profiling (as otherwise the running time goes unnecessarily high with a large number of 'raw' traces).

Finding the Optimized Parameter. At the beginning of ridge-based profiling, the adversary should first find the optimized parameter (i.e., the λ). We evaluate parameter optimization algorithm from Sect. 3.1. We consider the settings whose the degrees (of both leakage function and model) are fixed to 4 and under different signal-noise ratios (SNRs) (0.5, 0.1, 1). Let the set of parameter choices be $\Lambda = \{0.1, 1, 10, 50, 200, 800, 2000, 8000\}$, for which we conduct the parameter optimization algorithm 100 times (each time with a different random leakage function). For a fair comparison, we normalized[4] the averaged goodness-of-fits (of each experiment) and plot them in Fig. 3. We also highlight the mean of the averaged goodness-of-fits with red bold line. This confirms the intuition that the optimized parameter (which corresponds to each setting's minimum averaged goodness-of-fit) decreases with SNR.

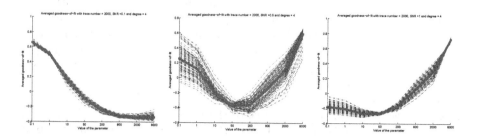

Fig. 3. Averaged goodness-of-fits and their mean values, with SNR = 0.1 (left-hand), 0.5 (middle), 1 (right-hand). (Color figure online)

---

[4] We apply the averaged goodness-of-fit for normalization, i.e., $\mathrm{norm}(R_\lambda) = (R_\lambda - \mathrm{mean}(R)/(\mathrm{max}(R) - \mathrm{min}(R)))$, where $\mathrm{mean}(R)$ is the average of $\{R_\lambda\}_{\lambda \in \Lambda}$ and $\mathrm{norm}(\cdot)$ is the normalization function.

A Comparison of Different Profilings in Simulation-Based Experiments. We compare different profilings (i.e., classical, LR-based and ridge-based profiling) using two metrics, namely, theoretical information and guessing entropy. The former computes the Perceived Information (PI) [6] between the secret variable and its leakage, and the latter combines the correlation DPA with the model built from one of three different profilings above and mounts the attack 100 times (each time with a different random leakage function) to compute the averaged ranking of the real key.

Figure 4 compares the Perceived Information and guessing entropies (as functions of the number of profiling traces) for different degrees of leakage function. The left-hand three sub-figures show the Perceived Information and the right-hand ones present the guessing entropies. The two sub-figures of the same row correspond to the Perceived Information and guessing entropy for leakage functions of the same degree respectively. Intuitively, the PI is an information theoretic metric that relates to the success rate of a profiled adversary using the estimated model obtained thanks to LR-based or ridge-based regression [5]. So it is the most revealing metric for comparing profiling phases [22]. In particular, the left parts of Fig. 4 exhibit both the informativeness of the model after sufficient profiling (i.e. the final Y axis values) and the efficiency of the profiling (i.e. how fast we converge towards this value). The guessing entropy metric is used as a confirmation that this intuition is matched and could be computed for any number of traces in the exploitation phase. In the profiling phase, we choose the same degree for the model and the leakage function. For all scenarios, the two metrics are consistent: the PI increases and the guessing entropies approaches to 1 with the increase of the number of traces. As clear from the PI figures, the ridge-based profiling performs better than the other two ones in all settings except for the $d = 1$ setting. More precisely, it generally has a better convergence speed, without any significant reduction of the final informativeness. Meanwhile the performance of LR-based profiling lies in between classical and ridge-based ones and it is largely affected by the degree of the leakage function. These observations confirm the theoretical analysis in Sect. 3.2. The guessing entropies computed in function of the number of profiling traces (for a fixed number of attack traces) confirm these trends.

Note that the typical scenario we are interested in is when the adversary has no knowledge about the actual degree of the leakage function for his profiling. In this case, our results show that he may use a conservative estimate about the degree of the model in the profiling phase without loosing efficiency (i.e. speed of convergence). To reflect this case, we also conduct experiments where the estimated degree of the model is higher that its actual value. That is, we simulate the traces with leakage functions of degrees 1 and 2 and then conduct the above experiments assuming a model of degree 4 for profiling. As shown in Fig. 5, the performance of ridge-based profiling is again significantly better. Therefore, our results show that an adversary (or an evaluation laboratory) can simply use a 'conservatively' estimated degree in ridge-based profiling, instead of running an enumeration on its possible values.

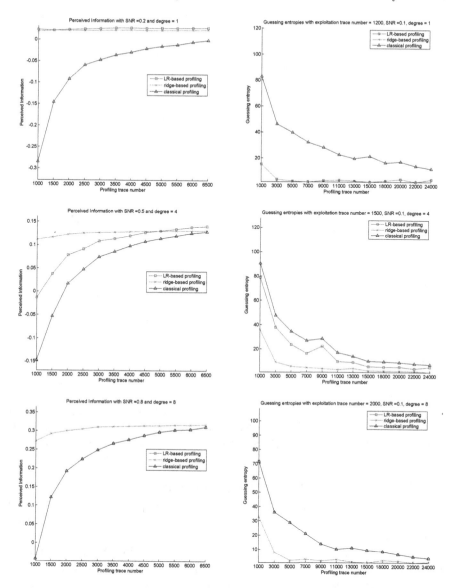

Fig. 4. A comparison of Perceived Information and guessing entropies (in functions of the number of profiling traces) for different degrees of leakage function, where the rows correspond to degrees 1, 4 and 8 respectively.

4.2 Experiments on Real FPGA Implementation

We carry out experiments on the SAKURA-X which running the AES on Xilinx FPGA devices Kintex-7 (XC7K70T/160T/325T). We amplified the signal using a (customized) LANGER PA 303N amplifier, providing 30 dB of gain. Then we measure the (absolute value of) power consumptions of the first round

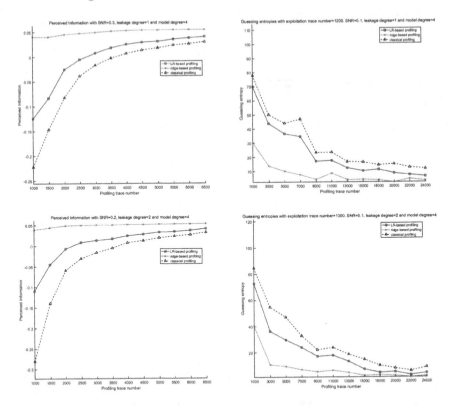

Fig. 5. The Perceived Information and guessing entropies with 'conservatively' degree of model for different numbers of exploitation traces, where the rows correspond to degrees of leakage function 1 and 2 respectively, and the degree of both models is 4.

S-box output, using a LeCroy waverunner 610Zi digital oscilloscope at a sampling rate of 1 GHz. Figure 6 shows the averaged trace[5] of the measurements of first round, we marked the leakage regions of the intermediate variable (i.e., the S-box output) in the figure and target them in our following attacks. We can see that the intermediate variable leaks in both region A and B similarly. Additionally, for each region, we apply the principal component analysis (PCA) to compact measurements [1,2,21], then only target the point of first principal component. And before the profiling, we perform the pre-processing, whose results are 256 mean traces of single point. In the following, to better illustrate the improvement of our new proposed method, we conduct two experiments for two different settings, in which we always profile on points of region A but attack (do the exploitation) on points of different regions.

---

[5] We shall not confuse the 'averaged trace' with the '256 mean power traces', where the former one is the mean of all the power traces which is only for the presentation of the measurements. And the latter one, as the result of pre-processing, is the means of the traces of same corresponding plaintext.

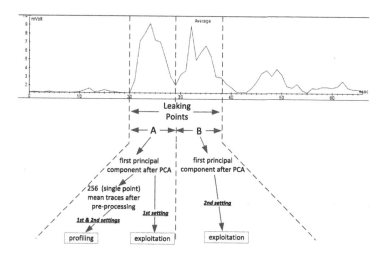

Fig. 6. The average trace of the measurements and the leaking points.

First, we assume a common setting (the 1st setting in Fig. 6) where the profiling and exploitation points are perfect aligned, thus we use the same region (i.e., region A) for both profiling and exploitation. Figure 7(a) shows the guessing entropies (as functions of the number of profiling traces) for ridge-based with different degrees power model in this setting. The parameter (i.e., $\lambda = 8000$) is chosen by means of the cross-validation as simulation-based experiments. We present the guessing entropies of the LR-based profiling with power model of degree 1 as the base line, since (in our attack scenario) it outperforms the LR-based profiling with higher degree as well as the classical one. We can see that the degree of the leakage function of our implementation is around 2. The result shows that (under this implementation) the ridge-based profiling with power model of degree 2 is the best one and perform better than the LR-based one (with power model of degree 1), which is consistent to the results of simulation-based experiments and theoretical analysis.

Further, we conduct another experiments to show that our new method can be used as a type of robust profiling [25], which can tolerate (some) differences between profiling and exploitation traces in a more realistic setting. As shown in Fig. 6 (the 2nd setting), we profile on the points in A and attack (do the exploitation) on the points in B. We aim to show how the miss-alignment of the points affects the ridge-based profiling. Figure 7(b) presents the guessing entropies (as functions of the number of profiling traces) for ridge-based with different degrees power model. We choose a larger parameter $\lambda = 500000$ by using the parameter optimization process in Sect. 3.1. We also add the LR-based profiling (with power model of degree 1) as the base line. The results show that the performance of ridge-based profiling is better than the LR-based one, which means that the performance of the new profiling method is better robust than LR-based one to the distortions between profiling and exploitation points.

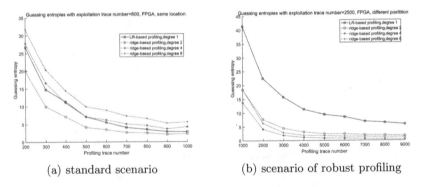

(a) standard scenario   (b) scenario of robust profiling

Fig. 7. A comparison of guessing entropy (in functions of the number of profiling traces) for FPGA implementation.

5   Conclusion

In this paper, we propose a new profiled differential power analysis based on ridge regression. Our theoretical analysis and experiments double confirm that the proposed profiling method has better performance than LR-based one by using a more stable (to avoid overfitting) and has a good potential to be a type of robust profiling. In view of the importance of profiled based side-channel analysis in security evaluations, these results show ridge-based profiling can allow laboratories to save significant factors in the number of traces they need to build a satisfying leakage model.

Acknowledgments. This work has been funded in parts by Major State Basic Research Development Program (973 Plan), the European Commission through the ERC project NANOSEC and by the INNOVIRIS project SCAUT. Yu Yu was supported by the National Natural Science Foundation of China Grant (Nos. 61472249, 61572192, 61572149), Science and Technology on Communication Security Laboratory (9140C110203140C11049), and International Science & Technology Cooperation & Exchange Projects of Shaanxi Province (2016KW-038). François-Xavier Standaert is a research associate of the Belgian Fund for Scientific Research (FNRS-F.R.S.). Dawu Gu was supported by National Natural Science Foundation of China (No. 61472250).

References

1. Archambeau, C., Peeters, E., Standaert, F., Quisquater, J.: Template attacks in principal subspaces. In: Goubin, L., Matsui, M. (eds.) [9], pp. 1–14
2. Batina, L., Hogenboom, J., Woudenberg, J.G.J.: Getting more from PCA: first results of using principal component analysis for extensive power analysis. In: Dunkelman, O. (ed.) CT-RSA 2012. LNCS, vol. 7178, pp. 383–397. Springer, Heidelberg (2012). doi:10.1007/978-3-642-27954-6_24
3. Chari, S., Rao, J.R., Rohatgi, P.: Template attacks. In: Kaliski, B.S., Koç, K., Paar, C. (eds.) CHES 2002. LNCS, vol. 2523, pp. 13–28. Springer, Heidelberg (2003). doi:10.1007/3-540-36400-5_3

4. Choudary, O., Kuhn, M.G.: Efficient template attacks. In: Francillon, A., Rohatgi, P. (eds.) [7], pp. 253–270. http://dx.doi.org/10.1007/978-3-319-08302-5

5. Duc, A., Faust, S., Standaert, F.-X.: Making masking security proofs concrete. In: Oswald, E., Fischlin, M. (eds.) EUROCRYPT 2015. LNCS, vol. 9056, pp. 401–429. Springer, Heidelberg (2015). doi:10.1007/978-3-662-46800-5_16

6. Durvaux, F., Standaert, F.-X., Veyrat-Charvillon, N.: How to certify the leakage of a chip? In: Nguyen, P.Q., Oswald, E. (eds.) EUROCRYPT 2014. LNCS, vol. 8441, pp. 459–476. Springer, Heidelberg (2014). doi:10.1007/978-3-642-55220-5_26

7. Francillon, A., Rohatgi, P. (eds.): CARDIS 2013. LNCS, vol. 8419. Springer, Cham (2014). http://dx.doi.org/10.1007/978-3-319-08302-5

8. Gierlichs, B., Lemke-Rust, K., Paar, C.: Templates vs. stochastic methods. In: Goubin, L., Matsui, M. (eds.) [9], pp. 15–29

9. Goubin, L., Matsui, M. (eds.): CHES 2006. LNCS, vol. 4249. Springer, Heidelberg (2006)

10. Güneysu, T., Handschuh, H. (eds.): CHES 2015. LNCS, vol. 9293. Springer, Heidelberg (2015). http://dx.doi.org/10.1007/978-3-662-48324-4

11. Hastie, T., Tibshirani, R., Friedman, J.: The Elements of Statistical Learning: Data Mining, Inference, and Prediction, 2nd edn., vol. 1, pp. 43–94. Springer, New York (2009)

12. Heuser, A., Zohner, M.: Intelligent machine homicide. In: Schindler, W., Huss, S.A. (eds.) COSADE 2012. LNCS, vol. 7275, pp. 249–264. Springer, Heidelberg (2012). doi:10.1007/978-3-642-29912-4_18

13. Hoerl, A.E., Kennard, R.W.: Ridge regression: biased estimation for nonorthogonal problems. Technometrics 12(1), 55–67 (1970)

14. Hospodar, G., Gierlichs, B., Mulder, E.D., Verbauwhede, I., Vandewalle, J.: Machine learning in side-channel analysis: a first study. J. Cryptographic Eng. 1(4), 293–302 (2011)

15. Kocher, P., Jaffe, J., Jun, B.: Differential power analysis. In: Wiener, M. (ed.) CRYPTO 1999. LNCS, vol. 1666, pp. 388–397. Springer, Heidelberg (1999). doi:10.1007/3-540-48405-1_25

16. Lerman, L., Bontempi, G., Markowitch, O.: Power analysis attack: an approach based on machine learning. IJACT 3(2), 97–115 (2014)

17. Lerman, L., Poussier, R., Bontempi, G., Markowitch, O., Standaert, F.-X.: Template attacks vs. machine learning revisited (and the curse of dimensionality in side-channel analysis). In: Mangard, S., Poschmann, A.Y. (eds.) COSADE 2014. LNCS, vol. 9064, pp. 20–33. Springer, Heidelberg (2015). doi:10.1007/978-3-319-21476-4_2

18. Martinasek, Z., Hajny, J., Malina, L.: Optimization of power analysis using neural network. In: Francillon, A., Rohatgi, P. (eds.) [7], pp. 94–107. http://dx.doi.org/10.1007/978-3-319-08302-5

19. Quisquater, J., Samyde, D.: Automatic code recognition for smartcards using a kohonen neural network. In: Proceedings of the Fifth Smart Card Research and Advanced Application Conference, CARDIS 2002, November 21–22, 2002, San Jose, CA, USA (2002)

20. Schindler, W., Lemke, K., Paar, C.: A stochastic model for differential side channel cryptanalysis. In: Rao, J.R., Sunar, B. (eds.) CHES 2005. LNCS, vol. 3659, pp. 30–46. Springer, Heidelberg (2005). doi:10.1007/11545262_3

21. Standaert, F.-X., Archambeau, C.: Using subspace-based template attacks to compare and combine power and electromagnetic information leakages. In: Oswald, E., Rohatgi, P. (eds.) CHES 2008. LNCS, vol. 5154, pp. 411–425. Springer, Heidelberg (2008). doi:10.1007/978-3-540-85053-3_26

22. Standaert, F.-X., Koeune, F., Schindler, W.: How to compare profiled side-channel attacks? In: Abdalla, M., Pointcheval, D., Fouque, P.-A., Vergnaud, D. (eds.) ACNS 2009. LNCS, vol. 5536, pp. 485–498. Springer, Heidelberg (2009). doi:10.1007/978-3-642-01957-9_30

23. Wang, W., Yu, Y., Liu, J., Guo, Z., Standaert, F., Gu, D., Xu, S., Fu, R.: Evaluation and improvement of generic-emulating DPA attacks. In: Güneysu, T., Handschuh, H. (eds.) [10], pp. 416–432. http://dx.doi.org/10.1007/978-3-662-48324-4

24. Whitnall, C., Oswald, E.: Profiling DPA: efficacy and efficiency trade-offs. In: Bertoni, G., Coron, J.-S. (eds.) CHES 2013. LNCS, vol. 8086, pp. 37–54. Springer, Heidelberg (2013). doi:10.1007/978-3-642-40349-1_3

25. Whitnall, C., Oswald, E.: Robust profiling for DPA-style attacks. In: Güneysu, T., Handschuh, H. (eds.) [10], pp. 3–21. http://dx.doi.org/10.1007/978-3-662-48324-4

26. Whitnall, C., Oswald, E., Standaert, F.-X.: The myth of generic DPA...and the magic of learning. In: Benaloh, J. (ed.) CT-RSA 2014. LNCS, vol. 8366, pp. 183–205. Springer, Cham (2014). doi:10.1007/978-3-319-04852-9_10

My Traces Learn What You Did in the Dark: Recovering Secret Signals Without Key Guesses

Si Gao[1,2], Hua Chen[1(✉)], Wenling Wu[1], Limin Fan[1], Weiqiong Cao[1,2], and Xiangliang Ma[1,2]

[1] Trusted Computing and Information Assurance Laboratory, Institute of Software, Chinese Academy of Sciences, Beijing 100190, People's Republic of China
{gaosi,chenhua,wwl,fanlimin,caowq,maxiangliang}@tca.iscas.ac.cn
[2] University of Chinese Academy of Sciences, Beijing 100049, People's Republic of China

Abstract. In side channel attack (SCA) studies, it is widely believed that unprotected implementations leak information about the intermediate states of the internal cryptographic process. However, directly recovering the intermediate states is not common practice in today's SCA study. Instead, most SCAs exploit the leakages in a "guess-and-determine" way, where they take a partial key guess, compute the corresponding intermediate states, then try to identify which one fits the observed leakages better. In this paper, we ask whether it is possible to take the other way around—directly learning the intermediate states from the side channel leakages. Under certain circumstances, we find that the intermediate states can be efficiently recovered with the well-studied Independent Component Analysis (ICA). Specifically, we propose several methods to convert the side channel leakages into effective ICA observations. For more robust recovery, we also present a specialized ICA algorithm which exploits the specific features of circuit signals. Experiments confirm the validity of our analysis in various circumstances, where most intermediate states can be correctly recovered with only a few hundred traces. Our approach brings new possibilities to the current SCA study, including building an alternative SCA distinguisher, directly attacking the middle encryption rounds and reverse engineering with fewer restrictions. Considering its potential in more advanced applications, we believe our ICA-based SCA deserves more research attention in the future study.

Keywords: Side channel analysis · Signal recovery · Independent component analysis

1 Introduction

Nowadays, Side Channel Attacks (SCA) pose a major threat for various cryptographic devices [1–4]. With some data-dependent leakage measurements, an SCA attacker can efficiently retrieve the secret key, even if the underlying cipher is cryptographically strong.

© Springer International Publishing AG 2017
H. Handschuh (Ed.): CT-RSA 2017, LNCS 10159, pp. 363–378, 2017.
DOI: 10.1007/978-3-319-52153-4_21

In a typical SCA context (illustrated in Fig. 1), the attacker Eve encrypts T plaintexts and measures the corresponding leakages $\mathbf{L}$. In SCA, it is widely believed that $\mathbf{L}$ depends on some key-related intermediate states, denoted as $\mathbf{x}_k = \{x_k(1), ..., x_k(T)\}$. With certain key guess k_i, Eve computes the intermediate state sequence $\mathbf{x}_{k_i} = \{x_{k_i}(1), ..., x_{k_i}(T)\}$ according to the encryption algorithm. Throughout this paper, we denote this sequence as a *signal*. Since the leakages $\mathbf{L}$ only depend on the correct signal $\mathbf{x}_k$, Eve combines all possible $\mathbf{x}_{k_i}$ with $\mathbf{L}$ and learns the most likely key guess.

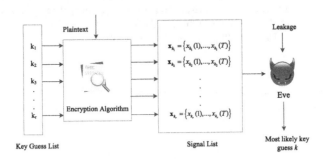

Fig. 1. A typical SCA procedure

In Fig. 1, Eve has a list of all possible intermediate state sequences (signals) and tries to find the correct one with the corresponding leakages. In SCA studies, such procedure is called a *side channel distinguisher*. The term *distinguisher* demonstrates its inherent limitation: such process only distinguishes the correct signal from the wrong ones, yet never directly retrieves any secret. A natural question to ask, is whether Eve can take one step further and directly learn the correct signal from the leakages, without an enumerative signal list.

It is not surprising that little SCA study answers this question. As the master key is the only secret in modern block ciphers, the "key-distingshers" above already present enough threat for unprotected chips. Nonetheless, recovering the intermediate states without a key guess may still be helpful in certain circumstances. For instance, in a typical SCA procedure, the computation cost strictly depends on the correlated key size. If the target intermediate state involves a large proportion of the secret key (>32 bits), enumerating all $\mathbf{x}_{k_i}$ becomes infeasible. Other examples include Side Channel Analysis for Reverse Engineering (SCARE), where the secret components baffle the computation of $\mathbf{x}_{k_i}$. Since the signal list cannot be efficiently computed, SCA in these cases needs a more general secret recovery technique.

Related Work. Despite a few *ad hoc* SPA-like attacks, in general, most previous non-profiled SCAs cannot directly recover the intermediate states. Collision attack may be the closest match in this direction [5]. In a collision attack, the attacker collects a few collisions and solves the intermediate states from the collision equations. More specifically, if the measurements from two Sbox computations match, we can reasonably predict they share the same input (a.k.a a *collision*). Since the leakages of the exact same Sbox computation are not always available, such "online-profiling" stage imposes restrictions on the implementations as well as the target ciphers. Indeed, none of the previous collision-based SCAREs gave realistic experiments to validate their results [6–8].

Our Contribution. In this paper, we aim to recover the secret intermediate states directly from the observed leakages. Our analysis shows that, under certain circumstances, recovering the intermediate states can be regarded as a noisy Blind Source Separation (BSS) problem. Following the study of BSS, we introduce the well-studied Independent Component Analysis (ICA) [9] to SCA. In signal processing, ICA is a widely used tool for recovering unknown sources in a blind context. Considering ICA takes at least n observations to recover n sources, we propose several methods to construct multi-channel observations from the side channel leakages. Moreover, since typical ICA algorithms are sensitive to noise, we present a more robust ICA algorithm based on Belouchrani and Cardoso's work on discrete ICA [10]. By exploiting the specific features of circuit signals, our specialized ICA gives efficient SCA recoveries: in our realistic experiments, our ICA-based SCA recovers over 80% of the intermediate states correctly, with only a few hundred traces. Furthermore, our ICA-based SCA brings several new possibilities to the current non-profiled SCA studies. In our experiments, our ICA-based SCA helps to improve the key-recovery result in a limited trace set, extend SCA to the middle encryption rounds as well as loosen the restrictions of current SCAREs. As a potentially powerful tool, we believe our ICA-based SCA deserves more research attention in the future.

Paper Organization. In the next section, we present a brief introduction of Independent Component Analysis, discussing its assumptions as well as limitations. Section 3 shows how to convert an SCA recovery to an ICA problem. Specifically, we propose several methods to construct multi-channel observations from the side channel leakages and build a specialized ICA algorithm for circuit signals. Section 4 demonstrates some advanced applications of our ICA-based SCA. With realistic measurements of an unprotected software implementation of DES, we confirm the validity of our approach. Further discussions about our ICA-based SCA and its promising prospects in the future are presented in Sect. 5.

2   Independent Component Analysis

2.1   Definition

Independent Component Analysis (ICA) [9] belongs to a boarder class of problems called *Blind Source Separation* (BSS) [11], which requires to separate a set of mixed signals, without the aid of information about the source signals or the mixing process. A common example is the *cocktail party problem*, which suggests a partygoer can focus on a single conversation in a noisy room.

Suppose we have n simultaneous conversations (sources) $\mathbf{S} = \{s_1, s_2, ..., s_n\}$ going on in the party room. Microphones are placed in different positions, recording m mixtures (observations) of the original sources $\mathbf{Y} = \{y_1, y_2, ..., y_m\}$. Assuming the observation y_j is a linear mixture of all sources, we have

$$y_j = a_{j,1}s_1 + a_{j,2}s_2 + ... + a_{j,n}s_n$$

where $a_{j,i}$ stands for the real-valued coefficient. The overall mixing procedure can be written as

$$\mathbf{Y} = \mathbf{AS}$$

where $\mathbf{A}$ is called the *mixing matrix*. In signal processing, such statistical model is called *Independent Component Analysis* [9]. With additional multivariate Gaussian noise $\mathbf{N}$, the noisy ICA model is defined as

$$\mathbf{Y} = \mathbf{AS} + \mathbf{N}$$

2.2   Assumptions and Ambiguities

Since both $\mathbf{S}$ and $\mathbf{A}$ are not given, in general, the BSS problem is highly underdetermined. In order to find useful solutions, BSS usually requires some additional assumptions. Specifically, ICA relies on the following assumptions [9]:

- **Independence:** The source signals are independent of each other.
- **Non-Gaussianity:** The source signals have non-Gaussian distributions.

In addition, typical ICA algorithms also assume the number of observations is no less than the number of sources ($m \geq n$) [9]. Like most noisy statistical models, noise significantly affects the effectiveness of ICA.

From the ICA model, it is not hard to see the following ambiguities hold [9]:

- **ICA cannot determine the amplitude of the sources.** As both $\mathbf{S}$ and $\mathbf{A}$ are unknown, any scalar multiplier in s_i can always be cancelled by dividing the corresponding column of $\mathbf{A}$ with the same scalar.
- **ICA cannot determine the order of the sources.** As both $\mathbf{S}$ and $\mathbf{A}$ are unknown, we can pick a random permutation matrix $\mathbf{P}$. In the ICA model, $\mathbf{Y} = \mathbf{AP}^{-1}\mathbf{PS}$, where $\mathbf{A}' = \mathbf{AP}^{-1}$ and $\mathbf{S}' = \mathbf{PS}$ is also a valid solution.

Besides, in ICA, the input signal usually enters a *whitening transformation* before any further analysis. Theoretically speaking, a *whitening transformation* is a decorrelation transformation that transforms a set of random variables into a set of new random variables with zero means and identity covariance. As a result, ICA only returns the estimates of the "whitened sources". In other words, the means of the original sources cannot be determined through ICA. For typical ICA applications like separating independent speeches, none of these ambiguities presents an obstacle in practice.

2.3   Existing ICA Algorithms

Despite the fact that the BSS problem is highly underdetermined, in the past 30 years, researchers have proposed many successful ICA algorithms, such like FastICA [12], JADE [13] and Infomax [14]. Although these algorithms use different measurements of independence, as Hyvärinen's explanation [9], they are indeed not "that" different. For efficiency, we simply choose FastICA as our primary ICA technique in this paper. Due to the space limit, here we omit further technical details. Interested readers can learn more from Hyvärinen's tutorial [9].

3   ICA-Based Signal Recovery

3.1   ICA versus SCA: Similarities and Differences

Throughout this paper, we assume the leakage follows the weighted Hamming Weight model. For an n-bit intermediate state X, the corresponding data-dependent leakage can be written as

$$L(x) = \alpha_0 + \alpha_1 x_1 + \alpha_2 x_2 + ... + \alpha_n x_n, \ \alpha_i \in \mathbb{R} \tag{1}$$

where x_i represents the i-th bit of x. With T times measurements, the sequence of the intermediate states $\mathbf{x} = \{\mathbf{x}(1), \mathbf{x}(2), ..., \mathbf{x}(T)\}$ forms a T-length signal. Apparently, $\mathbf{x}$ can also be regarded as a group of binary signals, where $\mathbf{x} = \{\mathbf{x}_1, ..., \mathbf{x}_n\}^\top$. As the leakages $\mathbf{y}$ capture the instantaneous mixtures of $\mathbf{x}$, SCA in this setting shares many similarities with ICA. Indeed, the basic assumptions of ICA—non-gaussianity and independence—are naturally satisfied: since all $\mathbf{x}_i$ are 1-bit 0–1 signals, the distribution of $\mathbf{x}_i$ is far from Gaussian. Considering the sources come from a cryptographic intermediate state, the cryptographic operation ensures each bit is statistically independent. The major difference is ICA requires at least n observations, whereas the SCA context provides only one. Although not explicitly stated, the additional noise may be another problem: since SCA exploits the unintended information leakage, the Signal-to-Noise-Ratio (SNR) in SCA is usually much lower than typical ICA contexts.

Table 1. Similarities and differences between ICA and SCA

	ICA	SCA
Sources	$\mathbf{s} = \{\mathbf{s}_1, \mathbf{s}_2, ..., \mathbf{s}_n\}$	$\mathbf{x} = \{\mathbf{x}_1, \mathbf{x}_2, ..., \mathbf{x}_n\}$
Distribution	non-Gaussian	Bernoulli
Independence	Independent	Independent
Observation	$\mathbf{Y} = \mathbf{AS} + \mathbf{N}$	$\mathbf{l} = \alpha\mathbf{x} + \mathbf{N}$
Number of observations	m	1
Level of noise	Low	High

Toy example. Assume our intermediate state $\mathbf{x}$ has only 2 bits (n=2). The leakages follow the standard Hamming Weight model, where both α_1 and α_2 in Equation (1) equals to 1 and $\alpha_0 = 0$. If the attacker takes 4 leakage measurements (T=4) with $\{\mathbf{x}(1) = 0, \mathbf{x}(2) = 1, \mathbf{x}(3) = 2, \mathbf{x}(4) = 3\}$, the resultant leakage measurements $\mathbf{l} = \{0, 1, 1, 2\}$ can be regarded as an observation in ICA. $\mathbf{x}_1 = \{0, 0, 1, 1\}$ and $\mathbf{x}_2 = \{0, 1, 0, 1\}$ are two blind sources in ICA and the mixing procedure is $\mathbf{l} = \mathbf{x}_1 + \mathbf{x}_2$.

3.2   Applying ICA in SCA: Obstacles and Solutions

As demonstrated in Table 1, the number of observations and the level of noise are two major obstacles for applying ICA in the SCA context. In the following, we further discuss these two obstacles and propose possible solutions.

Constructing multi-channel observations. The first difficulty we have to overcome is to construct multi-channel observations from the side channel leakages. In the following, our discussion focuses on the best solution we found for power leakage.

Our primary solution is based on a simple observation: if a binary source $\mathbf{s}$ is XORed with a constant k, the resultant source $\mathbf{s}'$ is

$$\mathbf{s}' = \begin{cases} \mathbf{s} & k = 0 \\ 1 - \mathbf{s} & k = 1 \end{cases}$$

In ICA, since the *whitening transformation* removes all constant terms in $\mathbf{s}'$, XORing $k = 1$ has the same effect as flipping the sign of the whitened source $\mathbf{s}$. According to our discussion in Sect. 2.2, we can move the flipping sign to the corresponding coefficients in the mixing matrix $\mathbf{A}$. In SCA, suppose we can measure the leakages of $X = S \oplus k$. With m different k_i, the overall model can be regarded as

$$\mathbf{x}' = \mathbf{s}$$

$$\mathbf{A}' = \begin{pmatrix} \alpha_1 & \cdots & \alpha_n \\ \alpha_1 & \cdots & \alpha_n \\ \vdots & \vdots & \vdots \\ \alpha_1 & \cdots & \alpha_n \end{pmatrix} \circ \begin{pmatrix} 2k_{1,1} - 1 & \cdots & 2k_{1,n} - 1 \\ 2k_{2,1} - 1 & \cdots & 2k_{2,n} - 1 \\ \vdots & & \vdots \\ 2k_{m,1} - 1 & \cdots & 2k_{m,n} - 1 \end{pmatrix}$$

where $k_{i,j}$ represents the j-th bit of k_i. If the resultant $\mathbf{A}'$ is non-singular, the attacker can simply uses n different k_i to collect input for ICA. On the other hand, if $\mathbf{A}'$ is singular, the attacker has to construct $m > n$ observations to ensure that ICA gets n linearly independent channels.

Toy example. In the previous example, assume the attacker can control a constant k that XORed to the intermediate state x. The attacker measures one observation l_1 when $k = 0$. Then, the attacker repeats his measurements with exactly the same plaintext inputs, using $k = 1$. In this case, the measured leakage signal becomes $l_2 = 1 - x_1 + x_2$. Considering the whitening stage, we can omit the constant term α_0 and write the model as

$$l = \begin{pmatrix} l_1 \\ l_2 \end{pmatrix} = \begin{pmatrix} 1 & 1 \\ -1 & 1 \end{pmatrix} \begin{pmatrix} x_1 \\ x_2 \end{pmatrix}$$

Apparently, l forms a valid input for ICA ($m = n = 2$).

Considering XOR is a common operation in symmetric cryptography, finding such constant should be easy. Since ICA does not need the actual value of k, in many block ciphers, the round key serves as a good candidate. It is worth mentioning that constructing multi-channel observations might be easier for certain implementations, such like software implementation of DES (Sect. 4).

Noise tolerance. As stated in Table 1, the Signal-to-Noise Ratio (SNR) in the SCA context is often much lower than the SNR in typical ICA applications. Thus, designing a more robust ICA is essential for ICA's application in SCA. Compared with the standard ICA context, ICA-based SCA does have some unusual *a priori* knowledge: all sources follow the Bernoulli distribution with $p = 0.5$. Taking the *a priori* distribution into consideration, ICA with the Maximum Likelihood Principle becomes a more robust choice. Specifically, Belouchrani and Cardoso had proposed an ICA algorithm specialized for discrete sources in 1994 [10]. Their approach estimates the unknown sources and the mixing matrix as well as the additional noise, then maximizes the likelihood via the Expectation-Maximization (EM) algorithm. With moderate adjustments, their EM-ICA can be a more robust candidate for circuit signal recovery.

3.3 Specialized ICA Algorithm

Taking the *a priori* distribution into consideration, we present a specialized ICA based on Belouchrani and Cardoso's EM-ICA [10]. In order to formally describe our ICA algorithm, we assume the sources X consist of T intermediate states $\{x(1), ..., x(T)\}$. Each bit of X forms an independent source, denoted as $x_i = \{x_i(1), ..., x_i(T)\}$. Similarly, the observations Y can also be written as $Y = \{y_1, ..., y_m\}^\top$, where $y_i = \{y_i(1), ..., y_i(T)\}$. The detailed algorithm is presented in Algorithm 1.

4   Applications in SCA

As our specialized ICA does not require any information from the plaintexts, it sheds light on several more advanced applications in SCA. Throughout this section, our experiments use leakages acquired from an unprotected software implementation of DES. The power consumptions were measured with a LeCroy WaveRunner 610Zi oscilloscope at a sampling rate of 20 MSa/s. The entire trace set contains 20 000 traces, with 80 000 sample points covering the first 3 rounds. An appealing property of this card, is that it performs DES's permutation P

Algorithm 1. Specialized ICA for SCA

Step 1: Collect m observations $\mathbf{Y} = \{\mathbf{y}_1, ..., \mathbf{y}_m\}^{\top}$

Step 2: Get n independent components $(\mathbf{IC}_1, \mathbf{IC}_2, ..., \mathbf{IC}_n)$ from FastICA

Step 3: Find the closest binary signals $\tilde{\mathbf{x}} = \{\tilde{\mathbf{x}}_1, \tilde{\mathbf{x}}_2, ..., \tilde{\mathbf{x}}_n\}$ for $\mathbf{IC}$

Step 4: Start EM-ICA with $\tilde{\mathbf{x}}$, estimate the initial parameter $\theta^{(0)}$

    while $\Delta L >$ threshold **do**

        E-Step: Compute the expectation of the log-likelihood $L\left(\theta^{(j)}\right)$ with current $\theta^{(j)}$

        M-Step: Compute the $\theta^{(j+1)}$ that maximize the expected log-likelihood function L

    end while

Step 5: Find the $\hat{\mathbf{x}}$ that maximize the expected log-likelihood function L

return binary signals $\hat{\mathbf{x}}$

---

bit-by-bit. As the influence of each signal bit is separated in time, the leakage trace naturally provides multi-channel observations. Most experiments in this section take advantage of this property. However, in Sect. 4.4, we present ICA analysis against the leakages of the Sboxes' input, where such property does not hold. Although the analysis becomes trickier, we can still perform a successful recovery with our XOR-constant method.

4.1 New SCA Distinguisher

Although key recovery is not our primary goal, surprisingly, in our experiments, our specialized ICA can serve as a competitor for common key recovery attacks. Specifically, let us focus on one of the Sboxes (S_5) in the first round. Figure 2(a) presents the performance of traditional univariate CPA (10 to 200 traces): the correct key (red line) stands out after 60 traces, although the distinguishing margin is quite small.

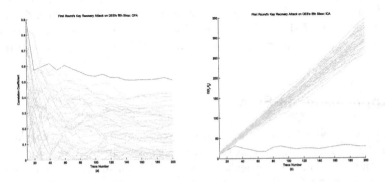

Fig. 2. Attacking S_5 in the first round: CPA v.s. ICA

On the other hand, in this particular implementation, we can also use our specialized ICA to recover the Sbox output sequence $\mathbf{X_r}$. In our experiments, we picked 5 points of interest ($m = 5$) from the leakages that correspond to S_5's output ($n = 4$). Similar to CPA, the attacker computes the guessed Sbox output sequences $\mathbf{X_{\hat{k}}}$ from the guessed key $\hat{k}$ and the known plaintext sequences $\{P_1, P_2, ..., P_T\}$:

$$\mathbf{X_{\hat{k}}} = \left\{ \mathrm{DES}_{S_5}(\hat{k}, P_1), \mathrm{DES}_{S_5}(\hat{k}, P_2), ..., \mathrm{DES}_{S_5}(\hat{k}, P_T) \right\}$$

where DES_{S_5} represents the corresponding DES encryption in the first round. In the following, the attacker need to decide which $\mathbf{X_{\hat{k}}}$ is the closest match for $\mathbf{X_r}$. Considering the ICA's ambiguity, if $\mathbf{X_r}$ is the signal that ICA returned, applying any bit permutation B or flipping any sign also gives a correct ICA result. Thus, we choose the distance between $\mathbf{X_{\hat{k}}}$ and its closest equivalent of $\mathbf{X_r}$ as our distinguish value. In the following, $\|\mathbf{v}\|_1$ stands for the L_1 norm (Manhattan norm) of $\mathbf{v}$, whereas $\mathbf{X_{\hat{k},i}}$ stands for the i-th bit of $\mathbf{X_{\hat{k}}}$.

$$D_{ICA}(\hat{k}) = \mathbf{D}\left(\mathbf{X_{\hat{k}}}, \mathbf{X_r}\right) = \min_B \left\{ \sum_{i=1}^{n} dist\left(\mathbf{X_{\hat{k},i}}, \mathbf{X_{r,B(i)}}\right) \right\}$$

$$dist\left(\mathbf{X_{\hat{k},i}}, \mathbf{X_{r,B(i)}}\right) = \min\left\{ \left\|\mathbf{X_{\hat{k},i}} - \mathbf{X_{r,B(i)}}\right\|_1, \left\|\mathbf{X_{\hat{k},i}} - \overline{\mathbf{X_{r,B(i)}}}\right\|_1 \right\}$$

Figure 2(b) presents the performance of our ICA-based SCA. The correct key stands out after 30 traces, which shows a slight advantage over CPA (60 traces). Besides, our ICA-based SCA provides a larger distinguishing margin: the distance with correct key stays stable after 50 traces, while the distances with incorrect keys increase linearly with the number of traces.

4.2 Extending SCA to the Middle Rounds

Considering ICA does not take a guess-and-determine procedure, in theory, our ICA-based SCA can be applied to the leakages of any encryption operation. This feature becomes crucial when the target implementation protects the first/last few rounds. In traditional SCAs, the attacker has to increase the guessed key space, in order to compute the middle round's intermediate states. This is hardly an issue for our ICA-based SCA: as long as the attacker can build effective observations, ICA can recover the intermediate states of any round.

Figure 3 demonstrates the results of recovering the 8 Sboxes' outputs ($n = 4$) in the second round. Following our discussion in Sect. 4.1, we further define the success rate of an ICA recovery as:

$$\mathbf{Succ}(\mathbf{X_r}) = \frac{\mathbf{D}\left(\mathbf{X_r}, \mathbf{X_c}\right)}{nT}$$

From this definition, it is not hard to see that the success rate is at least 0.5. If the success rate equals to 0.5, the ICA-returned signal is no better than

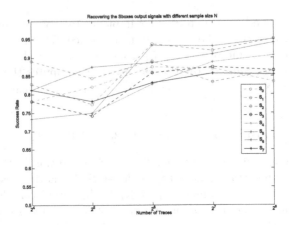

Fig. 3. Recovering the 8 Sboxes' outputs in the second round

a random guess. The success rates of all 8 attempts in Fig. 3 are over 0.8, which suggests our ICA-based attack learns most signal bits correctly. As our goal here is merely key recovery, 80% accuracy is enough for further cryptanalysis.

4.3   Reverse Engineering on Sbox

Despite the applications in key recovery, reverse engineering seems to be a more natural application for our ICA-based SCA. In order to infer the underlying cryptographic operations, SCARE usually requires to recover the secret intermediate states first. Indeed, the experiments in Sect. 4.1 already confirm our ICA-based SCA can recover some information about the intermediate states. In the following, let us first consider how to recover secret Sboxes.

Assume the target chip implements a customized DES with secret Sboxes. In the first round, the attacker can compute the output of the Expansion E. Similar to most SCAREs, the secret key is treated as a part of the secret Sbox $(S'(x) = S(x \oplus k))$, which means the output of E can be regarded as the input of S'. Since the Sboxes are secret, the attacker cannot take a key guess and analyze the leakages with traditional SCAs. Our ICA-based SCA works for this case, as long as the attacker can pick several independent leakage points. Figure 4(a) demonstrates the recovery of all 8 Sboxes' outputs ($n = 4$) in the first round with 5 ($m = 5$) manually picked leakage points. The analysis procedure is exactly the same as Sect. 4.1, except for the key distinguish step in the end. In all 8 attempts, our approach works very well with only 100 traces, recovering more than 80% of the intermediate states correctly. In Fig. 4(a), all 8 attempts report over 95% success rate after the number of traces reaches 400. Meanwhile, ICA with manually picked leakages always gives a few faulty recovery, even if the trace set is large (Fig. 4(a)).

The major drawback of picking leakages manually is the attacker may not know how to pick valid leakages in a non-profiled setting. In SCARE, since we

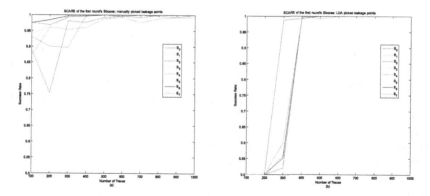

Fig. 4. Reverse engineering the first round's 8 Sboxes with ICA

have the Sboxes' input, Linear Discriminant Analysis (LDA) [15] can make the selection process easier. In our experiments, as the implementation performs the permutation bit-by-bit, the first 5 LDA components form a valid input for ICA. The benefit of LDA is twofold: on the one hand, the attacker does not have to choose leakage points himself. All he has to do is to estimate an approximate range on the trace corresponding to the target computation. On the other hand, in Fig. 4(b), our attack with LDA shows better recovery with larger trace sets. In fact, all 8 attempts achieve 100% accurate recovery with 600 traces. Noted in our traces set, the target operation lasts for 300–400 sample points. It is well known that LDA are not suitable for the cases where the number of traces is smaller than the range of interest [15]. This explains the fact that LDA gives poor results when the trace set is smaller than 400. Since the attacker has both the Sbox input and output now, writing the input and output in sequence gives the equivalent Sbox (up to ICA's ambiguity).

4.4 Reverse Engineering on Feistel Round Function

So far, all our experiments rely on the specific implementation of DES: as the implementation naturally provides multi-channel observations, the attacker can directly build ICA's input from the measured traces. For other ciphers or other implementations, this nice property does not always hold. In the following, our experiment demonstrates how the attacker can build his observations with our XOR-constant method and recover the output of the first round function.

Assume the attacker is reverse engineering a customized version of DES, where both the Sboxes and the linear permutation are altered. For convenience, we assume the attacker already knows the initial permutation IP and the expansion permutation E. Let F_{k1} denote the first round function of DES and (L_0, R_0) denote the initial input state (after IP). The first Sbox's input in the second round can be written as

$$X = E_0(L_0 \oplus F_{k1}(R_0)) = E_0(L_0) \oplus E_0(F_{k1}(R_0))$$

where E_0 stands for taking the 6 least significant bits after the expansion E. Clearly, we can use the corresponding bits in L_0 as our XORed constant and recover the intermediate states $E_0(F_{k1}(R_0))$ $(n = 6)$. Specifically, in our experiments, we choose $E_0(L_0) = \{0x01, 0x02, 0x04, 0x08, 0x10, 0x20\}$ $(m = 6)$. For each value of $E_0(L_0)$, we set the other bits in L_0 to random numbers and let R_0 take 64 fix values $\{R_0(1), ..., R_0(64)\}$ $(T = 64)$. Considering the following recovery, the attacker can choose 64 random values, or set $\{R_0(1), ..., R_0(64)\}$ to anything he likes. Thus, we have 6 groups of 64 traces, where the intermediate state (S_0's input) sequences are the same, except for $E_0(L_0)$. The attacker can then pick one leakage point on the trace, and build 6 64-length leakage observations. As our discussion in Sect. 3.2, these observations can be regarded as the results of 6 different leakage functions with the same sources $\{E_0 \circ F_{k1}(R_0(1)), ..., E_0 \circ F_{k1}(R_0(64))\}$. Considering the noise, the attacker can also repeat the above measurements several times: all settings stay the same, except for the random bits in L_0. In our experiments, we repeat the above measurements 10 times and get 6 trace sets, with 640 traces in each set.

Figure 5(a) presents the recovery with one manually picked leakage point. Our analysis works well with all 6 sets, recovering over 90% of the intermediate state bits correctly without any repetition. If the attacker repeats the measurement one more time and increases the trace set size to 128, our ICA-based SCA successfully recovers $\{E_0 \circ F_{k1}(R_0(1)), ..., E_0 \circ F_{k1}(R_0(64))\}$, up to ICA's ambiguity. On the other hand, if the attacker cannot build 6 trace sets, our analysis still works with inadequate observations. With 4 or 5 observations, our ICA-based SCA still learns about 80% of the intermediate state bits correctly.

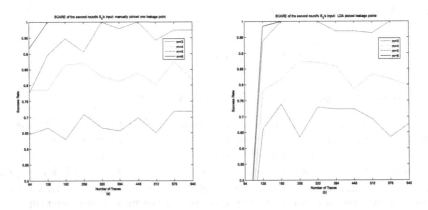

Fig. 5. Reverse engineering the first round's outputs of DES

As finding a valid leakage point for ICA might be difficult in practice, we also present our ICA-based SCA with LDA in Fig. 5(b). Unlike the previous section, LDA in this section simply takes the first component as our new observation[1].

---

[1] Here we simply use $\{1, ..., 64\}$ as LDA's labels.

Similar to the previous sections, LDA only gives valid results after the number of traces is much larger than the input range ($\gg 50$). In Fig. 5(b), LDA gives better recovery when the observations are not adequate ($m = 5$). As the first round's outputs are already recovered, the attacker can further perform other attacks to recover the inner structure of F_{k1}. It is worth mentioning that this attack is not specific to DES. Indeed, it works for any cipher with Feistel scheme, regardless of the inner structure or the specific implementation.

Remarks. Due to the ambiguities of ICA, our recovery cannot learn the actual intermediate states, only its ICA equivalent. This is not an issue in SCARE: as most cryptographic components in SCARE are secret, most SCAREs only learn an equivalent form of the original cipher [8]. The difference between the recovered components and the original are usually cancelled by the following recovery. Take our recovery above for instance, if the original intermediate state is X, our ICA-based SCA returns $\hat{X} = B(X) \oplus c$, where B is a bit permutation and c is a constant. The following Sbox computation can be written as $S'(\hat{X}) = S \circ B^{-1}\left(\hat{X} \oplus c\right)$. Since the attacker knows nothing about S, $\hat{X}$ and S' might as well be an valid form of the original cipher.

5 Discussions and Perspectives

5.1 Comparison with Other SCAREs

As a state recovery technique, our ICA-based SCA shares many inherent similarities with previous SCARE techniques. To date, Collision attack [5,7,8] is probably the most prevalent SCARE. Indeed, our attacker model shares many features with Collision Attacks: both attacks recover the intermediate state without key guesses. In theory, both attacks apply to any implementation; although in practice, most experimental verifications come from sequential software implementations [8]. Nonetheless, most applications in Sect. 4 cannot use collision attacks. The major difficulty comes from the "online profiling" step, where the attacker has to find another computation of the exact same Sbox for profiling. Furthermore, the attacker must know the Sboxes' input in this profiling set, at least up to its permutation equivalent. This assumption limits the profiling set to the first round, as the Sboxes' inputs of the second round are not accessible. Despite the fact that the secret cipher might use different Sboxes (like DES), studies also show that even the same Sbox computation can sometimes produce different leakages [5]. Since the attacker cannot build effective templates in these cases, our ICA-based SCA gives more stable recovery. However, our approach does have one major drawback: it only handles linear leakages. If the leakages contain significant non-linear components, collision attacks may be the only choice, thanks to the "online profiling" step (Table 2).

Apart from Collision Attacks, there are a few other SCAREs in the literature [16–18]. To our knowledge, these attacks are often restricted to certain cryptographic structures [16] or certain implementations [17]. Besides, their recovery

Table 2. Similarities and differences: collision attack v.s. ICA-based SCA

	Collision attack	ICA-based SCA
Target	Intermediate states	Intermediate states
Point of interest	Approximate	Approximate
Implementation	Sequential & Software	Sequential & Software
Profiling	"Online profilling"	None
Attacked round	First/Last	Any
Assumption on leakage	None	Linear

usually focuses on a single component, such as an Sbox [18]. These SCAREs may present more realistic threats in certain applications, while our approach recovers secret intermediate states in a more general way.

5.2   Future Improvements

In this section, we present some interesting extensions in this direction, which may further expand the applications of our ICA-based SCA.

- **Parallel hardware implementations:** In Sect. 4, our experiments mainly focus on unprotected software implementations. As hardware implementations are getting more and more popular, whether our ICA-based SCA works for hardware implementations is an interesting question. Theoretically, the common leakage model in hardware implementations—the Hamming Distance (HD) model—is still a linear model. Suppose the attacker knows the last state in the target register, the HD model alone should not hinder our ICA-based SCA. However, the assumption of knowing the last state is not always reasonable, especially for the SCARE applications. Another issue with hardware implementations is they usually involves parallel operations, which significantly reduces the Signal-to-Noise Ratio.
- **More convenient observation collection:** As stated in Sect. 4.2, the constant in our XOR-constant method should not affect the secret signal. For Feistel (or generalized Feistel) structures, we can choose part of the plaintext as our XOR-constant and attack the second round (or the first a few rounds). In the following rounds, since the target signal is affected by the chosen constant, constructing multi-channel observation is still an open problem. In SPN ciphers (like AES), with a few assumptions, our XOR-constant method can be applied to the second round. Although our XOR-constant method does ease the pain, in order to fully explore ICA's potential, we still need more convenient methods to construct multi-channel observations.

6   Conclusion

Despite their threat to various embedded devices, typical side channel attacks usually involve a "guess-and-determine" procedure. Instead of directly learning any secrets from the leakage, SCA takes key guesses and verifies them with the side channel leakages. In this paper, we propose an algorithm that directly learns the secret intermediate states from the observed leakages. Specifically, we show that under certain circumstances, the signal recovery problem can be regarded as a *Blind Source Separation* problem, and solved by the well-studied *Independent Component Analysis*. In order to find valid inputs for ICA, we propose several methods to construct multi-channel observations from the side channel leakages. In addition, to further exploit the specific features of circuit signals, we introduce a customized EM-ICA algorithm. Experiments show our analysis works well in certain ICA models, recovering most of the secret signal correctly with only a few hundred traces. Furthermore, our ICA-based SCA brings new possibilities to the current non-profiled SCA study, including attacking the middle round's encryption and reverse engineering with fewer restrictions. Considering ICA is a more aggressive tool than most previous SCA techniques, we believe our ICA-based SCA is a promising tool for the future SCA study.

Acknowledgements. We would like to thank Prof. Ming Tang and Dr. Carolyn Whitnall for the inspiring discussions on this topic. We would also like to thank the anonymous reviewers for providing valuable comments. This work is supported by the National Basic Research Program of China (No. 2013CB338002) and National Natural Science Foundation of China (No. 61272476, 61672509 and 61232009).

References

1. Kocher, P.C.: Timing attacks on implementations of diffie-hellman, RSA, DSS, and other systems. In: Koblitz, N. (ed.) CRYPTO 1996. LNCS, vol. 1109, pp. 104–113. Springer, Heidelberg (1996). doi:10.1007/3-540-68697-5_9
2. Kocher, P., Jaffe, J., Jun, B.: Differential power analysis. In: Wiener, M. (ed.) CRYPTO 1999. LNCS, vol. 1666, pp. 388–397. Springer, Heidelberg (1999). doi:10.1007/3-540-48405-1_25
3. Quisquater, J.-J., Samyde, D.: ElectroMagnetic analysis (EMA): measures and counter-measures for smart cards. In: Attali, I., Jensen, T. (eds.) E-smart 2001. LNCS, vol. 2140, pp. 200–210. Springer, Heidelberg (2001). doi:10.1007/3-540-45418-7_17
4. Genkin, D., Shamir, A., Tromer, E.: RSA key extraction via low-bandwidth acoustic cryptanalysis. In: Garay, J.A., Gennaro, R. (eds.) CRYPTO 2014. LNCS, vol. 8616, pp. 444–461. Springer, Heidelberg (2014). doi:10.1007/978-3-662-44371-2_25
5. Gérard, B., Standaert, F.X.: Unified and optimized linear collision attacks and their application in a non-profiled setting: extended version. J. Cryptographic Eng. **3**(1), 45–58 (2013)
6. Clavier, C.: An improved SCARE cryptanalysis against a secret A3/A8 GSM algorithm. In: McDaniel, P., Gupta, S.K. (eds.) ICISS 2007. LNCS, vol. 4812, pp. 143–155. Springer, Heidelberg (2007). doi:10.1007/978-3-540-77086-2_11
7. Clavier, C., Isorez, Q., Wurcker, A.: Complete SCARE of AES-like block ciphers by chosen plaintext collision power analysis. In: Paul, G., Vaudenay, S. (eds.) INDOCRYPT 2013. LNCS, vol. 8250, pp. 116–135. Springer, Heidelberg (2013). doi:10.1007/978-3-319-03515-4_8

8. Rivain, M., Roche, T.: SCARE of secret ciphers with SPN structures. In: Sako, K., Sarkar, P. (eds.) ASIACRYPT 2013. LNCS, vol. 8269, pp. 526–544. Springer, Heidelberg (2013). doi:10.1007/978-3-642-42033-7_27

9. Hyvärinen, A., Oja, E.: Independent component analysis: algorithms and applications. Neural Netw. **13**, 411–430 (2000)

10. Belouchrani, A., Cardoso, J.F.: Maximum likelihood source separation by the expectation-maximization technique: deterministic and stochastic implementation. In: Proceedings of the NOLTA, pp. 49–53 (1995)

11. Wiki: Blind signal separation. https://en.wikipedia.org/wiki/Blind_signal_separation

12. Hyvarinen, A.: Fast and robust fixed-point algorithms for independent component analysis. IEEE Trans. Neural Netw. **10**(3), 626–634 (1999)

13. Cardoso, J.: High-order contrasts for independent component analysis. Neural Comput. **11**(1), 157–192 (1999)

14. Bell, A., Sejnowski, T.: An information-maximization approach to blind separation and blind deconvolution. Neural Comput. **7**(6), 1129–1159 (1995)

15. Standaert, F.-X., Archambeau, C.: Using subspace-based template attacks to compare and combine power and electromagnetic information leakages. In: Oswald, E., Rohatgi, P. (eds.) CHES 2008. LNCS, vol. 5154, pp. 411–425. Springer, Heidelberg (2008). doi:10.1007/978-3-540-85053-3_26

16. Novak, R.: Side-channel attack on substitution blocks. In: Zhou, J., Yung, M., Han, Y. (eds.) ACNS 2003. LNCS, vol. 2846, pp. 307–318. Springer, Heidelberg (2003). doi:10.1007/978-3-540-45203-4_24

17. Daudigny, R., Ledig, H., Muller, F., Valette, F.: SCARE of the DES. In: Ioannidis, J., Keromytis, A., Yung, M. (eds.) ACNS 2005. LNCS, vol. 3531, pp. 393–406. Springer, Heidelberg (2005). doi:10.1007/11496137_27

18. Guilley, S., Sauvage, L., Micolod, J., Réal, D., Valette, F.: Defeating any secret cryptography with SCARE attacks. In: Abdalla, M., Barreto, P.S.L.M. (eds.) LATINCRYPT 2010. LNCS, vol. 6212, pp. 273–293. Springer, Heidelberg (2010). doi:10.1007/978-3-642-14712-8_17

Cryptographic Protocols

Actively Secure 1-out-of-N OT Extension with Application to Private Set Intersection

Michele Orrù[1]([⊠]), Emmanuela Orsini[2], and Peter Scholl[2]

[1] CNRS, ENS Paris, Paris, France
michele.orru@ens.fr
[2] Department of Computer Science, University of Bristol, Bristol, UK
{Emmanuela.Orsini,Peter.Scholl}@bristol.ac.uk

Abstract. This paper describes a 1-out-of-N oblivious transfer (OT) extension protocol with active security, which achieves very low overhead on top of the passively secure protocol of Kolesnikov and Kumaresan (Crypto 2011). Our protocol obtains active security using a consistency check which requires only simple computation and has a communication overhead that is independent of the total number of OTs to be produced. We prove its security in both the random oracle model and the standard model, assuming a variant of correlation robustness. We describe an implementation, which demonstrates our protocol only costs around 5–30% more than the passively secure protocol.

Random 1-out-of-N OT is a key building block in recent, very efficient, passively secure private set intersection (PSI) protocols. Our random OT extension protocol has the interesting feature that it even works when N is exponentially large in the security parameter, provided that the sender only needs to obtain polynomially many outputs. We show that this can be directly applied to improve the performance of PSI, allowing the core private equality test and private set inclusion subprotocols to be carried out using just a single OT each. This leads to a reduction in communication of up to 3 times for the main component of PSI.

Keywords: Oblivious transfer · Private set intersection · Multi-party computation

1 Introduction

Oblivious transfer (OT) is a fundamental primitive in cryptography, first introduced by Rabin [Rab81] and now employed in a variety of protocols, ranging from contract signing [EGL85] to special-purpose tasks such as private set intersection [PSZ14]. It plays a decisive role in protocols for secure two-party and multi-party computation, including those based on Yao's garbled circuits [Yao82] and secret-sharing [NNOB12, LOS14, KOS16]. The most commonly studied form

---

Full version available at http://eprint.iacr.org/2016/933.pdf
M. Orrù–Work done while visiting University of Bristol.

© Springer International Publishing AG 2017
H. Handschuh (Ed.): CT-RSA 2017, LNCS 10159, pp. 381–396, 2017.
DOI: 10.1007/978-3-319-52153-4_22

of oblivious transfer is 1-out-of-2 OT, where a sender has two messages (x_0, x_1) as input, and a receiver chooses a bit b; the goal of the protocol is for the receiver to learn x_b, but no information on x_{1-b}, whilst the sender learns nothing about b. This can be generalized to 1-out-of-N OT and k-out-of-N OT, in which the receiver learns k of the sender's N messages.

Unfortunately, due to a result of Impagliazzo and Rudich [IR89], oblivious transfer is highly unlikely to be possible without the use of public-key cryptography; consequently, even the most efficient oblivious transfer constructions [PVW08, CO15] come with a relatively high cost.

OT Extensions. In 1996, Beaver [Bea96] first showed that it is possible to extend OT starting with a small number (say, security parameter κ) of "base" OTs, to create $\mathsf{poly}(\kappa)$ additional OTs using only symmetric primitives, with computational security κ. This construction is very impractical as it requires the evaluation of pseudorandom generators within Yao's garbled circuits.

Later, in 2003, Ishai et al. [IKNP03] proposed a protocol for extending oblivious transfers: the passively secure version of this protocol (hereafter IKNP) only requires black-box use of a correlation robust hash function, and is very efficient. Concretely, an optimized version of IKNP for OT on random strings (described in [ALSZ13, KK13]) requires sending κ bits and computing three hash function evaluations per OT, after a one-time cost of κ base OTs, for computational security κ. With a carefully optimized implementation, the dominant cost of this is communication [ALSZ16].

Kolesnikov and Kumaresan [KK13] showed how to modify the IKNP protocol using Walsh-Hadamard error-correcting codes and obtain a passively secure protocol for 1-out-of-N OT on random strings. The cost is only a small constant factor more than the 1-out-of-2 IKNP for values of N up 256.

Several recent works have proposed increasingly efficient protocols for 1-out-of-2 OT extension with active security [NNOB12, ALSZ15, KOS15]. The latter work of Keller et al. [KOS15], which is proven secure in the random oracle model, brings the cost of actively secure 1-out-of-2 OT to essentially the same as the passive IKNP protocol by adding a simple consistency check.

1.1   Contributions

Actively Secure 1-out-of-N OT Extension. Our main contribution is a practical, actively secure 1-out-of-N OT extension protocol with very low overhead on top of the passively secure protocol of Kolesnikov and Kumaresan [KK13]. For the case of random OT, where the sender's strings are sampled at random, our protocol (proven secure in the random oracle model) improves upon [KK13] by allowing for much larger values of N with a suitable choice of binary linear code. Our protocol even works when N is *exponential in the security parameter*, provided that the sender is only required to learn polynomially many output strings. The protocol requires only κ base OTs, and the extension phase has an amortized communication cost of $O(\kappa)$ bits per random OT.

At a high level, our protocol starts with the passively secure [KK13] protocol and adds a simple consistency check to obtain active security (similar to [KOS15] for 1-out-of-2 OT). However, there are several technical challenges to solve on the way. In [KOS15], a check is used to verify that pairs of strings are of the form $(\mathbf{x}_i, \mathbf{x}_i + \mathbf{b})$ for a fixed correlation $\mathbf{b}$ (with addition modulo 2), when the receiver only knows one string from each pair. In the [KK13] protocol, however, we must ensure that strings are of the form $\mathbf{x}_i + \mathbf{b} \odot C(m_i)$, where C encodes a message m_i using an error-correcting code and $\odot$ denotes the component-wise product of bit vectors. The check of [KOS15] cannot be applied to this situation. We overcome this by adapting a check used previously in additively homomorphic UC commitments [FJNT16], which requires that C is a *linear* code with sufficiently large minimum distance.[1] The number of codewords in the binary linear code determines N in the 1-out-of-N OT, which gives a range of choices of N depending on the choice of code.

To be able to handle exponentially large N, it may seem that we just need to choose a suitable binary linear code of the right length. However, we need to take care that the security reduction does not contain any loss in security that scales with N: the reduction in [KK13] incurs a loss in $O(N^2)$, which would give a meaningless security result in this case. To ensure this, we modify the 1-out-of-N random OT functionality so that the sender can only obtain $N' = \mathsf{poly}(\kappa)$ of the output messages, and show that the loss in the resulting reduction is in $O(N')$.

Security in the Standard Model. For random OT extension, it is not known how to prove security without using a programmable random oracle as in [ALSZ13,KOS15]. However, for the case of non-random 1-out-of-N OT, we prove our protocol secure in the standard model, assuming a hash function that satisfies a variant of correlation robustness on high min-entropy secrets. This is a similar assumption to the protocol in [ALSZ15], but more general as we require the assumption to hold for a range of different parameters. This gives the first actively secure OT extension protocol needing only κ base OTs for security parameter κ and is proven secure without random oracles, even in the 1-out-of-2 case.[2]

Faster Private Set Intersection. We show that random 1-out-of-N OT with an exponentially large N can be directly applied to improve the efficiency of the previous fastest (semi-honest) private set intersection protocols. OT-based PSI protocols [PSZ14,PSSZ15] use random 1-out-of-N OT as a building block for a private equality test protocol, where two parties learn whether their inputs are

---

[1] We observe an interesting connection between our protocol and additively homomorphic UC commitment schemes [FJNT16,CDD+16]: our protocol essentially runs a homomorphic commitment protocol and hashes the resulting commitments to obtain random OTs. However, this mechanism seems very specific to the workings of these commitment schemes and appears unlikely to lead to a generic transformation.

[2] Note that our security reduction requires fixing the adversary's random coins, so is non-uniform. Obtaining a uniform reduction seems to need at least $\kappa + s$ base OTs, for statistical security parameter s.

equal (and nothing more). In that protocol, one random OT is used to perform an equality test on $\log N$-bit inputs. Since the random OT protocol of [KK13] only works for values of N up to 256 (due to the use of small Walsh-Hadamard codes) several OTs are XORed together to construct a protocol for comparing large (e.g. up to 128 bit) messages. Using our protocol with $N = 2^k$ gives a very simple private equality test on k-bit messages, for any $k = \mathsf{poly}(\kappa)$, using just a single 1-out-of-N random OT. This can be generalized to perform *private set inclusion*—where one party holds a single value and another party a set of m values—at the cost of one random OT and sending $m \cdot s$ bits, where s is the statistical security parameter. This results in a reduction in communication of around 2–5 times (depending on the bit-length of the input) for this component of the semi-honest PSI protocol in [PSSZ15].

Implementation. We have implemented and benchmarked our 1-out-of-N random OT extension protocol and compared its performance with the passive protocol of [KK13]. Although our implementation is not heavily optimized (it occupies around 800 lines of C in all), we show that the overhead of our consistency check for achieving active security is very low: the actively secure protocol takes only around 20% more time than the passive version, depending on parameters.

Towards Efficient Actively Secure PSI. Currently, the most efficient PSI protocols are the OT-based ones mentioned above, but these are only secure against a passive adversary. Since 1-out-of-N random OT is a key component in these protocols, our work can be seen as a step towards constructing more efficient PSI with active security. Actively secure PSI was recently studied by Lambæk [Lam16], who showed the protocol of [PSSZ15] can be modified to provide active security for one party, assuming the underlying OT protocol is actively secure; our protocol therefore provides an instantiation of this proposal.

Recent Work and Open Problems. In a very recent, independent work, Kolesnikov et al. [KKRT16] describe a batched oblivious PRF evaluation protocol with application to private set intersection. Although their protocol is phrased in the language of oblivious PRFs rather than 1-out-of-N OT, it is very similar to ours, only with passive security. Instead of using a traditional error-correcting code, they show that a random oracle has the necessary properties for passive security. In contrast, our protocol requires the linearity and erasure decoding properties of the binary code to achieve active security. They describe the same application to improved performance of PSI (with slightly better parameters than ours due to use of a random oracle) and give a thorough efficiency evaluation and implementation of the resulting PSI protocol. We note that it is still an interesting open problem to obtain a fully actively secure variant of the PSI protocol in [PSSZ15] with low overhead.

Regarding OT extension in general, there are still some interesting unsolved problems. Our 1-out-of-N OT extension cannot be used directly to improve performance of 1-out-of-2 OT on short secrets (as was done for the passive case in [KK13]), since the standard reduction from 1-out-of-N to 1-out-of-2 OT [NP99] is only passively secure. Therefore, it is still an open problem to construct a practical 1-out-of-2 OT extension on short strings with communication sublinear in the security parameter. Also, the case of constructing k-out-of-N OT with active security using OT extensions is still open; there is an elegant passively secure protocol [SSR08], but it seems difficult to make this actively secure.

2   Preliminaries

Notation. We denote by κ and s the computational, resp. statistical, security parameters. We use bold lower case letters for vectors. Given a matrix A, we let $\mathbf{a}_i$ denote the i-th row of A, and $\mathbf{a}^j$ denote the j-th column of A. When referring to a vector $\mathbf{v} \in \mathbb{F}^n$, we write $\mathbf{v}[i]$, with $1 \leq i \leq n$, to mean the i-th component of $\mathbf{v}$. We identify bit strings as vectors over the finite field $\mathbb{F}_2$, and use "$+$" and "$\cdot$" to mean addition and multiplication in this field. We use the notation $\mathbf{a} \odot \mathbf{b}$ to denote the component-wise product of vectors $\mathbf{a}, \mathbf{b} \in \mathbb{F}_2^n$. Given an integer N, we denote by $[N]$ the set of integers $\{1, \ldots, N\}$.

Error-Correcting Codes. Our protocol uses an $[n_C, k_C, d_C]$ binary linear code C, where n_C is the length, k_C the dimension and d_C the distance of C. So, $C : \mathbb{F}_2^{k_C} \to \mathbb{F}_2^{n_C}$ is a linear map such that for every pair of messages $\mathbf{m}_1, \mathbf{m}_2 \in \mathbb{F}_2^{k_C}$, the Hamming weight of the sum of the encodings of the messages satisfies $\mathsf{wt}_\mathsf{H}(C(\mathbf{m}_1) + C(\mathbf{m}_2)) \geq d_C$.

Oblivious Transfer Functionalities. We now recall some definitions of oblivious transfer. Following Even et al. [EGL85], 1-out-of-2 OT is a two-party protocol between a sender P_S, who inputs two messages v_0, v_1, and a receiver P_R who inputs a choice bit c and learns as output v_c and nothing about v_{1-c}, in such a way that P_S remains oblivious as what message was received by P_R. Formally, the general case of 1-out-of-N OT on κ-bit strings is defined as the functionality:

$$\mathcal{F}_{N\text{-OT}}((\mathbf{v}_0, \ldots, \mathbf{v}_{N-1}), c) = (\perp, \mathbf{v}_c),$$

where $\mathbf{x}_i \in \{0,1\}^\kappa$ are the sender's inputs and $c \in \{0, \ldots, N-1\}$ is the receiver's input. We denote by $\mathcal{F}_{N\text{-OT}}^{\kappa,m}$ the functionality that runs $\mathcal{F}_{N\text{-OT}}$ m times on messages in $\{0,1\}^\kappa$. For example, in $\mathcal{F}_{2\text{-OT}}^{\kappa,m}$, P_S inputs $(\mathbf{v}_{i,0}, \mathbf{v}_{i,1})$ and P_R inputs c_i for $i \in [m]$, and P_R receives the output $\mathbf{v}_{i,c_i}$.

Another important variant is the random OT functionality $\mathcal{F}_{N\text{-ROT}}^{\kappa,m}$, in which the sender provides no input, but receives random messages $(\mathbf{v}_0, \ldots, \mathbf{v}_{N-1})$ from the functionality as output.

2.1   Passively Secure OT Extension: The KK Protocol

We now recall the passively secure KK protocol for 1-out-of-N OT extension described in [KK13], which is a generalized version of the IKNP protocol for 1-out-of-2 OT [IKNP03].

Suppose the two parties wish to perform m sets of 1-out-of-N random OTs, where N is a power of two. There is a sender P_S with no input, and a receiver

P_R, who inputs the choices $w_1, \ldots, w_m \in \{0, \ldots, N-1\}$, which are represented as vectors $\mathbf{w}_i \in \mathbb{F}_2^{\log N}$. The two parties begin by performing n_c base 1-out-of-2 OTs on random inputs, with the roles of sender and receiver reversed. So, P_R obtains n_c pairs of random strings $(\mathbf{r}_0^j, \mathbf{r}_1^j)$ of length κ and P_S obtains $(b_j, \mathbf{r}_{b_j}^j)$, where $b_j \overset{\$}{\leftarrow} \{0, 1\}$, for $j \in [n_c]$.

Next, both parties locally extend their base OT outputs to length m using a pseudorandom generator, where m is the final number of OTs desired. This results in κ sets of 1-out-of-2 OTs on m-bit strings, which we represent as matrices $T_0, T_1 \in \mathbb{F}_2^{m \times n_c}$, held by P_R, whilst P_S holds the vector $\mathbf{b} = (b_1, \ldots, b_{n_c}) \in \mathbb{F}_2^{n_c}$ and the matrix

$$T_{\mathbf{b}} := \begin{pmatrix} \mathbf{t}_{b_1}^1 & \cdots & \mathbf{t}_{b_{n_c}}^{n_c} \end{pmatrix} \in \mathbb{F}_2^{m \times n_c},$$

where $\mathbf{t}_0^j, \mathbf{t}_1^j$ are the columns of T_0, T_1, for $j \in [n_c]$.

At this point P_R constructs a matrix $C \in \mathbb{F}_2^{m \times n_c}$, where each row $\mathbf{c}_i$ is the encoding $C(\mathbf{w}_i)$ of the input $\mathbf{w}_i \in \mathbb{F}_2^{k_c}$, where C is a binary code of length n_c, dimension $k_c = \log_2 N$ and minimum distance $d_c \geq \kappa$. Then P_R sends to P_S the matrix

$$U = T_0 + T_1 + C.$$

Note that for each column of U, all information on the receiver's encoded input is masked by the value $\mathbf{t}_{1-b_j}^j$, which is unknown to P_S.

After this step P_S defines an $m \times n_c$ matrix Q with columns $\mathbf{q}^j = b_j \cdot \mathbf{u}^j + \mathbf{t}_{b_j}^j = b_j \cdot \mathbf{c}^j + \mathbf{t}_0^j$ (where $\mathbf{c}^j$ are the columns of C). Notice that the rows of Q are given by

$$\mathbf{q}_i = \mathbf{c}_i \odot \mathbf{b} + \mathbf{t}_i,$$

where $\mathbf{t}_i$ are the rows of T_0. Here, P_R holds $\mathbf{t}_i$ and P_S holds $(\mathbf{q}_i, \mathbf{b})$, for $i \in [m]$. The key observation to turn these values into OTs is that for each of the possible receiver choices $\mathbf{w} \in \mathbb{F}_2^{k_c}$, P_S can compute the value $\mathbf{q}_i + C(\mathbf{w}) \odot \mathbf{b}$. If $\mathbf{w} = \mathbf{w}_i$ then this is equal to $\mathbf{t}_i$ so is known to P_R. Otherwise, for any $\mathbf{w} \neq \mathbf{w}_i$, P_R must guess κ bits of P_S's secret $\mathbf{b}$ to be able to compute $\mathbf{q}_i + C(\mathbf{w}) \odot \mathbf{b}$, since the minimum distance of C guarantees that $C(\mathbf{w})$ and $C(\mathbf{w}_i)$ are at least Hamming distance κ apart.

Therefore, the parties can convert these values to random 1-out-of-N OTs by simply hashing their outputs with a random oracle, H. P_S outputs the values $\mathbf{v}_{w,i} = \mathsf{H}(i, \mathbf{q}_i + C(\mathbf{w}) \odot \mathbf{b})$, for all $\mathbf{w} \in \mathbb{F}_2^{k_c}$, and P_R outputs $\mathbf{v}_{w_i,i} = \mathsf{H}(i, \mathbf{t}_i)$.

Instantiating the Code. As noticed in [KK13], if we instantiate the binary linear code C with the $[\kappa, 1, \kappa]$ binary repetition code, we obtain the 1-out-of-2 IKNP protocol [IKNP03]. In this case, each row of the matrix C constructed by the receiver is either 0^κ or 1^κ, depending on the receiver's choice bits. If instead C is chosen to be a Walsh-Hadamard code as in [KK13], then the result is a 1-out-of-2^{k_c} OT. This needs a code length of $N = 2^{k_c}$ with security parameter

$N/2$; this turns out to be more efficient than constructing 1-out-of-N OT from 1-out-of-2 OT for values of $N \leq 256$ with 128-bit security.

Security. The KK protocol (and hence IKNP) is actively secure against a corrupt sender, since after the base OTs, there is no opportunity for P_S to cheat. However, it only provides passive security against a corrupt receiver, since P_R may incorrect compute the encodings of their input in the matrix U. It was explained in [IKNP03, ALSZ15] that if P_R cheats in this way, and also learns (via a side-channel, for instance) the sender's outputs in just κ of the random OTs then P_R can compute the sender's secret $\mathbf{b}$, and thus learn all of the sender's outputs in every remaining OT.

3 Actively Secure Random 1-out-of-N OT Extension

In this section we present our actively secure OT extension protocol in the random oracle model. Since we want to construct 1-out-of-N random OT when N may be exponential in the security parameter, our protocol implements a modified random OT functionality $\mathcal{F}_{N\text{-ROT+}}$ (Fig. 1), which allows the sender to query the functionality to obtain their random OT outputs one at a time, so that all N need not be produced.

Functionality $\mathcal{F}_{N\text{-ROT+}}^{\kappa,m}$

Upon receiving (ROT) from P_S and (ROT, $(w_1, \ldots, w_m)$), where $\forall i \in [m], 0 \leq w_i < N$, from P_R, do the following:

  - Sample $\mathbf{v}_{j,i} \xleftarrow{\$} \mathbb{F}_2^\kappa$ for each $j \in \{0, \ldots, N-1\}$ and $i \in [m]$.
  - Send the outputs $\mathbf{v}_{w_i,i}$ to P_R for each $i \in [m]$
  - Upon receiving (SenderOutput, j, i) from P_S, where $i \in [m]$ and $j \in \{0, \ldots, N-1\}$, send $\mathbf{v}_{j,i}$ to P_S.

Fig. 1. Ideal functionality $\mathcal{F}_{N\text{-ROT+}}$ for m 1-out-of-($\leq N$) random OTs on κ-bit strings between a sender P_S and receiver P_R

The high-level idea of our protocol (Fig. 2) is that, to deal with a malicious receiver in the KK protocol, we add a consistency check that ensures P_R inputs codewords as rows of the matrix C when sending the matrix U in step 3. If the check passes then the protocol carries on and the correlated OTs are hashed to obtain random OTs. Otherwise, the protocol aborts.

The intuition behind security is that if not all the P_R's inputs $\mathbf{c}_i$ are codewords then to pass the check, the errors must 'cancel out' when taking the random linear combinations. However, the $x_i^{(\ell)}$ values used in the consistency check are unknown when P_R chooses $\mathbf{c}_i$ so this can only happen with negligible probability; since each $x_i^{(\ell)} \in \{0,1\}$, there is a $1/2$ probability that $\mathbf{c}_i$ is not

Protocol $N\text{-}ROT^{\kappa,m}$

COMMON INPUT: κ and s are the computational and statistical security parameters, respectively; C is an $[n_c, k_c, d_c]$ binary linear code such that $k_c = \log_2 N$ and $d_c \geq \kappa$.
INPUT OF P_R: m selection integers $(w_1, \dots, w_m)$, each in $\{0, \dots, N-1\}$, encoded as bit strings $\mathbf{w}_1, \dots, \mathbf{w}_m \in \mathbb{F}_2^{k_c}$.
INPUT OF P_S: m subsets $S_i \subseteq \mathbb{F}_2^{k_c}$, with $|S_i| = \text{poly}(\kappa)$, for $i \in [m]$.
REQUIRE: $H : [m] \times \mathbb{F}_2^{n_c} \rightarrow \mathbb{F}_2^\kappa$ is a random oracle and $PRG : \{0,1\}^\kappa \rightarrow \{0,1\}^{m'}$, $m' = m + s$, is a pseudorandom generator.

Init: Both parties call $\mathcal{F}_{2\text{-}OT}^{\kappa, n_c}$, with P_R playing the role of the sender and P_S playing the role of the receiver, inputing n_c random bits. P_S receives $\{(b_j, \mathbf{r}_{b_j}^j)\}_{j \in [n_c]} \in \mathbb{F}_2 \times \mathbb{F}_2^\kappa$ and P_R receives $\{(\mathbf{r}_0^j, \mathbf{r}_1^j)\}_{j \in [n_c]}$.

Extend: Let $m' = m + s$.

1. P_R constructs matrices $T_0, T_1 \in \mathbb{F}_2^{m' \times n_c}$ from seeds $\{(\mathbf{r}_0^j, \mathbf{r}_1^j)\}$ so that the respective columns are:

$$\mathbf{t}_0^j = PRG(\mathbf{r}_0^j) \in \mathbb{F}_2^{m'}, \qquad \mathbf{t}_1^j = PRG(\mathbf{r}_1^j) \in \mathbb{F}_2^{m'}, \quad j \in [n_c].$$

In the same way P_S produces $\mathbf{t}_{b_j}^j$, for each $j \in [n_c]$. Summarizing, P_R holds $\{(\mathbf{t}_0^j, \mathbf{t}_1^j)\}_{j \in [n_c]}$ and P_S holds $\{\mathbf{t}_{b_j}^j\}_{j \in [n_c]}$

2. P_R samples random $\mathbf{w}_{m+\ell} \xleftarrow{\$} \mathbb{F}_2^{k_c}$, for $\ell \in [s]$, and then constructs a matrix $C \in \mathbb{F}_2^{m' \times n_c}$ such that each row $\mathbf{c}_i$ is the codeword $C(\mathbf{w}_i)$. Then, P_R sends to P_S the values

$$\mathbf{u}^j = \mathbf{t}_0^j + \mathbf{t}_1^j + \mathbf{c}^j, \quad j \in [n_c], \tag{1}$$

where $\mathbf{c}^j$ is the j-th column of C.

3. P_S receives $\mathbf{u}^j \in \mathbb{F}_2^{m'}$ and computes

$$\mathbf{q}^j = b_j \cdot \mathbf{u}^j + \mathbf{t}_{b_j}^j = b_j \cdot \mathbf{c}^j + \mathbf{t}_0^j, \tag{2}$$

that form the columns of an $(m' \times n_c)$ matrix Q. Denoting the rows of T_0, Q by $\mathbf{t}_i, \mathbf{q}_i$, P_R now holds $\mathbf{c}_i, \mathbf{t}_i$ and P_S holds $\mathbf{b}, \mathbf{q}_i$ so that

$$\mathbf{q}_i = \mathbf{c}_i \odot \mathbf{b} + \mathbf{t}_i.$$

4. *Consistency check:*
   - P_S samples s random strings $\{(x_1^{(\ell)}, \dots, x_m^{(\ell)}) \in \mathbb{F}_2^m\}_{\ell \in [s]}$ and sends them to P_R.
   - P_R computes and sends, for $\ell \in [s]$:

$$\mathbf{t}^{(\ell)} = \sum_{i \in [m]} \mathbf{t}_i \cdot x_i^{(\ell)} + \mathbf{t}_{m+\ell}, \qquad \mathbf{w}^{(\ell)} = \sum_{i \in [m]} \mathbf{w}_i \cdot x_i^{(\ell)} + \mathbf{w}_{m+\ell}$$

   - P_S computes $\mathbf{q}^{(\ell)} = \sum_{i \in [m]} \mathbf{q}_i \cdot x_i^{(\ell)} + \mathbf{q}_{m+\ell}$, and checks that:

$$\mathbf{t}^{(\ell)} + \mathbf{q}^{(\ell)} = C(\mathbf{w}^{(\ell)}) \odot \mathbf{b}, \quad \forall \ell \in [s]. \tag{3}$$

If the check fails, P_S sends **Abort**, otherwise continue.

Output: P_S sets $\forall i \in [m]$ and $\mathbf{w} \in S_i$: $\mathbf{v}_{w,i} = H(i, \mathbf{q}_i + C(\mathbf{w}) \odot \mathbf{b})$ and P_R sets $\forall i \in [m]$: $\mathbf{v}_{w_i, i} = H(i, \mathbf{t}_i)$.

Fig. 2. An actively secure protocol for $\mathcal{F}_{N\text{-}ROT+}^{\kappa, m}$, extending $\mathcal{F}_{2\text{-}OT}^{\kappa, n_c}$.

included in the linear combination, so s sets of checks are needed to ensure a negligible cheating probability.

Compared with the consistency check of [KOS15] (for the 1-out-of-2 case), our check is simpler as we only require XOR operations instead of multiplications in the finite field $\mathbb{F}_{2^\kappa}$. However, being over $\mathbb{F}_2$ means that we must repeat the check s times, whereas [KOS15] only needs one check; in our case, working in $\mathbb{F}_{2^\kappa}$ would not allow the linear encoding relation to be verified, which is why we use $\mathbb{F}_2$.

We observe that our protocol, minus the final hashing step, is essentially the same as the additively homomorphic commitment protocol from [FJNT16] (which inspired our consistency check). Although our security proof requires quite some extra work to implement OT instead of commitments, it is interesting to see how the same construction can lead to two very different applications with just a small modification. More recently, another scheme [CDD+16] improved upon [FJNT16] by using a linear-time computable consistency check based on a special class of universal hash functions, and constructing a linear-time encodable error-correcting code. These changes can also be applied to our protocol, but it is not clear how efficient these would be in practice, and since we aim for practical (rather than asymptotic) efficiency we do not present this here.

Theorem 1. *Assuming that* H *is a random oracle and* PRG *a pseudo-random generator, the protocol* N-ROT$^{\kappa,m}$ *in Fig. 2 securely implements* $\mathcal{F}^{\kappa,m}_{\text{N-ROT}+}$ *(Fig. 1) in the* $\mathcal{F}_{\text{2-OT}}$*-hybrid model with computational security parameter* κ *and statistical security parameter* s *against a static malicious adversary.*

Proof of this result can be found in the full version of this work.

The case of a corrupt sender is straightforward and reduces to the security of PRG, similar to previous works [ALSZ16,KOS15]. For a corrupt receiver, the first main challenge is for the simulator to extract the receiver's inputs, $\mathbf{w}_i$, to send to the functionality $\mathcal{F}_{\text{N-ROT}+}$. This is done by using the values sent during the check to identify a set of positions where the receiver has attempted to 'guess' some bits of the sender's secret, $\mathbf{b}$. Removing these positions from the $\mathbf{c}_i$ values used by P_R leaves behind an incomplete codeword, which can be erasure decoded to recover the message.

After decoding the inputs, the simulator must then respond to random oracle queries made by the environment. We do this in an *optimistic* manner, meaning, we do not abort if conflicting queries are made, but answer at random in that case; the environment may not always notice this inaccurate behaviour if the sender did not learn all N outputs from the OTs. This allows us to obtain a security bound that depends on N', the *maximum* number of outputs learnt by P_S in any OT, rather than N, which may be exponential in κ.

Instantiating the Code. It remains to instantiate the binary linear code, $\mathcal{C}$, to obtain a 1-out-of-N random OT protocol for a desired power of two choice of N. As well as the repetition code (for 1-out-of-2 OT), we suggest a more efficient form of the Walsh-Hadamard code for $N \leq 512$; a binary Golay code

Table 1. Parameters for various choices of code

Code	N	Length	Distance/Security
Repetition [IKNP03]	2	128	128
Walsh-Hadamard [KK13]	≤ 256	256	128
Punctured Walsh-Hadamard	≤ 512	256	128
Binary Golay	2048	384	128
BCH-511	2^{76}	511	171
BCH-1023	2^{443}	1023	128

for $N = 2048$; and BCH codes for values of N that are exponential in the security parameter. The parameters of these codes are presented in Table 1; note that the code length determines exactly the amount of communication required per extended OT. We obtained the generator matrices for all of these codes using Sage[3]. For further details of the constructions, see the full version.

4   Security in the Standard Model

In this section we consider the case of non-random 1-out-of-N OT. In this protocol (Fig. 3), we remove the random oracle assumption and prove security in the standard model. Similarly to [ALSZ15], we need a stronger version of correlation robustness than that given in [IKNP03], and require that the secret correlation **b** comes from a distribution of min-entropy k and in addition is multiplied by a codeword in the binary linear code $\mathcal{C}$.

Protocol N-OT$^{\kappa,m}$ (Standard Model)

INPUT: As in Protocol N-ROT$^{\kappa,m}$.
REQUIRE: $\mathsf{H} : \mathbb{F}_2^{n_c} \to \mathbb{F}_2^{\kappa}$ a k-min-entropy strongly $\mathcal{C}$-correlation-robust and PRG : $\{0,1\}^{\kappa} \to \{0,1\}^{m'}$.

Init: As in Protocol N-ROT$^{\kappa,m}$.
Extend: As in Protocol N-ROT$^{\kappa,m}$.
Output: P$_\mathsf{S}$ sends, $\forall i \in [m]$ and $0 \leq w < N - 1$: $\mathbf{y}_{w,i} = \mathbf{v}_{w,i} + \mathsf{H}(\mathbf{q}_i + \mathcal{C}(\mathbf{w}) \odot \mathbf{b})$.
    P$_\mathsf{R}$ recovers, $\forall i \in [m]$: $\mathbf{v}_{w_i,i} = \mathbf{y}_{w_i,i} + \mathsf{H}(\mathbf{t}_i)$.

Fig. 3. An implementation of $\mathcal{F}_{N\text{-OT}}^{\kappa,m}$ extending $\mathcal{F}_{2\text{-OT}}^{\kappa,n_c}$ in the Standard Model.

---

[3] http://www.sagemath.org.

Definition 1 (k-min-entropy code correlation robustness). *Let χ be a distribution on $\mathbb{F}_2^{n_C}$ with min-entropy k and C be an $[n_C, k_C, d_C]$ binary linear code. An efficiently computable function $\mathsf{H} : \mathbb{F}_2^{n_C} \to \mathbb{F}_2^{\kappa}$ is said to be k-min-entropy C -correlation robust if it holds that:*

$$\{\mathbf{t}_i, \mathsf{H}(\mathbf{t}_i + \mathbf{c} \odot \mathbf{b})\}_{i \in [m], \mathbf{c} \in C} \overset{c}{\equiv} \mathcal{U}^{m \cdot n_C + (m + |C|) \cdot \kappa},$$

where $\mathbf{b} \overset{\$}{\leftarrow} \chi$ and $\mathbf{t}_1, \ldots, \mathbf{t}_m \in \mathbb{F}_2^{n_C}$ are independent and uniformly distributed.

Similarly, $\mathsf{H}(\cdot)$ is said to be k-min-entropy strongly C -correlation robust if it holds that:

$$\{\mathsf{H}(\mathbf{t}_i + \mathbf{c} \odot \mathbf{b})\}_{i \in [m], \mathbf{c} \in C} \overset{c}{\equiv} \mathcal{U}^{(m + |C|) \cdot \kappa},$$

where $\mathbf{b} \overset{\$}{\leftarrow} \chi$, for any distribution of the $\{\mathbf{t}_i\}_{i \in [m]}$.

Notice that in the values used to mask P_S's inputs, it is the receiver that effectively chooses the $\mathbf{t}_j$'s, and they can not only choose these values non-uniformly, but even maliciously. This is the reason why we need a *strong* code-correlation robust hash function.

We claim that if H is a k-min-entropy strongly correlation robust function for all $n_C - s \leq k \leq n_C$, then the protocol is secure in the standard model. For further discussion on parameter choices regarding this assumption, see the full version.

Theorem 2. *Assuming that H is k-min-entropy strongly code-correlation robust for all $k \in \{n_C - s, \ldots, n_C\}$, and PRG is a pseudo-random generator, the protocol N-$\mathsf{OT}^{\kappa,m}$ in Fig. 3 securely implements $\mathcal{F}_{N\text{-}\mathsf{OT}}^{\kappa,m}$ in the $\mathcal{F}_{2\text{-}\mathsf{OT}}$-hybrid model against a static malicious adversary.*

Proof of this result can be found in the full version.

5   Application to Private Set Intersection

We now show how to apply the 1-out-of-N random OT extension protocol to increase the efficiency and obtain stronger security guarantees in existing private set intersection (PSI) protocols. We describe a simpler and more efficient private set inclusion protocol with active security, which is used as a key component of the most efficient passively secure PSI protocols.

5.1   Private Set Inclusion

A core building block of OT-based PSI protocols is a *private set inclusion* protocol, where party P_A has input $a \in \{0, 1\}^k$, party P_B has input a set $B \subset \{0, 1\}^k$ and the parties wish to learn whether $a \in B$. Note that the special case of $|B| = 1$ is a private equality test.

The previous most efficient protocol [PSSZ15, Sect. 6.1] requires $t = k/8$ executions of 1-out-of-256 random OT, and uses the KK protocol with length

256 Walsh-Hadamard codes. However, with the observation that our random OT protocol can be used for exponentially large values of N, we can in fact choose $N = 2^k$ and perform a private set inclusion with just a single 1-out-of-N random OT. This is possible because the OT sender is only required to learn one of the random OT outputs in order to run the set inclusion protocol.

The protocol, shown in Fig. 4, is very simple: P_A inputs their value a as the receiver's choice in a 1-out-of-N random OT, and P_B inputs each of their values $b \in B$ to obtain $|B|$ of the sender's random outputs. Thus, P_A learns a random value r_a and P_B learns a set of random values $R = \{r_b\}_{b \in B}$. P_B randomly permutes R and sends this to P_A, who checks whether $r_a \in R$ to determine the result (and can send this to P_B if desired).

Since P_A only learns one of the random OT outputs initially, all other possible elements of the set R are uniformly random so do not leak any information on P_B's input. Note that because our 1-out-of-N OT protocol is actively secure, we actually obtain an actively secure private set inclusion protocol (although this does not seem to suffice to make the PSI protocol of [PSSZ15] actively secure).

Protocol Π_{SetInc}

Inputs: P_A has input $a \in \{0,1\}^k$ and P_B has input $B \subset \{0,1\}^k$.

1. The parties run $\mathcal{F}^{s,1}_{N\text{-ROT+}}$ with $N = 2^k$, where P_A plays receiver with input a.
2. P_A receives $r_a \in \{0,1\}^s$.
3. For each $b \in B$, P_B calls $\mathcal{F}_{N\text{-ROT+}}$ with $(\mathsf{SenderOutput}, 0, b)$ to obtain r_b.
4. P_B randomly permutes and sends $R = \{r_b\}_{b \in B}$ to P_A.
5. P_A outputs 1 if $r_a \in R$ and 0 otherwise.

Fig. 4. Private set inclusion protocol

The complete security proof and functionality that we implement is given in the full version.

Efficiency. The cost of the above protocol is precisely the cost of 1-out-of-N random OT, plus sending $s \cdot |B|$ bits. If the protocol is run in a large batch

Table 2. Comparing the communication cost of private set inclusion subprotocols on k-bit strings and size $|B|$ sets with statistical security s.

k	Cost with BCH (bit)	Cost with W-H (bit)
32	$467 + s \cdot |B|$	$1024 + s \cdot |B|$
64	$499 + s \cdot |B|$	$2048 + s \cdot |B|$
128	$708 + s \cdot |B|$	$4096 + s \cdot |B|$

using $\mathcal{F}_{N\text{-ROT}+}^{\kappa,m}$ for large m (which is possible for the application to private set intersection) then this gives an amortized cost of $n_C + s \cdot |B|$ bits per execution, where n_C is the length of the code. The costs for this when instantiated with BCH codes (as described previously) are illustrated for various choices of k in Table 2, and compared with the Walsh-Hadamard code used in [PSSZ15]. In practice, the set size used in the set inclusion subprotocol for PSI in [PSSZ15] is around $|B| = 20$. For a large item length of $k = 128$ bits, and $s = 40$-bit statistical security, this gives a 3.3× reduction in communication for the dominant component of PSI.

6 Implementation

We now evaluate the complexity of our random OT protocol, and compare its performance to a passively secure variant by analysing implementation results.

Complexity Analysis. The main overhead introduced by our protocol to produce m OTs, compared with the passively secure KK protocol, is the computation of $m \cdot s$ XORs (on n_C-bit strings) by each party, and the communication of $s \cdot m$ random bits from the sender to receiver, followed by $s \cdot (n_C + k_C)$ bits in the other direction, in the consistency check. However, the $s \cdot m$ bits can be reduced to κ by having P_S send only a single random seed for these values, and expanding the seed using a PRG.

Outside of the consistency check, the main protocol costs are the encodings, hash function evaluations and the n_C bits that are sent by P_R for each extended OT. Of course, the sender's computational cost also highly depends on the number of random OT outputs that are desired.

Implementation. We evaluated our protocol on two machines running over a 1 Gbps local network, and also simulated a WAN environment with 50 Mbps bandwidth and 100 ms round-trip latency to model a real-life scenario over the Internet. All benchmarks have been run on modern Core i7 machines at 2–3 GHz.

Our implementation is in plain C, and uses the SimpleOT [CO15] oblivious transfer software[4] to run the base OTs, and BLAKE2 [ANWOW13] for hashing, as this provides fast hashing on short inputs. Otherwise, it does not rely on any other software and is available in the public domain. The executable occupies $280K$.

The core protocol covers roughly 200 lines of C code. It mostly runs on single thread, except we use OpenMP to parallelize the encodings and hash function evaluations, which are the computational bottlenecks of the protocol. We fix the computational security parameter $\kappa = 128$ and statistical security parameter $s = 40$. We used Intel AVX instructions to efficiently implement vector addition, componentwise product, and matrix transposition. Encoding of the binary linear code is implemented with multiplication by the generator matrix.

---

[4] http://users-cs.au.dk/orlandi/simpleOT/.

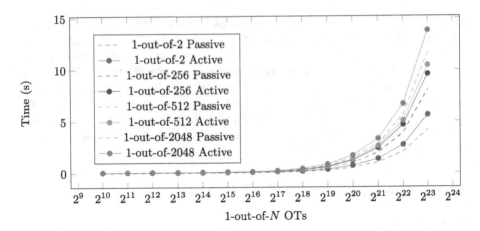

Fig. 5. Benchmarking different 1-out-of-N random OTs in LAN environment; average time for 20 runs.

Table 3. Data transmitted per OT and runtimes in seconds for 2^{23} OTs (LAN) or 2^{20} OTs (WAN), for several choices of N

Setting	$N = 2$	256	512	2048	2^{76}
Comms. (bit)	128	256	256	384	512
LAN, passive	4.1812	8.0260	8.1193	11.6642	23.4738
LAN, active	5.6191	9.5693	10.4379	13.8065	25.4001
WAN, passive	27.3982	54.2414	54.274	81.0548	108.89
WAN, active	27.882	54.7445	54.8189	81.6644	109.44

Our results for 1-out-of-N random OT for the small sized codes (repetition, Walsh-Hadamard, punctured Walsh-Hadamard and binary Golay) in the LAN setting can be seen in Fig. 5, for varying numbers of OTs. Table 3 compares the performance of the active and passively secure variants in both LAN and WAN settings, including the BCH-511 code, which could be used for the private set intersection application. We see that the overhead of active security is around 20–30% of the passive protocol over a LAN, and less than 5% in the WAN setting. This fits with the fact that the main cost of the check is computation, which is less significant in a WAN. The table also gives the amount of communication required for each choice of N, which shows that this reflects the main total cost of the protocol. Encoding of larger BCH codes (for $N = 2^{76}$) does have a noticeable effect in a LAN, though: here, BCH runtimes are 3 times higher than Walsh-Hadamard, but only have twice the communication cost. We expect that this could be improved by a more sophisticated encoding algorithm, rather than naive multiplication by the generator matrix.

Acknowledgements. We thank Ranjit Kumaresan for providing us with an extended version of [KK13].

The work in this paper has been partially supported by the ERC via Advanced Grant ERC-2010-AdG-267188-CRIPTO and the Defense Advanced Research Projects Agency (DARPA) and Space and Naval Warfare Systems Center, Pacific (SSC Pacific) under contract No. N66001-15-C-4070.

References

[ALSZ13] Asharov, G., Lindell, Y., Schneider, T., Zohner, M.: More efficient oblivious transfer and extensions for faster secure computation. In: ACM Conference on Computer and Communications Security, CCS 2013, pp. 535–548 (2013)

[ALSZ15] Asharov, G., Lindell, Y., Schneider, T., Zohner, M.: More efficient oblivious transfer extensions with security for malicious adversaries. In: Proceedings of Advances in Cryptology - EUROCRYPT 2015, Sofia, Bulgaria, Part I, pp. 673–701, 26–30 April 2015

[ALSZ16] Asharov, G., Lindell, Y., Schneider, T., Zohner, M.: More efficient oblivious transfer extensions. J. Cryptol. 1–54 (2016)

[ANWOW13] Aumasson, J.-P., Neves, S., Wilcox-O'Hearn, Z., Winnerlein, C.: BLAKE2: simpler, smaller, fast as MD5. In: Jacobson, M., Locasto, M., Mohassel, P., Safavi-Naini, R. (eds.) ACNS 2013. LNCS, vol. 7954, pp. 119–135. Springer, Heidelberg (2013). doi:10.1007/978-3-642-38980-1_8

[Bea96] Beaver, D.: Correlated pseudorandomness and the complexity of private computations. In: Proceedings of the Twenty-Eighth Annual ACM Symposium on Theory of Computing, pp. 479–488. ACM (1996)

[CDD+16] Cascudo, I., Damgård, I., David, B., Döttling, N., Nielsen, J.B.: Rate-1, linear time and additively homomorphic UC commitments. In: Proceedings of Advances in Cryptology - CRYPTO 2016, Santa Barbara, CA, USA, Part III, pp. 179–207, 14–18 August 2016

[CO15] Chou, T., Orlandi, C.: The simplest protocol for oblivious transfer. In: Progress in Cryptology - LATINCRYPT 2015, Guadalajara, Mexico, pp. 40–58, 23–26 August 2015

[EGL85] Even, S., Goldreich, O., Lempel, A.: A randomized protocol for signing contracts. Commun. ACM **28**(6), 637–647 (1985)

[FJNT16] Frederiksen, T.K., Jakobsen, T.P., Nielsen, J.B., Trifiletti, R.: On the complexity of additively homomorphic UC commitments. In: Proceedings of Theory of Cryptography - TCC 2016-A, Tel Aviv, Israel, 10–13 January 2016, Part I, pp. 542–565 (2016)

[IKNP03] Ishai, Y., Kilian, J., Nissim, K., Petrank, E.: Extending oblivious transfers efficiently. In: Boneh, D. (ed.) CRYPTO 2003. LNCS, vol. 2729, pp. 145–161. Springer, Heidelberg (2003). doi:10.1007/978-3-540-45146-4_9

[IR89] Impagliazzo, R., Rudich, S.: Limits on the provable consequences of one-way permutations. In: Proceedings of the Twenty-First Annual ACM Symposium on Theory of Computing, pp. 44–61. ACM (1989)

[KK13] Kolesnikov, V., Kumaresan, R.: Improved OT extension for transferring short secrets. In: Canetti, R., Garay, J.A. (eds.) CRYPTO 2013. LNCS, vol. 8043, pp. 54–70. Springer, Heidelberg (2013). doi:10.1007/978-3-642-40084-1_4

[KKRT16] Kolesnikov, V., Kumaresan, R., Rosulek, M., Trieu, N.: Efficient batched oblivious PRF with applications to private set intersection. In: ACM Conference on Computer and Communications Security, CCS (2016)

[KOS15] Keller, M., Orsini, E., Scholl, P.: Actively secure OT extension with optimal overhead. In: Advances in Cryptology - CRYPTO Santa Barbara, CA, USA, pp. 724–741, 16–20 August 2015

[KOS16] Keller, M., Orsini, E., Scholl, P.: MASCOT: faster malicious arithmetic secure computation with oblivious transfer. In: ACM Conference on Computer and Communications Security, Vienna, Austria, pp. 830–842 (2016)

[Lam16] Lambæk, M.: Breaking and fixing private set intersection protocols. Cryptology ePrint Archive, Report 2016/665 (2016)

[LOS14] Larraia, E., Orsini, E., Smart, N.P.: Dishonest majority multi-party computation for binary circuits. In: Garay, J.A., Gennaro, R. (eds.) CRYPTO 2014. LNCS, vol. 8617, pp. 495–512. Springer, Heidelberg (2014). doi:10.1007/978-3-662-44381-1_28

[NNOB12] Nielsen, J.B., Nordholt, P.S., Orlandi, C., Burra, S.S.: A new approach to practical active-secure two-party computation. In: Safavi-Naini, R., Canetti, R. (eds.) CRYPTO 2012. LNCS, vol. 7417, pp. 681–700. Springer, Heidelberg (2012). doi:10.1007/978-3-642-32009-5_40

[NP99] Naor, M., Pinkas, B.: Oblivious transfer and polynomial evaluation. In: Proceedings of the Thirty-First Annual ACM Symposium on Theory of Computing, Atlanta, GA, USA, pp. 245–254 (1999)

[PSSZ15] Pinkas, B., Schneider, T., Segev, G., Zohner, M.: Phasing: private set intersection using permutation-based hashing. In: 24th USENIX Security Symposium, Washington, D.C., USA, 12–14 August 2015, pp. 515–530 (2015)

[PSZ14] Pinkas, B., Schneider, T., Zohner, M.: Faster private set intersection based on OT extension. In: 23rd USENIX Security Symposium, San Diego, CA, pp. 797–812, August 2014

[PVW08] Peikert, C., Vaikuntanathan, V., Waters, B.: A framework for efficient and composable oblivious transfer. In: Wagner, D. (ed.) CRYPTO 2008. LNCS, vol. 5157, pp. 554–571. Springer, Heidelberg (2008). doi:10.1007/978-3-540-85174-5_31

[Rab81] Rabin, M.O.: How to exchange secrets with oblivious transfer (1981)

[SSR08] Shankar, B., Srinathan, K., Rangan, C.P.: Alternative protocols for generalized oblivious transfer. In: 9th International Conference on Distributed Computing and Networking, ICDCN (2008)

[Yao82] Yao, A.C.-C.: Protocols for secure computations (extended abstract). In: 23rd Annual Symposium on Foundations of Computer Science, pp. 160–164 (1982)

Low-Leakage Secure Search for Boolean Expressions

Fernando Krell[1], Gabriela Ciocarlie[2], Ashish Gehani[2],
and Mariana Raykova[3(✉)]

[1] Dreamlab Technologies, Bern, Switzerland
fernando.krell@dreamlab.net
[2] SRI International, Menlo Park, USA
{gabriela,gehani}@csl.sri.com
[3] Yale University, New Haven, USA
mariana.raykova@yale.edu

Abstract. Schemes for encrypted search face inherent trade-offs between efficiency and privacy guarantees. Whereas search in plaintext can leverage efficient structures to achieve sublinear query time in the data size, similar performance is harder to achieve for secure search. Oblivious RAM (ORAM) techniques can provide the desired efficiency for simple look-ups, but do not address the needs of complex search protocols. Several recent works achieve efficiency at the price of revealing the access pattern. We propose a new encrypted search scheme that reduces the leakage of current Boolean queries solutions, while introducing limited overhead and preserving the sublinear efficiency properties for the search protocol in the semi-honest model. Our scheme achieves a privacy-efficiency trade-off that lies between highly optimized systems such as Blind Seer [18] and OXT-OSPIR [15], which exhibit significant access pattern leakage, and the secure search solution of Gentry et al. [8], which has no leakage, but a much higher efficiency cost.

Our solution is based on a hybrid approach, which integrates ORAM techniques with the efficient search index structure of the Blind Seer system. We reduce the leakage to the server to only the number of nodes visited in the search tree during query execution. Queries that execute in sublinear time in Blind Seer execute also in sublinear time in our scheme.

To enable delegated queries, we develop a new protocol for oblivious PRF sum evaluation and perform secure Boolean queries in a Bloom filter that reveals only the match result. We also enable oblivious-search token generation to hide the specifics of the delegated query from the data owner issuing the search tokens.

We evaluated our system by implementing a prototype and testing it on a 100,000-record database. Our results indicate that the index can be traversed at a rate of a few seconds per matching record for both conjunction and small Disjunctive Normal Form queries.

F. Krell—Work described here was carried out while this author was at SRI International and partially at Columbia University

M. Raykova—Work described here was mainly carried out while this author was at SRI International.

H. Handschuh (Ed.): CT-RSA 2017, LNCS 10159, pp. 397–413, 2017.
DOI: 10.1007/978-3-319-52153-4_23

Keywords: Private search · Boolean queries · Bloom filters · ORAM

1   Introduction

The ability to search over encrypted data provides critical capabilities for database systems that need to guarantee privacy protection for data and queries; examples include information sharing between law enforcement agencies; electronic discovery in private databases such as log files, bank records, during lawsuits, and private queries to census data; police investigations using data from automated license plate readers [4, 18]. Such functionality enables data outsourcing, where a client stores its data on a remote server and later sends queries that the server executes without learning them. An extension to this functionality is the setting of delegated search, where the data owner can generate query tokens for third parties that enable them to execute only the authorized query requests with the storage server without learning any other information about the outsourced data. An immediate application of this extension is the ability to audit cloud applications and allow auditors only issue to authorized queries.

However, there are two main relevant questions of privacy and efficiency for encrypted search schemes, whose answers are also related. The privacy question considers how much the storage server learns about the queries it executes. While encryption techniques can help hide the content stored on the server, they do not protect the client's access pattern which becomes a privacy leakage to the server. A recent work [14] has demonstrated that this leakage can be substantial even in the simplest search scenario of exact match such as keyword search. For more complex types of queries, such as Boolean queries, this leakage can have even more serious security implications. The second question is related to efficiency. Search algorithms in plaintext usually have sublinear efficiency in the size of the database, and this efficiency guarantee is crucial for the usability of an algorithm. The sublinear efficiency of a search algorithm implies that it does not access all data. At the same time, the ability of the server to know what data has been accessed translates into access pattern leakage. Thus, an inherent trade-off between the privacy and efficiency questions for secure search emerges.

There have been several approaches to secure search in previous works that achieve various trade-offs between the efficiency and the privacy guarantees of their schemes. Searchable encryption [3, 6, 22] provides the capability to encrypt data items such as keywords, issue search tokens for particular items using the private parameters for the scheme, and support matching functionality to check whether a ciphertext contains the same item as a search token. Using this primitive, one can implement a search protocol with sublinear efficiency, which completely reveals the access pattern into the database for each query. Private information retrieval (PIR) [5] and symmetric PIR [9] techniques allow a client to retrieve a record stored on a server without revealing what record is being retrieved or allowing the client to learn anything more about the data. Such techniques can be employed for secure search, but they require computation proportional to the size of the database.

While earlier work on secure search considered mostly single keyword search queries [3,6,17,19,21,22], recent approaches address more complex queries over databases, such as Boolean and range queries, and model queries over databases of records, which are described by several attributes and a main payload [4,7,13,15,18][1]. These last solutions provide surprisingly good efficiency at the price of access pattern leakage. Unfortunately, the access pattern is not uniquely defined. In fact, each of these solutions reveals an access pattern that is specific to the structure of the scheme's data storage and goes beyond the records that match each query. Hence, it is difficult to analyze this leakage and to precisely define what is protected, making it impossible to compare the leakage of different schemes that employ different underlying encrypted search structures.

A different line of work [20] proposes a solution for encrypted search that can handle SQL queries. However, it considers a different adversarial model that aims to protect the data against curious database administrators, but assumes fully trusted proxy that encrypts the queries. This model does not match the guarantees that we want to achieve. This solution also reveals the access patterns for the query terms.

A completely different approach for the secure search problem is to employ secure computation techniques to implement the search functionality. While generic secure computation techniques require computation time at least linear in the size of its inputs, the works of Gordon et al. [12] and Afshar et al. [1] manage to achieve sublinear (amortized) time for sublinear RAM computations in the semi-honest and malicious setting respectively. These approaches leverage the random access machine computation (RAM) model together with a special structure for memory storage called oblivious RAM (ORAM) [11], which provides access patterns hiding with only polylogarithmic overhead for memory accesses. This approach was further pursued in the work of Gentry et al. [8] focusing on the database query functionality and employing somewhat-homomorphic encryption to implement the small secure computation steps. This work handles keyword database queries and limited conjunction queries.

Our work is motivated by the lack of a good grasp on analyzing leakage in the Boolean search protocols mentioned earlier. We propose a construction that adopts the approach of combining ORAM together with small secure computation steps. We focus on functionality for Boolean search queries and we develop tailored solutions for that. Our solution for secure search enables the same functionality for Boolean queries as Blind Seer [7,18], but it diminishes the access pattern leakage, while preserving the sublinear efficiency overhead for queries that are executed in sublinear time in these protocols. As expected, when compared with these solutions that reveal complete access patterns, the concrete efficiency overhead for our protocol increases. Although the direct comparison with the work of Gentry et al. [8] is hard, since their work implements much simpler queries compared to our protocols, our protocols achieve a much better efficiency for comparable functionality.

---

[1] Interesting is the single-keyword range-query solution of [13] which provide a tunable privacy-efficiency trade-off.

1.1 Setting

The setting we are interested in is called Outsourced Symmetric Private Information Retrieval (OSPIR) [15]. It captures the scenario in which the data owner outsources the data to a server, and gives search capabilities to clients. Such a scheme can be defined by two phases OSPIRSetup and OSPIRSearch.

In the OSPIRSetup phase, the data owner (Owner) on input DB, does some preliminary computation on the data and produces an *encrypted* database EDB and access parameters params. EDB is then given to the server (Server). In the OSPIRSearch phase, a client (Client) inputs a query $\mathbf{q}$, Owner inputs params, and Server inputs EDB. After protocol execution, Client obtains database records satisfying its query. We provide a formal definition next, allowing a tunable false positive rate on the records returned to Client.

Definition 1. *We define an* OSPIR Scheme *as a pair of interactive algorithms*

- (params, EDB)←OSPIRSetup(1^λ, DB, fp). Owner *inputs database* DB = $\{(D_i, W_i)\}_{i=1}^D$, *and gets back* params. Server *gets* EDB.
- (records)←OSPIRSearch(params, EDB, $\mathbf{q}$). Owner *inputs* params, Server *inputs* EDB *and* Client *inputs* $\mathbf{q}$. Client *gets* records.

such that for all λ *and for all* DB, $\mathbf{q}$, *if* (params, EDB)←OSPIRSetup(1^λ, DB, fp), *and* (records)←OSPIRSearch(params, EDB, $\mathbf{q}$), *then* $DB_{fp}(q)$ = records, *where* $DB_{fp}(q)$ *denotes the records of* DB *satisfying the query* $\mathbf{q}$ *plus each* DB *record with probability* fp.

The OSPIR scheme described in this work assumes a semi-honest behavior of the participants. That is, we assume that every participant honestly follows the description of the protocol, and we define and prove security in such setting.

1.2 Related Work

As mentioned earlier, the problem of privately searching a database can be solved by generic secure computation schemes [10,23]. However, these generic protocols require a computation time that is at least linear in the size of the participant's inputs. A scenario closer to out setting is the one of PIR and SPIR protocols, where a client obliviously selects an item from the server's database. Although PIR-like protocols provide sublinear communication, they do require linear-time computation. Ad hoc solutions [8,12] provide sublinear computation time for look-ups and single keyword search in the private DB setting.

A more practical approach is taken in the searchable encryption schemes [3,4,6,22] and OSPIR protocols [7,15,18]. These schemes achieve an efficiency close to plaintext solutions, but at the cost of revealing access patterns to the database records and the underlying search structure. The OSPIR solution of [13] provides a tunable efficiency-privacy trade-off solution. They achieve efficiency comparable to [7,15,18] with virtually no access pattern leakage for a tunable number of queries. After this threshold is reached, the index need to be rebuild to avoid incurring access pattern leakage.

Among the works just mentioned, the HE-over-ORAM approach [8], and the Blind Seer and OXT-OSPIR [15] are of particular interest. First, these schemes focus on the delegated query scenario. Secondly, while HE-over-ORAM aims for a secure asymptotically sublinear solution for single keyword search, the Blind Seer and OXT-OSPIR systems focus on practicality: they both support a rich set of queries and their efficiency is close to the plaintext database case. Our goal is to build a system that lies in-between these systems in terms of the privacy vs. efficiency trade-off. Hence, we borrow techniques from all these solutions.

OSPIR-OXT and Blind Seer. The first solution for the OSPIR setting was proposed by Jarecki et al. [15] and Pappas et al. [18]. Although they solve the same problem, they provide very different approaches. OSPIR-OXT [15] is an extension of the OXT searchable encryption scheme [4]. This solution allows for Boolean queries in Searchable Normal Form $(t_1 \wedge \phi(t_2, ..., t_n))$, and runs in time proportional to the number of records satisfying the term t_1. The solution is based on an inverted index approach, which is used to search information about the leading term t_1. This information is used then to search for the records satisfying the sub-queries $t_1 \wedge t_i$. A completely different approach was taken in Blind Seer [7,18]. Instead of using an inverted index, Blind Seer builds a Bloom filter tree on the searchable keywords of the database. Each leaf of the tree is associated with a record in the database, and each internal node corresponds to a *masked* Bloom filter containing the searchable keywords of the records in its subtree. Hence, a Boolean formula is answered by following root-to-leaves paths, where the nodes' Bloom filter satisfy the query. To evaluate each node, the server and the client engage in a secure two-party computation protocol (implemented using Yao's protocol [23]), where the server inputs masked Bloom filter bits, and the client inputs the mask. The big advantage of OSPIR-OXT over Blind Seer is efficiency. This is due to the interactive nature of Blind Seer. In terms of leakage, these systems are incomparable since their underlying data structures are completely different. Blind Seer, though, has the advantage that the search procedure does not reveal the partial evaluation results. In addition, Blind Seer can answer *any* Boolean query in sublinear time.

HE-over-ORAM Database Search. Gentry et al. [8] recently proposed a private DB system with no leakage based on ORAM and Somewhat Homomorphic Encryption Scheme. ORAM is used to protect the client's access patterns and the owner's data from the server. To protect the database information from the client, data is also encrypted using a variation of a Somewhat Homomorphic Encryption Scheme that enables Equal-to-Zero and Comparison operations. These operations enable the client to blindly perform ORAM operations until the requested value is found. Although this work shows the feasibility of the HE-over-ORAM approach, it has significant limitations in efficiency and functionality. In terms of efficiency, their experimental results shows that it requires 30 min to execute a single keyword query on a 2^{22} record database. In terms of functionality, the system only allows single keyword queries, and conjunction may be enabled by a trivial addition of the keywords into the database index.

1.3   Approach

Our approach is to use the Bloom filter Search Tree of Blind Seer as our search structure, while storing the encrypted data and its index in ORAM structures at the server. We give the ORAM access parameters to the client, as done in the HE-over-ORAM scheme. To avoid the case where the client learns more information than necessary, the actual data held by the ORAM should be encrypted in a special way. While this is done using Somewhat Homomorphic Encryption by Gentry et al. [8], we provide a new encoding scheme that allows parties to securely evaluate an index node, revealing to the client only the necessary information to continue the search procedure. We accomplish this with a novel protocol for conjunctive query evaluation on specially encrypted Bloom filters. This protocol is then extended to handle queries in Disjunctive Normal Form.

The use of ORAM eliminates all important leakage to the index server of Blind Seer [18]. ORAM protocols, however, do leak the number of queries performed by the client; hence, our solution reveals the amount of work done by the client (which is unavoidable if we require sublinear time). In particular, the server can infer the number of records retrieved by the client. It also learns the relation between the amount of work in the index and the amount of records retrieved. Nevertheless, the server is unable to link the work done in the index and the specific record retrieved.

2   Preliminaries

Bloom Filter. A Bloom filter [2] is a data structure that allows for set membership queries. The structure is composed by a bit array $B[1..n]$ and a set of hash functions $\mathcal{H} = \{H^{(i)} : \{0,1\}^* \rightarrow [n]\}_{i=1}^{h}$. An element is inserted by turning on the bits at the positions indicated by the hash values of the element. Hence, if an element e is in the filter, then $B[H^{(i)}(e)] = 1$ for all $i \in \{1..h\}$. A Bloom filter is parametrized by a *false positive* rate since elements not in the set may hash to positions that are all set in B. Given a false positive probability fp and the number of elements N in the filter the optimal length n of B and the optimal number of hash functions h to use can be approximately computed as $n = \lceil \frac{N \ln \mathsf{fp}}{\ln \frac{1}{2^{\ln 2}}} \rceil$, $h = \lceil \ln 2 \frac{n}{N} \rfloor \approx \log_2(1/\mathsf{fp})$.

Bloom Filter Search Tree. A Bloom Filter Search Tree (BFT) is an index for a database $(D_i, W_i)_{i=1}^{D}$, where D_i is an arbitrary document and W_i is D_i's associated set of searchable keywords. Given a parameter fp (false positive), a Bloom filter tree is constructed by building a b-ary tree of D leaves. Each leaf of this tree is associated with a document D_i and holds a Bloom filter with the corresponding set of keywords W_i. Each internal node contains a Bloom filter having inserted all keywords held at its children nodes. A search procedure for documents containing a keyword w starts by querying the Bloom filter at the root node. If the keyword is present, we continue recursively querying its children until reaching the leaf nodes whose associated document contains w.

Oblivious RAM. An Oblivious RAM protocol [11] is a two-party protocol that allows a client to outsource its data and completely hide the access pattern of future queries. It is composed by an algorithm OSetup and by a protocol OAccess. OSetup is run by the client (or data owner) and outputs parameters param (including an initial state state), and data structure struct. OAccess is a two-party protocol in which the client inputs an operation op $\in$ {ORead, OWrite}, an index i, data D^* (if op is OWrite), and parameters param. The server inputs struct. At the end of the protocol execution, the server obtains an updated structure, struct', while the client obtains the updated state state', and data D_i (if op was ORead). It is well known that any ORAM holding n elements and simulating m RAM accesses requires $\Omega(m \log n)$ accesses [11].

Participants. The system supports three actors: Owner, Server, and Client. The Owner knows the database $(D_i, W_i)_{i=1}^D$, builds an index and outsources the list of documents (D_i) and the index to the Server. The Client has a *query* $\mathbf{q} = \phi(W)$ composed by a Boolean formula $\phi(\cdot)$ over a set of keywords W. The Client gets the set of documents $\{D_i : W_i \text{ satisfies } \phi(W)\}$.

Notation. We use λ to denote a security parameter, and fp a false positive rate. The set $\{1, 2, ..., i\}$ will be denoted as $[i]$. Let $\mathcal{G}$ be a group of generator g and prime order p, where the Decisional Diffie-Hellman (DDH) assumption holds. We use multiplicative notation for the group operations. Let $H : \{0,1\}^\lambda \times \{0,1\}^* \to \{0,1\}^\lambda$ be a keyed hash function (or MAC) having keys in $\{0,1\}^\lambda$, in which $H(k, w)$ is denoted as $H_k(w)$. Similarly, let $F : \{0,1\}^\lambda \times \{0,1\}^\lambda \to \mathcal{G}$ be a pseudo random function (PRF) indexed by keys in $\{0,1\}^\lambda$, having domain in $\{0,1\}^\lambda$, and image in $\mathcal{G}$. We denote $F(k, r)$ as $F_k(r)$. Let $\mathcal{E} = \langle \text{Gen}, \text{Enc}, \text{Dec} \rangle$ be a semantically secure encryption scheme. For a query $\mathbf{q}$ corresponding to a DNF formula $\phi(\cdot)$, we let $|\mathbf{q}| = |\phi|$ be the number of conjunctive clauses in ϕ. For a clause $C_i \in \phi$, we let $|C_i|$ be the number of terms in C. The *topology* of $\mathbf{q}$, denoted as $\text{topo}(\mathbf{q})$ (or $\text{topo}(\phi)$), correspond to $|\mathbf{q}|$ and $|C_i|$ for each $i \in [|\mathbf{q}|]$. We denote by $x \xleftarrow{\$} S$ the process of sampling a uniformly random element x from set S. For a tree node v, we let Children(v) be the set of children nodes of v. We let $\text{BFBuild}(S, \text{fp})$ denote the process of building a Bloom filter for set S with false positive rate fp, and $\text{BFMatch}(\text{BF}, w)$ denotes the process of matching a keyword w in Bloom filter BF. For a set of hash functions $\mathcal{H}$, we let $\mathcal{H}(w)$ denote the set $\{H(w) : H \in \mathcal{H}\}$. Finally, we abuse ORAM notation and let $D_i \leftarrow \text{ORead}(i, \text{struct})$ denote a read ORAM access on address i at ORAM structure struct held by the server. That is, we omit in the notation the client's parameters and the updated structure given to the server.

3   Cryptographic Primitives

In this section, we introduce the necessary cryptographic primitives for the construction of our private search scheme (Sect. 4). These primitives are presented in a modular way, and can be of independent interest.

Oblivious PRF. First, our solution uses an Oblivious Pseudorandom Function (OPRF). It involves two parties, C having input m and S having input k, who jointly evaluate a pseudorandom function $F_k(m)$, keeping k, m private to the respective party. A simple construction proposed by Jarecki and Liu [16] uses the Hashed Diffie-Hellman PRF $(F_k(m) = \mathsf{Hash}(m)^k)$. The protocol is described in Fig. 1. C starts by sampling a uniformly random invertible exponent α and sends $X = \mathsf{Hash}(m)^\alpha$ to S. S responds with $Y = X^k$. Finally, C outputs $Z = Y^{\alpha^{-1}} = \mathsf{Hash}(m)^k$.

MUL-OPRF. In a simple variation of the above primitive, C inputs a set $\{m_1, ..., m_n\}$, S inputs the secret key k, and C receives as output $\prod_i F_k(m_i)$. We call this new primitive MUL-OPRF. We obtain a secure protocol for this primitive by using $\prod_i \mathsf{Hash}(m_i)$, as the random hash function in the protocol of Fig. 1.

Masked MOPRF. For the purpose of the construction in Sect. 4, we require a slight modification on the above MUL-OPRF functionality. We call this new primitive a Masked MOPRF. In this primitive, C gets the result of the MUL-OPRF protocol masked with a random value R, while S obtains the mask R. This simple modification is achieved by adding one extra message in the protocol (Fig. 2). The server starts by sampling a uniformly random exponent β, and sending $W = g^\beta$ back to C. C responds with $X = (W \cdot \prod_i \mathsf{Hash}(m_i))^\alpha$ for the uniformly random invertible exponent α. S replies with $Y = X^k$, and outputs $R = g^{\beta \cdot k}$. C outputs $Z = Y^{\alpha^{-1}} = R \cdot \prod_i \mathsf{Hash}(m_i)^k$.

Security. The security of the MUL-OPRF protocol follows directly form the security of the Hashed DH Oblivious PRF protocol [16] by using $\prod \mathsf{Hash}(\cdot)$ as the random function in the random oracle model. The security and correctness of the Masked MUL-OPRF protocol follows directly from DDH assumption since it implies that the value $g^{\beta \cdot k} \times \prod_i F_k(m_i)$ is pseudo-random even given g^β (and even if the adversary somehow knows $\prod_i F_k(m_i)$).

4   Scheme

In this section, we present our private search scheme. Our ultimate goal is a secure search functionality that enables oblivious delegated queries on outsourced

---

Two-party Protocol OPRF

Parameters. A random hash function $\mathsf{Hash} : \{0,1\}^* \to \mathcal{G}$.
Inputs. C: $w \in \{0,1\}^*$. S: k.

1. C samples $\alpha \xleftarrow{\$} \mathbb{Z}_p^*$, and sends back $X = \mathsf{Hash}(w)^\alpha$.
2. S replies with $Y = X^k$.
3. C outputs $Z = Y^{\alpha^{-1}}$.

Fig. 1. The two-party protocol OPRF

Two-party Protocol Masked-MOPRF

Parameter: A random hash function $\mathsf{Hash} : \{0,1\}^* \to \mathcal{G}$.
Input $\mathcal{C}: \{m_i\}_{i \in [n]}$. $\mathcal{S}: k$.

1. $\mathcal{S}$ samples $\beta \xleftarrow{\$} \mathbb{Z}_p$ and sends $W = g^\beta$.
2. $\mathcal{C}$ samples $\alpha \xleftarrow{\$} \mathbb{Z}_p^*$, and sends back $X = (W \cdot \prod_{i=1}^n (\mathsf{Hash}(m_i)))^\alpha$.
3. $\mathcal{S}$ replies with $Y = X^k$.
4. $\mathcal{C}$ outputs $Z = Y^{\alpha^{-1}}$.
5. $\mathcal{S}$ outputs $R = g^{\beta \cdot k}$.

Fig. 2. The two-party protocol Masked-MOPRF

data to a server (Server), where the data owner (Owner) can obliviously issue a search token to a client (Client) for a query that remains hidden from the owner. Given this search token, Client should only learn the data matching the query, while minimizing the information that the server (Server) learns about the issued queries (we analyze what Server learns formally in the full version of this work).

Recall from Sect. 1.3 that our search structure is a Bloom filter tree in which documents are associated with the leaves of the tree and each node contains a Bloom filter holding the searchable keywords of the documents associated with the leaves of its subtree. In the simple two-party setting, where Owner is the querier (or client), Owner can build a plaintext Bloom filter tree storing it as an ORAM at the server. Then, for each query, Owner traverses the Bloom filter tree (via ORAM accesses), to find the documents that satisfy its query (which it retrieves and decrypts also via ORAM accesses).

In the delegated queries scenario (i.e., where Client is not the database owner), if complete ORAM access is allowed to Client, information beyond what is strictly necessary is revealed. First, since each ORAM access may retrieve several elements, Client gets bits of the index that do not correspond to its query. Equally important, Client learns partial evaluation information, such as which keywords of the formula are satisfied at each node, and which Bloom filter bits are set. Ideally, Client should only learn if the complete query is satisfied by the index node being evaluated. These two problems are addressed by *specially* encrypting the index bits and introducing an oblivious protocol that allows Client to only learn whether the formula is satisfied by an index node, but nothing more.

In Sect. 4.1, we introduce techniques that allow for secure delegated queries leveraging Bloom filter tree and ORAM approaches. We first show how to generate query tokens without revealing the client's query to either party. We then describe how to securely evaluate single term queries, conjunctions and DNF queries on each Bloom filter, allowing Client to traverse the tree and find the documents satisfying its query. Finally, we describe how Client can decrypt the retrieved documents without any party knowing the identifiers of these documents. In Sect. 4.2, we present the complete construction of our private search scheme.

4.1   Building Block Techniques

Obliviously Generating Search Tokens. Before Client can evaluate its query, it needs to be able to compute Bloom filter indices corresponding to the terms in the query for each Bloom filter in the tree. These indices need to be derived from a PRF, whose key, s_{bf}, is held by the database owner. For this purpose, each term is mapped through the use of this PRF to a *search token*, which is then hashed to get the Bloom filter indices. Similarly to Jarecki et al. [15], we use the Hashed Diffie-Hellman PRF $F_{s_{bf}}(w) = \mathsf{Hash}(w)^{s_{bf}}$ as our PRF to compute search tokens for each term. This PRF can be obliviously computed via the protocol in Fig. 1.

Single Term Queries. For single keyword queries, $\mathbf{q} = \phi(w) = w$, the client needs to learn if all the bits queried in a Bloom filter are set. For this purpose, we leverage the Masked-MOPRF protocol making use of the underlying PRF to encrypt each bit. We encode a bit to an arbitrary element in the range group of the PRF F, and use F to encrypt the bit. Let g be a group generator (that we keep secret from the client); we map a bit b_i to g^{b_i} and encrypt it as $\langle g^{b_i} \cdot F_k(r_i), r_i \rangle$ for position i in the Bloom filter[2]. The client and server use the Masked-MOPRF primitive described in Fig. 2 to evaluate a Bloom filter query that reveals no additional information to the client as follows. They execute the Masked-MOPRF protocol with inputs a set of $\{r_i\}_{i \in S}$, for the client, and a PRF key k for the server, where S is the set of BF indices corresponding to the query. At the end of the protocol, the client obtains $R \cdot \prod_{i \in S} F_k(r_i)$, while the server obtains a random value R. Next, the client computes $\prod_{i \in S} (g^{b_i} \cdot F_k(r_i))$, and, using its output from the Masked-MOPRF protocol, obtains

$$\prod_{i \in S} \left(g^{b_i} \cdot F_k(r_i) \right) \left(R \cdot \prod_{i \in S} F_k(r_i) \right)^{-1} = R^{-1} \cdot g^{\sum_{i \in S} b_i}$$

The server now provides $H(R^{-1} \cdot g^h)$ so that the client can do the matching evaluating $H(R^{-1} \cdot g^{\sum_{i \in S} b_i})$ and the comparison. The random element R binds together the values from all BF indices corresponding to a query, and does not allow the client to learn any information about subsets of the BF bits in the corresponding positions. The hash over the server-side matching key $R^{-1} \cdot g^h$ hides R from the client. Hence, the protocol completely hides the value $\sum_{i \in S} b_i$ for a mismatching query.

Conjunction Queries. The method described above can be trivially extended to conjunctions since the single term case is in fact a conjunction on the corresponding Bloom filter bits. We can treat a conjunction as a bigger single term query. Let C be a conjunction, and let $|C|$ denote the number of terms in C, then the number of bits to be checked is $h \times |C|$.

Disjunctive Normal Form Queries. In the case of single term queries (and conjunctions), a match requires that all the bits at the query indices of the

---

[2] The values r_i across different Bloom filters are independent.

Bloom filter be set to one. Therefore, it suffices that the server provides the hash of a single "randomized matching key" $H(R \cdot g^h)$ to the client. In the case of disjunctions, on the other hand, there are many settings for the bit values of the query BF indices that can satisfy the query; hence, there are many possible matching keys. In fact, there can be as many as $|C| \cdot 2^{h \cdot (|C|-1)}$ different satisfying bit value assignments for the BF query indices. However, in our construction, we consider the expression $g^{\sum_{i \in S} b_i}$ for each term in the conjunction, which has h different possible values which depend only on the number of ones in the set of bits. Hence, there are only $|C| \cdot h^{|C|-1}$ possible matching evaluation values for the client formula. With this observation in mind, we construct the following protocol:

1. For each conjunctive clause C the client and the server execute the protocol for the single query matching (without the final stage where server reveals the hashed matching key), and the client learns the value $R_C \cdot g^{\sum_{i \in S_C} b_i}$, where S_C denotes the set of Bloom filter positions to be checked for clause C.
2. Each of the resulting values is blinded by a public random exponent L_C, and the final matching evaluation key is computed as $\prod_{C \in \phi} (R_C \cdot g^{\sum_{i \in S_C} b_i})^{L_C}$.
3. The server computes the set Matching of all the possible matching values, and the client *obliviously* does the matching. There are several ways to do the matching. One possibility is to hash and permute all the matching keys, before sending them to the client. Another approach is through a Bloom filter.

The purpose of the exponent L_C is to separate the space of possible values of each clause evaluation, such that there are no overlaps that could (with high probability) make a set of unsatisfying clauses evaluate to a matching key.

Record Decryption. After finding the list of identifiers of records satisfying the query, Client can actually retrieve them by querying the ORAM that contains the records. However, as mentioned earlier, in the case of the index ORAM, each ORAM access can potentially reveal records that do not satisfy Client's query. Hence, each document should be encrypted under a key unknown to Client. However, the client should be allowed to decrypt the satisfying records. For this purpose, Owner samples a secret key s_r and, using again the Hashed Diffie-Hellman PRF, it derives for each document D_i an encryption key $k_i \leftarrow F_{s_r} = \mathsf{Hash}(i)^{s_r}$. For each document identifier obtained by Client, Owner and Client execute the OPRFprotocol in Fig. 1 to derive the decryption keys.

4.2   Final Scheme

Preprocessing. The procedure is parametrized by a false positive rate fp and a security parameter λ. The database owner starts by choosing a key s_{bf} for the PRF F and keys s_k, s_r for the keyed hash function H. It then proceeds by building a Bloom filter Search Tree with false positive rate fp for the database $\mathsf{DB} = (D_i, W_i)_{i=1}^D$, where each keyword $w \in W_i$ is mapped to $H_{s_k}(w)$ forming set $\tilde{W}_i$. Each record D_i is encrypted using the derived key $k_i \leftarrow H_{s_r}(i)$,

$\tilde{D}_i = \mathsf{Enc}_{k_i}(D_i)$. The Bloom filter tree is then *encrypted* by encoding each Bloom filter bit b as g^b and encrypting it as $\mathsf{bEnc}_{s_{\mathsf{bf}}}(g^b) = \langle g^b \cdot F_{s_{\mathsf{bf}}}(r), r \rangle$, where r is sampled uniformly random from $\{0,1\}^\lambda$. The owner continues by preparing an ORAM structure $(\mathsf{param}_I, \mathsf{struct}_I)$ holding the encrypted index, and the ORAM structure $(\mathsf{param}_D, \mathsf{struct}_D)$ holding the records. In principle, each encrypted Bloom filter bit can be an ORAM block. However, this can be optimized to pack several bits in the same ORAM block to reduce the number of ORAM lookups. We can choose, for example, to hold an entire Bloom filter in one ORAM block, or to pack together bits in the same position across sibling Bloom filters.

We describe next the basic procedures used by the setup phase:

○ $\mathsf{BFTBuild}(\{\tilde{W}_i\}_{i=1}^D, \mathsf{fp}, d)$: Let BFT be a balanced d-ary tree of D leafs. Let $L = \lceil \log_d D \rceil$ be the height of the tree. We build the tree level by level, starting from the bottom level L. We then proceed recursively until reaching the root of the tree. Let $N_L = \max |W_i|$. Using N_L and fp, compute Bloom filters length n_L and number of Bloom filter hash function h_L. Then, we sample h_L independent hash function $\mathcal{H}_L = \{H^{(1)}, ..., H^{(h_L)}\}$ with image $\{0, 1, ..., n_L - 1\}$. For each $i \in [D]$, we build a Bloom filter B_i (using $\mathcal{H}_L$) inserting the elements of $\tilde{W}_i$. We maintain each Bloom filter in a unique leaf of BFT. The internal nodes of the tree are built recursively as follows: we associate each node at level ℓ with the keywords held in its children. That is, for each internal node, we build a Bloom filter that contains the elements from all its d children. Return $\mathcal{H} = \{\mathcal{H}_1, \mathcal{H}_2, ..., \mathcal{H}_L\}$ and tree BFT. We force the sets of hash functions to be of the same size $h = h_L = |\mathcal{H}_L|$, such that, for each query, the number of lookups in every node is the same. This will prevent the server from learning the level in the tree of the nodes being evaluated.

○ $\mathsf{BFTEncrypt}(\mathsf{BFT}, 1^\lambda)$: Sample a uniformly random key s_{bf} for PRF F. Build a tree EBFT by: (a) encoding each bit b of BFT as g^b, (b) encrypting g^b as $\mathsf{bEnc}_{s_{\mathsf{bf}}}(g^b) = \langle g^b \cdot F_{s_{\mathsf{bf}}}(r), r \rangle$, where r is uniformly random in $\{0,1\}^\lambda$. Return key s_{bf} and tree EBFT.

Search. Client inputs a DNF formula $\mathbf{q} = \phi(\boldsymbol{W}) = C_1 \vee C_2 \vee \cdots \vee C_{|\mathbf{q}|}$ on keywords in $\boldsymbol{W}$. The client reveals the query *topology* (number of clauses and size of each clause) to Server. Client and Owner then execute the protocol in Fig. 1 to obtain search tokens for each keyword in each clause. For each clause C in the query, Client (or Server) uniformly samples L_C from $[|\mathbf{q}|]$ and sends it to Server (Client). Client and Server then start the tree traversal protocol. For each node being evaluated, both parties proceed as follows:

1. For each clause C of the query, the client computes the Bloom filter positions of the clause's hashed keywords for the node being evaluated, and performs the ORAM queries to get the corresponding encrypted bits $\langle g^{b_i} \cdot F_{s_{\mathsf{bf}}}(r_i), r_i \rangle$.
2. To get each clause evaluation key, Client and Server engage in the Masked-MOPRF protocol, where Client inputs the encryption randomness r_i of each encrypted bit, and Server inputs the PRF secret key s_{bf}. Client obtains $\pi_C = R_C \cdot \prod_{t \in S_C} F_{s_{\mathsf{bf}}}(r_i)$, and Server obtains the random mask R_C. Client computes each clause C evaluation key as $\prod_{i \in S_C}(g^{b_i} \cdot F_{s_{\mathsf{bf}}}(r_i)) \cdot (\pi_C)^{-1}$. The key obtained is $\zeta_C = R^{-1} \cdot g^{\sum b_i}$.

3. The client computes each clause evaluation key $K_C = \zeta_C^{L_C}$, and multiplies all keys together to obtain the final evaluation key FinalKey: $\prod_{C \in \phi} K_C = \prod_{C \in \phi} (R_C^{-1} \cdot g^{\sum_{i \in S_C} b_i})^{L_C}$

4. The server computes all possible matching keys. That is, for each clause C, Server computes the set $\text{Matching}_C = \left\{ (R_C \cdot g^{|C| \cdot h})^{L_C} \cdot \prod_{C' \neq C} (R_{C'} \cdot g^{\nu_{C'}})^{L_C} \right\}$, where each $\nu_{C'} \in \{0, \ldots, |C'| \cdot h\}$.

5. Each node evaluation finishes by checking if Client's FinalKey belongs to the set $\text{Matching} = \bigcup_C \text{Matching}_C$. This can be done securely by computing a Bloom filter with all matching keys and sending the filter to the client, or by sending a permutation of all hashed keys.

After the tree traversal, Client gets the indices of all documents satisfying the query. It can obtain the documents by querying the documents ORAM structure. To obtain the document decryption keys, Client and Owner execute protocol OPRF, where Owner inputs key s_r and Client inputs the document identifiers. A formal description of the protocol is presented in the full version of this work.

5   Evaluation

In this section, we quantify the performance of the encrypted index traversal of our OSPIR protocol by both showing the results of running our prototype on datasets of 1K, 10K, and 100K records, and providing an asymptotic analysis of performance.

Experimental Setup. Motivated by the audit log application on cloud services, we collected provenance data from an Ubuntu 14.04 system running Apache. From this data, we built a single table database containing on each row a node from the provenance graph and its annotation. We set up two Intel Xeon E5-2430 2.2 Ghz (2 cores of 12 threads), 100 GB RAM machines with Broadcom 1 GB Ethernet. Server and Owner run on the same machine. Our system parameters were set so that the index for the 100K records database fits in 100 GB of RAM. Specifically, we fixed the degree of the tree to 10, the Bloom filter false positive to 10^{-5}, and the number of searchable keywords per record to 4.

Queries. We ran SELECT-id queries that match a single record. The performance of the queries that return one result provides the worst-case latency per record, since queries returning several records do not need to inspect already-evaluated nodes. Additionally, by returning just the record identifier, we can evaluate exactly the cost of the search procedure. The types of queries covered were single term, conjunctions, disjunctions and 3-DNFs.

Conjunctions vs. Disjunctions. Figure 3 shows, in $\log_{10}$ scale, the latency time for conjunctions and disjunctions of sizes 1, 2, 3, 4, and 5 on a 100 K records database. We observe that while conjunctive queries run in a few seconds, disjunctive queries are exponentially more expensive. It is interesting to note that the number of ORAM queries performed by both types of queries is exactly the

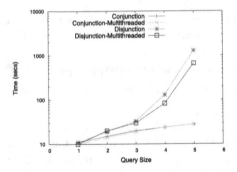

Fig. 3. Latency of conjunctions and disjunctions of sizes 1, 2, 3, 4 and 5 for 100K records DB.

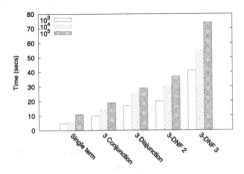

Fig. 4. Query latency time for different-size DNF queries for databases of sizes 1K, 10K and 100K records.

same; hence, the latency time is dominated by the cryptographic operations and the data transfer of the matching keys set. In the case of disjunctive queries, we also note that the use of multiple cores does not reduce the latency significantly (at most a factor of two for 24 cores). In the case of conjunctions, the evaluation is entirely sequential and the use of multiple cores has no effect.

Varying Database Size. Figure 4 shows the latency for different DNF queries across databases of sizes 1K, 10K, and 100K. The difference between running a query on databases of varying sizes is captured in the number of nodes to be evaluated and the potentially larger ORAM size for larger databases. Observe the sub-linearity of our system's running time: an increase in the database size by a factor of 10 increases the running time by a comparatively small amount, which is due to a single extra evaluation node and a larger ORAM structure.

ORAM vs. Node Evaluation. In Table 1, the second and third columns illustrate the time our prototype spent in ORAM read queries and node evaluation, once ORAM queries have been performed. Since same-size queries require the same number of ORAM operations, the ORAM time is identical for same-size queries. Disjunctive queries, however, exhibit a much more expensive node

Table 1. Latency in seconds of tasks in protocol and network usage per query on a 100K records DB.

Query	ORAM	Eval	Network	Query	ORAM	Eval	Network
Single term	4	6	26 MB				
2-Conjunction	9	6	52 MB	4-Conjunction	18	6	105 MB
2-Disjunction	9	10	52 MB	4-Disjunction	18	90	140MB
3-Conjunction	14	6	78 MB	5-Conjunction	18	6	131 MB
3-Disjunction	14	20	80 MB	5-Disjunction	18	90	932 MB
				Size 2 3-DNF	25	11	158 MB
				Size 3 3-DNF	35	35	249 MB
				Size 4 3-DNF	50	680	1173 MB

evaluation execution, since they involve an exponentially large number of possible matching keys, which Server has to compute and hash individually. Moreover, the fourth column indicates that the network usage increases significantly with bigger disjunctions. The reason is that Server also needs to send the set of possible matching keys to Client. In particular, for size-4 3-DNF, the network usage raises to 1 GB, and we can infer that for these queries the index traversal will dominate the running time for queries that also return the records' payload.

Index Size. One of the drawbacks of our solution is the space utilization of the index. Each bit of a plaintext version of our index is encoded using 140 bytes. Moreover, the index is stored as is in an ORAM structure, which multiplies the space by a non-small constant factor. In our evaluation, each record was associated with 4 searchable keywords. Consequently, for our 100K records dataset, the encrypted index uses 75 GB of RAM.

6 Conclusions

We proposed an private search scheme that supports Boolean queries and delegated queries. Our system diminishes the leakage of existent solutions, while preserving sublinear search efficiency. Our construction integrates ORAM techniques with efficient search index structures, and leaks to the server only the number of nodes visited in the search tree during the execution of a query. We proposed a new protocol for oblivious PRF evaluation that allows to securely evaluate Bloom filters. This enables the delegated-query feature by disclosing only the match result. Finally, protect the client's queries from the data owner. We implemented our system prototype and ran it on a 100,000-record database. We showed that our system can handle conjunctive queries and small DNF formulas in 10–30 s. The sublinearity of our solution, also experimentally illustrated in Fig. 4, allows us to extrapolate that queries on much larger databases (10^6, 10^7, and 10^8 records) will run in a few minutes. The cost of eliminating leakage is substantial; the Blind Seer and OXT-OSPIR systems manage to return records in less than a second for databases of size 10^8 records with a much larger number of searchable keywords. On the other hand, our system outperforms the secure single-keyword search of the HE-over-ORAM solution whose experimental results showed that their system answers a query in 30 min for 4×10^6 record databases. Therefore, our scheme provides a new tradeoff mark between privacy and efficiency.

Acknowledgments. This work was funded by the US Department of Homeland Security (DHS) Science and Technology (S&T) Directorate under contract no. HSHQDC-10-C-00144. The views and conclusions contained herein are the authors' and should not be interpreted as necessarily representing the official policies or endorsements, either expressed or implied, of DHS or the US government.

While at Columbia University, Fernando Krell was supported by NSF awards #CNS-1445424 and #CCG-1423306.

Mariana Raykova is supported by NSF grants CNS-1633282, 1562888, 1565208, and DARPA W911NF-15-C-0236, W911NF-16-1-0389.

References

1. Afshar, A., Hu, Z., Mohassel, P., Rosulek, M.: How to efficiently evaluate RAM programs with malicious security. In: Oswald, E., Fischlin, M. (eds.) EUROCRYPT 2015. LNCS, vol. 9056, pp. 702–729. Springer, Heidelberg (2015). doi:10.1007/978-3-662-46800-5_27

2. Bloom, B.H.: Space/time trade-offs in hash coding with allowable errors. Commun. ACM **13**, 422–426 (1970)

3. Boneh, D., Crescenzo, G.D., Ostrovsky, R., Persiano, G.: Public key encryption with keyword search. In: Cachin, C., Camenisch, J.L. (eds.) EUROCRYPT 2004. LNCS, vol. 3027, pp. 506–522. Springer, Heidelberg (2004). doi:10.1007/978-3-540-24676-3_30

4. Cash, D., Jarecki, S., Jutla, C., Krawczyk, H., Roşu, M.-C., Steiner, M.: Highly-scalable searchable symmetric encryption with support for Boolean queries. In: Canetti, R., Garay, J.A. (eds.) CRYPTO 2013. LNCS, vol. 8042, pp. 353–373. Springer, Heidelberg (2013). doi:10.1007/978-3-642-40041-4_20

5. Chor, B., Kushilevitz, E., Goldreich, O., Sudan, M.: Private information retrieval. J. ACM **45**(6), 965–981 (1998)

6. Curtmola, R., Garay, J., Kamara, S., Ostrovsky, R.: Searchable symmetric encryption: improved definitions and efficient constructions. In: CCS (2006)

7. Fisch, B., Vo, B., Krell, F., Kumarasubramanian, A., Kolesnikov, V., Malkin, T., Bellovin, S.M.: Malicious-client security in blind seer: a scalable private DBMS. Cryptology ePrint Archive, Report 2014/963 (2014). http://eprint.iacr.org/

8. Gentry, C., Halevi, S., Jutla, C., Raykova, M.: Private database access with HE-over-ORAM architecture. In: Malkin, T., Kolesnikov, V., Lewko, A.B., Polychronakis, M. (eds.) ACNS 2015. LNCS, vol. 9092, pp. 172–191. Springer, Heidelberg (2015). doi:10.1007/978-3-319-28166-7_9

9. Gertner, Y., Ishai, Y., Kushilevitz, E., Malkin, T.: Protecting data privacy in private information retrieval schemes. J. Comput. Syst. Sci. **60**(3), 592–629 (2000)

10. Goldreich, O., Micali, S., Wigderson, A.: How to play any mental game. In: STOC (1987)

11. Goldreich, O., Ostrovsky, R.: Software protection and simulation on oblivious rams. J. ACM **43**(3), 431–473 (1996)
12. Gordon, S.D., Katz, J., Kolesnikov, V., Krell, F., Malkin, T., Raykova, M., Vahlis, Y.: Secure two-party computation in sublinear (amortized) time. In: CCS (2012)
13. Ishai, Y., Kushilevitz, E., Lu, S., Ostrovsky, R.: Private large-scale databases with distributed searchable symmetric encryption. In: Sako, K. (ed.) CT-RSA 2016. LNCS, vol. 9610, pp. 90–107. Springer, Heidelberg (2016). doi:10.1007/978-3-319-29485-8_6
14. Islam, M.S., Kuzu, M., Kantarcioglu, M.: Access pattern disclosure on searchable encryption: ramification, attack and mitigation. In: NDSS (2012)
15. Jarecki, S., Jutla, C.S., Krawczyk, Rosu, H., Steiner, M.: Outsourced symmetric private information retrieval. In: CCS (2013)
16. Jarecki, S., Liu, X.: Fast secure computation of set intersection. In: Garay, J.A., Prisco, R. (eds.) SCN 2010. LNCS, vol. 6280, pp. 418–435. Springer, Heidelberg (2010). doi:10.1007/978-3-642-15317-4_26
17. Kamara, S., Papamanthou, C., Roeder, T.: Dynamic searchable symmetric encryption. In: CCS (2012)
18. Pappas, V., Krell, F., Vo, B., Kolesnikov, V., Malkin, T., Choi, S., George, W., Keromytis, A., Bellovin, S.: Blind seer: a scalable private DBMS. In: IEEE S&P (2014)
19. Pappas, V., Raykova, M., Vo, B., Bellovin, S.M., Malkin, T.: Private search in the real world. In: ACSAC 2011, pp. 83–92 (2011)
20. Popa, R.A., Redfield, C.M.S., Zeldovich, N., Balakrishnan, H.: CryptDB: protecting confidentiality with encrypted query processing. In: SOSP (2011)
21. Raykova, M., Vo, B., Bellovin, S., Malkin, T.: Secure anonymous database search. In: CCSW 2009 (2009)
22. Song, D.X., Wagner, D., Perrig, A.: Practical techniques for searches on encrypted data. In: IEEE S&P (2000)
23. Yao, A.C.: Protocols for secure computations. In: FOCS (1982)

Public Key Algorithms

Constructions Secure Against Receiver Selective Opening and Chosen Ciphertext Attacks

Dingding Jia[1,2,3]($\boxtimes$), Xianhui Lu[1,2,3], and Bao Li[1,2,3]

[1] State Key Laboratory of Information Security,
Institute of Information Engineering, CAS, Beijing, China
{ddjia,xhlu,lb}@is.ac.cn
[2] Data Assurance and Communication Security Research Center,
CAS, Beijing, China
[3] University of Chinese Academy of Sciences, Beijing, China

Abstract. In this paper we study public key encryption schemes of indistinguishability security against receiver selective opening (IND-RSO) attacks, where the attacker can corrupt some receivers and get the corresponding secret keys in the multi-party setting. Concretely:

- We present a general construction of RSO security against chosen ciphertext attacks (RSO-CCA) by combining any RSO secure scheme against chosen plaintext attacks (RSO-CPA) with any regular CCA secure scheme, along with an appropriate non-interactive zero-knowledge proof.
- We show that the leakage-resistant construction given by Hazay *et al.* in Eurocrypt 2013 from weak hash proof system (wHPS) is RSO-CPA secure.
- We further show that the CCA secure construction given by Cramer and Shoup in Eurocrypt 2002 based on the universal HPS is RSO-CCA secure, hence obtain a more efficient paradigm for RSO-CCA security.

Keywords: Receiver selective opening · Chosen ciphertext security · Hash proof system

1 Introduction

Indistinguishability against chosen plaintext and chosen ciphertext attacks (IND-CPA, IND-CCA) are widely accepted security notions for public key encryption (PKE). However, in the multi-party situation, when attacks such as selective opening [7,11] are possible, the above security requirements are not enough.

Generally, in selective opening attacks the adversary may corrupt a fraction of parties and get the plaintext messages together with internal randomness for

This work is Supported by the National Basic Research Program of China (973 project) (No. 2013CB338002), the National Nature Science Foundation of China (No. 61502484, No. 61379137, No. 61572495).

H. Handschuh (Ed.): CT-RSA 2017, LNCS 10159, pp. 417–431, 2017.
DOI: 10.1007/978-3-319-52153-4_24

encryption or decryption, while it is hoped that messages for uncorrupted parties remain protected. The notion of selective opening attacks is considered in two settings: sender selective opening (SSO), where part of senders are corrupted and messages together with randomness for encryption are revealed; and receiver selective opening (RSO), where part of receivers are corrupted and messages together with secret keys for decryption are revealed [8].

Formal study of selective opening in PKE scenario was initiated by Bellare, Hofheinz and Yilek [4,5] in 2009. They gave rigorous definitions with two styles: indistinguishability-based (IND) and simulation-based (SIM). Considering that in the selective opening scenario, part of random coins or secret keys are opened, whether the ciphertext is consistant with the plaintext can be checked. In security proof this restricts the way how the target ciphertext generated, thus whether the ordinary IND security implies SO security and relations of SO security of different styles attracts much attention [1,3,12,21–23,31].

Earlier constructions of SO security either depended on erasures, updating secret keys, with long secret keys or were in the random oracle model [7,8,30]. As to the result in the random oracle model, Heuer et al. [17] proved that the practical schemes RSA-OAEP and DHIES were SIM-SSO-CCA secure. Next we review constructions that are stateless, non-interactive and without erasures in the standard model.

For constructions secure in the SSO setting a lot of works have been done in recent years [4,13,17–19,27–29]. Up to now constructions secure in the RSO setting [8,23] are relatively less, and these constructions are only RSO-CPA secure. In this paper we will focus on the constructions that are secure against RSO of the indistinguishability style and CCA attacks simultaneously.

1.1 Our Contribution

In this paper we show the existence of IND-RSO-CCA secure schemes by giving a construction from a variant of the Noar-Yung paradigm [6]. The construction is a combination of any IND-RSO-CPA secure scheme, any IND-CCA secure scheme and an appropriate non-interactive zero-knowledge proof (NIZK). And we prove that the leakage-resistant construction from weak hash proof systems (wHPS) in [20] is actually IND-RSO-CPA secure. For more efficient constructions, we prove that the Cramer-Shoup paradigm [9,10] from universal HPS is IND-RSO-CCA secure. In the following we outline the main idea of the construction.

To modify an IND-RSO-CPA secure scheme to be IND-RSO-CCA secure, one should handle decryption queries appropriately. We observe that when applying the Noar-Yung paradigm (or its variant), it is possible to keep secret keys unchanged by taking only the first copy of the secret key of the IND-RSO-CPA secure scheme as the secret key for the whole encryption scheme. Our first construction, which is constructed from an IND-RSO-CPA secure scheme, an IND-CCA secure scheme, an appropriate NIZK and a one-time signature, is inspired by the paradigm to achieving key-dependent message security against chosen ciphertext attacks (KDM-CCA) [6]. The proof sketch is shown in Fig. 5.

Besides, we prove the IND-RSO-CPA security for the leakage-resistant construction from wHPS given by Hazay *et al.* [20]. Since wHPS can be constructed from any CPA secure scheme, our result shows that IND-RSO-CPA secure PKE can be built from any IND-CPA secure PKE. Considering that IND-CCA secure PKE can be get from any IND-CPA secure PKE and an appropriate NIZK, we get that IND-RSO-CCA security can be built from any IND-CPA, an appropriate NIZK and a one-time signature. Generally speaking, a wHPS is a key encapsulation mechanism (KEM) along with a fake encapsulation algorithm. The fake encapsulation algorithm can generate a fake ciphertext, which is indistinguishable from the real ciphertext even given the secret key and is non-committing to any message when given the public key. In fact, the construction from wHPS, which adds to the encryption and decryption algorithm a bitwise XOR with the message, is IND-RSO-CPA secure. The security proof is straightforward, since when the adversary gets fake ciphertexts, messages are completely hidden, while fake ciphertexts are indistinguishable from real ciphertexts.

Although the framework we give above implies the existence of IND-RSO-CCA secure PKE, the use of NIZK makes it less efficient. In the final part, we prove that the construction from universal hash proof system (HPS) [9], which is more efficient, is IND-RSO-CCA secure. Here we give a general explaination. Hazay *et al.* demonstrated that smooth HPS implies tNCER, which leads to IND-RSO-CPA security [21]. Although the CCA construction from universal HPS adds elements in secret key for ciphertext verification compared with construction for CPA security, this does not affect the non-committing property, for the simulator is able to open messages along with secret keys which it holds.

One may notice that constructions in this paper can only achieve single-message security, while a more reliable requirement for practice is security for multi-message. In the full version [24] we give a reduction from multi-message security to single-message case through a hybrid argument. The reduction leads to a security loss related to the number of messages. We leave constructions that are secure for multi-messages with a tight reduction as an open problem.

Organization. The rest of our paper is organized as follows: in Sect. 2 we give definitions and preliminaries; in Sect. 3 we give a variant of the Noar-Yung paradigm to build IND-RSO-CCA secure encryption and prove that the leakage-resistant construction given by Hazay *et al.* from wHPS is IND-RSO-CPA secure; in Sect. 4 we prove that the construction in [9] is IND-RSO-CCA secure.

2   Preliminaries and Definitions

2.1   Preliminaries

Notations. In this paper we use PPT to represent probabilistic polynomial time for short. Let $[n]$ be the set of $\{1, 2, ..., n\}$. $a \leftarrow A$ is to denote choosing a random element from A when A is a set, and to denote picking a uniformly distributed randomness, running A with the randomness and assigning the output to a when A is a PPT algorithm. we use the lower case boldface to denote vectors. $Enc(\boldsymbol{pk}, \boldsymbol{m}) := (Enc(pk_1, m_1), ..., Enc(pk_n, m_n))$ when $\boldsymbol{pk}, \boldsymbol{m}$ are vectors

of dimension n. The statistical distance of two distributions $\mathcal{X}, \mathcal{Y}$ is defined as
$SD(\mathcal{X}, \mathcal{Y}) := \frac{1}{2} \Sigma_x |\Pr[\mathcal{X} = x] - \Pr[\mathcal{Y} = x]|$.

Besides efficiently samplable, the message space is required to be efficiently conditional resamplable to accompany the security definition we will give later.

Definition 1 (Efficiently Conditional Resamplable [4]). *Let dist be a joint distribution over* $\mathbb{M}^n$, *where* $\mathbb{M}$ *is the message space, then dist is efficiently conditional resamplable if there is a PPT algorithm Redist such that for any* $I \subset [n]$ *and any* $\mathbf{m}_I := (m_i)_{i \in I}$, *where* $\mathbf{m} = (m_i)_{i \in [n]}$ *is sampled from dist, the output* $\mathbf{m}' \leftarrow Redist(\mathbf{m}_I)$ *satisfies that* $\mathbf{m}'$ *is distributed according to dist and* $m_i' = m_i$ *for* $i \in I$.

2.2 Security Definitions

Public Key Encryption (PKE). A PKE scheme supported ciphertexts with labels consists of three algorithms: $Keygen(1^\lambda) \rightarrow (pk, sk)$, $Enc(pk, m, l) \rightarrow c$, $Dec(sk, c, l) \rightarrow m$ or $\perp$, where $Keygen$ is the key generation algorithm, Enc is the encryption algorithm with label l and Dec is the decryption algorithm.

Correctness. A PKE scheme satisfies correctness, if for all $(pk, sk) \leftarrow Keygen(1^\lambda)$, $m \in \mathbb{M}$, $Dec(sk, Enc(pk, m, l), l) = m$.

Clearly, an ordinary PKE scheme can be seen as a PKE scheme with empty label spaces.

Security. Here we give the definition of indistinguishability based security against receiver selective opening chosen ciphertext attacks (IND-RSO-CCA) as in [21] and IND-CCA security definition for ciphertexts with labels in Fig. 1. As in [4,19], we require the message space be efficiently conditional resamplable. The security experiment proceeds as follows:

Note that in $Exp^{\text{ind-rso-cca}}(\mathcal{A})$, the decryption query is of the form (c, j) satisfying that $c \neq c_j^*$, and is answered by $Dec(sk_j, c)$. And after the adversary gets sk_I, it is required that $j \notin I$. The advantage is defined as $Adv_{\mathcal{A}}^{\text{IND-RSO-CCA}} =$

Experiment. $Exp^{\text{ind-rso-cca}}(\mathcal{A})$:	Experiment. $Exp^{\text{ind-cca}}(\mathcal{A})$:
$b \leftarrow \{0,1\}$	
$(\boldsymbol{pk}, \boldsymbol{sk}) := (pk_i, sk_i)_{i \in [n]} \leftarrow Setup(1^\lambda)$	$b \leftarrow \{0,1\}$
$(dist, Redist, state_1) \leftarrow \mathcal{A}^{Dec(\cdot, \cdot)}(\boldsymbol{pk})$	$(pk, sk) \leftarrow Setup(1^\lambda)$
$\boldsymbol{m}_0 \leftarrow dist$	$(m_0, m_1, l^*, state_1) \leftarrow \mathcal{A}^{Dec(\cdot, \cdot)}(pk)$
$\boldsymbol{c}^* \leftarrow Enc(\boldsymbol{pk}, \boldsymbol{m}_0)$	$c^* \leftarrow Enc(pk, m_b, l^*)$
$(I, state_2) \leftarrow \mathcal{A}^{Dec(\cdot, \cdot)}(\boldsymbol{c}^*, state_1)$	$b' \leftarrow \mathcal{A}^{Dec(\cdot, \cdot)}(c^*, state_1)$
$\boldsymbol{m}_1 \leftarrow Redist(\boldsymbol{m}_{0I})$	Return 1 if $b' = b$ and 0 else .
$b' \leftarrow \mathcal{A}^{Dec(\cdot, \cdot)}(\boldsymbol{sk}_I, \boldsymbol{m}_b, state_2)$	
Return 1 if $b' = b$ and 0 else.	

Fig. 1. The IND-RSO-CCA and IND-CCA experiment

$\left| 2 \Pr[Exp^{\text{ind-rso-cca}}(\mathcal{A}) = 1] - 1 \right|$. In $Exp^{\text{ind-cca}}(\mathcal{A})$, the decryption query is of the form (c, l) such that $(c, l) \neq (c^*, l^*)$, where l is a label, and the query is answered by $Dec(sk, c, l)$. The advantage is defined as $Adv_{\mathcal{A}}^{\text{IND-CCA}} = \left| 2 \Pr[Exp^{\text{ind-cca}}(\mathcal{A}) = 1] - 1 \right|$. When omitting the decryption oracle, the above experiment gives a definition of IND-RSO-CPA and IND-CPA security respectively.

Definition 2 (IND-RSO-CCA/CPA Security). *A PKE scheme is IND-RSO-CCA secure if for any PPT adversary $\mathcal{A}$, $Adv_{\mathcal{A}}^{IND\text{-}RSO\text{-}CCA}$ is negligible in λ. And it is IND-RSO-CPA secure if for any PPT adversary $\mathcal{A}$, $Adv_{\mathcal{A}}^{IND\text{-}RSO\text{-}CPA}$ is negligible in λ. IND-CCA/CPA security are defined similarly.*

One-Time Signature. A signature scheme consists of three PPT algorithms satisfying that for all: $Sig.Kg(1^\lambda) \rightarrow (vk, sigk), m \in \mathbb{M}, Ver(vk, m, Sign(sigk, m)) = 1$, where $Sig.Kg$ is the key generation algorithm, $Sign$ is the signature algorithm and Ver is the verification algorithm.

Security. Here we give the security notion of strong existential unforgeability under one-time chosen message attack in the following experiment between a challenger $\mathcal{C}$ and a PPT adversary $\mathcal{A}$ (Fig. 2):

Definition 3 (One-time Unforgeable Security). *A signature scheme is strongly existential unforgeable under one-time chosen message attack if for any PPT adversary $\mathcal{A}$, $Adv_{\mathcal{A}}^{ots} := \Pr[Exp_{sig}^{uf\text{-}ot}(\mathcal{A}) = 1]$ is negligible in λ.*

2.3 Non-interactive Zero-Knowledge Proofs

Let R be a binary relation that is efficiently computable. Let $\mathcal{L} := \{x : \exists w, \ s.t. \ (x, w) \in R\}$. A non-interactive zero-knowledge (NIZK) proof system for R consists of three PPT algorithms $(CRSGen, P, V)$ satisfying the completeness property such that: for all $\mathfrak{C} \leftarrow CRSGen$, all $(x, w) \in R$, and $\mathfrak{p} \leftarrow P(\mathfrak{C}, x, w)$, $V(\mathfrak{C}, x, \mathfrak{p}) = 1$ where $CRSGen$ generates a common reference string (CRS), P is the proof algorithm and V is the verification algorithm.

Experiment. $Exp_{\text{sig}}^{\text{uf-ot}}(\mathcal{A})$:

$(vk, sigk) \leftarrow Sig.kg(1^\lambda)$
$(m, st) \leftarrow \mathcal{A}(vk)$
$\sigma \leftarrow Sign(sigk, m)$
$(m', \sigma') \leftarrow \mathcal{A}(st, \sigma)$
if $(m', \sigma') \neq (m, \sigma)$ and $Ver(vk, m', \sigma') = 1$, outputs 1, and 0 else

---

Fig. 2. One-time unforgeable for signatures

Definition 4 (NIZK [2,14]). $(CRSGen, P, V)$ *is an NIZK proof system for* R *if it satisfies the following properties:*

Computational Soundness: *For any PPT* $\mathcal{A}$, $Adv^{cs}_{nizk,\mathcal{A}} = \Pr[\mathcal{A}(\mathfrak{C}) \to (x, \mathfrak{p}) \wedge$ $x \notin \mathcal{L} \wedge V(\mathfrak{C}, x, \mathfrak{p}) = 1]$ *is negligible, where* $\mathfrak{C} \leftarrow CRSGen$ *is given to* $\mathcal{A}$.

Computational Zero-knowledge: *There exists a simulator* $\mathcal{S}$ *such that for any PPT adversary* $\mathcal{A}$, $Adv^{czk}_{nizk,\mathcal{A}} = |\Pr[Exp^{real}(\mathcal{A}) = 1] - \Pr[Exp^{sim}(\mathcal{A}) = 1]|$ *is negligible, where* $Exp^{real}(\mathcal{A})$ *and* $Exp^{sim}(\mathcal{A})$ *are defined in Fig. 3, in which* ϵ *denotes an empty string and* $\mathcal{E}$ *denotes an empty set.*

Experiment. $Exp^{real}(\mathcal{A})$:	Experiment. $Exp^{sim}(\mathcal{A})$:
$\mathfrak{C} \leftarrow CRSGen$, $st = \epsilon$, $\mathfrak{P} = \mathcal{E}$	$(\mathfrak{C}, t) \leftarrow \mathcal{S}$, $st = \epsilon$, $\mathfrak{P} = \mathcal{E}$
for $i = 1, ..., n$	for $i = 1, ..., n$
$\mathcal{A}(\mathfrak{C}, st, \mathfrak{P}) \to (x_i, w_i, st_i)$	$\mathcal{A}(\mathfrak{C}, st, \mathfrak{P}) \to (x_i, w_i, st_i)$
$P(\mathfrak{C}, x_i, w_i) \to \mathfrak{p}_i$	$\mathcal{S}(t, x_i) \to \mathfrak{p}_i$
$st \leftarrow st_i$, $\mathfrak{P} \leftarrow \mathfrak{P} \cup \mathfrak{p}_i$	$st \leftarrow st_i$, $\mathfrak{P} \leftarrow \mathfrak{P} \cup \mathfrak{p}_i$
end for	end for
$b \leftarrow \mathcal{A}(st, \mathfrak{P})$	$b \leftarrow \mathcal{A}(st, \mathfrak{P})$
outputs b	outputs b

Fig. 3. Computational zero-knowledge

Loosely speaking, CZK means that with the help of the secret information t generated with $\mathfrak{C}$, the simulator $\mathcal{S}$ can produce a proof that is indistinguishable from the real proof without the witness for $x \in \mathcal{L}$. For the construction in this paper, although only one message is encrypted for each public key, there are multi public keys, the one-time definition of computational zero-knowledge given by Blum *et al.* [2] is not enough.

3   An IND-RSO-CCA Secure Construction

In this section, we give an IND-RSO-CCA secure construction analogous to that in [6] with the following building blocks: a PKE $\mathbf{E}_1$ with IND-RSO-CPA security, a regular CCA secure PKE $\mathbf{E}_2$ that supports ciphertexts with labels, an NIZK proof system for the language consisting of the set of all pairs that encrypt the same message using $\mathbf{E}_1$ and $\mathbf{E}_2$, and a strong existential unforgeable one-time signature scheme. Then we prove that the construction from wHPS [20] is IND-RSO-CPA secure.

3.1   Preliminaries for Section 3

Tweaked Non-committing Encryption for Receivers (tNCER). In [21], Hazay *et al.* defined tNCER and proved that a tNCER is IND-RSO-CPA secure. A tweaked PKE (tPKE) consists of five algorithms $(tKeygen, tEnc, tEnc^*, tDec, tOpen)$, where $(tKeygen, tEnc, tDec)$ form a regular PKE and the tweaked encryption algorithm $tEnc^*$ outputs a fake ciphertext $c^* \leftarrow tEnc^*(pk, sk, m)$ and the (possibly inefficient)open algorithm $tOpen$ outputs a secret key $sk^* \leftarrow tOpen(pk, c^*, m)$, satisfying that $tDec(sk^*, c^*) = m$.

Experiment. $Exp_{tpke}^{ind\text{-}tcipher}(\mathcal{A})$:
$b \leftarrow \{0, 1\}$ $(pk, sk) \leftarrow tKeygen(1^\lambda)$ $(m, st) \leftarrow \mathcal{A}(pk)$ $c_0 \leftarrow tEnc(pk, m)$ $c_1 \leftarrow tEnc^*(pk, sk, m)$ $b' \leftarrow \mathcal{A}(sk, c_b, st)$ if $b = b'$, outputs 1, else outputs 0

Experiment. $Exp_{tpke}^{ind\text{-}tncer}(\mathcal{A})$:
$b \leftarrow \{0, 1\}$ $(pk, sk_0) \leftarrow tKeygen(1^\lambda)$ $(m, st) \leftarrow \mathcal{A}(pk)$ $c_0 \leftarrow tEnc^*(pk, sk_0, m)$ $m' \leftarrow \mathbb{M}$ $c_1 \leftarrow tEnc^*(pk, sk_0, m')$ $sk_1 \leftarrow tOpen(pk, c_1, m)$ $b' \leftarrow \mathcal{A}(sk_b, c_b, st)$ if $b = b'$, outputs 1, else outputs 0

Fig. 4. Tweaked NCER

Definition 5 *(tNCER). A tPKE is a tweaked NCER (Fig. 4) if:*

- *for any PPT adversary $\mathcal{A}$, $Adv_{tpke,\mathcal{A}}^{ind\text{-}tcipher} := |2 \Pr[Exp_{tpke}^{ind\text{-}tcipher}(\mathcal{A}) = 1] - 1|$ is negligible.*
- *for any unbounded adversary $\mathcal{A}$, $Adv_{tpke,\mathcal{A}}^{ind\text{-}tncer} := |2 \Pr[Exp_{tpke}^{ind\text{-}tncer}(\mathcal{A}) = 1] - 1|$ is negligible.*

Weak Hash Proof System (wHPS). Weak hash proof system, which can be seen as a generalization of HPS, was proposed by Hazay *et al.* to provide leakage resistant security from CPA secure schemes [20]. Here we give a brief review. A wHPS is an ordinary KEM in addition with a fake encryption algorithm Enc^* that takes as input pk, outputs an invalid ciphertext. $c^* \leftarrow Enc^*(pk)$.

It should satisfy indistinguishability and smoothness properties.

Indistinguishability. Given $(pk, sk) \leftarrow Keygen(1^\lambda)$, any PPT adversary $\mathcal{A}$ cannot distinguish a valid ciphertext from an invalid ciphertext. That is, for any PPT adversary $\mathcal{A}$, $Adv_{\mathcal{A}, \text{wHPS}}^{\text{CI}}$ is negligible, where

$$Adv_{\mathcal{A}, \text{wHPS}}^{\text{CI}} = |\Pr[\mathcal{A}(pk, sk, c|(c, K) \leftarrow Enc(pk)) = 1]$$
$$- \Pr[\mathcal{A}(pk, sk, c^*|c^* \leftarrow Enc^*(pk)) = 1]|.$$

Smoothness. For any invalid ciphertext c^*, the distribution of (pk, c^*, K^*) and (pk, c^*, K) are identical, where $K^* = Dec(sk, c^*)$ and K is chosen randomly from the session key space.

3.2 Construction

Let $\mathbf{E}_1 := (Keygen_1, Enc_1, Dec_1)$ be IND-RSO-CPA secure, and $\mathbf{E}_2 := (Keygen_2, Enc_2, Dec_2)$ be IND-CCA secure and supports ciphertext with labels, $\mathbf{S} := (Sig.Kg, Sign, Ver)$ be strong existential unforgeable under one-time chosen message attack, $\mathcal{L}_{eq} := \{(c_1, c_2, l) | \exists m, r_1, r_2, s.t. c_1 = Enc_1(pk_1, m; r_1), c_2 = Enc_2(pk_2, m, l; r_2)\}$. Let $\mathbf{P} := (CRSGen, P, V)$ be an NIZK proof for $\mathcal{L}_{eq}$. The scheme is described as follows:

Keygen: Generate $(pk_i, sk_i) \leftarrow Keygen_i(1^\lambda)$ for $i = 1, 2$, run $CRSGen$ to get the CRS $\mathfrak{C}$ of the NIZK $\mathbf{P}$. Set $pk := (pk_1, pk_2, \mathfrak{C}), sk := sk_1$.

Enc: Generate $(vk, sigk) \leftarrow Sig.Kg(1^\lambda)$, randomly choose r_1, r_2 and compute $c_1 = Enc_1(pk_1, m; r_1), c_2 = Enc_2(pk_2, m, vk; r_2), \mathfrak{p} \leftarrow P(\mathfrak{C}, (c_1, c_2, vk), (m, r_1, r_2)), \sigma = Sign(Sigk, c_1 \| c_2 \| \mathfrak{p})$. The ciphertext $c = (vk, c_1, c_2, \mathfrak{p}, \sigma)$.

Dec: Verifies whether $V(\mathfrak{C}, c_1 \| c_2 \| vk, \mathfrak{p}) = 1$ and $Ver(vk, c_1 \| c_2 \| \mathfrak{p}, \sigma) = 1$, if both equations hold, output $m = Dec_1(sk, c_1)$, otherwise reject.

Correctness of the decryption algorithm is trivially follows from the completeness of NIZK, correctness of the signature scheme and correctness of the IND-RSO-CPA scheme.

Theorem 1. *Let* $\mathbf{E}_1$ *be IND-RSO-CPA secure,* $\mathbf{E}_2$ *be IND-CCA secure that supports ciphertext with labels,* $\mathbf{S}$ *be existential unforgeable under one-time chosen message attack,* $\mathbf{P}$ *be an NIZK proof for* $\mathcal{L}_{eq}$, *then the scheme constructed above is IND-RSO-CCA secure. Concretely,*

$$Adv_{pke}^{IND\text{-}RSO\text{-}CCA} \le 2q(Adv_{nizk}^{cs} + nAdv_{sig}^{uf\text{-}ot}) + 2nAdv_{pke}^{cca} + 2Adv_{nizk}^{czk} + Adv_{pke}^{IND\text{-}RSO\text{-}CPA}$$

Proof. The proof is through a sequence of games depicted in Fig. 5, where the boxed item is the change from the former game.

Next we give the formal description of the games. Let W_i denote the event that the adversary outputs 1 in Game$_i$.

Game$_0$: the real security game when $b = 0$.

Game$_1$: the same as Game$_0$, except that when responding to a decryption query (c, j), the challenger computes $m = Dec_2(sk_{2j}, c_2)$ instead of $m = Dec_1(sk_{1j}, c_1)$. From the soundness property of $\mathbf{P}$, one can get that $\Pr[W_1] - \Pr[W_0]$ is negligible.

Game$_2$: the same as Game$_1$, except that $\mathfrak{C}$ is generated by a simulator $\mathcal{S}$ and when responding to the encryption query $dist$, the challenger produce simulated proofs $\mathfrak{p} \leftarrow \mathcal{S}(t, (c_1, c_2, vk))$ instead of a real $\mathfrak{p}$. From the zero-knowledge property of $\mathbf{P}$, one can get that $\Pr[W_2] - \Pr[W_1]$ is negligible.

Game3: the same as Game2, except that when responding to a decryption oracle (c, j), where $c = (vk, c_1, c_2, \mathfrak{p}, \sigma)$, the challenger checks whether $vk = vk_j^*$, if the equation holds, then it just rejects. From the existential unforgeable property of **S**, one can get that $\Pr[W_3] - \Pr[W_2]$ is negligible.

Game4: the same as Game3 except that when responding to the encryption query $dist$, the challenger samples $\boldsymbol{m_0} \leftarrow dist$, and random $\boldsymbol{m_R}$ from the message space, generates $(\boldsymbol{vk}, \boldsymbol{sigk}) \leftarrow Sig.Kg^n(1^\lambda)$, computes $\boldsymbol{c_1^*} = Enc_1(\boldsymbol{pk_1}, \boldsymbol{m_0})$, $\boldsymbol{c_2^*} = Enc_2(\boldsymbol{pk_2}, \boldsymbol{m_R}, \boldsymbol{vk^*})$, and other parts of the ciphertext vector as in Game3. From the CCA security of $\mathbf{E_2}$, by a hybrid argument one can get that $\Pr[W_4] - \Pr[W_3]$ is negligible.

Game5: the same as Game4, except that in the open phase, the adversary resamples $\boldsymbol{m_1} \leftarrow Redist(\boldsymbol{m_{0I}})$ and responds with $(\boldsymbol{sk_I}, \boldsymbol{m_1})$. From the RSO-CPA security of $\mathbf{E_1}$, one can get that $\Pr[W_5] - \Pr[W_4]$ is negligible.

Game6: the same as Game5, except that when responding to the encryption query $dist$, the challenger computes $\boldsymbol{c_2} = Enc_2(\boldsymbol{pk_2}, \boldsymbol{m_0}, \boldsymbol{vk^*})$, with the real sampled message vector instead of randomly chosen one. From the CCA security of $\mathbf{E_2}$, one can get that $\Pr[W_6] - \Pr[W_5]$ is negligible.

Game7: the same as Game6, except that when responding to a decryption query (c, j), the challenger no longer rejects when $vk = vk_j^*$. From the existential unforgeable property of **S**, one can get that $\Pr[W_7] - \Pr[W_6]$ is negligible.

Game8: the same as Game7, except that $\mathfrak{C}$ is normally generated and when responding to the encryption query $dist$, the challenger produce real proofs $\mathfrak{p}$. From the zero-knowledge property of **P**, one can get that $\Pr[W_8] - \Pr[W_7]$ is negligible.

Game9: the real security game when $b = 1$. From the soundness property of **P**, one can get that $\Pr[W_9] - \Pr[W_8]$ is negligible.

Combining the above game sequences, we get that $\Pr[W_9] - \Pr[W_0]$ is negligible. $\square$

Game	Enc	Dec	Open	Remarks
0	m_0, m_0,real $\mathfrak{p}$	sk_1	m_0, sk_I	
1	m_0, m_0,real $\mathfrak{p}$	sk_2	m_0, sk_I	soundness of **P**
2	$m_0, m_0,$ fake $\mathfrak{p}$	sk_2	m_0, sk_I	NIZK of **P**
3	m_0, m_0,fake $\mathfrak{p}$	$sk_2,$ reject vk^*	m_0, sk_I	unforgeable Signature **S**
4	$m_0,$ m_R ,fake $\mathfrak{p}$	sk_2,reject vk^*	m_0, sk_I	cca security of $\mathbf{E_2}$
5	m_0, m_R,fake $\mathfrak{p}$	sk_2,reject vk^*	m_1, sk_I	rso-cpa security of $\mathbf{E_1}$
6	$m_0,$ m_0 ,fake $\mathfrak{p}$	sk_2,reject vk^*	m_1, sk_I	cca security of $\mathbf{E_2}$
7	m_0, m_0,fake $\mathfrak{p}$	$sk_2,$ no reject vk^*	m_1, sk_I	unforgeable Signature **S**
8	$m_0, m_0,$ real $\mathfrak{p}$	$sk_2,$	m_1, sk_I	NIZK of **P**
9	m_0, m_0,real $\mathfrak{p}$	sk_1	m_1, sk_I	soundness of **P**

Fig. 5. Game transform for RSO-CCA security from RSO-CPA security

3.3 IND-RSO-CPA Secure PKE from wHPS

Up to now there are instantiations of RSO-CPA secure PKE [21], CCA secure scheme with labeled ciphertext [6], NIZK for equal message relations [6,16], one-time signatures [15]. Here we prove that the leakage-resistant construction from wHPS [20] is IND-RSO-CPA secure. Since in [20] Hazay *et al.* showed that wHPS can be realized from CPA secure PKE schemes, our result implies that IND-RSO-CPA secure PKE can be constructed from any IND-CPA secure PKE.

Lemma 1 ([21]). *For any PPT adversary $\mathcal{A}$ attacking tPKE in the IND-RSO-CPA scheme, there exists a PPT adversary $\mathcal{B}$ and an unbounded adversary $\mathcal{C}$, such that $Adv_{tpke}^{ind\text{-}rso\text{-}cpa}(\mathcal{A}) \leq 2n(Adv_{tpke}^{ind\text{-}tcipher}(\mathcal{B}) + Adv_{tpke}^{ind\text{-}tncer}(\mathcal{C}))$.*

Construction. Next we show that the PKE constructed from wHPS [20] is a tNCER. The scheme is described as follows.

$tKeygen(1^\lambda)$: The key generation algorithm is the generation algorithm of wHPS. $(pk, sk) \leftarrow wHPS.Keygen(1^\lambda)$.

$tEnc(pk, m)$: $c = (c_1, c_2)$, where $(c_1, K) \leftarrow wHPS.Enc(pk), c_2 = K + m$, here we assume that the encrypted messages are in an additive group.

$tDec(sk, c)$: $K \leftarrow wHPS.Dec(sk, c_1), m \leftarrow c_2 - K$.

$tEnc^*(pk, sk, m)$: $c^* = (c_1^*, c_2^*), c_1^* \leftarrow wHPS.Enc^*(pk), K^* \leftarrow wHPS.Dec(sk, c_1^*)$,
$c_2^* = K^* + m$.

$tOpen(pk, c^*, m)$: Parse c^* as $c^* = (c_1^*, c_2^*)$, compute $K^* = c_2^* - m$, find an sk^* such that $wHPS.Dec(sk^*, c^*) = m$.

Correctness can be easily verified from the correctness property of wHPS. It is obvious that the decryption of a fake ciphertext c^* outputs the encrypted message m. Since c_1^* is an output of $wHPS.Enc^*(pk)$, from the smooth property of wHPS, $(pk, c_1^*, wHPS.Dec(sk, c_1^*))$ is distributed as (pk, c_1^*, K) for randomly chosen K. Hence for a given K^*, there exists a sk^* corresponding to pk such that $wHPS.Dec(sk^*, c_1^*) = K^*$, an unbounded algorithm can find it. The ciphertext indistinguishability of tPKE easily follows from the indistinguishability of wHPS. And the non-committing property for fake ciphertexts follows from the smoothness property of wHPS.

4 IND-RSO-CCA Secure PKE from Universal HPS

The construction of the above section implies the existence of IND-RSO-CCA secure scheme. However, due to the employment of NIZK (pairing), the construction is less efficient, and the ciphertext is not compact. In this section we prove that the compact and efficient CCA secure scheme in [9] based on HPS is IND-RSO-CCA secure.

4.1 Universal Hash Proof System

Projective Hash Family. Firstly we recall the concept of hash proof system (HPS) introduced by Cramer and Shoup [9]. A projective hash family consists of $(\Lambda, \mathcal{SK}, \mathcal{X}, \mathcal{L}, \mathcal{W}, \mathcal{Y}, \mathcal{PK}, \mu)$, where $\mathcal{X}, \mathcal{Y}, \mathcal{L}, \mathcal{W}, \mathcal{SK}, \mathcal{PK}$ are sets and $\mathcal{L} \subset \mathcal{X}$ is a language, Let Λ be a family of hash functions indexed by $sk \in \mathcal{SK}$ mapping from $\mathcal{X}$ to $\mathcal{Y}$. Let μ be a polynomial time function mapping from $\mathcal{SK}$ to $\mathcal{PK}$. A hash family $\mathbf{H} = (\Lambda, \mathcal{SK}, \mathcal{X}, \mathcal{L}, \mathcal{W}, \mathcal{Y}, \mathcal{PK}, \mu)$ is projective if for all $sk \in \mathcal{SK}$, the action of Λ_{sk} on $\mathcal{L}$ is determined by $\mu(sk)$.

Definition 6 (ϵ-smoothness [9]). *The projective hash family is ϵ-smooth if for randomly chosen $sk \leftarrow \mathcal{SK}$, $X \leftarrow \mathcal{X} \backslash \mathcal{L}, pk = \mu(sk)$, given pk, X, the distribution of $Y = \Lambda_{sk}(X)$ and randomly chosen $\tilde{Y} \in \mathcal{Y}$ are statistically indistinguishable,*

$$SD((pk, X, Y), (pk, X, \tilde{Y})) \le \epsilon.$$

Definition 7 (ι-related ϵ-smoothness). *The projective hash family is ι-related ϵ-smooth if for ι randomly chosen $\boldsymbol{sk} = (sk_1, ..., sk_\iota) \leftarrow \mathcal{SK}^\iota$, $\boldsymbol{X} = (X_1, ..., X_\iota) \leftarrow (a\mathcal{L})^\iota, a \leftarrow \mathcal{X} \backslash \mathcal{L}$, compute $\boldsymbol{pk} = (\mu(sk_1), ..., \mu(sk_\iota))$, $\boldsymbol{Y} = (\Lambda_{sk_1}(X_1), ..., \Lambda_{sk_\iota}(X_\iota))$, for randomly chosen $\tilde{\boldsymbol{Y}} \in \mathcal{Y}^\iota$,*

$$SD((\boldsymbol{pk}, \boldsymbol{X}, \boldsymbol{Y}), (\boldsymbol{pk}, \boldsymbol{X}, \tilde{\boldsymbol{Y}})) \le \epsilon.$$

ι-related ϵ-smoothness property can be easily deduced from the ordinary smoothness property of hash family with a hybrid proof argument.

As in [9], we introduce a finite set $\mathcal{E}$ to extend the sets $\mathcal{X}$ and $\mathcal{L}$ to define a universal$_2$ extended projective hash family $\mathbf{H} = (\Lambda, \mathcal{SK}, \mathcal{X} \times \mathcal{E}, \mathcal{L} \times \mathcal{E}, \mathcal{W}, \mathcal{Y}, \mathcal{PK}, \mu)$.

Definition 8 (universal$_2$ [9,25]). *The extended projective hash family is universal$_2$ if for all $pk \in \mathcal{PK}$, $X_1, X_2 \in \mathcal{X} \backslash \mathcal{L}, E_1, E_2 \in \mathcal{E}$, $(X_1, E_1) \ne (X_2, E_2)$, for all $Y_1, Y_2 \in \mathcal{Y}$,*

$$\Pr[\Lambda_{sk}(X_2, E_2) = Y_2 | \mu(sk) = pk, \Lambda_{sk}(X_1, E_1) = Y_1] = \frac{1}{|\mathcal{Y}|}.$$

Subset Membership Problem (SMP). An SMP specifies an instance ensembles $\{I_n\}_n$ such that for each n, I_n specifies a distribution over instance $\Gamma = (\mathcal{X}, \mathcal{L}, \mathcal{W}, \mathcal{R})$, where $\mathcal{X}, \mathcal{L}, \mathcal{W}$ are non-empty sets and $\mathcal{L} \subset \mathcal{X}$ and $\mathcal{R} \subset \mathcal{X} \times \mathcal{W}$ is a binary relation such that $x \in \mathcal{L}$ iff there exists a w satisfying $(x, w) \in \mathcal{R}$.

We assume that there are efficient algorithms to sample instances from I_n, elements from $\mathcal{X}, \mathcal{X} \backslash \mathcal{L}$ and elements L from $\mathcal{L}$ together with its witness $w \in \mathcal{W}$. Also we require that $\mathcal{X}, \mathcal{Y}$ being abelian groups (with computational symbol "$+$") and $\mathcal{L}$ being subgroup of $\mathcal{X}$.

Definition 9 (Subset Membership (SM) Problem [9]). *The advantage of an adversary $\mathcal{A}$ in breaking SMP is defined as:*

$$Adv_{\mathcal{A}}^{SM} = |\Pr[\mathcal{A}(\Gamma, Z_0) = 1] - \Pr[\mathcal{A}(\Gamma, Z_1) = 1]|,$$

where the probability is taken over the randomness of choosing instance Γ and elements Z_0, Z_1, the internal randomness of $\mathcal{A}$. We say that the SM problem is hard if for every PPT $\mathcal{A}, Adv_{\mathcal{A}}^{SM}$ is negligible.

Hash Proof System (HPS). An HPS associates each SM instance Γ with a projective hash family $\mathbf{H} = (\Lambda, \mathcal{SK}, \mathcal{X}, \mathcal{L}, \mathcal{W}, \mathcal{Y}, \mathcal{PK}, \mu)$. In addition, it provides PPT algorithms to choose $sk \in \mathcal{SK}$ and $X \in \mathcal{X}$ uniformly at random, PPT algorithm to compute $\mu(sk)$, and PPT algorithms $(Priv, Pub)$ to compute $\Lambda_{sk}(L)$ for $L \in \mathcal{L}$ with witness w :

$$\Lambda_{sk}(L) = Priv(sk, L) = Pub(\mu(sk), L, w).$$

HPS with Trapdoor. Following [25,26], we also require that the SM problem can be efficiently solved with a master trapdoor, which will be used not in the actual scheme but in the security proof. In fact, all known hash proof systems have such a trapdoor.

4.2   Construction

Let $\mathbf{H}_1 = (\Lambda_1, \mathcal{SK}_1, \mathcal{X}, \mathcal{L}, \mathcal{W}, \mathcal{Y}_1, \mathcal{PK}_1, \mu_1)$ be a smooth projective hash proof system, $\mathbf{H}_2 = (\Lambda_2, \mathcal{SK}_2, \mathcal{X} \times \mathcal{Y}_1, \mathcal{L} \times \mathcal{Y}_1, \mathcal{W}, \mathcal{Y}_2, \mathcal{PK}_2, \mu_2)$ be an extended universal$_2$ projective hash proof system. Public parameters are set as $pp = (\mathbf{H}_1, \mathbf{H}_2)$.

$Keygen(pp)$: The key generation algorithm chooses random secret key $sk_1 \leftarrow \mathcal{SK}_1, sk_2 \leftarrow \mathcal{SK}_2$ and computes the public key as $pk = (pk_1 = \mu_1(sk_1), pk_2 = \mu_2(sk_2))$.

$Enc(pk, m)$: The encryption algorithm samples random $L \in \mathcal{L}$ with witness w, and computes the ciphertext $c = (c_0, c_1, c_2)$ as:

$$c_0 = L, Y_1 = Pub(pk_1, L, w), c_1 = Y_1 + m, c_2 = Pub(pk_2, L, c_1, w).$$

$Dec(sk, c)$: The decryption algorithm first verifies whether $c_2 = Priv(sk_2, c_0, c_1)$, if the equation does not hold, it just rejects, else it computes the message as:
$$Y_1 = Priv(sk_1, c_0), m = c_1 - Y_1.$$

Correctness can be easily verified from the projective property of the HPS.

4.3   Security Proof

Theorem 2. *If $\mathbf{H}_1$ is a ϵ_1-smooth projective HPS with the corresponding SM problem hard, $\mathbf{H}_2$ is an extended universal$_2$ projective hash proof system with the same corresponding SM problem hard, then our PKE scheme is IND-RSO-CCA secure. Concretely,*

$$Adv_{\mathcal{A}}^{IND\text{-}RSO\text{-}CCA} \leq Adv_{\mathcal{B}}^{SM,HPS} + q(\frac{1}{(|\mathcal{X}| - |\mathcal{L}|) \cdot |\mathcal{Y}_1|} + \frac{1}{|\mathcal{Y}_2|}) + n\epsilon_1.$$

where q is the number of decryption queries, n is the number of key pairs.

Proof. A ciphertext c is invalid if $c_0 \notin \mathcal{L}$. The master trapdoor mt is used to solve the SM problem.

To prove the security of our scheme, we define a sequence of games whereby any PPT adversary can not tell the difference between consecutive games.

$Game_0$: the real security game.

$Game_1$: the same as $Game_0$ except that the challenge ciphertexts are generated using the secret keys. That is $Y_{i1}^* = Priv_1(sk_{i1}, c_{i0}^*), c_{i2}^* = Priv_2(sk_{i2}, c_{i0}^*, c_{i1}^*)$.

$Game_2$: the same as $Game_1$ except that the challenge ciphertexts are invalid. Concretely, $\{c_{i0}^*\}_{i \in [n]}$ are chosen uniformly from a random coset of $\mathcal{L}$, that is $a\mathcal{L}, a \leftarrow \mathcal{X} \backslash \mathcal{L}$.

$Game_3$: the same as $Game_2$ except that the decryption oracle rejects all queries (c, j) that satisfy $c_0 \notin \mathcal{L}$. This can be achieved with the help of the master trapdoor mt.

Let $Adv_{\mathcal{A}}^i$ denote $\mathcal{A}$'s advantage in $Game_i$ for $i = 0, 1, 2, 3$.

It is clear to see $Adv_{\mathcal{A}}^0 = Adv_{\mathcal{A}}^1$ from the projective property of HPS.

Lemma 2. *Suppose that there exists a PPT adversary $\mathcal{A}$ such that $Adv_{\mathcal{A}}^1 - Adv_{\mathcal{A}}^2 = \epsilon$, then there exists a PPT adversary $\mathcal{B}$ with advantage ϵ in solving the SM problem.*

Lemma 3. $Adv_{\mathcal{A}}^2 - Adv_{\mathcal{A}}^3 \le \epsilon$ *if the projective HPS H_2 satisfies the universal$_2$ property, where $\epsilon = q(\frac{1}{(|\mathcal{X}| - |\mathcal{L}|) \cdot |\mathcal{Y}_1|} + \frac{1}{|\mathcal{Y}_2|})$.*

Lemma 4. $Adv_{\mathcal{A}}^3 \le n\epsilon_1$, *if the underlying projective HPS H_1 is ϵ_1-smooth.*

Concrete proofs for Lemmas 2, 3 and 4 are deferred to the full version. $\square$

Instantiations. The instantiations are the same as that in [9] from the DDH, DCR and QR assumptions.

Acknowledgments. We are grateful to Yamin Liu and Haiyang Xue for helpful discussions and advice. We also thank the anonymous reviewers of CT-RSA 2017 for their useful comments.

References

1. Bellare, M., Dowsley, R., Waters, B., Yilek, S.: Standard security does not imply security against selective-opening. In: Pointcheval, D., Johansson, T. (eds.) EUROCRYPT 2012. LNCS, vol. 7237, pp. 645–662. Springer, Heidelberg (2012). doi:10.1007/978-3-642-29011-4_38

2. Blum, M., Feldman, P., Micali, S.: Non-interactive zero-knowledge and its applications (extended abstract). In: STOC 1988, pp. 103–112 (1988)

3. Böhl, F., Hofheinz, D., Kraschewski, D.: On definitions of selective opening security. In: Fischlin, M., Buchmann, J., Manulis, M. (eds.) PKC 2012. LNCS, vol. 7293, pp. 522–539. Springer, Heidelberg (2012). doi:10.1007/978-3-642-30057-8_31

4. Bellare, M., Hofheinz, D., Yilek, S.: Possibility and impossibility results for encryption and commitment secure under selective opening. In: Joux, A. (ed.) EUROCRYPT 2009. LNCS, vol. 5479, pp. 1–35. Springer, Heidelberg (2009). doi:10.1007/978-3-642-01001-9_1

5. Bellare, M., Yilek, S.: Encryption schemes secure under selective opening attack. IACR Cryptology ePrint Archive 2009, 101 (2009)

6. Camenisch, J., Chandran, N., Shoup, V.: A public key encryption scheme secure against key dependent chosen plaintext and adaptive chosen ciphertext attacks. In: Joux, A. (ed.) EUROCRYPT 2009. LNCS, vol. 5479, pp. 351–368. Springer, Heidelberg (2009). doi:10.1007/978-3-642-01001-9_20

7. Canetti, R., Feige, U., Goldreich, O., Naor, M.: Adaptively secure multi-party computation. In: Twenty-Eighth Annual ACM Symposium on Theory of Computing, Proceedings of STOC 1995, pp. 639–648. ACM Press (1996)

8. Canetti, R., Halevi, S., Katz, J.: Adaptively-secure, non-interactive public-key encryption. In: Kilian, J. (ed.) TCC 2005. LNCS, vol. 3378, pp. 150–168. Springer, Berlin (2005). doi:10.1007/978-3-540-30576-7_9

9. Cramer, R., Shoup, V.: Universal hash proofs and a paradigm for adaptive chosen ciphertext secure public-key encryption. In: Knudsen, L.R. (ed.) EUROCRYPT 2002. LNCS, vol. 2332, pp. 45–64. Springer, Heidelberg (2002). doi:10.1007/3-540-46035-7_4

10. Cramer, R., Shoup, V.: Design and analysis of practical public-Key encryption schemes secure against adaptive chosen ciphertext attack. SIAM J. Compt. **33**(1), 167–226 (2003)

11. Dwork, C., Naor, M., Reingold, O., Stockmeyer, L.: Magic functions. J. ACM **50**(6), 852–921 (2003)

12. Fuchsbauer, G., Heuer, F., Kiltz, E., Pietrzak, K.: Standard security does imply security against selective opening for Markov distributions. In: Kushilevitz, E., Malkin, T. (eds.) TCC 2016. LNCS, vol. 9562, pp. 282–305. Springer, Heidelberg (2016). doi:10.1007/978-3-662-49096-9_12

13. Fehr, S., Hofheinz, D., Kiltz, E., Wee, H.: Encryption schemes secure against chosen-ciphertext selective opening attacks. In: Gilbert, H. (ed.) EUROCRYPT 2010. LNCS, vol. 6110, pp. 381–402. Springer, Heidelberg (2010). doi:10.1007/978-3-642-13190-5_20

14. Feige, U., Lapidot, D., Shamir, A.: Multiple non-interactive zero knowledge proofs based on a single random string (extended abstract). In: FOCS 1990, pp. 308–317 (1990)

15. Groth, J.: Simulation-sound NIZK proofs for a practical language and constant size group signatures. In: Lai, X., Chen, K. (eds.) ASIACRYPT 2006. LNCS, vol. 4284, pp. 444–459. Springer, Heidelberg (2006). doi:10.1007/11935230_29

16. Groth, J., Sahai, A.: Efficient non-interactive proof systems for bilinear groups. In: Smart, N. (ed.) EUROCRYPT 2008. LNCS, vol. 4965, pp. 415–432. Springer, Heidelberg (2008). doi:10.1007/978-3-540-78967-3_24

17. Heuer, F., Jager, T., Kiltz, E., Schäge, S.: On the selective opening security of practical public-key encryption schemes. In: Katz, J. (ed.) PKC 2015. LNCS, vol. 9020, pp. 27–51. Springer, Heidelberg (2015). doi:10.1007/978-3-662-46447-2_2

18. Huang, Z., Liu, S., Qin, B., Chen, K.: Fixing the sender-equivocable encryption scheme in Eurocrypt 2010. In: INCOS, pp. 366–372 (2013)

19. Hemenway, B., Libert, B., Ostrovsky, R., Vergnaud, D.: Lossy encryption: constructions from general assumptions and efficient selective opening chosen ciphertext security. In: Lee, D.H., Wang, X. (eds.) ASIACRYPT 2011. LNCS, vol. 7073, pp. 70–88. Springer, Heidelberg (2011). doi:10.1007/978-3-642-25385-0_4

20. Hazay, C., López-Alt, A., Wee, H., Wichs, D.: Leakage-resilient cryptography from minimal assumptions. In: Johansson, T., Nguyen, P.Q. (eds.) EUROCRYPT 2013. LNCS, vol. 7881, pp. 160–176. Springer, Heidelberg (2013). doi:10.1007/978-3-642-38348-9_10

21. Hazay, C., Patra, A., Warinschi, B.: Selective opening security for receivers. In: Iwata, T., Cheon, J.H. (eds.) ASIACRYPT 2015. LNCS, vol. 9452, pp. 443–469. Springer, Heidelberg (2015). doi:10.1007/978-3-662-48797-6_19

22. Hofheinz, D., Rupp, A.: Standard versus selective opening security: separation and equivalence results. In: Lindell, Y. (ed.) TCC 2014. LNCS, vol. 8349, pp. 591–615. Springer, Heidelberg (2014). doi:10.1007/978-3-642-54242-8_25

23. Hofheinz, D., Rao, V., Wichs, D.: Standard security does not imply indistinguishability under selective opening. In: Hirt, M., Smith, A. (eds.) TCC 2016. LNCS, vol. 9986, pp. 121–145. Springer, Heidelberg (2016). doi:10.1007/978-3-662-53644-5_5

24. Jia, D., Lu, X., Li, B.: Constructions secure against receiver selective opening and chosen ciphertext attacks. IACR Cryptology ePrint Archive 2016, 1083 (2016)

25. Kurosawa, K., Desmedt, Y.: A new paradigm of hybrid encryption scheme. In: Franklin, M. (ed.) CRYPTO 2004. LNCS, vol. 3152, pp. 426–442. Springer, Heidelberg (2004). doi:10.1007/978-3-540-28628-8_26

26. Kiltz, E., Pietrzak, K., Stam, M., Yung, M.: A new randomness extraction paradigm for hybrid encryption. In: Joux, A. (ed.) EUROCRYPT 2009. LNCS, vol. 5479, pp. 590–609. Springer, Heidelberg (2009). doi:10.1007/978-3-642-01001-9_34

27. Lai, J., Deng, R.H., Liu, S., Weng, J., Zhao, Y.: Identity-based encryption secure against selective opening chosen-ciphertext attack. In: Nguyen, P.Q., Oswald, E. (eds.) EUROCRYPT 2014. LNCS, vol. 8441, pp. 77–92. Springer, Heidelberg (2014). doi:10.1007/978-3-642-55220-5_5

28. Liu, S., Paterson, K.G.: Simulation-based selective opening CCA security for PKE from key encapsulation mechanisms. In: Katz, J. (ed.) PKC 2015. LNCS, vol. 9020, pp. 3–26. Springer, Heidelberg (2015). doi:10.1007/978-3-662-46447-2_1

29. Liu, S., Zhang, F., Chen, K.: Public-key encryption scheme with selective opening chosen-ciphertext security based on the Decisional Diffie-Hellman assumption. Concurrency Comput. Pract. Experience 26(8), 1506–1519 (2014)

30. Nielsen, J.B.: Separating random oracle proofs from complexity theoretic proofs: the non-committing encryption case. In: Yung, M. (ed.) CRYPTO 2002. LNCS, vol. 2442, pp. 111–126. Springer, Heidelberg (2002). doi:10.1007/3-540-45708-9_8

31. Ostrovsky, R., Rao, V., Visconti, I.: On selective-opening attacks against encryption schemes. In: Abdalla, M., Prisco, R. (eds.) SCN 2014. LNCS, vol. 8642, pp. 578–597. Springer, Heidelberg (2014). doi:10.1007/978-3-319-10879-7_33

New Revocable IBE in Prime-Order Groups: Adaptively Secure, Decryption Key Exposure Resistant, and with Short Public Parameters

Yohei Watanabe[1,2]([⊠]), Keita Emura[3], and Jae Hong Seo[4]

[1] The University of Electro-Communications, Tokyo, Japan
watanabe@uec.ac.jp
[2] National Institute of Advanced Industrial Science and Technology (AIST),
Tokyo, Japan
[3] National Institute of Information and Communications Technology (NICT),
Tokyo, Japan
k-emura@nict.go.jp
[4] Myongji University, Yongin, Korea
jaehongseo@mju.ac.kr

Abstract. Revoking corrupted users is a desirable functionality for cryptosystems. Since Boldyreva, Goyal, and Kumar (ACM CCS 2008) proposed a notable result for scalable revocation method in identity-based encryption (IBE), several works have improved either the security or the efficiency of revocable IBE (RIBE). Currently, all existing scalable RIBE schemes that achieve adaptively security against decryption key exposure resistance (DKER) can be categorized into two groups; either with long public parameters or over composite-order bilinear groups. From both practical and theoretical points of views, it would be interesting to construct adaptively secure RIBE scheme with DKER and short public parameters in prime-order bilinear groups.

In this paper, we address this goal by using Seo and Emura's technique (PKC 2013), which transforms the Waters IBE to the corresponding RIBE. First, we identify necessary requirements for the input IBE of their transforming technique. Next, we propose a new IBE scheme having several desirable properties; satisfying all the requirements for the Seo-Emura technique, constant-size public parameters, and using prime-order bilinear groups. Finally, by applying the Seo-Emura technique, we obtain the first adaptively secure RIBE scheme with DKER and constant-size public parameters in prime-order bilinear groups.

Keywords: Revocable identity-based encryption · Static assumptions · Asymmetric pairings

1 Introduction

Identity-Based Encryption (IBE) scheme is a public key cryptosystem enabling one to use arbitrary bit-string as her/his public key. In dynamic cryptosystem,

The first author is supported by JSPS Research Fellowships for Young Scientists.

H. Handschuh (Ed.): CT-RSA 2017, LNCS 10159, pp. 432–449, 2017.
DOI: 10.1007/978-3-319-52153-4_25

user registration and revocation are important functionalities. When Boneh and Franklin proposed the first realization of IBE [4], they already explained how to revoke corrupted users; for an identity I of a non-revoked user at time T, $I\|T$ is regarded as the identity, and Key Generation Center (KGC) issues a secret key for $I\|T$ to a non-revoked user I for each time period. Even though this simple identity-encoding method can successfully revoke users from the system, KGC's huge overhead (linear computational complexity in the number of users per each time period) is an inherent problem. To resolve this problem, Boldyreva, Goyal, and Kumar [2] proposed a scalable revocation method by using the symmetric key broadcast encryption technique, so-called the Complete Subtree (CS) method [21]. They called IBE with such the efficient revocation Revocable IBE (RIBE).

After the seminal work by Boldyreva, Goyal, and Kumar [2], several RIBE schemes have been proposed so far. Almost all such subsequent works basically follow Boldyreva et al.'s revocation methodology. Let us briefly explain Boldyreva et al.'s approach; as in IBE, each user has a (long-term) secret key sk_I. At each time T, KGC broadcasts key update information ku_T which is constructed by the Complete Subtree (CS) method [21]. Remark that no secure channel is required to send ku_T to users. A user can compute a decryption key $dk_{I,T}$ from ku_T and own sk_I if the user is not revoked at T. Due to the CS method, the size of ku_T is $O(r\log(n/r))$, where n is the number of maximum users and r is the number of revoked users. Thus, Bolyreva et al. RIBE scheme is scalable. The first adaptively secure RIBE scheme was proposed by Libert and Vergnaud [20]. Seo and Emura extended the Boldyreva et al.'s security notion to consider more practical threats; *decryption key exposure resistance (DKER)* [28,30]. Intuitively, this notion considers the case where several decryption keys $dk_{I^*,T}$ for the target identity I^* are leaked to an adversary but the target decryption key dk_{I^*,T^*} is not exposed. This notion is important where the secret key is stored in physically secure devices such as USB pen drives to be isolated from the Internet but decryption keys are stored in weaker device such as a smart phone. They also proposed the first scalable RIBE scheme with adaptive security with DKER. The Seo-Emura RIBE is based on the Waters IBE [35], so that long public parameters are inevitable. Since there exist several efficient IBE schemes, it is quite natural to ask

whether we attain an adaptively secure RIBE scheme with DKER, which achieves similar performance to efficient IBE schemes, in particular, short public parameters in prime-order groups.

Although several RIBE schemes are proposed so far [5,7,8,12,16,22,34], none of them achieves adaptive security against decryption key exposure and short parameters (in the sense of constant public parameters and prime-order groups) at the same time. We found that the answer is not trivial due to the following reasons. Basically, there are two approaches to achieving constant-size public parameter IBE: One is to use strong assumptions such as static ones in composite-order groups and q-type ones (e.g., [11,37]); and the other is to apply the dual system encryption methodology [36] in either prime-order or composite-order groups. Therefore, if we want to realize an RIBE scheme with constant-size public parameter under static assumptions in prime-order groups, it is quite natural to apply the latter approach for our purpose.

Unfortunately, there exists a subtle obstacle in applying the dual system encryption methodology for adaptive security with decryption key exposure resistance. In fact, Lee observed such an obstacle [15] and also, basing on his observation, pointed out a flaw of an Revocable Hierarchical IBE (RHIBE) scheme [31]. Let us briefly review such an obstacle. In the dual system encryption framework, ciphertexts and secret keys can be transformed into semi-functional ones. Normal ciphertexts can be decrypted with either a normal or semi-functional key, whereas semi-functional ciphertexts can be decrypted with only a normal secret key. In the security proof, a normal challenge ciphertext and secret keys are transformed into their semi-functional forms one by one. In the process of changing some normal key (called a *target key*) into its semi-functional form, a simulator has to embed some function f into public parameters. Thus, the simulator can generate randomness $r_C := f(\mathtt{I}^*)$ for the challenge ciphertext, as well as randomness $r_K := f(\mathtt{I})$ for the target key, where $\mathtt{I}^*$ is the target identity and $\mathtt{I}$ is an identity such that $\mathtt{I} \neq \mathtt{I}^*$. The proof goes well since f is a pairwise independent function and $\mathtt{I} \neq \mathtt{I}^*$, i.e., r_C is independent of r_K from an adversarial view in the information-theoretic sense. To the best of our knowledge, such a pairwise independent function f is necessary for proving security of all of the currently-known IBE schemes using the dual system encryption methodology. On the other hand, an adversary against the security game of RIBE can get not only a challenge ciphertext for $\mathtt{I}^*$ but also a secret key for $\mathtt{I}^*$ (see Definition 1). Therefore, we cannot argue that randomness r_C for the challenge ciphertext and randomness r_K for the secret key are independent of each other from the view point of the adversary, since it holds $r_C = r_K = f(\mathtt{I}^*)$.

Lee [15] introduced a way to circumvent the above obstacle and also proposed provably secure RHIBE scheme in the adaptive adversary model. Since we can consider a 1-level HIBE as an IBE scheme, Lee's RHIBE can be considered as an adaptively secure RIBE with DKER and short public parameters. We note that, however, his approach essentially used composite-order bilinear groups. Moreover, there are other RHIBE schemes [9,17,26,27,29,32,33], but none of them satisfies both adaptive security with decryption key exposure resistance and short parameters (i.e., short public parameters in prime-order groups) at the same time. Therefore, designing an adaptively secure RIBE scheme with DKER and short parameters (possibly through the dual system encryption approach) is still open.

1.1   Our Contribution

In this paper, we propose the first adaptively secure RIBE scheme with constant size public parameters in asymmetric bilinear groups of prime order. Our RIBE scheme also supports decryption key exposure resistance (i.e., our scheme meets the strong security notion for RIBE). The security of our scheme is proved under static assumptions, which are mild variants of the symmetric external Diffie-Hellman (SXDH) assumption.

We overcome the difficulty mentioned above by the following strategy: Taking the Seo-Emura approach [28]. Seo and Emura proposed an adaptively secure RIBE scheme based on the Waters IBE [35], and showed a security reduction from the Waters IBE to their RIBE scheme. Note that the Waters IBE does not use the dual system encryption methodology, and requires long public parameters which depend on the bit-length of identities. Therefore, by taking the Seo-Emura approach we want to avoid the randomness correlation problem specific to dual system encryption-based RIBE schemes. Namely, we want to make a security reduction from some IBE scheme using the dual system encryption methodology to our RIBE scheme. However, the Seo-Emura technique essentially requires *the secret-key re-randomization*[1] of the underlying IBE scheme, but almost all of the dual system encryption-based IBE schemes in prime-order groups (e.g., [6,19,36]) do not have this property.

Therefore, we employ the Jutla-Roy IBE [13] (and its variant [25]) as a promising candidate of our basic IBE scheme since it allows one to publicly re-randomize the secret key. However, the public parameter of the Jutla-Roy IBE lacks some important elements for simulating secret keys in the security proof. In the security proof taking the Seo-Emura approach, a simulator extracts the master key of the underlying IBE scheme by using the Boneh-Boyen technique [3], and creates a secret key sk_{I^*} or decryption key $dk_{I^*,T}$ for any T, where I^* is the challenge identity and T is a time period such that it is not the challenge one. The Boneh-Boyen technique requires some group elements that contain the master key in the exponent in the public parameter of the underlying IBE, however the original Jutla-Roy IBE does not contain them (For details, see Sect. 3). Hence, we modify the Jutla-Roy IBE so that the Seo-Emura technique can be applied to it, and we prove the security under the Augmented Decisional Diffie-Hellman on $\mathbb{G}_1$ (ADDH1), which is a new static assumption, and Decisional Diffie-Hellman on $\mathbb{G}_2$ (DDH2) assumptions. The ADDH1 assumption is newly introduced in this paper, and therefore it is a non-standard one. However, this assumption is not so complicated and similar to the previously used assumption in [24]. The security of the ADDH1 assumption is proved in the generic bilinear group model.

We then propose an RIBE scheme based on the Jutla-Roy IBE, and the security is proved by making a security reduction from the modified Jutla-Roy IBE to the RIBE scheme.[2] As a result, we obtain the first RIBE scheme that achieves adaptive security with decryption key exposure resistance and constant-size public parameters in prime-order asymmetric bilinear groups. Furthermore, our proof technique provides a better reduction loss, which is elaborated in the next paragraph.

---

[1] It means that each secret key can be re-randomized with fresh randomness.

[2] This situation is the same as that of Ishida et al.'s construction [12]. Since the Kiltz-Galindo IB-KEM [14] is not directly applicable due to the same reason, they constructed a variant of the Kiltz-Galindo IB-KEM, and then showed a security reduction from the variant scheme to their scheme.

Table 1. Efficiency comparison among adaptively secure RIBE schemes with decryption key exposure resistance.

Scheme	#mpk	#msk	#C
Seo-Emura [28, 30]	$(6+\ell)|\mathbb{G}_p|$	$|\mathbb{G}_p|$	$3|\mathbb{G}_p| + |\mathbb{G}_T^{sym}|$
Lee [15] ($L=1$)	$8|\mathbb{G}_N| + 3|\mathbb{G}_T^{comp}|$	$|\mathbb{G}_N|$	$4|\mathbb{G}_N| + |\mathbb{G}_T^{comp}|$
Proposed Scheme	$7|\mathbb{G}_1| + 11|\mathbb{G}_2| + |\mathbb{G}_T^{asym}|$	$2|\mathbb{G}_2|$	$4|\mathbb{G}_1| + |\mathbb{G}_T^{asym}| + |\mathbb{Z}_p|$

Scheme	#sk	#ku	#dk	Assumption
Seo-Emura [28, 30]	$(2\log n)|\mathbb{G}_p|$	$(2r\log(n/r))|\mathbb{G}_p|$	$3|\mathbb{G}_p|$	DBDH
Lee [15] ($L=1$)	$(2\log n)|\mathbb{G}_N|$	$(2r\log(n/r))|\mathbb{G}_N| + 2|\mathbb{Z}_N|$	$4|\mathbb{G}_N|$	Static
Proposed Scheme	$(5\log n)|\mathbb{G}_2|$	$(3r\log(n/r))|\mathbb{G}_2|$	$6|\mathbb{G}_1|$	ADDH1, DDH2

Let $|\mathbb{G}_1|$, $|\mathbb{G}_2|$, and $|\mathbb{G}_T^{asym}|$ be the bit-length of an element of asymmetric bilinear groups $\mathbb{G}_1$, $\mathbb{G}_2$, and $\mathbb{G}_T$ respectively. Let $|\mathbb{G}_p|$ and $|\mathbb{G}_T^{sym}|$ be the bit-length of an element of symmetric bilinear groups $\mathbb{G}_p$ and $\mathbb{G}_T$ employed in [28, 30], respectively. Let $|\mathbb{G}_N|$ and $|\mathbb{G}_T^{comp}|$ be the bit-length of an element of symmetric bilinear groups $\mathbb{G}_N$ and $\mathbb{G}_T$ of composite order $N = p_1 p_2 p_3$, where p_1, p_2, and p_3 are distinct prime numbers, employed in [15], respectively. Let $|\mathbb{Z}_p|$ and $|\mathbb{Z}_N|$ be the bit-length of an element of $\mathbb{Z}_p$ and $\mathbb{Z}_N$, respectively. On 256-bit Barreto-Naehrig curve [1], $|\mathbb{G}_1| = 256$, $|\mathbb{G}_2| = |\mathbb{G}_p| = 512$, and $|\mathbb{G}_T^{asym}| = |\mathbb{G}_T^{sym}| = 3072$ due to [6]. Note that $|\mathbb{G}_N|$ and $|\mathbb{G}_T^{comp}|$ should be much larger so that N cannot be factored. L is the hierarchy depth, n is the maximum number of users, r is the number of revoked users, and ℓ is the bit-length of identity. For example, if 32 byte e-mail address is regarded as identity, then $\ell = 256$.

Efficiency Comparison. We give an efficiency comparison in Table 1. All of the schemes meet adaptive security with decryption key exposure resistance. We use the KUNode algorithm for efficient revocation as in previous RIBE schemes (For details, see Sect. 2 or [21]). Therefore, the sizes of secret keys and key updates in every scheme are $O(\log n)$ and $O(r\log(n/r))$, respectively, due to the KUNode algorithm. Lee's scheme [15] is less efficient than the others since it is constructed over composite-order bilinear groups. Our scheme is more efficient than the Seo-Emura RIBE in terms of constant-size public parameters and asymmetric pairings, and other parameters are comparable to those of the Seo-Emura RIBE. In addition, our proof technique provides a better reduction loss than that of the Seo-Emura RIBE. More precisely, the reduction loss of our scheme is $O(q_1 q|\mathcal{T}|)$, whereas that of the Seo-Emura RIBE is $O(\ell q^2 |\mathcal{T}|)$, where ℓ is the bit-length of identity, q is the maximum number of queries in the security game, q_1 is the maximum number of queries *before the challenge phase* in the security game, and $|\mathcal{T}|$ is the number of time periods in the schemes.

1.2 Paper Organization

In Sect. 2, we describe notation and definitions throughout this paper. In Sect. 3, we propose an IBE scheme, which is used as the underlying IBE scheme of our RIBE scheme, based on the Jutla-Roy IBE. In Sect. 4, we show the first adaptively secure RIBE scheme with DKER and short public parameters in prime-order groups, and we conclude in Sect. 5.

2 Preliminaries

Notation. In this paper, "probabilistic polynomial-time" is abbreviated as "PPT". For a prime p, let $\mathbb{Z}_p := \{0, 1, \ldots, p-1\}$ and $\mathbb{Z}_p^{\times} := \mathbb{Z}_p \setminus \{0\}$. If we write $(y_1, y_2, \ldots, y_m) \leftarrow \mathcal{A}(x_1, x_2, \ldots, x_n)$ for an algorithm $\mathcal{A}$ having n inputs and m outputs, it means to input $x_1, x_2, \ldots, x_n$ into $\mathcal{A}$ and to get the resulting output $y_1, y_2, \ldots, y_m$. We write $(y_1, y_2, \ldots, y_m) \leftarrow \mathcal{A}^{\mathcal{O}}(x_1, x_2, \ldots, x_n)$ to indicate that an algorithm $\mathcal{A}$ that is allowed to access an oracle $\mathcal{O}$ takes $x_1, x_2, \ldots, x_n$ as input and outputs $(y_1, y_2, \ldots, y_m)$. If $\mathcal{X}$ is a set, we write $x \xleftarrow{\$} \mathcal{X}$ to mean the operation of picking an element x of $\mathcal{X}$ uniformly at random. We use λ as a security parameter. $\mathcal{M}$, $\mathcal{I}$, and $\mathcal{T}$ denote sets of plaintexts, IDs, and time periods, respectively, which are determined by the security parameter λ.

Bilinear Groups. A bilinear group generator $\mathcal{G}$ is an algorithm that takes a security parameter λ as input and outputs a bilinear group $(p, \mathbb{G}_1, \mathbb{G}_2, \mathbb{G}_T, g_1, g_2, e)$, where p is a prime, $\mathbb{G}_1$, $\mathbb{G}_2$, and $\mathbb{G}_T$ are multiplicative cyclic groups of order p, g_1 and g_2 are (random) generators of $\mathbb{G}_1$ and $\mathbb{G}_2$, respectively, and e is an efficiently computable and non-degenerate bilinear map $e : \mathbb{G}_1 \times \mathbb{G}_2 \to \mathbb{G}_T$ with the following bilinear property: For any $u, u' \in \mathbb{G}_1$ and $v, v' \in \mathbb{G}_2$, $e(uu', v) = e(u, v)e(u', v)$ and $e(u, vv') = e(u, v)e(u, v')$.

A bilinear map e is called symmetric or a "Type-1" pairing if $\mathbb{G}_1 = \mathbb{G}_2$. Otherwise, it is called asymmetric. In the asymmetric setting, e is called a "Type-2" pairing if there is an efficiently computable isomorphism from $\mathbb{G}_2$ to $\mathbb{G}_1$. If no efficiently computable isomorphism between $\mathbb{G}_1$ and $\mathbb{G}_2$ is known, then it is called a "Type-3" pairing. Throughout this paper, we focus on the Type-3 pairing. Type-3 is the most efficient setting since compared to Type-1, the size of representation of $\mathbb{G}_1$ in the Type-3 setting is smaller and whole operations in the Type-3 setting are more efficient; and compared to Type-2, the size of representation of $\mathbb{G}_2$ in the Type-3 setting is smaller and group operations in $\mathbb{G}_2$ in the Type-3 are more efficient. For details, see [10].

KUNode Algorithm. To reduce costs of a revocation process, we use a binary tree structure and apply the following KUNode algorithm as in the previous RIBE schemes [2,20,28]. KUNode(BT, RL, T) takes as input a binary tree BT, a revocation list RL, and a time period $T \in \mathcal{T}$, and outputs a set of nodes. When η is a non-leaf node, then we write η_L and η_R as the left and right child of η, respectively. When η is a leaf node, Path(BT, η) denotes the set of nodes on the path from η to the *root*. Each user is assigned to a leaf node. If a user who is assigned to η is revoked on a time period $T \in \mathcal{T}$, then $(\eta, T) \in RL$. KUNode(BT, RL, T) is executed as follows. It sets $\mathcal{X} := \emptyset$ and $\mathcal{Y} := \emptyset$. For any $(\eta_i, T_i) \in RL$, if $T_i \le T$ then it adds Path(BT, η_i) to $\mathcal{X}$ (i.e., $\mathcal{X} := \mathcal{X} \cup$ Path(BT, η_i)). That is, KUNode adds at most $r \log n$ nodes to $\mathcal{X}$ where $r = |RL|$ and n is the number of leaves of BT. Then, for any $\eta \in \mathcal{X}$, if $\eta_L \notin \mathcal{X}$, then it adds η_L to $\mathcal{Y}$. If $\eta_R \notin \mathcal{X}$, then it adds η_R to $\mathcal{Y}$. That is, KUNode adds at most $r \log n$ nodes to $\mathcal{Y}$. Actually, due to the result of [21], the size of $\mathcal{Y}$ is $O(r \log(n/r))$, and the time complexity is $O(\log \log n)$. Finally, it outputs $\mathcal{Y}$ if $\mathcal{Y} \ne \emptyset$. If $\mathcal{Y} = \emptyset$, then it adds *root* to $\mathcal{Y}$ and outputs $\mathcal{Y}$.

Revocable Identity-Based Encryption. An RIBE scheme Π consists of seven-tuple algorithms (Setup, SKGen, KeyUp, DKGen, Enc, Dec, Revoke) defined as follows: For simplicity, we omit a public parameter in the input of all algorithms except for the Setup algorithm.

- $(mpk, msk, RL, st) \leftarrow$ Setup(λ, N): A probabilistic algorithm for setup. It takes a security parameter λ and the maximum number of users N as input and outputs a public parameter mpk, a master secret key msk, an initial revocation list $RL = \emptyset$ and a state st.
- $(sk_\mathrm{I}, st) \leftarrow$ SKGen(st, I): An algorithm for private key generation. It takes st and an identity $\mathrm{I} \in \mathcal{I}$ as input and outputs a secret key sk_I and updated state information st.[3]
- $ku_T \leftarrow$ KeyUp(msk, st, RL, T): An algorithm for key update generation. It takes msk, state st, a current revocation list RL, and a time period T as input, and then outputs key update ku_T.
- $dk_{\mathrm{I},T}$ or $\perp \leftarrow$ DKGen(sk_I, ku_T): A probabilistic algorithm for decryption key generation. It takes sk_I and ku_T as input and then outputs a decryption key $dk_{\mathrm{I},T}$ at T or $\perp$ if I has been revoked by T.
- $C_{\mathrm{I},T} \leftarrow$ Enc(M, I, T): A probabilistic algorithm for encryption. It takes $M \in \mathcal{M}$, $\mathrm{I} \in \mathcal{I}$, and $T \in \mathcal{T}$ as input and then outputs a ciphertext $C_{\mathrm{I},T}$.
- M or $\perp \leftarrow$ Dec($dk_{\mathrm{I},T}, C_{\mathrm{I},T}$): A deterministic algorithm for decryption. It takes $dk_{\mathrm{I},T}$ and $C_{\mathrm{I},T}$ as input and then outputs M or $\perp$.
- $RL \leftarrow$ Revoke(I, T, RL, st): An algorithm for revocation. It takes $(\mathrm{I}, T) \in \mathcal{I} \times \mathcal{T}$, the current revocation list RL, and a state st as input and then outputs an updated revocation list RL.

In the above model, we assume that Π meets the following correctness property: For all security parameter $\lambda \in \mathbb{N}$, all $(mpk, msk, RL, st) \leftarrow$ Setup(λ, N), all $M \in \mathcal{M}$, all $\mathrm{I} \in \mathcal{I}$, all $T \in \mathcal{T}$, if I is not revoked on $T \in \mathcal{T}$, it holds that $M =$ Dec(DKGen(SKGen(st, I), KeyUp(msk, st, RL, T)), Enc(M, I, T)).

We describe the notion of indistinguishability against chosen plaintext attack (IND-RID-CPA). Note that this notion also captures decryption key exposure resistance, which was introduced by Seo and Emura [28], and this security model is the strongest known one. Let $\mathcal{A}$ be a PPT adversary, and $\mathcal{A}$'s advantage against

---

[3] We consider the SKGen algorithm in the sense of history-free RHIBE [32,33], i.e., the algorithm takes st, rather than msk, as input.

IND-RID-CPA security is defined by

$$Adv_{\Pi,\mathcal{A}}^{IND\text{-}RID\text{-}CPA}(\lambda, N) :=$$

$$\left| \Pr \left[b' = b \left| \begin{array}{l} (mpk, msk, RL, st) \leftarrow \mathsf{Setup}(\lambda, N), \\ (M_0^*, M_1^*, \mathtt{I}^*, T^*, state) \leftarrow \mathcal{A}^{\mathcal{O}}(\mathsf{find}, mpk), \\ b \xleftarrow{\$} \{0,1\}, \\ C_{\mathtt{I}^*, T^*}^* \leftarrow \mathsf{Enc}(M_b^*, \mathtt{I}^*, T^*), \\ b' \leftarrow \mathcal{A}^{\mathcal{O}}(\mathsf{guess}, C_{\mathtt{I}^*, T^*}^*, state) \end{array} \right. \right] - \frac{1}{2} \right|.$$

Here, $\mathcal{O}$ is a set of oracles $\{SKGen(\cdot),\ KeyUp(\cdot),\ Revoke(\cdot,\cdot),\ DKGen(\cdot,\cdot)\}$ defined as follows.

SKGen($\cdot$): For a query $\mathtt{I} \in \mathcal{I}$, it stores and returns $\mathsf{SKGen}(st, \mathtt{I})$.

KeyUp($\cdot$): For a query $T \in \mathcal{T}$, it stores and returns $\mathsf{KeyUp}(msk, RL, st, T)$.

Revoke($\cdot,\cdot$): For a query $(\mathtt{I}, T) \in \mathcal{I} \times \mathcal{T}$, it updates a revocation list RL by running $\mathsf{Revoke}(\mathtt{I}, T, RL, st)$.

DKGen($\cdot,\cdot$): For a query $(\mathtt{I}, T) \in \mathcal{I} \times \mathcal{T}$, it finds $sk_\mathtt{I}$ and ku_T generated by the $SKGen$ and $KeyUp$ oracles, respectively (If $sk_\mathtt{I}$ has not been generated yet, $DKGen$ executes $(sk_\mathtt{I}, st) \leftarrow \mathsf{SKGen}(st, \mathtt{I})$).[4] $DKGen$ returns $\mathsf{DKGen}(sk_\mathtt{I}, ku_T)$ and stores it unless it is $\bot$.

The above oracles represent the following realistic threats and situations: $SKGen$ represents the collusion among users as in ordinary IBE. $\mathcal{A}$ can access $KeyUp$ since key updates are broadcasted by the KGC. The reason why $\mathcal{A}$ can access $Revoke$ is an RIBE scheme should be secure against any situations in terms of the revocation list. $DKGen$ represents decryption key exposure.

We then impose the following restrictions on $\mathcal{A}$. Specifically, the first three restrictions are placed to take into account practical situations, and we circumvent some trivial attacks by the other restrictions.

1. $KeyUp(\cdot)$ and $Revoke(\cdot,\cdot)$ can be queried at a time period which is later than or equal to that of all previous queries.
2. $Revoke(\cdot,\cdot)$ cannot be queried at a time period T after issuing T to $\mathsf{KeyUp}(\cdot)$.
3. $DKGen(\cdot,\cdot)$ cannot be queried at T before issuing T to $\mathsf{KeyUp}(\cdot)$.
4. If $\mathtt{I}^*$ was issued to $SKGen(\cdot)$ at T', then $(\mathtt{I}^*, T)$ must be issued to $\mathsf{Revoke}(\cdot,\cdot)$ such that $T' \leq T \leq T^*$.
5. $(\mathtt{I}^*, T^*)$ cannot be issued to $DKGen(\cdot,\cdot)$.

Definition 1. *An RIBE scheme Π is said to be IND-RID-CPA secure if for all PPT adversaries $\mathcal{A}$, $Adv_{\Pi,\mathcal{A}}^{IND\text{-}RID\text{-}CPA}(\lambda, N)$ is negligible in λ.*

---

[4] Contrary to $sk_\mathtt{I}$, ku_T is already stored by the $KeyUp$ oracle due to the restrictions on the oracles.

3   The Basic IBE Scheme

We begin with reviewing Seo and Emura's approach for transforming IBE to RIBE [28]. Although their approach is not generic, it seems quite broadly applicable to the other IBE schemes. We find some requirements for applying their technique. Then, we propose an IBE scheme satisfying such the requirements, which has short public parameters and over prime-order bilinear groups.

Seo and Emura constructed an RIBE scheme based on the Waters IBE [35] and provided as security reduction to the Waters IBE. In the reduction, almost all queries can be easily simulated due to the adaptive security of the underlying IBE. The most non-trivial part in the reduction is simulating decryption keys for (I^*, T), where I^* is the target identity, since the security of usual IBE scheme does not handle this case related to I^*. To this end, Seo and Emura employed two techniques; the Boneh-Boyen technique [3] and secret-key *re-randomization*.

The Boneh-Boyen technique is originally for *selectively secure* scheme[5]; that is, if the simulator knows the target (time T^* in our case) in advance, then the simulator embeds it into public parameters so that the simulator can simulate all the other queries not related to T^*.[6] The Boneh-Boyen technique enables the simulator to compute decryption keys for (I^*, T) with biased distribution, where T is not the target time. The secret-key re-randomization can resolve the biased distribution by forcing that all decryption keys have uniform randomness.

From the above interpretation, we find two requirements for the input IBE; (1) the secret-key re-randomization property and (2) applicability of the Boneh-Boyen technique. The latter requirement can be further segmentalized. (2-1) Each component of a secret key contains at most one component of a master key and (2-2) each component of the master-key is available in the public parameters in some form of elements in source-groups (of bilinear groups). The former is due to that the Boneh-Boyen technique can extract at most one master-key component from each secret-key component. The latter is due to that in the security reduction the master-key is embedded into key updates that consist of only elements in source-groups by using the master-key-related public parameters.[7]

The Waters IBE satisfies all the above requirements, but most of dual-system-encryption-based IBE schemes in prime-order groups do not. For example, the first scheme by Waters [36] and almost all of the IBE schemes using dual pairing vector spaces (DPVS) (e.g.,[6,19]) do not satisfy any requirement, in particular, the public re-randomization requirement.

---

[5] Although our goal is adaptive security, the polynomial reduction loss enables one to use the selective security technique in terms of (polynomial-size) time period.

[6] Although the decryption key (I^*, T) is related to the target identity I^*, it is not related to T^* so that the Boneh-Boyen technique is applicable.

[7] In (usual-but-not-all) pairing-based IBE schemes, private keys consist of elements in source-groups. Since both key updates and secret keys of RIBE are materials for decryption keys, they also should consist of source-group elements.

3.1 Modified Jutla-Roy IBE

We employ a modified version of the Jutla-Roy IBE [13] (and its variant [25]). The original scheme satisfies two requirements (1) and (2-1). In this subsection, we modify the Jutla-Roy IBE to additionally satisfy the requirement (2-2).

The master key of the Jutla-Roy IBE is $(y_0, x_0) \in \mathbb{Z}_p^2$. To get a basic IBE scheme for our RIBE scheme based on the Jutla-Roy IBE, we add the master key in the forms of elements in $\mathbb{G}_1$ and $\mathbb{G}_2$ with a random mask $\beta \in \mathbb{Z}_p^\times$, respectively, to the public parameters. Specifically, we add four group elements $(\chi_1 := g_1^{\beta(-x_0\alpha+y_0)}, g_2^{x_0\beta}, g_2^{y_0\beta}, g_2^{1/\beta})$ to the original public parameter. However, we then cannot apply the original security proof of the Jutla-Roy IBE, and so we add a new twist to the proof. The modified Jutla-Roy IBE $\Pi_{\mathrm{JR}} =$(Init, KeyGen, IBEnc, IBDec) is constructed as follows.[8]

- Init(λ): It runs $(\mathbb{G}_1, \mathbb{G}_2, \mathbb{G}_T, p, g_1, g_2, e) \leftarrow \mathcal{G}$. It chooses $x_0, y_0, x_1, y_1, x_2, y_2, x_3, y_3 \xleftarrow{\$} \mathbb{Z}_p$ and $\alpha, \beta \xleftarrow{\$} \mathbb{Z}_p^\times$, and sets

$$z = e(g_1, g_2)^{-x_0\alpha+y_0}, \quad u_1 := g_1^{-x_1\alpha+y_1}, \quad w_1 := g_1^{-x_2\alpha+y_2},$$
$$h_1 := g_1^{-x_3\alpha+y_3}, \quad \chi_1 := g_1^{\beta(-x_0\alpha+y_0)}.$$

It outputs $PP := (g_1, g_1^\alpha, u_1, w_1, h_1, \chi_1, g_2, g_2^{x_1}, g_2^{x_2}, g_2^{x_3}, g_2^{y_1}, g_2^{y_2}, g_2^{y_3}, z, g_2^{x_0\beta}, g_2^{y_0\beta}, g_2^{\frac{1}{\beta}})$, $MK := (g_2^{y_0}, g_2^{-x_0})$.

- KeyGen($PP, MK, \mathtt{I}$): Parse MK as (d_1', d_2'). It chooses $r \xleftarrow{\$} \mathbb{Z}_p$ and computes

$$D_1 := (g_2^{y_2})^r, \quad D_1' := d_1'\left((g_2^{y_1})^{\mathtt{I}}g_2^{y_3}\right)^r,$$
$$D_2 := (g_2^{x_2})^{-r}, \quad D_2' := d_2'\left((g_2^{x_1})^{\mathtt{I}}g_2^{x_3}\right)^{-r}, \quad D_3 := g_2^r.$$

It outputs $SK_{\mathtt{I}} := (D_1, D_1', D_2, D_2', D_3)$.

- IBEnc($PP, \mathtt{I}, M$): It chooses $t, \mathtt{tag} \xleftarrow{\$} \mathbb{Z}_p$. For $M \in \mathbb{G}_T$, it computes

$$C_0 := Mz^t, \quad C_1 := g_1^t, \quad C_2 := (g_1^\alpha)^t, \quad C_3 := \left(u_1^{\mathtt{I}}w_1^{\mathtt{tag}}h_1\right)^t.$$

It outputs $C := (C_0, C_1, C_2, C_3, \mathtt{tag})$.

- IBDec($PP, SK_{\mathtt{I}}, C$): Parse $SK_{\mathtt{I}}$ and C as $(D_1, D_1', D_2, D_2', D_3)$ and $(C_0, C_1, C_2, C_3, \mathtt{tag})$, respectively. It computes

$$M = \frac{C_0 e(C_3, D_3)}{e(C_1, D_1^{\mathtt{tag}}D_1')e(C_2, D_2^{\mathtt{tag}}D_2')}.$$

---

[8] Due to space limitation, we omit the syntax of IBE.

We show the correctness of Π_{JR}. Suppose that $sk_{\text{I}} = (D_1, D_1', D_2, D_2', D_3)$ and $C = (C_0, C_1, C_2, C_3, \text{tag})$ are correctly generated. Then, we have

$$\frac{C_0 e(C_3, D_3)}{e(C_1, D_1^{\text{tag}} D_1') e(C_2, D_2^{\text{tag}} D_2')}$$

$$= M e(g_1, g_2)^{(-x_0\alpha + y_0)t} \frac{e(g_1^{t(I(-x_1\alpha + y_1) + \text{tag}(-x_2\alpha + y_2) - x_3\alpha + y_3)}, g_2^r)}{e(g_1^t, g_2^{y_2 r \text{tag} + y_0 + r(y_1 I + y_3)}) e(g_1^{\alpha t}, g_2^{-x_2 r \text{tag} - x_0 - r(x_1 I + x_3)})}$$

$$= M e(g_1, g_2)^{(-x_0\alpha + y_0)t} \frac{1}{e(g_1^t, g_2^{y_0}) e(g_1^{\alpha t}, g_2^{-x_0})} = M.$$

3.2   Proof of Security

We describe complexity assumptions used for proving the security proof of the modified Jutla-Roy IBE.

First, we give the definition of the decisional Diffie-Hellman (DDH) assumption in $\mathbb{G}_1$ and $\mathbb{G}_2$, which are called the DDH1 and DDH2 assumptions, respectively. We say that the SXDH assumption holds if both the DDH1 and DDH2 assumptions hold. Let $\mathcal{A}$ be a PPT adversary and we consider $\mathcal{A}$'s advantage against the DDHi problem ($i = 1, 2$) as follows.

$$Adv_{\mathcal{G},\mathcal{A}}^{DDHi}(\lambda) := \left| \Pr\left[b' = b \; \middle| \; \begin{array}{l} D := (p, \mathbb{G}_1, \mathbb{G}_2, \mathbb{G}_T, g_1, g_2, e) \leftarrow \mathcal{G}, \\ c_1, c_2 \xleftarrow{\$} \mathbb{Z}_p, \; b \xleftarrow{\$} \{0,1\}, \\ \text{if } b = 0 \text{ then } Z := g_i^{c_1 c_2}, \text{ else } Z \xleftarrow{\$} \mathbb{G}_i, \\ b' \leftarrow \mathcal{A}(\lambda, D, g_i^{c_1}, g_i^{c_2}, Z) \end{array} \right] - \frac{1}{2} \right|.$$

Definition 2 (DDHi Assumption). *The DDHi assumption relative to a generator $\mathcal{G}$ holds if for all PPT adversaries $\mathcal{A}$, $Adv_{\mathcal{G},\mathcal{A}}^{DDHi}(\lambda)$ is negligible in λ.*

We then introduce a new assumption based on the DDH1 assumption, which is called *Augmented DDH1 (ADDH1) assumption*. Let $\mathcal{A}$ be a PPT adversary and we consider $\mathcal{A}$'s advantage against the ADDH1 problem as follows.

$$Adv_{\mathcal{G},\mathcal{A}}^{ADDH1}(\lambda) :=$$

$$\left| \Pr\left[b' = b \; \middle| \; \begin{array}{l} D := (p, \mathbb{G}_1, \mathbb{G}_2, \mathbb{G}_T, g_1, g_2, e) \leftarrow \mathcal{G}(\lambda), \\ d, c_1, c_2 \xleftarrow{\$} \mathbb{Z}_p, c_3 \xleftarrow{\$} \mathbb{Z}_p^{\times}, \; b \xleftarrow{\$} \{0,1\}, \\ \text{if } b = 0 \text{ then } Z := g_1^{c_1 c_2}, \text{ else } Z \xleftarrow{\$} \mathbb{G}_1, \\ b' \leftarrow \mathcal{A}(\lambda, D, g_1^{c_1}, g_1^{c_2}, g_1^{dc_3}, g_2^d, g_2^{c_2 c_3}, g_2^{dc_3}, g_2^{\frac{1}{c_3}}, Z) \end{array} \right] - \frac{1}{2} \right|.$$

Definition 3 (ADDH1 Assumption). *The ADDH1 assumption relative to a generator $\mathcal{G}$ holds if for all PPT adversaries $\mathcal{A}$, $Adv_{\mathcal{G},\mathcal{A}}^{ADDH1}(\lambda)$ is negligible in λ.*

This assumption is similar to the DDH2v assumption ("v" stands for "variant"), which was used for constructing the Lewko-Waters IBE [18] in prime-order groups in [24]. Similarly, we can also consider the DDH1v assumption.[9]

---

[9] We give the formal definition of the DDH2v and DDH1v assumptions in the full version of this paper.

The authors of [24] argued that the DDH2v (resp., DDH1v) assumption is the minimal assumption when one tries to put some information about c_1 or c_2 in an instance of DDH1 (resp., DDH2) while staying in the hardness of the problem. We define the ADDH1 problem by removing g_1^d from the DDH1v problem and adding $g_1^{dc_3}$ and g_2^{1/c_3}. Therefore, we may say this new assumption is also a not-so-strange one. Actually, we prove the security of this assumption in the generic bilinear group model as follows (For the formal proof, see the full version of this paper).

Theorem 1 (Informal). *Let $\mathcal{A}$ be an algorithm that attempts to solve the ADDH1 problem in the generic group model. $\mathcal{A}$ makes at most q queries to the oracles computing the group actions in $\mathbb{G}_1$, $\mathbb{G}_2$, and $\mathbb{G}_T$, and the bilinear map e. Then, the advantage ϵ of $\mathcal{A}$ in solving the problem is bounded by $\epsilon \leq 3(q+11)^2/4p$.*

We prove the security of Π_{JR} under the above assumptions.

Theorem 2. *If the ADDH1 and DDH2 assumptions hold, then the resulting Jutla-Roy IBE Π_{JR} is IND-ID-CPA secure.*

Proof (Sketch). Our security proof is the same as that of the Jutla-Roy IBE except that we have to care the extra terms $(\chi_1, g_2^{x_0\beta}, g_2^{y_0\beta}, g_2^{1/\beta})$ that were added to their scheme. We replace the DDH1 assumption of Jutla-Roy's proof with "DDH1 with the additional instance", the ADDH1 assumption, in order to treat these extra terms. More specifically, we need the ADDH1 assumption in the proof of indistinguishability of the semi-functional challenge ciphertext and the random element in the ciphertext space. In the proof, the simulator $\mathcal{B}$ receives the DDH1 instance $(g_1^{c_1}, g_1^{c_2}, Z)$ with the additional instance $(g_1^{dc_3}, g_2^d, g_2^{c_2c_3}, g_2^{dc_3}, g_2^{1/c_3})$, where $Z = g_1^{c_1c_2}$ or $Z \xleftarrow{\$} \mathbb{G}_1$. $\mathcal{B}$ chooses $\alpha \xleftarrow{\$} \mathbb{Z}_p^{\times}$ and (implicitly) sets $x_0 := c_2$, $y_0' := d$, $y_0 := \alpha x_0 + y_0'$, and $\beta := c_3$. $\mathcal{B}$ then can create the elements as follows: $\chi_1 := g_1^{dc_3}$, $g_2^{\beta x_0} := g_2^{c_2c_3}$, $g_2^{\beta y_0} := (g_2^{c_2c_3})^{\alpha} g_2^{dc_3}$, and $g_2^{1/\beta} := g_2^{1/c_3}$. Furthermore, g_2^d is used for creating $z := e(g_1, g_2^d)$. For the full proof, see the full version. $\qquad\square$

4 Our Construction

We construct an RIBE scheme based on the original Jutla-Roy IBE, and prove that the security of the proposed scheme relies on that of the modified Jutla-Roy IBE. An RIBE scheme $\Pi = (\text{Setup}, \text{SKGen}, \text{KeyUp}, \text{DKGen}, \text{Enc}, \text{Dec}, \text{Revoke})$ is constructed as follows.

- Setup(λ, N): It runs $(\mathbb{G}_1, \mathbb{G}_2, \mathbb{G}_T, p, g_1, g_2, e) \leftarrow \mathcal{G}$. It chooses $x_0, y_0, x_1, y_1,$ $x_2, y_2, x_3, y_3, x_4, y_4, x_5, y_5 \xleftarrow{\$} \mathbb{Z}_p$ and $\alpha \xleftarrow{\$} \mathbb{Z}_p^{\times}$, and sets

$$z = e(g_1, g_2)^{-x_0\alpha+y_0}, \quad u_1 := g_1^{-x_1\alpha+y_1}, \quad w_1 := g_1^{-x_2\alpha+y_2},$$
$$h_1 := g_1^{-x_3\alpha+y_3}, \quad v_1 := g_1^{-x_4\alpha+y_4}, \quad \hat{v}_1 := g_1^{-x_5\alpha+y_5},$$

Let BT be a binary tree that has N leaves, where N is a power of two for simplicity. It outputs $mpk := (g_1, g_1^\alpha, u_1, w_1, h_1, v_1, \hat{v}_1, g_2, g_2^{x_1}, g_2^{x_2}, \ldots, g_2^{x_5}, g_2^{y_1}, g_2^{y_2}, \ldots, g_2^{y_5}, z)$, $msk := (g_2^{y_0}, g_2^{-x_0})$, $st := \mathtt{BT}$, and $RL := \emptyset$.

- SKGen(st, I): Parse st as BT. It randomly chooses an unassigned leaf η from BT, and stores I in the node η. For each node $\theta \in \mathsf{Path}(\mathtt{BT}, \eta)$, it recalls P_θ if it was defined. Otherwise, it chooses $P_\theta \xleftarrow{\$} \mathbb{G}_2$ and stores P_θ in the node θ. Then, it chooses $r_\theta \xleftarrow{\$} \mathbb{Z}_p$ and it computes

$$\mathsf{SK}_{1,\theta} := (g_2^{y_2})^{r_\theta}, \ \mathsf{SK}'_{1,\theta} := P_\theta\left((g_2^{y_1})^{\mathtt{I}} g_2^{y_3}\right)^{r_\theta},$$

$$\mathsf{SK}_{2,\theta} := (g_2^{x_2})^{-r_\theta}, \ \mathsf{SK}'_{2,\theta} := P_\theta\left((g_2^{x_1})^{\mathtt{I}} g_2^{x_3}\right)^{-r_\theta}, \ \mathsf{SK}_{3,\theta} := g_2^{r_\theta}.$$

It outputs $sk_{\mathtt{I}} := \{(\mathsf{SK}_{1,\theta}, \mathsf{SK}'_{1,\theta}, \mathsf{SK}_{2,\theta}, \mathsf{SK}'_{2,\theta}, \mathsf{SK}_{3,\theta})\}_{\theta \in \mathsf{Path}(\mathtt{BT}, \eta)}$.

- KeyUp(msk, st, RL, T): Parse msk as $(\mathtt{MK}_1, \mathtt{MK}_2)$. For each node $\theta \in \mathsf{KUNode}(\mathtt{BT}, RL, T)$, it recalls P_θ if it was defined. Otherwise, it chooses $P_\theta \xleftarrow{\$} \mathbb{G}_2$ and stores P_θ in the node θ. It chooses $s_\theta \xleftarrow{\$} \mathbb{Z}_p$ and computes

$$\mathsf{KU}'_{1,\theta} := P_\theta^{-1}\mathtt{MK}_1\left((g_2^{y_4})^T g_2^{y_5}\right)^{s_\theta}, \ \mathsf{KU}'_{2,\theta} := P_\theta^{-1}\mathtt{MK}_2\left((g_2^{x_4})^T g_2^{x_5}\right)^{-s_\theta}, \ \mathsf{KU}_{3,\theta} := g_2^{s_\theta}.$$

It outputs $ku_T := \{(\mathsf{KU}'_{1,\theta}, \mathsf{KU}'_{2,\theta}, \mathsf{KU}_{3,\theta})\}_{\theta \in \mathsf{KUNode}(\mathtt{BT}, RL, T)}$.

- DKGen($sk_{\mathtt{I}}$, ku_T): Parse $sk_{\mathtt{I}}$ and ku_T as $\{(\mathsf{SK}_{1,\theta}, \mathsf{SK}'_{1,\theta}, \mathsf{SK}_{2,\theta}, \mathsf{SK}'_{2,\theta}, \mathsf{SK}_{3,\theta})\}_{\theta \in \Theta_{\mathsf{SK}}}$ and $\{(\mathsf{KU}'_{1,\theta}, \mathsf{KU}'_{2,\theta}, \mathsf{KU}_{3,\theta})\}_{\theta \in \Theta_{\mathsf{KU}}}$, respectively. It outputs $\perp$ if $\Theta_{\mathsf{SK}} \cap \Theta_{\mathsf{KU}} = \emptyset$. Otherwise, for some $\theta \in \Theta_{\mathsf{SK}} \cap \Theta_{\mathsf{KU}}$, it computes as follows. It chooses $R, S \xleftarrow{\$} \mathbb{Z}_p$ and computes

$$\mathsf{DK}_1 := \mathsf{SK}_{1,\theta}(g_2^{y_2})^R, \ \mathsf{DK}'_1 := \mathsf{SK}'_{1,\theta}\mathsf{KU}'_{1,\theta}\left((g_2^{y_1})^{\mathtt{I}} g_2^{y_3}\right)^R \left((g_2^{y_4})^T g_2^{y_5}\right)^S,$$

$$\mathsf{DK}_2 := \mathsf{SK}_{2,\theta}(g_2^{x_2})^{-R}, \ \mathsf{DK}'_2 := \mathsf{SK}'_{2,\theta}\mathsf{KU}'_{2,\theta}\left((g_2^{x_1})^{\mathtt{I}} g_2^{x_3}\right)^{-R} \left((g_2^{x_4})^T g_2^{x_5}\right)^{-S},$$

$$\mathsf{DK}_3 := \mathsf{SK}_{3,\theta}g_2^R, \ \mathsf{DK}_4 := \mathsf{KU}_{3,\theta}g_2^S.$$

It outputs $dk_{\mathtt{I},T} := (\mathsf{DK}_1, \mathsf{DK}'_1, \mathsf{DK}_2, \mathsf{DK}'_2, \mathsf{DK}_3, \mathsf{DK}_4)$.

- Enc(M, I, T): It chooses $t, \mathtt{tag} \xleftarrow{\$} \mathbb{Z}_p$. For $M \in \mathbb{G}_T$, it computes

$$C_0 := Mz^t, \ C_1 := g_1^t, \ C_2 := (g_1^\alpha)^t, \ C_3 := \left(u_1^{\mathtt{I}} w_1^{\mathtt{tag}} h_1\right)^t, \ C_4 := (v_1^T \hat{v}_1)^t.$$

It outputs $C_{\mathtt{I},T} := (C_0, C_1, C_2, C_3, C_4, \mathtt{tag})$.

- Dec($dk_{\mathtt{I},T}$, $C_{\mathtt{I},T}$): Parse $dk_{\mathtt{I},T}$ and $C_{\mathtt{I},T}$ as $(\mathsf{DK}_1, \mathsf{DK}'_1, \mathsf{DK}_2, \mathsf{DK}'_2, \mathsf{DK}_3, \mathsf{DK}_4)$ and $(C_0, C_1, C_2, C_3, C_4, \mathtt{tag})$, respectively. It computes

$$M = \frac{C_0 e(C_3, \mathsf{DK}_3)e(C_4, \mathsf{DK}_4)}{e(C_1, \mathsf{DK}_1^{\mathtt{tag}}\mathsf{DK}'_1)e(C_2, \mathsf{DK}_2^{\mathtt{tag}}\mathsf{DK}'_2)}.$$

- Revoke(I, T, RL, st): Output $RL := RL \cup \{(\mathtt{I}, T)\}$.

Due to space limitation, we give the correctness of our RIBE scheme Π in the full version. The security of the above construction is given as follows.

Theorem 3. *If the ADDH1 and DDH2 assumptions holds, then the resulting RIBE scheme Π is IND-RID-CPA secure.*

We show the following lemma, and we obtain Theorem 3 as a corollary of the lemma.

Lemma 1. *The proposed RIBE scheme Π is IND-RID-CPA secure as long as the modified Jutla-Roy IBE Π_{JR}, which is described in Section 3.1, is IND-ID-CPA secure.*

Proof (Sketch). Due to space limitation, we here give a sketch of the proof. For the full proof, see the full version. We construct a PPT algorithm $\mathcal{B}$ which breaks the IND-ID-CPA security of the modified Jutla-Roy IBE Π_{JR} using a PPT adversary $\mathcal{A}$ which breaks the IND-RID-CPA security of Π.

At the beginning, $\mathcal{B}$ receives a public parameter PP of Π_{JR}. $\mathcal{B}$ guesses what time period T^* will be submitted from $\mathcal{A}$ in the challenge phase, and it holds with probability $1/|\mathcal{T}|$. We assume $\mathcal{B}$'s guess is right. $\mathcal{B}$ chooses $\tilde{x}, \hat{x}, \tilde{y}, \hat{y} \xleftarrow{\$} \mathbb{Z}_p$ and (implicitly) sets

$$x_4 := \beta x_0 + \tilde{x}, \quad x_5 := -T^*\beta x_0 + \hat{x}, \quad y_4 := \beta y_0 + \tilde{y}, \quad y_5 := -T^*\beta y_0 + \hat{y},$$
$$- x_4\alpha + y_4 = -(\beta x_0 + \tilde{x})\alpha + \beta y_0 + \tilde{y} = \beta(-x_0\alpha + y_0) - \alpha\tilde{x} + \tilde{y},$$
$$- x_5\alpha + y_5 = -(-T^*\beta x_0 + \hat{x})\alpha - T^*\beta y_0 + \hat{y} = -T^*\beta(-x_0\alpha + y_0) - \alpha\hat{x} + \hat{y}.$$

Then, $\mathcal{B}$ computes

$$g_2^{x_4} := g_2^{\beta x_0}g_2^{\tilde{x}}, \quad g_2^{x_5} := (g_2^{\beta x_0})^{-T^*}g_2^{\hat{x}}, \quad g_2^{y_4} := g_2^{\beta y_0}g_2^{\tilde{y}}, \quad g_2^{y_5} := (g_2^{\beta y_0})^{-T^*}g_2^{\hat{y}},$$
$$v_1 := g_1^{-x_4\alpha + y_4} = \chi_1(g_1^{\alpha})^{-\tilde{x}}g_1^{\tilde{y}}, \quad \hat{v}_1 := g_1^{-x_5\alpha + y_5} = \chi_1^{-T^*}(g_1^{\alpha})^{-\hat{x}}g_1^{\hat{y}},$$

and sends mpk to $\mathcal{A}$.

Since $\mathcal{B}$ changes a way to simulate oracles based on A's behavior, $\mathcal{B}$ has to guess whether $\mathcal{A}$ will issue the target identity I^* to the $SKGen$ oracle, and when it will first issue I^* to the ($SKGen$ and) $DKGen$ oracle. The probability that $\mathcal{B}$'s guess is right is $1/2(q_1 + 1)$, where q_1 is the maximum number of identities issued to the $SKGen$ and $DKGen$ oracles *before the challenge phase*. In this sketch, we only show the case that $\mathcal{A}$ never issues the target identity I^* to the $SKGen$ oracle, and that $\mathcal{B}$ knows when the target identity I^* is first issued to the $DKGen$ oracle. Although $\mathcal{B}$ does not know the master key $(g_2^{y_0}, g_2^{-x_0})$, it can easily return a secret key sk_I and a decryption key $dk_{\mathrm{I},T}$ for any I ($\neq \mathrm{I}$) and T by using the $KeyGen$ oracle of the modified Jutla-Roy IBE Π_{JR}. However, $\mathcal{B}$ has to respond to the decryption-key query (I^*, T) for any T without the knowledge of the master key. We show how $\mathcal{B}$ creates the decryption key $dk_{\mathrm{I}^*,T}$ as follows.

$\mathcal{B}$ chooses $r, s \xleftarrow{\$} \mathbb{Z}_p$ and computes

$$\mathrm{DK}_1 := (g_2^{y_2})^r, \quad \mathrm{DK}'_{1,\theta} := ((g_2^{y_1})^{\mathrm{I}^*} g_2^{y_3})^r ((g_2^{y_4})^T g_2^{y_5})^s (g_2^{\frac{1}{\beta}})^{-\frac{T\tilde{y}+\hat{y}}{T-T^*}},$$

$$\mathrm{DK}_2 := (g_2^{x_2})^{-r}, \quad \mathrm{DK}'_2 := ((g_2^{x_1})^{\mathrm{I}^*} g_2^{x_3})^{-r} ((g_2^{x_4})^T g_2^{x_5})^{-s} (g_2^{\frac{1}{\beta}})^{\frac{T\tilde{x}+\hat{x}}{T-T^*}},$$

$$\mathrm{DK}_3 := g_2^r, \quad \mathrm{DK}_4 := g_2^s (g_2^{\frac{1}{\beta}})^{-\frac{1}{T-T^*}}.$$

$\mathcal{B}$ sends $dk_{\mathrm{I}^*,T} := (\mathrm{DK}_1, \mathrm{DK}'_1, \mathrm{DK}_2, \mathrm{DK}'_2, \mathrm{DK}_3, \mathrm{DK}_4)$ to $\mathcal{A}$. The simulation goes well ~~since it holds that~~

$$((g_2^{y_1})^{\mathrm{I}^*} g_2^{y_3})^r ((g_2^{y_4})^T g_2^{y_5})^s (g_2^{\frac{1}{\beta}})^{-\frac{T\tilde{y}+\hat{y}}{T-T^*}}$$

$$= g_2^{y_0} ((g_2^{y_1})^{\mathrm{I}^*} g_2^{y_3})^r (g_2^{(T-T^*)\beta y_0 + T\tilde{y}+\hat{y}})^s g_2^{-\frac{T\tilde{y}+\hat{y}}{(T-T^*)\beta}} g_2^{-y_0}$$

$$= g_2^{y_0} ((g_2^{y_1})^{\mathrm{I}^*} g_2^{y_3})^r (g_2^{(T-T^*)\beta y_0 + T\tilde{y}+\hat{y}})^s (g_2^{(T-T^*)\beta y_0 + T\tilde{y}+\hat{y}})^{-\frac{1}{(T-T^*)\beta}}$$

$$= g_2^{y_0} ((g_2^{y_1})^{\mathrm{I}^*} g_2^{y_3})^r ((g_2^{y_4})^T g_2^{y_5})^{s'},$$

$$((g_2^{x_1})^{\mathrm{I}^*} g_2^{x_3})^{-r} ((g_2^{x_4})^T g_2^{x_5})^{-s} (g_2^{\frac{1}{\beta}})^{\frac{T\tilde{x}+\hat{x}}{T-T^*}}$$

$$= g_2^{-x_0} ((g_2^{x_1})^{\mathrm{I}^*} g_2^{x_3})^{-r} (g_2^{(T-T^*)\beta x_0 + T\tilde{x}+\hat{x}})^{-s} g_2^{\frac{T\tilde{x}+\hat{x}}{(T-T^*)\beta}} g_2^{x_0}$$

$$= g_2^{-x_0} ((g_2^{x_1})^{\mathrm{I}^*} g_2^{x_3})^{-r} (g_2^{(T-T^*)\beta x_0 + T\tilde{x}+\hat{x}})^{-s} (g_2^{(T-T^*)\beta x_0 + T\tilde{x}+\hat{x}})^{\frac{1}{(T-T^*)\beta}}$$

$$= g_2^{-x_0} ((g_2^{x_1})^{\mathrm{I}^*} g_2^{x_3})^{-r} ((g_2^{x_4})^T g_2^{x_5})^{-s'},$$

$$g_2^s (g_2^{\frac{1}{\beta}})^{-\frac{1}{T-T^*}} = g_2^{s - \frac{1}{(T-T^*)\beta}} = g_2^{s'},$$

where $s' = s - \frac{1}{(T-T^*)\beta}$.

In the challenge phase, $\mathcal{B}$ receives $(M_0^*, M_1^*, \mathrm{I}^*, T^*)$ from $\mathcal{A}$. It then sends $(M_0^*, M_1^*, \mathrm{I}^*)$ to the challenger in the IND-ID-CPA game of Π_{JR}. After receiving $(C_0^*, C_1^*, C_2^*, C_3^*, \mathtt{tag}^*)$ from the challenger, $\mathcal{B}$ sets $C_4^* := (C_2^*)^{-(T^*\tilde{x}+\hat{x})}(C_1^*)^{T^*\tilde{y}+\hat{y}}$. Since $C_4^* = (v_1^{T^*}\hat{v}_1)^t = g_1^{t(-T^*\tilde{x}\alpha + T^*\tilde{y} - \hat{x}\alpha + \hat{y})} = g_1^{-t\alpha(T^*\tilde{x}+\hat{x})+t(T^*\tilde{y}+\hat{y})}$, this is well-formed. $\mathcal{B}$ sends $(C_0^*, C_1^*, C_2^*, C_3^*, C_4^*, \mathtt{tag}^*)$ to $\mathcal{A}$. When $\mathcal{A}$ outputs b', then $\mathcal{B}$ transfer it.

Thus, we have $Adv_{\Pi,\mathcal{A}}^{IND\text{-}RID\text{-}CPA}(\lambda) = 2|T|(q_1+1)Adv_{\Pi_{\mathrm{JR}},\mathcal{B}}^{IND\text{-}ID\text{-}CPA}(\lambda)$. Therefore, we have $Adv_{\Pi,\mathcal{A}}^{IND\text{-}RID\text{-}CPA}(\lambda) \leq 8|T|(q_1+1)Adv_{\mathcal{G},\mathcal{B}}^{ADDH1}(\lambda) + 2|T|q(q_1+1)Adv_{\mathcal{G},\mathcal{B}}^{DDH2}(\lambda)$, where q is the maximum number of queries issued to the $KeyGen$ oracle in the IND-ID-CPA game of Π_{JR}. $\square$

5   Concluding Remarks

In the context of identity-based encryption schemes, it is natural to employ dual system encryption methodology. However, as aforementioned in the introduction, if we consider revocation functionality in the identity-based cryptosystem, there is a subtle obstacle in an approach using dual system encryption methodology, in particular, in prime-order groups. To circumvent this obstacle, we revisited the proof of Seo-Emura RIBE scheme [28], which does not uses dual system encryption methodology, but give a reduction to the IND-CPA security of the underlying IBE scheme. We extract several important requirements for Seo-Emura approach, and then construct a new IBE scheme satisfying all such the requirements, based on Jutla-Roy IBE scheme. Then, we construct an RIBE based on the proposed modified Jutla-Roy IBE scheme. We prove the IND-RID-CPA security of the proposed scheme under mild variants of the SXDH assumption, which are static and generically secure. Furthermore, we can extend the proposed scheme to guarantee the CCA security by using the technique in [12] and also to a server-aided scheme [23] (For details, see the full version).

Acknowledgments. We would like to thank anonymous reviewers for valuable comments. Yohei Watanabe was supported by Grant-in-Aid for JSPS Fellows Grant Number JP16J10532. Keita Emura was supported by JSPS KAKENHI Grant Number JP16K00198.

References

1. Barreto, P.S.L.M., Naehrig, M.: Pairing-friendly elliptic curves of prime order. In: Preneel, B., Tavares, S. (eds.) SAC 2005. LNCS, vol. 3897, pp. 319–331. Springer, Heidelberg (2006). doi:10.1007/11693383_22
2. Boldyreva, A., Goyal, V., Kumar, V.: Identity-based encryption with efficient revocation. In: Proceedings of the 15th ACM Conference on Computer and Communications Security, CCS 2008, pp. 417–426. ACM, New York (2008)
3. Boneh, D., Boyen, X.: Efficient selective-ID secure identity-based encryption without random oracles. In: Cachin, C., Camenisch, J.L. (eds.) EUROCRYPT 2004. LNCS, vol. 3027, pp. 223–238. Springer, Heidelberg (2004). doi:10.1007/978-3-540-24676-3_14
4. Boneh, D., Franklin, M.: Identity-based encryption from the weil pairing. In: Kilian, J. (ed.) CRYPTO 2001. LNCS, vol. 2139, pp. 213–229. Springer, Heidelberg (2001). doi:10.1007/3-540-44647-8_13
5. Chen, J., Lim, H.W., Ling, S., Su, L., Wang, H.: Anonymous and adaptively secure revocable IBE with constant size public parameters (2012). http://arxiv.org/abs/1210.6441
6. Chen, J., Lim, H.W., Ling, S., Wang, H., Wee, H.: Shorter identity-based encryption via asymmetric pairings. Des. Codes Crypt. **73**(3), 911–947 (2014)
7. Chen, J., Lim, H.W., Ling, S., Wang, H., Nguyen, K.: Revocable identity-based encryption from lattices. In: Susilo, W., Mu, Y., Seberry, J. (eds.) ACISP 2012. LNCS, vol. 7372, pp. 390–403. Springer, Heidelberg (2012). doi:10.1007/978-3-642-31448-3_29. http://eprint.iacr.org/2011/583
8. Cheng, S., Zhang, J.: Adaptive-ID secure revocable identity-based encryption from lattices via subset difference method. In: Lopez, J., Wu, Y. (eds.) ISPEC 2015. LNCS, vol. 9065, pp. 283–297. Springer, Heidelberg (2015). doi:10.1007/978-3-319-17533-1_20
9. Emura, K., Seo, J.H., Youn, T.: Semi-generic transformation of revocable hierarchical identity-based encryption and its DBDH instantiation. IEICE Trans. **99-A**(1), 83–91 (2016)
10. Galbraith, S.D., Paterson, K.G., Smart, N.P.: Pairings for cryptographers. Discrete Appl. Math. **156**(16), 3113–3121 (2008)

11. Gentry, C.: Practical identity-based encryption without random oracles. In: Vaudenay, S. (ed.) EUROCRYPT 2006. LNCS, vol. 4004, pp. 445–464. Springer, Heidelberg (2006). doi:10.1007/11761679_27

12. Ishida, Y., Watanabe, Y., Shikata, J.: Constructions of CCA-secure revocable identity-based encryption. In: Foo, E., Stebila, D. (eds.) ACISP 2015. LNCS, vol. 9144, pp. 174–191. Springer, Heidelberg (2015). doi:10.1007/978-3-319-19962-7_11

13. Jutla, C.S., Roy, A.: Shorter quasi-adaptive NIZK proofs for linear subspaces. In: Sako, K., Sarkar, P. (eds.) ASIACRYPT 2013. LNCS, vol. 8269, pp. 1–20. Springer, Heidelberg (2013). doi:10.1007/978-3-642-42033-7_1

14. Kiltz, E., Galindo, D.: Direct chosen-ciphertext secure identity-based key encapsulation without random oracles. Theor. Comput. Sci. 410(47–49), 5093–5111 (2009)

15. Lee, K.: Revocable hierarchical identity-based encryption with adaptive security. Cryptology ePrint Archive, Report 2016/749 (2016)

16. Lee, K., Lee, D.H., Park, J.H.: Efficient revocable identity-based encryption via subset difference methods. Cryptology ePrint Archive, Report 2014/132 (2014). http://eprint.iacr.org/

17. Lee, K., Park, S.: Revocable hierarchical identity-based encryption with shorter private keys and update keys. Cryptology ePrint Archive, Report 2016/460 (2016). http://eprint.iacr.org/

18. Lewko, A., Waters, B.: New techniques for dual system encryption and fully secure HIBE with short ciphertexts. In: Micciancio, D. (ed.) TCC 2010. LNCS, vol. 5978, pp. 455–479. Springer, Heidelberg (2010). doi:10.1007/978-3-642-11799-2_27

19. Lewko, A.: Tools for simulating features of composite order bilinear groups in the prime order setting. In: Pointcheval, D., Johansson, T. (eds.) EUROCRYPT 2012. LNCS, vol. 7237, pp. 318–335. Springer, Heidelberg (2012). doi:10.1007/978-3-642-29011-4_20

20. Libert, B., Vergnaud, D.: Adaptive-ID secure revocable identity-based encryption. In: Fischlin, M. (ed.) CT-RSA 2009. LNCS, vol. 5473, pp. 1–15. Springer, Heidelberg (2009). doi:10.1007/978-3-642-00862-7_1

21. Naor, D., Naor, M., Lotspiech, J.: Revocation and tracing schemes for stateless receivers. In: Kilian, J. (ed.) CRYPTO 2001. LNCS, vol. 2139, pp. 41–62. Springer, Heidelberg (2001). doi:10.1007/3-540-44647-8_3

22. Park, S., Lee, K., Lee, D.H.: New constructions of revocable identity-based encryption from multilinear maps. IEEE Trans. Inf. Forensics Secur. 10(8), 1564–1577 (2015)

23. Qin, B., Deng, R.H., Li, Y., Liu, S.: Server-aided revocable identity-based encryption. In: Pernul, G., Ryan, P.Y.A., Weippl, E. (eds.) ESORICS 2015. LNCS, vol. 9326, pp. 286–304. Springer, Heidelberg (2015). doi:10.1007/978-3-319-24174-6_15

24. Ramanna, S.C., Chatterjee, S., Sarkar, P.: Variants of Waters' dual system primitives using asymmetric pairings. In: Fischlin, M., Buchmann, J., Manulis, M. (eds.) PKC 2012. LNCS, vol. 7293, pp. 298–315. Springer, Heidelberg (2012). doi:10.1007/978-3-642-30057-8_18

25. Ramanna, S.C., Sarkar, P.: Efficient (Anonymous) compact HIBE from standard assumptions. In: Chow, S.S.M., Liu, J.K., Hui, L.C.K., Yiu, S.M. (eds.) ProvSec 2014. LNCS, vol. 8782, pp. 243–258. Springer, Heidelberg (2014). doi:10.1007/978-3-319-12475-9_17

26. Ryu, G., Lee, K., Park, S., Lee, D.H.: Unbounded hierarchical identity-based encryption with efficient revocation. In: Kim, H., Choi, D. (eds.) WISA 2015. LNCS, vol. 9503, pp. 122–133. Springer, Heidelberg (2016). doi:10.1007/978-3-319-31875-2_11

27. Seo, J.H., Emura, K.: Efficient delegation of key generation and revocation functionalities in identity-based encryption. In: Dawson, E. (ed.) CT-RSA 2013. LNCS, vol. 7779, pp. 343–358. Springer, Heidelberg (2013). doi:10.1007/978-3-642-36095-4_22

28. Seo, J.H., Emura, K.: Revocable identity-based encryption revisited: security model and construction. In: Kurosawa, K., Hanaoka, G. (eds.) PKC 2013. LNCS, vol. 7778, pp. 216–234. Springer, Heidelberg (2013). doi:10.1007/978-3-642-36362-7_14

29. Seo, J.H., Emura, K.: Revocable hierarchical identity-based encryption. Theor. Comput. Sci. **542**, 44–62 (2014)

30. Seo, J.H., Emura, K.: Revocable identity-based cryptosystem revisited: security models and constructions. IEEE Trans. Inf. Forensics Secur. **9**(7), 1193–1205 (2014)

31. Seo, J.H., Emura, K.: Adaptive-ID secure revocable hierarchical identity-based encryption. In: Tanaka, K., Suga, Y. (eds.) IWSEC 2015. LNCS, vol. 9241, pp. 21–38. Springer, Heidelberg (2015). doi:10.1007/978-3-319-22425-1_2

32. Seo, J.H., Emura, K.: Revocable hierarchical identity-based encryption: history-free update, security against insiders, and short ciphertexts. In: Nyberg, K. (ed.) CT-RSA 2015. LNCS, vol. 9048, pp. 106–123. Springer, Heidelberg (2015). doi:10.1007/978-3-319-16715-2_6

33. Seo, J.H., Emura, K.: Revocable hierarchical identity-based encryption via history-free approach. Theor. Comput. Sci. **615**, 45–60 (2016)

34. Su, L., Lim, H.W., Ling, S., Wang, H.: Revocable IBE systems with almost constant-size key update. In: Cao, Z., Zhang, F. (eds.) Pairing 2013. LNCS, vol. 8365, pp. 168–185. Springer, Heidelberg (2014). doi:10.1007/978-3-319-04873-4_10

35. Waters, B.: Efficient identity-based encryption without random oracles. In: Cramer, R. (ed.) EUROCRYPT 2005. LNCS, vol. 3494, pp. 114–127. Springer, Heidelberg (2005). doi:10.1007/11426639_7

36. Waters, B.: Dual system encryption: realizing fully secure IBE and HIBE under simple assumptions. In: Halevi, S. (ed.) CRYPTO 2009. LNCS, vol. 5677, pp. 619–636. Springer, Heidelberg (2009). doi:10.1007/978-3-642-03356-8_36

37. Wee, H.: Déjà Q: Encore! un petit IBE. In: Kushilevitz, E., Malkin, T. (eds.) TCC 2016. LNCS, vol. 9563, pp. 237–258. Springer, Heidelberg (2016). doi:10.1007/978-3-662-49099-0_9

Author Index